CONTINUUM THERMODYNAMICS

Part I: Foundations

Series on Advances in Mathematics for Applied Sciences – Vol. 77

CONTINUUM THERMODYNAMICS

Part I: Foundations

Krzysztof Wilmanski
University of Zielona Gora, Poland

World Scientific

NEW JERSEY • LONDON • SINGAPORE • BEIJING • SHANGHAI • HONG KONG • TAIPEI • CHENNAI

Published by

World Scientific Publishing Co. Pte. Ltd.
5 Toh Tuck Link, Singapore 596224
USA office: 27 Warren Street, Suite 401-402, Hackensack, NJ 07601
UK office: 57 Shelton Street, Covent Garden, London WC2H 9HE

Library of Congress Cataloging-in-Publication Data
Wilmanski, Krzysztof.
Continuum thermodynamics / by Krzysztof Wilmanski.
p. cm. -- (Series on advances in mathematics for applied sciences ; v. 77)
Includes bibliographical references and index.
Contents: pt. 1. Foundations
ISBN-13: 978-981-283-556-7 (hardcover : alk. paper)
ISBN-10: 981-283-556-3 (hardcover : alk. paper)
1. Thermodynamics. I. Title.
QC311.W489 2008
536'.7--dc22

2008033325

British Library Cataloguing-in-Publication Data
A catalogue record for this book is available from the British Library.

Printed in Singapore by World Scientific Printers

Preface

This book consists of three parts which complement each other

– Part I: **Foundations**
KRZYSZTOF WILMANSKI
(*University of Zielona Gora,* http://www.mech-wilmanski.de),

– Part II: **Applications and Exercises**
BETTINA ALBERS
(*Technical University of Berlin,* http://www.mech-albers.de),

– Part III: **Numerical Methods**
MIECZYSLAW KUCZMA
(*University of Zielona Gora,* http://www.uz.zgora.pl/~mkuczma).

For convenience, they appear separately but studying the subject of continuum thermodynamics should be simultaneous in all three aspects. Therefore, in Parts II and III we indicate interrelations with Part I of the book as well as the mutual relations between those two Parts. Each Chapter of Part II with exercises begins with a short recall of foundations which are needed for a particular problem and each numerical problem begins with the indication of both theoretical foundations of Part I and the corresponding problem in Part II.

Part II contains in addition the presentation of many general problems which were not included in Part I as they simultaneously form excellent exercises of skills for continuum models. For instance, the debate about the class of universal solutions, such as tension, shear, bending, torsion, inversion, etc. yields questions of material stability which will be discussed in this Part. For similar reasons, the discussion of fundamental rheological flows and their applications in experimental investigation of non-Newtonian fluids is discussed here. This also concerns the dynamic problems such as propagation of linear acoustic waves in nonlinear media or the propagation of various surface waves.

In Part III, techniques of approximation and numerical solutions of problems presented in Parts I and II will be introduced. Mainly the treatment of these problems by the finite element method will be presented. However, it will also contain a discussion of such fundamental problems as convergence of algorithms, stability of numerical analysis, or remarks on new methods designed for continua, for instance the method of finite volumes. The classical method of finite differences and, developed for flow problems in continua, the method of finite volumes will be particularly emphasized. The method of finite volumes allows for unstructured meshes, uses directly balance equations of continuous media and, for this reason, is particularly useful in problems of continua. Consequently, notions of Delaunay triangulation, Voronoi grid, MUSCL (Monotone Upstream-centered Schemes for Conservation Laws) applicable to hyperbolic field equations will be presented in Part III.

Zielona Gora, in June 2008

Krzysztof Wilmanski

Contents

CONTINUUM THERMODYNAMICS

Part I: Foundations

Chapter 1

Introduction

This volume contains the presentation of main topics of the field theory known as a *continuum thermomechanics*. It means that we present the fundamental notions of the nonlinear continuum mechanics as well as the foundations of continuum thermodynamics in application to thermomechanical processes. We limit the attention to Newtonian thermodynamics which covers all those subject of interest which are related to solids and the majority of fluids. Relativistic thermodynamics of gases will not be considered in this book and we mention only very briefly this topic in Chapter 14.

In contrast to many other books on this subject, we do not expose full theories of particular substances but rather concentrate on the *strategy of thermodynamics* in constructing models of thermomechanical processes in various materials. This means that material theories presented in this book, i.e.

rigid heat conductors,
thermoelastic materials, in particular: compressible Mooney-Rivlin materials,
various solids with constraints,
nonrelativistic ideal gases,
viscoelastic materials, in particular viscous fluids, viscoelastic solids, Maxwell fluids, Rivlin-Ericksen fluids,
elasto-viscoplastic solids,
mixtures of fluids,
poroelastic materials,

are considered to be examples of applications of this strategy rather than full material theories of these substances. For details, in particular, for the experimental verification of various models we refer to the literature.

This volume consists of four main parts. The first one, Chapters 2, 3, and 4, contains the presentation of the classical continuum mechanics. In particular, we present the description of large deformations, kinematics, and balance equations in their Lagrangian and Eulerian formulations. The second part, Chapters 5, 6, 7, is devoted to the formulation of the second law of thermodynamics by means of the entropy inequality. We motivate this law in a heuristic way as well as using the transition from microscopic kinetic theories. The exploitation of the entropy inequality is based on two theorems proved by I-Shih Liu. The first one, concerning the existence of Lagrange multipliers yields fundamental thermodynamical identities and the dissipation functions for various continuous systems. The second one, on the proportionality of fluxes of heat and entropy, leads to the evaluation of thermodynamical identities. Most of the results are obtained for isotropic materials satisfying the principle of material objectivity. For this reason, both topics, isotropy and objectivity, are extensively presented. To appreciate the nonequilibrium extension, the modicum of equilibrium Gibbs thermodynamics is also enclosed. The third part contained in Chapters 8 and 9 concerns a particular form of the theory called extended thermodynamics. Apart from some fundamental features of this approach we discuss in details the nonrelativistic extended thermodynamics of ideal gases, viscoelastic materials and, in particular, non-Newtonian fluids. Additionally, within the frame of classical thermodynamic approach we present the basic features of viscoplastic materials. Finally, the fourth part concerns the thermodynamical theory of multicomponent systems. Chapters 11, 12 and 13 contain the thermodynamics of mixtures of fluids as well as a nonlinear thermodynamical theory of porous materials.

The second half of the twentieth century was witness of the vehement research in the field theories of mechanical systems. The renaissance of these continuum theories begun during the World War II and was motivated by practical needs in the description of rubber, napalm and many other complicated materials with essentially nonlinear behavior. The papers of R. S. Rivlin [270], [264], [271], [272], [265], [266], [268], [267] from the years 1948-1955 (jointly a memoir of the volume of some 250 pages! see: the collected works of Rivlin [26]), J. G. Oldroyd [237] from 1950, C. A. Truesdell [326] from 1952, T. C. Doyle and J. Ericksen [89] from 1956, started this intensive research. Herman F. Nied, a former student of Rivlin's said: "Ronald Rivlin was pretty much universally recognized as being the father of nonlinear mechanics. Before World War II, it was widely considered to be virtually impossible to work in that field. But Rivlin really surprised people with the rigor and simplicity of his solutions." Extensive quotations from these works can be found in the books of C. A. Truesdell and R. A. Toupin [335], C. A. Truesdell and W. Noll [334] as well as Stuart S. Antman [13].

Ronald S. Rivlin
1915 – 2005

Jerald L. Ericksen
1924–

James G. Oldroyd
1921 – 1982

Since the beginning of sixties, the mechanical field theories were extended to cover also thermodynamics of continua. This step was even more difficult than the formulation of nonlinear continuum mechanics as thermodynamics developed in XIX century was mainly oriented to describe homogeneous chemical systems. The development of such an approach was practically completed by ingenious works of W. J. Gibbs [120]. It was very daring to go beyond the scheme of his research. In spite of contemporary criticism of some points of his research, it was primarily B. D. Coleman who started to develop the field theoretical approach to thermodynamics (e.g. [80], [77]). In the mid of sixties numerous papers were published extending and improving his ideas. The review of this development can be found in the books of Truesdell (e.g. [332], [334]).

The development of the modern continuum thermodynamics in Russia whose scientists had important contributions to continuum mechanics was slowed down for political reasons. In the sixties they worked on numerous important nonlinear problems of fluids (e.g. see the book of L. I. Sedov [290], [288], [289] which appeared in Russian in 1968) but did almost no research in continuum thermodynamics. This could be an influence of a single man – L. I. Sedov, who was in charge of the Russian space program and, simultaneously, decided about scientific programs of the country. For many years his influence stopped even the distribution of western publications on thermodynamics in Soviet Union. It was first in the mid of seventies that, for instance, translations of books of C. Truesdell started to appear on the Russian market.

In this book we have chosen one of many possible approaches to thermodynamics. This has been advocated by I. Müller (e.g. [221], [222]). For this reason, we do not go into any details of, for instance, Coleman's approach or Šilhavý's approach [297].

Clifford A. Truesdell
1919 – 2000

Walter Noll
1925–

Leonid I. Sedov
1907 – 1999

Thermodynamics of continua presented in this book is based on four fundamental principles

1. **Continuity,**
2. **Balance equations,**
3. **Local action,**
4. **Thermodynamical admissibility.**

The principle of continuity means that we consider functions on a three-dimensional manifold $\mathcal{B}_0$ called *a body* which satisfy certain mathematical assumptions on a continuity with respect to the volume measure defined on this manifold. These assumptions yield the existence of densities. For example, instead of mass of material points of the classical mechanics, we deal with masses of *subbodies* which are certain three-dimensional subsets of $\mathcal{B}_0$. Such masses are given by integrals of mass densities over subbodies. In continua it does not make any sense talking about a mass of a material point. The material point $\mathbf{X} \in \mathcal{B}_0$ is only a geometrical notion and densities (fields) of a continuum are functions of these points and of the time. Values of these functions have no direct physical meaning known from the classical mechanics. We speak about mass density, momentum density, energy density, etc. but we measure in laboratories their integrals over finite volumes.

Continuity means that densities are continuous functions of the point $\mathbf{X}$ of the body and of time t except of sets of volume measure zero. This means that these functions may possess finite discontinuities on surfaces, lines and at separate points. This is, for example, the case when we consider the propagation of waves. We return to this point later.

The continuity assumption means as well that we consider a special form of changes of the shape of the body due to *motions*. The motion is defined by a differentiable global mapping (*diffeomorphism*) $\mathbf{f}$ of the manifold $\mathcal{B}_0$ on the three-dimensional Euclidean space $\Re^3$. This space is called the *space of configurations*. For our purposes we can identify the body $\mathcal{B}_0$ with a domain in this space, occupied by the body at a chosen *reference time*, say t_0[1]. In this case we call $\mathcal{B}_0$ the *reference configuration*. *The function of motion*

$$\mathbf{f}(.,.) : \mathcal{B}_0 \times \mathcal{T} \rightarrow \Re^3, \tag{1.1}$$

defines for each instant of time $t \in \mathcal{T}$ a *current configuration* $\mathcal{B}_t \subset \Re^3$ of the body. The derivative of this function with respect to $\mathbf{X}$ is called the *deformation gradient* $\mathbf{F}(\mathbf{X}, \mathbf{t})$ (see Fig. 1) and the derivative with respect to time is the *velocity* $\mathbf{v}(\mathbf{X}, t)$ at the material point $\mathbf{X}$. We discuss these notions further in details. The deformation gradient is a linear mapping defined on the so-called *tangent space* to the material manifold and it defines *material vectors* essential for the description of deformations of the body.

We assume that the mapping $\mathbf{f}(., t)$ is *invertible* which means that each position $\mathbf{x}$ can be occupied only by one material point $\mathbf{X}$. This requires

$$\det \mathbf{F} \neq 0. \tag{1.2}$$

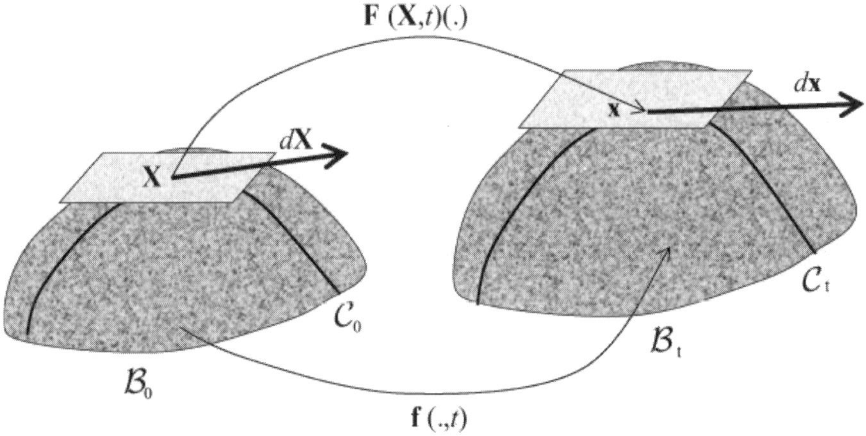

Fig. 1: *Local configuration of a continuum*

The existence of a continuous function of motion $\mathbf{f}$ imposes severe limitations on possible motions of the body. For instance, a creation of new surfaces (opening of a crack in solids, tearing or a creation of vortices) is forbidden by the topological continuity. Also the description of strong mixing (e.g. cigarette smoke in the air) is not possible.

[1]Certain models do not admit such an identification. For example, there are continuous models of dislocations which require a more general structure of the manifold than this indicated by the Euclidean space (e.g. [231]).

The second principle – *balance equations* – means that some most fundamental quantities appearing in models of the continuum satisfy relations describing their time changes in terms of surface and volume supplies. In thermomechanical models which we consider in this book these quantities are: mass density, momentum density, angular momentum density, energy density and entropy density. In particular cases, balance equations become *conservation laws* of mass, momentum, angular momentum and energy. We discuss further the detailed structure of these equations. Apart from balance equations a particular model may contain additional equations such as *evolution equations* of internal variables but we assume that the above listed conservation laws are unconditionally satisfied in any model. The violation of conservation laws of mass, momentum or energy leads to a *perpetuum mobile,* i.e. the system may do a useful work without any time limit and without any supply from the surrounding. Even though it may not be excluded in a microscopic world described by a quantum theory, the existence of perpetuum mobile contradicts our macroscopic experience.

The principle of *local action* requires that a reaction of the body on external actions is transmitted to material points by interactions of parts of the body through *surfaces of contacts*, i.e. a reaction of each material point is limited to an influence of its infinitesimal neighborhood. Direct interactions of two or more material points at finite distances are not possible. Consequently, such actions as gravitational forces between parts of the body or Coulomb electromagnetic interactions are not modelled by a continuum. Attempts to include these *nonlocal* interactions failed and only some approximations of such actions by the so-called higher gradient theories are possible without the violation of some basic mathematical assumptions of the continuum. We discuss this problem within the subject of constitutive (material) relations.

Finally, the principle of thermodynamical admissibility reflects the requirement that the *second law of thermodynamics* and certain *thermodynamical stability conditions* are satisfied. These will be the main subject of this book.

I would like to express my gratitude to my teachers, coworkers and colleagues for their contributions to this work either directly in common research which we did in the last 30 years or by their lectures, seminars and discussions on various occasions. Among them, I would like to mention in particular Prof. Erich Bauer (Technische Universität Graz), Prof. Renato Lancellotta (Politecnico di Torino), Prof. Benjamin Landkof, Prof. Miles Rubin (Technion, Haifa) and Prof. Claudio Tamagnini (Università di Perugia) who in recent years encouraged me to give courses on the subjects appearing in this volume:

(1) *Continuum Mechanics and Thermodynamics*, Politecnico di Torino, Turin, April 2001,
(2) *Continuum Theories of Mixtures*, La Sapienza, Università di Roma, Rome, June 2005,
(3) *Continuum Thermodynamics*, Technion, Haifa, Spring 2006,
(4) *Continuum Thermodynamics*, Technische Universität Graz, September 2007.

I have incorporated in this volume many parts of my notes for these courses.

Chapter 2

Geometry

2.1 Deformation gradient, material vectors

We proceed to discuss the geometry of the body changing in time due to motion. As already mentioned the current configuration of the body is defined by the function $\mathbf{f}$. Let us choose an arbitrary smooth curve $\mathcal{C}_0$ in the initial configuration $\mathcal{B}_0$ and investigate its current image $\mathcal{C}_t := \mathbf{f}(\mathcal{C}_0, t) \equiv \left\{\mathbf{x} \in \Re^3 \middle| \mathbf{x} = \mathbf{f}(\mathbf{X}, t), \mathbf{X} \in \mathcal{C}_0\right\}$. It is convenient to write the equation of $\mathcal{C}_0$ in the parametric form

$$\mathbf{X} = \mathbf{X}(S), \tag{2.1}$$

where S is the parameter defining the distance along the curve. Then the vector

$$\mathbf{T} = \frac{d\mathbf{X}}{dS}, \tag{2.2}$$

is a unit vector (i.e. $|\mathbf{T}|^2 = \mathbf{T} \cdot \mathbf{T} = 1$) tangent to the curve. The infinitesimal vector

$$d\mathbf{X} = \mathbf{T}dS, \tag{2.3}$$

is tangent to the curve $\mathcal{C}_0$ as well. According to the definition of the current image $\mathcal{C}_t$ its tangent infinitesimal vector $d\mathbf{x}$ is given by the relation [175], [108], [109], [176]

$$\forall \mathbf{X} \in \mathcal{C}_0 : \quad d\mathbf{x} = (\operatorname{Grad} \mathbf{f})\, d\mathbf{X} \equiv \mathbf{t}dS, \quad \mathbf{t} := \mathbf{FT}, \quad \mathbf{F} := \operatorname{Grad} \mathbf{f}, \tag{2.4}$$

where $\mathbf{F}$ is the deformation gradient at the point $\mathbf{X}$ and the instant of time t. Hence, the infinitesimal vector $d\mathbf{x}$ tangent to the curve $\mathcal{C}_t$ which deforms with the body is given by a *linear transformation* of the infinitesimal vector $d\mathbf{X}$. This transformation is defined by the quadratic matrix which is given by components of the deformation gradient $\mathbf{F}$. It is easy to be seen in the representation in Cartesian coordinates which are admissible due to the assumption that configuration spaces are Euclidean. If we choose the unit orthogonal basis vectors $\{\mathbf{e}_K\}, K = 1, 2, 3$ for the initial configuration $\mathcal{B}_0$ and $\{\mathbf{e}_k\}, k = 1, 2, 3$ for the current configuration $\mathcal{B}_t$ then the above relations can be written in the form

$$
\begin{aligned}
d\mathbf{X} &= dX^K \mathbf{e}_K = T^K \mathbf{e}_K dS, \\
d\mathbf{x} &= dx^k \mathbf{e}_k = t^k \mathbf{e}_k dS, \quad t^k := F^k_K T^K, \quad \mathbf{F} = F^k_K \mathbf{e}_k \otimes \mathbf{e}_K.
\end{aligned} \tag{2.5}
$$

The tangent vector $\mathbf{t}$ is the current image of the vector $\mathbf{T}$ and it is given by the rule defined by the relation (2.4). This rule of transformation defines the so-called *material* vectors. Relations (2.5) have been discovered by Euler [109]. Not all vectors transform according to this rule and we see an example of a different rule of transformation in the sequel.

The most important property of the above transformation is that it is independent of the choice of curve going through a chosen point $\mathbf{X}$. The deformation gradient $\mathbf{F}$ depends only on $\mathbf{X}$ and t and defines the transformation of an arbitrary vector $\mathbf{T}$ tangent to a material curve crossing the point $\mathbf{X}$. We say that the gradient $\mathbf{F}$ considered as a mapping maps a tangent space at the point $\mathbf{X}$ into the tangent space at the point $\mathbf{x} = \mathbf{f}(\mathbf{X}, t)$.

Further we consider also Eulerian description where spatial position $\mathbf{x}$ is considered to be the independent variable and the point $\mathbf{X}$ is given by the inverse of the function of motion, i.e.

$$
\mathbf{X} = \mathbf{f}^{-1}(\mathbf{x}, t), \quad \mathbf{x} \in \mathcal{B}_t, \quad t \in \mathcal{T}. \tag{2.6}
$$

As we have already mentioned the existence of this mapping follows from the assumption (1.2). In this description, we can define the inverse of the deformation gradient as the matrix of partial derivatives of $\mathbf{f}^{-1}$ with respect to $\mathbf{x}$. In Cartesian coordinates we have

$$
\begin{aligned}
x^k &= f^k\left(X^1, X^2, X^3, t\right) = f^k\left(X^K, t\right), \quad X^K = f^{-1K}\left(x^k, t\right), \\
F^k_K &= \frac{\partial f^k}{\partial X^K} \equiv f^k_{;K}, \quad {F^{-1K}}_k = \frac{\partial f^{-1K}}{\partial x^k} \equiv {f^{-1K}}_{;k},
\end{aligned} \tag{2.7}
$$

$$
f^k_{;K} {f^{-1K}}_{;m} = \delta^k_m, \quad {f^{-1K}}_{;k} f^k_{;M} = \delta^K_M.
$$

In these relations we indicate the shorthand notation for the variables – instead of writing all three spatial variables we write often the joint symbol for coordinates. Simultaneously, the semicolon denotes in the case of Cartesian coordinates the partial derivatives. However, the relations hold true also in arbitrary curvilinear coordinates and then this symbol denotes the covariant derivative (see Appendix A).

The third group of relations (2.7) is the set of nine equations either for F^k_K or for ${F^{-1K}}_k$. The solution can be constructed by Cramer's rule and it has been done by Euler [109]

$$
{f^{-1K}}_{;k} = \frac{\frac{1}{2}\epsilon^{KMP}\epsilon_{kmp}f^m_{;M}f^p_{;P}}{\frac{1}{6}\epsilon^{QRS}\epsilon_{qrs}f^q_{;Q}f^r_{;R}f^s_{;S}}, \tag{2.8}
$$

where $\epsilon_{kmp}, \epsilon_{KMP}$, etc. are permutation symbols (i.e. they are either ± 1 for even and, respectively, odd transformation of different indices, $1, 2, 3$, and zero for two equal indices).

Integrals over material surfaces appearing in balance equations of continua contain surface elements oriented by unit vectors perpendicular to such elements. A material surface $\mathcal{S}$ is defined as a collection of material curves and, for simplicity, we assume that it is parametrized by two orthogonal families of such curves, say $\mathbf{X}(\Xi^\alpha)$, $\alpha = 1, 2$. At a chosen point $\mathbf{X} \in \mathcal{S}$, we consider two orthogonal parametric curves whose tangent vectors are $\mathbf{T}_\alpha = \partial\mathbf{X}/\partial\Xi^\alpha$, $\alpha = 1, 2$. Then the surface element $d\mathcal{S}$ is defined through the vector product

$$
\left(\frac{\partial\mathbf{X}}{\partial\Xi^1}d\Xi^1\right) \times \left(\frac{\partial\mathbf{X}}{\partial\Xi^2}d\Xi^2\right) = \mathbf{N}\left|\mathbf{T}_1 \times \mathbf{T}_2\right| d\Xi^1 d\Xi^2 = \mathbf{N}d\mathcal{S}, \tag{2.9}
$$

where

$$
\mathbf{N} = \frac{\mathbf{T}_1 \times \mathbf{T}_2}{\left|\mathbf{T}_1 \times \mathbf{T}_2\right|}, \quad d\mathcal{S} = \left|\mathbf{T}_1 \times \mathbf{T}_2\right| d\Xi^1 d\Xi^2, \tag{2.10}
$$

and $\mathbf{N}$ is the unit vector perpendicular to the surface $\mathcal{S}$ at the point $\mathbf{X}$.

This surface in the current configuration, $\mathcal{S}_t = \{\mathbf{x} | \, \mathbf{x} = \mathbf{f}(\mathbf{X}, t), \mathbf{X} \in \mathcal{S}\}$, has at the point $\mathbf{x} = \mathbf{f}(\mathbf{X},t)$ the following tangent and unit orthogonal vectors

$$
\mathbf{t}_1 = \mathbf{F}\mathbf{T}_1, \quad \mathbf{t}_2 = \mathbf{F}\mathbf{T}_2, \quad \mathbf{n} = \frac{\mathbf{t}_1 \times \mathbf{t}_2}{\left|\mathbf{t}_1 \times \mathbf{t}_2\right|}, \tag{2.11}
$$

i.e.

$$
\left(\frac{\partial\mathbf{x}}{\partial\Xi^1}d\Xi^1\right) \times \left(\frac{\partial\mathbf{x}}{\partial\Xi^2}d\Xi^2\right) = \mathbf{n}\left|\mathbf{t}_1 \times \mathbf{t}_2\right| d\Xi^1 d\Xi^2 = \mathbf{n}d\mathcal{S}_t \tag{2.12}
$$

$$
\Rightarrow \quad d\mathcal{S}_t = \left|\mathbf{t}_1 \times \mathbf{t}_2\right| d\Xi^1 d\Xi^2.
$$

Simultaneously, we have

$$\begin{aligned}(\mathbf{t}_1\times\mathbf{t}_2)\cdot\mathbf{e}_k &= \epsilon_{kmn}t_1^m t_2^n = \epsilon_{kmn}F^m_M T_1^M F^n_N T_2^N \\ &= \epsilon_{imn}F^i_P F^{-1P}_{\;\;k} F^m_M F^n_N T_1^M T_2^N = \epsilon_{PMN} J F^{-1P}_{\;\;k} T_1^M T_2^N \\ &= J N_P F^{-1P}_{\;\;k}, \quad J := \det\mathbf{F} = \frac{1}{6}\epsilon^{QRS}\epsilon_{qrs}F^q_Q F^r_R F^s_S > 0.\end{aligned} \tag{2.13}$$

The above relations follow easily from the representation of the vector product in the form of the determinant. Consequently

$$\mathbf{n} = \frac{\boldsymbol{\nu}}{|\boldsymbol{\nu}|}, \quad \boldsymbol{\nu} = \mathbf{F}^{-T}\mathbf{N}. \tag{2.14}$$

This is the rule of transformation for unit vectors perpendicular to material surfaces (i.e. *normal vectors*).

Relation (2.12) yields immediately (*Nanson formula*)

$$d\mathbf{s} = J\mathbf{F}^{-T}d\mathbf{S}, \quad d\mathbf{s} = \mathbf{n}d\mathcal{S}_t, \quad d\mathbf{S} = \mathbf{N}d\mathcal{S}. \tag{2.15}$$

Hence, the rules of transformation (2.14) for $\mathbf{N}$ and (2.15) for $d\mathbf{S}$ differ from the rule (2.4) for material vectors. However, it is easy to prove that vectors

$$\mathbf{k} = \mathbf{F}\mathbf{K}, \quad \mathbf{m} = \mathbf{F}^{-T}\mathbf{K}, \tag{2.16}$$

the first one – material (i.e. tangent to a material curve), and the second one not material but perpendicular to a material surface, are parallel to each other. Namely, let us construct the vector product

$$\mathbf{k}\times\mathbf{m} = \epsilon_{ijk}F^j_P K^P F^{-1Q}_{\;\;k} K_Q \mathbf{e}_i. \tag{2.17}$$

Bearing Euler's relation (2.8) in mind, we obtain

$$\begin{aligned}\mathbf{k}\times\mathbf{m} &= \frac{1}{2J}\epsilon^{i}_{\cdot jk}\epsilon^{k}_{\cdot mr}\epsilon^{QMR}F^j_P F^m_M F^r_R K^P K_Q \mathbf{e}_i \\ &= \frac{1}{2J}\left(\delta^i_m\delta_{jr} - \delta^i_r\delta_{jm}\right)\epsilon^{QMR}F^j_P F^m_M F^r_R K^P K_Q \mathbf{e}_i \\ &= \frac{1}{2J}F_{iP}\left(F^i_M F^j_R - F^i_R F^j_M\right)\epsilon^{QMR}K^P K_Q \mathbf{e}_i = \mathbf{0},\end{aligned} \tag{2.18}$$

which proves the statement. We have used the relation (A.35) quoted in Appendix A.

By means of the relation (2.15) we can also define the change of the infinitesimal volume element, dV, into its current value, dv. We have

$$\begin{aligned}dV &= \left[\left(\mathbf{T}_1 dX^1\right)\times\left(\mathbf{T}_2 dX^2\right)\right]\cdot\left(\mathbf{T}_3 dX^3\right) = \left[(\mathbf{T}_1\times\mathbf{T}_2)\cdot\mathbf{T}_3\right]dX^1 dX^2 dX^3, \\ dv &= \left[\left(\mathbf{t}_1 dx^1\right)\times\left(\mathbf{t}_2 dx^2\right)\right]\cdot\left(\mathbf{t}_3 dx^3\right) = \left[(\mathbf{t}_1\times\mathbf{t}_2)\cdot\mathbf{t}_3\right]dx^1 dx^2 dx^3,\end{aligned} \tag{2.19}$$

where (of course, if we choose $|\mathbf{T}_1| = |\mathbf{T}_2| = |\mathbf{T}_3| = 1$ then $dV = dX^1 dX^2 dX^3$)

$$\mathbf{t}_1 = \mathbf{F}\mathbf{T}_1, \quad \mathbf{t}_2 = \mathbf{F}\mathbf{T}_2, \quad \mathbf{t}_3 = \mathbf{F}\mathbf{T}_3. \tag{2.20}$$

Scalar multiplication by $\mathbf{t}_3$ instead of $\mathbf{e}_3$ in (2.13) yields

$$dv = JdV. \tag{2.21}$$

Hence, the Jacobian J determines changes of infinitesimal volume elements caused by the transformation from the reference to current configuration. Its value for the identical mapping is equal to one. According to the condition (1.2) it cannot cross the line of zero values and consequently, due to continuity, it must be positive: $J > 0$.

2.2 Measures of deformation

As already mentioned the transformation of vectors $d\mathbf{X}$ caused by the motion determines local deformations of the body. We need only changes of length of infinitesimal vectors in an arbitrary direction in order to find the local changes of the size and shape of material elements. These length changes follow from the relation

$$ds^2 = d\mathbf{x} \cdot d\mathbf{x} = (\mathbf{F}d\mathbf{X}) \cdot (\mathbf{F}d\mathbf{X}) = d\mathbf{X} \cdot \mathbf{C}d\mathbf{X}, \quad \mathbf{C} := \mathbf{F}^T\mathbf{F} = \mathbf{C}^T, \tag{2.22}$$

$$\det \mathbf{C} = (\det \mathbf{F})^2 = J^2 > 0, \tag{2.23}$$

where the symmetric tensor $\mathbf{C}$ is called the *right Cauchy-Green deformation tensor*.

There arises the question of what happens to the nine components of the deformation gradient $\mathbf{F}$ if six components of $\mathbf{C}$ are sufficient to describe the deformation. The answer is given by the *polar decomposition theorem*: under the assumption of nonsingularity of motion (1.2) there exists a unique decomposition of the deformation gradient of the following form

$$\mathbf{F} = \mathbf{R}\mathbf{U}, \quad \mathbf{R}^{-1} = \mathbf{R}^T, \quad \mathbf{U}^T = \mathbf{U}, \tag{2.24}$$

i.e. there exist a unique orthogonal tensor $\mathbf{R}$ (it rotates vectors without changing their length) and a unique symmetric *right stretch tensor* $\mathbf{U}$ whose product is equal to the deformation gradient.

Augustin Louis Cauchy
1789 – 1857

Leonhard Euler
1707 – 1783

George Green
1793 – 1841

The proof of the theorem is easy and, simultaneously, it shows the procedure of calculating these two tensors. In order to take the square root of a tensor, we have to represent it in the diagonal form. We show further that $\mathbf{U}$ is the square root of $\mathbf{C}$. The diagonal representation of $\mathbf{C}$ is provided by the solution of the eigenvalue problem. Namely, for the right Cauchy-Green tensor we have the following eigenvalue problem

$$(\mathbf{C}-\lambda_C\mathbf{1})\,\mathbf{K}_C=\mathbf{0}, \tag{2.25}$$

where the eigenvalues λ_C satisfy the characteristic equation

$$\det(\mathbf{C}-\lambda_C\mathbf{1})=0,$$

i.e.

$$\lambda_C^3-I\lambda_C^2+II\lambda_C-III=0,\quad I=\operatorname{tr}\mathbf{C},\quad II=\frac{1}{2}\left(I^2-\operatorname{tr}\mathbf{C}^2\right),\quad III=\det\mathbf{C}, \tag{2.26}$$

and I, II, III are the so-called principal invariants of $\mathbf{C}$. The tensor $\mathbf{1}$ denotes the metric tensor of the reference configuration. Its components are as follows

$$\mathbf{1}=\delta^{KL}\mathbf{e}_K\otimes\mathbf{e}_L=G^{KL}\mathbf{G}_K\otimes\mathbf{G}_L, \tag{2.27}$$

where the first relation holds for Cartesian frames and the second one – for arbitrary frames.

Hence, there exist three eigenvalues $\lambda_C^{(\alpha)}, \alpha=1,2,3$, and due to the symmetry of $\mathbf{C}$ they are all real. They are called *principal stretches*. The corresponding three unit eigenvectors $\mathbf{K}_C^\alpha$ are linearly independent and this yields the following *spectral representation* of the deformation tensor $\mathbf{C}$

$$\mathbf{C} = \sum_{\alpha=1}^{3} \lambda_C^{(\alpha)} \mathbf{K}_C^\alpha \otimes \mathbf{K}_C^\alpha. \tag{2.28}$$

Simultaneously, for the stretch tensor $\mathbf{U}$ we have the following eigenvalue problem

$$(\mathbf{U} - \lambda_U \mathbf{1}) \mathbf{K}_U = \mathbf{0}. \tag{2.29}$$

If we multiply this relation by $\mathbf{U}$ from the left and use (2.25) we obtain

$$\left(\mathbf{C} - \lambda_U^2 \mathbf{1}\right) \mathbf{K}_U = \mathbf{0} \quad \Longrightarrow \quad \lambda_U = \sqrt{\lambda_C}, \quad \mathbf{K}_U = \mathbf{K}_C, \tag{2.30}$$

where we have used the relation

$$\mathbf{C} = \mathbf{F}^T \mathbf{F} = \mathbf{F}^T \mathbf{R} \mathbf{R}^{-1} \mathbf{F} = \left(\mathbf{R}^{-1} \mathbf{F}\right)^T \left(\mathbf{R}^{-1} \mathbf{F}\right) = \mathbf{U}^2. \tag{2.31}$$

This means that the spectral representation of the stretch tensor is as follows

$$\mathbf{U} = \sum_{\alpha=1}^{3} \sqrt{\lambda_C^{(\alpha)}} \mathbf{K}_C^\alpha \otimes \mathbf{K}_C^\alpha. \tag{2.32}$$

As both the determination of $\mathbf{C}$ as the product of the deformation gradient $\mathbf{F}$ with itself and the solution of the eigenvalue problem for $\mathbf{C}$ are straightforward, the above relation determines easily the stretch tensor $\mathbf{U} = \mathbf{C}^{1/2}$. It remains to find the inverse of $\mathbf{U}$ and we have

$$\mathbf{R} = \mathbf{F}\mathbf{U}^{-1} \quad \Longrightarrow \quad \mathbf{R}^T \mathbf{R} = \mathbf{U}^{-1} \mathbf{F}^T \mathbf{F} \mathbf{U}^{-1} = \mathbf{U}^{-1} \mathbf{C} \mathbf{U}^{-1} = \mathbf{1}, \tag{2.33}$$

and, consequently, $\mathbf{R}$ is orthogonal.

Similarly one can also prove the dual decomposition

$$\mathbf{F} = \mathbf{V}\mathbf{R}, \quad \mathbf{R}^{-1} = \mathbf{R}^T, \quad \mathbf{V}^T = \mathbf{V}, \tag{2.34}$$

$$\mathbf{V} = \sum_{\alpha=1}^{3} \sqrt{\lambda_B^{(\alpha)}} \mathbf{k}_B^\alpha \otimes \mathbf{k}_B^\alpha, \quad \mathbf{B} = \mathbf{V}^2, \tag{2.35}$$

where $\mathbf{V}$ is the *left stretch tensor* and $\mathbf{k}_B^\alpha$ denote eigenvectors of the left Cauchy-Green deformation tensor $\mathbf{B}$

$$\mathbf{B} = \mathbf{F}\mathbf{F}^T = \sum_{\alpha=1}^{3} \lambda_B^{(\alpha)} \mathbf{k}_B^\alpha \otimes \mathbf{k}_B^\alpha. \tag{2.36}$$

It can be shown immediately that eigenvalues λ_B and λ_C are identical. Namely

$$
\begin{aligned}
(\mathbf{B}-\lambda_B\mathbf{1})\,\mathbf{k}_B=\mathbf{0} \quad &\Rightarrow \quad \mathbf{F}\left(\mathbf{1}-\lambda_B\mathbf{F}^{-1}\mathbf{F}^{-T}\right)\mathbf{F}^T\mathbf{k}_B=\mathbf{0} \\
\Rightarrow \quad \mathbf{F}\mathbf{C}^{-1}\left(\mathbf{C}-\lambda_B\mathbf{1}\right)\mathbf{F}^T\mathbf{k}_B &= \mathbf{0} \quad \Rightarrow \quad \lambda_B=\lambda_C=\lambda, \\
\mathbf{F}^T\mathbf{k}_B \| \mathbf{K}_C \quad &\Rightarrow \quad \mathbf{k}_B=\frac{\mathbf{F}\mathbf{K}_C}{|\mathbf{F}\mathbf{K}_C|},
\end{aligned} \tag{2.37}
$$

where the relation (2.18) has been used.

Hence, the invariants of $\mathbf{C}$ and $\mathbf{B}$ are identical and the invariants for stretches $\mathbf{U}$ and, consequently also for $\mathbf{V}$, satisfy the relations

$$
\begin{aligned}
i &= \operatorname{tr}\mathbf{U} = \sum_{\alpha=1}^{3}\sqrt{\lambda^{(\alpha)}}, \\
ii &= \frac{1}{2}\left(i^2-\operatorname{tr}\mathbf{U}^2\right) = \sqrt{\lambda^{(1)}\lambda^{(2)}}+\sqrt{\lambda^{(2)}\lambda^{(3)}}+\sqrt{\lambda^{(1)}\lambda^{(3)}}, \\
iii &= \det\mathbf{U} = \sqrt{\lambda^{(1)}\lambda^{(2)}\lambda^{(3)}}.
\end{aligned} \tag{2.38}
$$

The above considerations show that local changes of geometry are given only by the tensor $\mathbf{U}$ and, consequently, by the tensor $\mathbf{C}$. This is the most important contribution of A.-L. Cauchy to the theory of finite deformations. The orthogonal tensor $\mathbf{R}$ possesses, of course, 3 independent components (e.g. Euler angles) and it determines local rotations as an infinitesimal material element were a rigid body.

Depending on a particular application, there are many possibilities to define deformation tensors. They are all equivalent. Some most frequently used deformation measures are quoted in Table 1 (for the extensive history of the subject of deformation measures see: [328], [335]).

In some works, for instance, on plasticity, some other measures of deformation are occasionally applied as well. In Table 1 (cont.) which is in part a modified version of Table 3a of Truesdell and Toupin [335] we quote some of them.

Obviously, the last two measures reduce to the Green-St.Venant measure and, respectively, to Almansi-Hamel measure for $m=2$, and the first one – to Biot's measure for $m=1$.

The logarithmic measure is motivated by the following simple consideration. The extension of a cylindrical bar from the initial length l_0 to the final length l_f can be defined by the formula

$$
e_{axial} = \int_{l_0}^{l_f}\frac{dl}{l} = \ln\left(\frac{l_f}{l_0}\right), \tag{2.39}
$$

where l is the current length of the bar. This strain measure is additive – if we extend the bar from l_0 to l_{int} and then from l_{int} to l_f then the full strain is the sum of two corresponding partial strains.

Table 1: *Measures of deformation*

Name		Def.	Eigen-values	Eigen-vectors	Author
right Cauchy – Green	$\mathbf{C}$	$\mathbf{F}^T\mathbf{F}$	$\lambda \equiv \lambda_C$	$\mathbf{K} \equiv \mathbf{K}_C$	G. Green 1841
left Cauchy – Green (Finger)	$\mathbf{B}$	$\mathbf{F}\mathbf{F}^T$	λ	$\mathbf{k} \equiv \frac{\mathbf{FK}}{\lvert\mathbf{FK}\rvert}$	J. Finger 1894
right stretch	$\mathbf{U}$	$\mathbf{C}^{1/2}$	$\lambda^{1/2}$	$\mathbf{K}$	L. Euler?
left stretch	$\mathbf{V}$	$\mathbf{B}^{1/2}$	$\lambda^{1/2}$	$\mathbf{k}$	L. Euler?
Cauchy	$\mathbf{c}$	$\mathbf{B}^{-1}$	$1/\lambda$	$\mathbf{k}$	L. A. Cauchy 1827
Green – St.Venant (Lagrange)[311]	$\mathbf{E}$	$\frac{\mathbf{C}-\mathbf{1}}{2}$	$\frac{\lambda-1}{2}$	$\mathbf{K}$	A. de St.Venant 1844
Almansi – Hamel (Euler)	$\mathbf{e}$	$\frac{\mathbf{1}-\mathbf{c}}{2}$	$\frac{1-1/\lambda}{2}$	$\mathbf{k}$	E. Almansi 1911
Piola	$\mathbf{C}^{-1}$		$1/\lambda$	$\mathbf{K}$	G. Piola 1833

This logarithmic axial strain can be generalized to the three-dimensional case in the following way

$$\mathbf{U}_{(l)} = \sum_{\alpha=1}^{3} \frac{1}{2} \ln \lambda^{(\alpha)} \mathbf{K}_C^{\alpha} \otimes \mathbf{K}_C^{\alpha}, \quad \lambda^{(\alpha)} \equiv \lambda_C^{(\alpha)}, \tag{2.40}$$

for the right logarithmic stretch tensor. In the three-dimensional case the additivity is not preserved anymore. The relation (2.40) yields expansions presented in Table 1 (cont.). Similarly, for the left logarithmic stretch tensor we have

$$\mathbf{V}_{(l)} = \sum_{\alpha=1}^{3} \frac{1}{2} \ln \lambda^{(\alpha)} \mathbf{k}_B^{\alpha} \otimes \mathbf{k}_B^{\alpha}, \quad \lambda^{(\alpha)} \equiv \lambda_B^{(\alpha)}. \tag{2.41}$$

Let us return to formula (2.2) and its geometrical interpretation. Obviously, the right Cauchy-Green deformation tensor can be treated as a transformation of the metric tensor in the reference configuration – from the metric $\delta_{KL} = \mathbf{e}_K \cdot \mathbf{e}_L$ (or, generally, $G_{KL} = \mathbf{G}_K \cdot \mathbf{G}_L$, where $\mathbf{G}_1, \mathbf{G}_2, \mathbf{G}_3$

are the vectors of the basis in the reference state) to C_{KL} with the corresponding change of the infinitesimal line element – from dS to ds after the deformation. Similarly, we can consider the transformation of the line element in the current configuration

Table 1 (cont.): *Measures of deformation*

Signorini [295]	$\frac{1}{2}\left(\mathbf{c}^{-1}-\mathbf{1}\right)$
Biot [35]	$\mathbf{C}^{1/2}-\mathbf{1}$
Murnaghan [216]	$\mathbf{c}^{-1/2}$
Swainger according to Herschey [144]	$\mathbf{1}-\mathbf{c}^{-1/2}$
Mooney [214]	$(III)^{-1/2}\,\mathbf{c}^{-1/2}$
Oldroyd [236]	$(III)^{1/2}\,\mathbf{c}$
Hencky [143]	$\mathbf{H}=\frac{1}{2}\ln\mathbf{C}=\frac{1}{2}\sum_{k=1}^{\infty}\frac{(-1)^{k-1}}{k!}\left(\mathbf{C}-\mathbf{1}\right)^{k}$
Hencky [143]	$\mathbf{h}=-\frac{1}{2}\ln\mathbf{c}=-\frac{1}{2}\sum_{k=1}^{\infty}\frac{(-1)^{k-1}}{k!}\left(\mathbf{c}-\mathbf{1}\right)^{k}$
generalized Green-St. Venant (Ogden?)	$\mathbf{E}^{(m)}=\frac{1}{m}\left(\mathbf{U}^{m}-\mathbf{1}\right),\quad m\neq 0$
generalized Almansi-Hamel (Ogden?)	$\mathbf{e}^{(m)}=\frac{1}{m}\left(\mathbf{1}-\mathbf{v}^{-m}\right)$

$$dS^2 = d\mathbf{X} \cdot d\mathbf{X} = \left(\mathbf{F}^{-1} d\mathbf{x}\right) \cdot \left(\mathbf{F}^{-1} d\mathbf{x}\right) = d\mathbf{x} \cdot \left(\mathbf{F}^{-T}\mathbf{F}^{-1}\right) d\mathbf{x} \qquad (2.42)$$
$$= d\mathbf{x} \cdot \left(\mathbf{F}\mathbf{F}^T\right)^{-1} d\mathbf{x} = d\mathbf{x} \cdot \mathbf{c} d\mathbf{x},$$

i.e. the Cauchy deformation tensor $\mathbf{c}$ (the inverse of the left Cauchy-Green deformation tensor $\mathbf{B}$) can be treated as the transformation of the metric tensor in the current configuration – from the metric $\delta_{kl} = \mathbf{e}_k \cdot \mathbf{e}_l$ (or, generally, $\mathbf{g}_{kl} = \mathbf{g}_k \cdot \mathbf{g}_l$, where $\mathbf{g}_1, \mathbf{g}_2, \mathbf{g}_3$ are the vectors of the basis in the current state) to c_{kl} with the corresponding change of the infinitesimal line element – from ds to dS.

In addition to these two interpretations one can consider a relative change of the line element which is usually measured by the difference of squares

$$ds^2 - dS^2 = d\mathbf{X} \cdot (\mathbf{C} - \mathbf{1}) \, d\mathbf{X} = 2d\mathbf{X} \cdot \mathbf{E} d\mathbf{X}, \qquad (2.43)$$
$$dS^2 - ds^2 = d\mathbf{x} \cdot (\mathbf{c} - \mathbf{1}) \, d\mathbf{x} = -2d\mathbf{x} \cdot \mathbf{e} d\mathbf{x}.$$

These two measures become identical in the theories of small deformations. Such theories are defined in terms of eigenvalues of the right Cauchy-Green tensor in the following way

$$\|\mathbf{C} - \mathbf{1}\| = \max\left\{\left|\lambda^{(1)} - 1\right|, \left|\lambda^{(2)} - 1\right|, \left|\lambda^{(3)} - 1\right|\right\} \ll 1, \qquad (2.44)$$

where $\lambda^{(1)}, \lambda^{(2)}, \lambda^{(3)}$ are eigenvalues of $\mathbf{C}$ (or, identically, $\mathbf{B}$). Then eigenvalues of the Green-St.Venant measure $\mathbf{E}$ and Almansi-Hamel measure $\mathbf{e}$ are approximately identical and equal to (comp. Table 1)

$$\frac{1}{2}\left(\lambda^{(1)} - 1\right), \quad \frac{1}{2}\left(\lambda^{(2)} - 1\right), \quad \frac{1}{2}\left(\lambda^{(3)} - 1\right). \qquad (2.45)$$

This means that in such linear theories we do not have to distinguish between metrics of the reference state and the current state. The Lagrangian and Eulerian descriptions which we discuss further in details become identical.

As we have indicated above, the change of length and relative direction caused by deformation is loosely called *strain* [260].

Now, we present some geometrical properties of the strain. Let us consider again the line element $d\mathbf{X}$ at the point $\mathbf{X} \in \mathcal{B}_0$ satisfying the equation

$$d\mathbf{X} \cdot \mathbf{dX} = \text{const.}, \qquad (2.46)$$

which defines a sphere of radius $|d\mathbf{X}|$ and center $\mathbf{X}$. According to (2.42) we have in the current configuration

$$d\mathbf{x} \cdot \mathbf{c} d\mathbf{x} = \text{const.}, \tag{2.47}$$

and, since $\mathbf{c}$ is positive definite this equation defines an ellipsoid centered on the point $\mathbf{x} = \mathbf{f}(\mathbf{X},t) \in \mathcal{B}_t$. The material contained in the sphere (2.46) is deformed into this ellipsoid. It is called *Eulerian strain ellipsoid*. Its principal axes (*Eulerian principal axes*) coincide with eigenvectors $\mathbf{k}^\alpha, \alpha = 1,2,3$ of $\mathbf{c}$ (and, hence, of the left stretch $\mathbf{V}$; see: Table 1) and the semi-axes of the ellipsoid have the lengths proportional to $\sqrt{\lambda^{(\alpha)}}, \alpha = 1,2,3$.

Perpendicular diameters of the sphere (2.46) along the line elements $d\mathbf{X}$ and $d\mathbf{X}'$ satisfy the relation $d\mathbf{X} \cdot d\mathbf{X}' = 0$. Hence, for these two diameters

$$d\mathbf{X} = \mathbf{F}^{-1} d\mathbf{x}, \quad d\mathbf{X}' = \mathbf{F}^{-1} d\mathbf{x}' \quad \Rightarrow \quad d\mathbf{x}' \cdot \mathbf{c} d\mathbf{x} = 0. \tag{2.48}$$

However, the vector $\mathbf{c}d\mathbf{x}$ is normal to the Eulerian strain ellipsoid because the equation of an ellipsoid $\phi = \mathbf{x} \cdot \mathbf{c}\mathbf{x} =$const. yields grad $\phi = 2\mathbf{c}\mathbf{x}$ which is orthogonal to ϕ. Hence, $d\mathbf{x}$ and $d\mathbf{x}'$ are conjugate diameters of the ellipsoid. If, simultaneously, $d\mathbf{x} \cdot d\mathbf{x}' = 0$ then they must coincide with the principal axes of the Eulerian strain ellipsoid. Only for such directions angles between the line elements do not change during the deformation.

On the other hand, the material contained in the sphere of radius $|d\mathbf{x}|$ with the center at $\mathbf{x} \in \mathcal{B}_t$

$$d\mathbf{x} \cdot d\mathbf{x} = \text{const.}, \tag{2.49}$$

comes from the ellipsoid

$$d\mathbf{X} \cdot \mathbf{C} d\mathbf{X} = \text{const.}, \tag{2.50}$$

centered at $\mathbf{X} = \mathbf{f}^{-1}(\mathbf{x},t) \in \mathcal{B}_0$. It is called the *reciprocal Lagrangian strain ellipsoid*. Its principal axes (*Lagrangian principal axes*) coincide with the eigenvectors $\mathbf{K}^\alpha, \alpha = 1,2,3$ of $\mathbf{C}$ (and, hence, of the right stretch $\mathbf{U}$; see: Table 1).

2.3 Extension, dilatation, shear

Now we consider some special cases of deformation which often appear in further considerations. In general, the deformation gradient $\mathbf{F}$ varies from point to point in $\mathcal{B}_0$. Such a deformation is called *inhomogeneous*. In special cases, when $\mathbf{F}$ is independent of $\mathbf{X}$ we say that the deformation of the body is *homogeneous*. The function of motion for such deformations has the general form

$$\mathbf{x} = \mathbf{F}\mathbf{X} + \mathbf{c}, \quad \mathbf{X} \in \mathcal{B}_0, \tag{2.51}$$

where the deformation gradient $\mathbf{F}$ and the vector of *rigid translation* $\mathbf{c}$ are independent of $\mathbf{X}$. We proceed to present three special cases of homogeneous deformations.

Let us consider a circular cylindrical body extended uniformly along its axis. Due to the symmetry the axes of Lagrangian reference configuration and Eulerian current configuration coincide – one of them is aligned with the axis and the other two are normal to this axis. Let L and l be the reference and the current lengths of the cylinder. Then, again due to the symmetry,

$$\begin{aligned} x^1 &= \sqrt{\lambda^{(1)}}X^1, \quad \lambda^{(1)} = \left(\frac{l}{L}\right)^2, \\ x^2 &= \sqrt{\lambda^{(2)}}X^2, \quad x^3 = \sqrt{\lambda^{(3)}}X^3, \quad \lambda^{(2)} = \lambda^{(3)}, \end{aligned} \tag{2.52}$$

where we have set $\mathbf{c} = \mathbf{0}$ for convenience. In the chosen coordinates the deformation gradient has the form of the following matrix

$$\left[F^k_K\right] = \begin{bmatrix} \sqrt{\lambda^{(1)}} & 0 & 0 \\ 0 & \sqrt{\lambda^{(2)}} & 0 \\ 0 & 0 & \sqrt{\lambda^{(3)}} \end{bmatrix}. \tag{2.53}$$

Expressed in terms of cylindrical polar coordinates (R, Θ, Z) and (r, θ, z) the function of motion becomes

$$r = \sqrt{\lambda^{(2)}}R, \quad \theta = \Theta, \quad z = \sqrt{\lambda^{(1)}}Z. \tag{2.54}$$

In the case of *isochoric motions* (i.e. the volume does not change: $J = 1$ $\Rightarrow \lambda^{(1)}\lambda^{(2)}\lambda^{(3)} = 1$) we have $\lambda^{(1)} = 1/\sqrt{\lambda^{(2)}} = 1/\sqrt{\lambda^{(3)}}$.

Obviously, only line elements in principal directions, identical with the chosen axes, do not rotate. Other elements change orientations as indicated in the left panel of Fig. 2.

A related simple form of the deformation appears for a cube whose edges have the initial length L and after the deformation the length changes to l. This deformation is called the *dilatation*. Due to the symmetry, we have $\lambda^{(1)} = \lambda^{(2)} = \lambda^{(3)} = (l/L)^2$. The deformation gradient has the following matrix representation in Cartesian coordinates

$$\left[F^k_K\right] = \begin{bmatrix} l/L & 0 & 0 \\ 0 & l/L & 0 \\ 0 & 0 & l/L \end{bmatrix}. \tag{2.55}$$

The deformation is isochoric if and only if $\lambda^{(1)} = 1$.

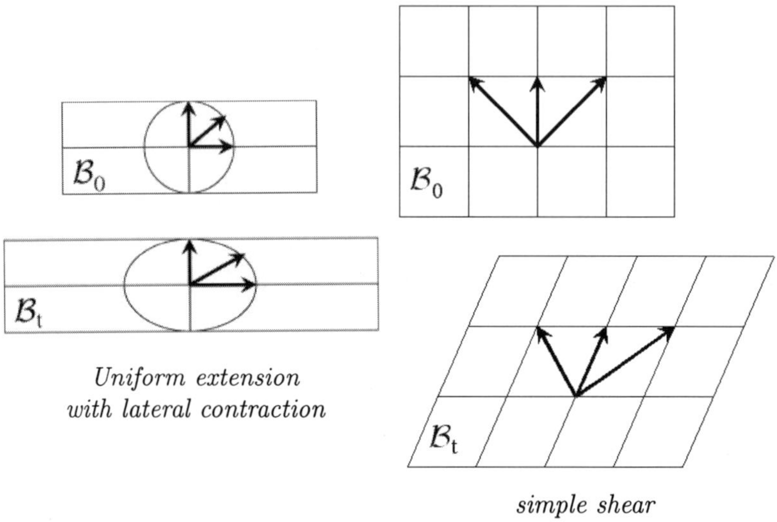

Fig. 2: *Uniform deformations*

The third example of the homogeneous deformation which we show in this part of the book is the *simple shear*. It is the plane isochoric deformation of a rectangular block defined in Lagrangian and Eulerian Cartesian coordinates by the following function of motion

$$x^1 = X^1 + \kappa X^2, \quad x^2 = X^2, \quad x^3 = X^3, \quad \kappa = \tan\varphi, \tag{2.56}$$

$$\text{i.e.} \quad \mathbf{x} = \mathbf{X} + \kappa\,(\mathbf{X}\cdot\mathbf{e}_2)\,\mathbf{e}_1. \tag{2.57}$$

The coefficient κ is called the *amount of shear*. This deformation is illustrated in the right panel of Fig. 2. The deformation gradient is represented in the chosen Cartesian coordinates by the matrix

$$\left[F^k_K\right] = \begin{bmatrix} 1 & \kappa & 0 \\ 0 & 1 & 0 \\ 0 & 0 & 1 \end{bmatrix}, \quad J = \det\mathbf{F} = 1. \tag{2.58}$$

For this deformation gradient, we can easily find the polar decomposition, the rotation $\mathbf{R}$ and, say, the right stretch tensor $\mathbf{U}$ such that $\mathbf{F} = \mathbf{RU}$. However, calculations are rather tedious and it is easier to perform the following geometrical analysis.

The right Cauchy-Green tensor has the following form

$$[C_{KL}] = \begin{bmatrix} 1 & \kappa & 0 \\ \kappa & 1+\kappa^2 & 0 \\ 0 & 0 & 1 \end{bmatrix}, \quad \text{i.e.} \tag{2.59}$$

$$\mathbf{C} = \mathbf{1} + \kappa\left(\mathbf{e}_1 \otimes \mathbf{e}_2 + \mathbf{e}_2 \otimes \mathbf{e}_1\right) + \kappa^2 \mathbf{e}_2 \otimes \mathbf{e}_2. \tag{2.60}$$

The characteristic equation

$$\det\left(\mathbf{C} - \lambda \mathbf{1}\right) = 0, \tag{2.61}$$

reduces to

$$\lambda + \frac{1}{\lambda} = 2 + \kappa^2, \tag{2.62}$$

since, for the isochoric plane motion ($J = 1$), we have, without loss of generality, $\lambda = \lambda^{(1)} = 1/\lambda^{(2)} \geq 1$.

Let θ denote the orientation (in the anticlockwise sense relative to $\mathbf{e}_1$ and $\mathbf{e}_2$) of the Lagrangian principal axes (i.e. eigenvectors $\mathbf{K}$ of the deformation tensor $\mathbf{C}$). Then the change of reference from $(\mathbf{e}_1, \mathbf{e}_2)$ to $\left(\mathbf{K}^1, \mathbf{K}^2\right)$ yields in the matrix representation

$$\begin{bmatrix} 1 & \kappa & 0 \\ \kappa & 1+\kappa^2 & 0 \\ 0 & 0 & 1 \end{bmatrix} = \begin{bmatrix} \cos\theta & -\sin\theta & 0 \\ \sin\theta & \cos\theta & 0 \\ 0 & 0 & 1 \end{bmatrix} \begin{bmatrix} \lambda & 0 & 0 \\ 0 & 1/\lambda & 0 \\ 0 & 0 & 1 \end{bmatrix} \begin{bmatrix} \cos\theta & \sin\theta & 0 \\ -\sin\theta & \cos\theta & 0 \\ 0 & 0 & 1 \end{bmatrix}, \tag{2.63}$$

which leads to the following set of equations

$$\begin{aligned} \lambda \cos^2\theta + \frac{1}{\lambda}\sin^2\theta &= 1, \\ \lambda \sin^2\theta + \frac{1}{\lambda}\cos^2\theta &= 1 + \kappa^2, \\ \left(\lambda - \frac{1}{\lambda}\right)\sin\theta\cos\theta &= \kappa. \end{aligned} \tag{2.64}$$

Obviously, the first and second equations are identical due to the relation (2.62). The solution of the remaining set of two equations can be written in the form

$$\tan 2\theta = -\frac{2}{\kappa}, \qquad \lambda = 1 + \kappa \tan\theta. \tag{2.65}$$

For a small amount of shear: $\tan\varphi \approx \varphi$, $\kappa^2 \approx 0$, $\theta \approx \pi/4$, $\lambda - 1 \approx \kappa$.

Many further examples of homogeneous deformations can be found in Part II of the book.

2.4 Displacement

It is often the case in the classical textbooks of continuum mechanics (e.g. [116], [199]) that the motion of the body is described by the *displacement vector* $\mathbf{u}$ defined by the relation

$$\mathbf{u}(\mathbf{X},\mathbf{t}) = \mathbf{f}(\mathbf{X},t) - \mathbf{X}. \tag{2.66}$$

This definition as the difference of two position vectors: $\mathbf{x} - \mathbf{X}$ is meaningful in the Euclidean translation space $\mathcal{E}^3$ (see Appendix A) and, from the operational point of view, it makes sense if we choose the same frame of coordinates in the reference and current configurations, $\mathcal{B}_0$ and $\mathcal{B}_t$, i.e. we have, for instance, in Cartesian coordinates

$$u^k(\mathbf{X},t) = f^k(\mathbf{X},t) - \delta^k_K X^K. \tag{2.67}$$

Such a definition does not possess invariance properties required from vectors on tangent spaces. We can write

$$\mathbf{u}(\mathbf{X},t) = u^K(\mathbf{X},t)\,\mathbf{G}_K(\mathbf{X}), \quad \mathbf{u}(\mathbf{x},t) = u^k(\mathbf{x},t)\,\mathbf{g}_k(\mathbf{x}), \tag{2.68}$$

and, as both vectors are supposed to be the same,

$$\begin{aligned} u^K(\mathbf{X},t) &= g^K_k(\mathbf{x},\mathbf{X})\,u^k(\mathbf{x},t), \quad g^K_k(\mathbf{x},\mathbf{X}) = \mathbf{G}^K(\mathbf{X})\cdot\mathbf{g}_k(\mathbf{x}), \\ u^k(\mathbf{x},t) &= g^k_K(\mathbf{x},\mathbf{X})\,u^K(\mathbf{X},t), \quad g^k_K(\mathbf{x},\mathbf{X}) = \mathbf{G}_K(\mathbf{X})\cdot\mathbf{g}^k(\mathbf{x}), \end{aligned} \tag{2.69}$$

where, obviously, g^K_k, g^k_K are *shifters* (see Appendix A). These relations yield proper invariance of the displacement vector but they are practically not very useful as shifters depend on two variables: $\mathbf{x},\mathbf{X}$ (they are the so-called *two-point tensors*) which are connected in an implicit manner by the displacement vector again. Only in fully linear models where not only deformations but also displacements are small in some sense these relations are useful and commonly used (e.g. [136]).

According to the relation (2.67) the deformation gradient $\mathbf{F}$ can be written in the Cartesian coordinates in the following form

$$F^k_K = \delta^k_K + u^k_{;K}, \quad \mathbf{F} = F^k_K \mathbf{e}_k \otimes \mathbf{e}_K. \tag{2.70}$$

Hence, the right Cauchy-Green tensor $\mathbf{C}$ is in these coordinates as follows

$$\begin{aligned} C_{KL} &= F^k_K F_{kL} = \delta_{KL} + g_{kK}u^k_{;L} + g_{kL}u^k_{;K} + u^k_{;K}u_{k;L}, \\ g_{kK} &= g_{kl}g^l_K \stackrel{*}{=} g^k_K \stackrel{*}{=} \delta^k_K, \end{aligned} \tag{2.71}$$

where $\overset{*}{=}$ means that the relation holds only in a chosen (in our case, Cartesian) frame of coordinates. Consequently, for the Green-St.Venant deformation tensor we obtain

$$E_{KL} = \frac{1}{2}\left(g_{kK}u^k_{;L} + g_{kL}u^k_{;K} + u^k_{;K}u_{k;L}\right). \tag{2.72}$$

For small deformations one can neglect the quadratic contribution and, identifying basis vectors in the reference and current configurations, we obtain

$$E_{kl} \approx e_{kl} \approx \frac{1}{2}\left(u_{k;l} + u_{l;k}\right) \equiv u_{(k;l)}. \tag{2.73}$$

This is indeed the measure of relative deformations commonly used in the linear elasticity.

Due to the above presented invariance problems we shall not refer in this book to the displacement vector $\mathbf{u}(\mathbf{X},t)$ in the Lagrangian description of nonlinear problems and use rather the function of motion $\mathbf{f}(\mathbf{X},t)$. In the Eulerian description used, for instance, in various models of fluids, the displacement vector as a field does not appear anyway.

2.5 Geometrical compatibility conditions

The assumption that the space of motion is Euclidean imposes certain restrictions on the deformation measures which are called *geometrical compatibility conditions*. They follow from the fact that the Euclidean space is *flat*. We proceed to discuss this notion.

According to the definition, the n-dimensional space is flat if one can introduce a frame of reference $\{\xi^\alpha\}_{\alpha=1}^n$ such that the so-called metric form Φ of this space has the form

$$\Phi = \varepsilon_1\left(d\xi^1\right)^2 + \varepsilon_2\left(d\xi^2\right)^2 + \ldots\varepsilon_n\left(d\xi^n\right)^2, \tag{2.74}$$

where $\varepsilon_1, \ldots, \varepsilon_n$ have values ± 1. Obviously, the space $\mathcal{E}^3$ possesses this property: for the Cartesian frame of reference $\varepsilon_1 = \varepsilon_2 = \varepsilon_3 = 1$ and $\Phi = ds^2$. In this space, we check the effect of *curvature* of the space (i.e. a deviation from the flat form of the space) by differentiating an arbitrary vector field, say, $\mathbf{v} = v_k\mathbf{g}^k$. For the second covariant derivative, one obtains

$$v_{k;mn} - v_{k;nm} = R^l_{.kmn}v_l, \tag{2.75}$$

where

$$R^l_{.kmn} = \frac{\partial\Gamma^l_{km}}{\partial y^n} - \frac{\partial\Gamma^l_{kn}}{\partial y^m} + \Gamma^p_{kn}\Gamma^l_{pm} - \Gamma^p_{km}\Gamma^l_{pn}, \tag{2.76}$$

is the so-called *mixed curvature tensor*. From the relation (2.75) it follows immediately that this object is a tensor – the left-hand side of (2.75) is the tensor because covariant derivatives have tensorial character. It means that if it is vanishing in one frame it must be vanishing in all. Of course, this is the case in the Euclidean space in which Christoffel symbols are zero in the Cartesian frame of reference. One can prove that this is both the necessary and sufficient condition for the space to be flat (for details, see, e.g. [320]).

The definition (2.76) yields the following notion of the *covariant curvature tensor* or *Riemann tensor*

$$R_{klmn} = g_{kp} R^{p}_{.lmn}, \tag{2.77}$$

and, directly from the definition, one can prove its following properties

$$R_{klmn} = -R_{lkmn} = -R_{klnm} = R_{mnkl},$$

$$R_{klmn} + R_{kmnl} + R_{knlm} = 0. \tag{2.78}$$

The last relation is called the *Bianchi identity*. These identities yield $\frac{1}{12}n^2(n^2-1)$ independent components of the Riemann tensor. For $n = 3$ we have 6 independent components.

After this introduction, we can return to the analysis of deformations in the Euclidean space. We follow the analysis proposed by Jarić [156] and investigate the integrability condition for the right Cauchy-Green tensor $\mathbf{C}$, i.e. the existence of the function of motion $\mathbf{f}(\mathbf{X}, t)$ and of the deformation gradient $\mathbf{F}(\mathbf{X}, t)$ for a chosen $\mathbf{C}$. In chosen frames we have

$$C_{KL} - g_{kl} F^k_K F^l_L = 0, \tag{2.79}$$

and we want to check the existence of the function of motion $f^k(\mathbf{X}, t)$ such that

$$F^k_K = \frac{\partial f^k}{\partial X^K}. \tag{2.80}$$

The integrability conditions of these relations have the form

$$\frac{\partial F^k_K}{\partial X^L} = \frac{\partial F^k_L}{\partial X^K}. \tag{2.81}$$

Combination of (2.79) and (2.81) yields then after some calculations

$$\overset{\mathbf{C}}{R}_{KLMN} = \overset{g}{R}_{klmn} F^k_K F^l_L F^m_M F^n_N, \tag{2.82}$$

where $\overset{\mathbf{C}}{R}_{KLMN}$ is the Riemann tensor with respect to the metric C_{KL} and $\overset{g}{R}_{klmn}$ with respect to the metric g_{kl}. As the latter is zero we obtain

$$\overset{\mathbf{C}}{R}_{KLMN} = 0. \tag{2.83}$$

As we have already mentioned there are only six independent components of the Riemann tensor. Hence, we have

$$\begin{aligned} &\overset{\mathbf{C}}{R}_{1212} = 0, \quad \overset{\mathbf{C}}{R}_{1213} = 0, \quad \overset{\mathbf{C}}{R}_{1223} = 0, \\ &\overset{\mathbf{C}}{R}_{1313} = 0, \quad \overset{\mathbf{C}}{R}_{1323} = 0, \quad \overset{\mathbf{C}}{R}_{2323} = 0. \end{aligned} \tag{2.84}$$

It is convenient to transform the condition (2.83) to the Green-St.Venant deformation measure $\mathbf{E}$. We only quote the result obtained by Graiff in 1958 [126]

$$\begin{aligned} &E_{KM,LN} + E_{LN,KM} - E_{KL,MN} - E_{NM,KL} \\ &+ \overset{-1}{C}{}^{PR}_{\cdot} \left[\left(E_{PM,K} + E_{KP,M} - E_{KM,P} \right) \left(E_{LR,N} + E_{NR,L} - E_{LN,R} \right) \right. \\ &\left. - \left(E_{PM,N} + E_{NP,M} - E_{NM,P} \right) \left(E_{LR,K} + E_{KR,L} - E_{LK,R} \right) \right] = 0. \end{aligned} \tag{2.85}$$

This is the *geometrical compatibility condition for deformations* in the fully nonlinear case. In this relation, the comma denotes the partial differentiation.

Finally, in the case of small deformations, when the relation (2.73) is satisfied the above relation simplifies to the following [67]

$$e_{ij,kl} + e_{kl,ij} - e_{ik,jl} - e_{jl,ik} = 0, \tag{2.86}$$

where the comma is again the partial differentiation. These are the compatibility relations of the classical linear elasticity. They form the basis for the derivation of the so-called Beltrami-Michell equations specifying stress components in the linear elasticity.

2.6 Geometrical compatibility condition for singular surfaces

The motion $\mathbf{f}$ in a continuum cannot be discontinuous. However, its derivatives with respect to the spatial variable (deformation gradient), and the time derivatives as well as other fields such as temperature may be discontinuous on some sets. These sets must have volume measure zero, i.e. they may form surfaces, lines or they may be isolated points. In this Section, we consider some geometrical properties of surfaces. Such surfaces model in continua vortex sheets, shock waves, acoustic waves, and, as a particular case, boundaries (for details, see Section 3.3).

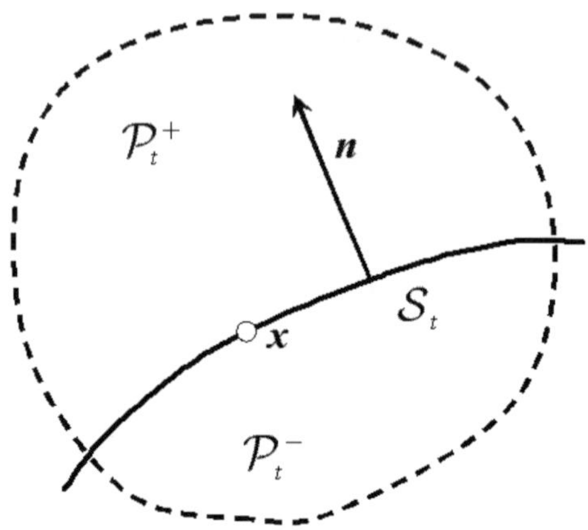

Fig. 3: *Singular surface in the current configuration*

Let us consider an orientable regular surface $\mathcal{S}_t$ which is the common boundary of two regions $\mathcal{P}_t^+, \mathcal{P}_t^-$ with a unit normal vector $\mathbf{n}$ oriented to $\mathcal{P}_t^+$ (see Fig. 3). Let $\varphi(\mathbf{x})$ be a function continuously differentiable in the interiors of $\mathcal{P}_t^+$ and $\mathcal{P}_t^-$ and which approaches definite limit values φ^+ and φ^- as $\mathbf{x}$ approaches a point $\mathbf{x}_0$ on $\mathcal{S}_t$. The *jump* of φ across $\mathcal{S}_t$ at $\mathbf{x}_0$ is denoted by

$$[[\varphi]] = \varphi^+ - \varphi^-. \tag{2.87}$$

The quantity $[[\varphi]]$ is the function of position $\mathbf{x}_0$ on the surface $\mathcal{S}_t$. If $[[\varphi]] \neq 0$ we say that $\mathcal{S}_t$ is the *singular surface* with respect to φ.

Hadamard's Lemma [138] Let the function φ and its derivative $\partial\varphi/\partial x^k$ approach finite limits φ^+ and $\left(\partial\varphi/\partial x^k\right)^+$ as $\mathcal{S}_t$ is approached upon paths interior to $\mathcal{P}_t^+$. Let $\mathbf{x} = \mathbf{x}(l)$ be a smooth curve on $\mathcal{S}_t$ and let φ^+ be differentiable on this curve. Then

$$\frac{d\varphi^+}{dl} = \left(\frac{\partial\varphi}{\partial x^k}\right)^+ \frac{dx^k}{dl}, \quad x^k = \mathbf{x} \cdot \mathbf{e}^k. \tag{2.88}$$

The proof is identical as in the case of the proof of the classical chain rule of differentiation (theorem of the total differential) as the path between two points of the curve can be chosen through interior points of $\mathcal{P}_t^+$. Consequently, under the same conditions the Lemma holds also for limits on the other side of the surface $\mathcal{S}_t$. Subtracting these results we obtain

$$\frac{d}{dl}[[\varphi]] = \left[\left[\frac{\partial\varphi}{\partial x^k}\frac{dx^k}{dl}\right]\right] = \left[\left[\frac{\partial\varphi}{\partial x^k}\right]\right]\frac{dx^k}{dl}. \tag{2.89}$$

This relation can be formulated as the following statement: the jump of a tangential derivative (right-hand side) is the tangential derivative of the jump (left-hand side).

Since values of φ in $\mathcal{P}_t^+$ and $\mathcal{P}_t^-$ are unrelated, the jump of the normal derivative of φ is not specified

$$\left[\left[\frac{d\varphi}{dn}\right]\right] = \left[\left[\frac{\partial\varphi}{\partial x^k}n^k\right]\right] \quad \text{is unrestricted.} \tag{2.90}$$

In this relation n^k are components of the normal vector $\mathbf{n}$.

Choosing the curve on the surface $\mathcal{S}_t$ as a parametric curve of the surface coordinates $\{\xi^\alpha\}, \alpha = 1,2$, where $x^k = \mathbf{s}\cdot\mathbf{e}^k = s^k(\xi^\alpha)$ are equations of the surface (i.e. $\mathbf{x} = \mathbf{s}(\xi^\alpha)$) we write the relations (2.89) in the form

$$\frac{\partial}{\xi^\alpha}[[\varphi]] = [[\operatorname{grad}\varphi]]\cdot\frac{\partial\mathbf{s}}{\partial\xi^\alpha} = \left[\left[\frac{\partial\varphi}{\partial x^k}\right]\right]s^k_{.;\alpha}. \tag{2.91}$$

Obviously, derivatives $s^k_{.;\alpha}$ define two vectors tangent to parametric curves on the surface $\mathcal{S}_t$.

The conditions (2.90) and (2.91) are called the *superficial conditions of compatibility.*

In the important special case when the jump of φ is zero, the relation (2.91) yields

$$[[\varphi]] = 0 \quad \Rightarrow \quad \left[\left[\frac{\partial\varphi}{\partial x^k}\right]\right] = bn_k, \quad \text{i.e.} \quad [[\operatorname{grad}\varphi]] = b\mathbf{n}, \tag{2.92}$$

as $n_k s^k_{.;\alpha} = \mathbf{n}\cdot(\partial\mathbf{s}/\partial\xi^\alpha) = 0$. In (2.92) b is a factor of proportionality. This yields the

Maxwell Theorem: [203] When $[[\varphi]] = 0$ then

$$\left[\left[\frac{\partial\varphi}{\partial x^k}\right]\right] = bn_k, \quad b = \left[\left[n^k\frac{\partial\varphi}{\partial x^k}\right]\right] = \left[\left[\frac{d\varphi}{dn}\right]\right], \tag{2.93}$$

i.e.

$$[[\operatorname{grad}\varphi]] = b\mathbf{n}, \quad b = \mathbf{n}\cdot[[\operatorname{grad}\varphi]]. \tag{2.94}$$

Obviously, the above results hold for an arbitrary tensorial quantity. In a particular case, when φ is a vector, say, $\mathbf{w} = w_i \mathbf{e}^i$ then we obtain

$$\left[\left[\frac{\partial w_i}{\partial x^k}\right]\right] = b_i n_k, \quad b_i = \left[\left[n^m \frac{\partial w_i}{\partial x^m}\right]\right] = \left[\left[\frac{d w_i}{dn}\right]\right], \tag{2.95}$$

i.e.

$$[[\operatorname{grad} \mathbf{w}]] = \mathbf{b} \otimes \mathbf{n}, \quad \mathbf{b} = [[(\mathbf{n} \cdot \operatorname{grad})\, \mathbf{w}]]\,. \tag{2.96}$$

Consequently, we obtain the following

Weingarten's first Theorem: [347] The longitudinal and transversal jumps of the gradient of a continuous vector $\mathbf{w}$ are the jumps of its divergence and rotation (curl)[1] , respectively:

$$[[\operatorname{rot} \mathbf{w}]] = \mathbf{n} \times \mathbf{b}, \quad [[\operatorname{div} \mathbf{w}]] = \mathbf{n} \cdot \mathbf{b}, \quad \mathbf{b} = \left[\left[\frac{d\mathbf{w}}{dn}\right]\right]. \tag{2.97}$$

Weingarten's second Theorem concerns the structure of the jump of displacement in the kinetic theory and we shall not quote it here (see [335]).

Iterating the above results for the vector field one can easily obtain conditions for second derivatives, $\left[\left[\partial^2 \varphi / \partial x^i \partial x^k\right]\right]$ in terms of $\left[\left[n^j n^l \partial^2 \varphi / \partial x^j \partial x^l\right]\right]$, $\left[\left[n^j \partial \varphi / \partial x^j\right]\right]$, $\left[\left[n^j \partial \varphi / \partial x^j\right]\right]_{;\alpha}$, $[[\varphi]]$, $[[\varphi]]_{;\alpha}$ and $[[\varphi]]_{;\alpha\beta}$. Such relations hold also for higher derivatives. We skip here the detailed discussion and quote from [335] only one important result

$$[[\varphi]] = 0, \quad \left[\left[\frac{\partial \varphi}{\partial x^k}\right]\right] = 0 \quad \Rightarrow \quad \left[\left[\frac{\partial^2 \varphi}{\partial x^i \partial x^k}\right]\right] = b n_i n_k, \tag{2.98}$$

i.e.

$$[[\varphi]] = 0, \quad [[\operatorname{grad} \varphi]] = 0 \quad \Rightarrow \quad [[\operatorname{grad} \otimes \operatorname{grad} \varphi]] = b\mathbf{n} \otimes \mathbf{n}, \tag{2.99}$$

where

$$b = \mathbf{n} \cdot [[\operatorname{grad} \otimes \operatorname{grad} \varphi]]\, \mathbf{n}, \tag{2.100}$$

for a scalar φ and analogously for tensorial functions $\boldsymbol{\varphi}$.

---

[1]We use the notation rot and curl as equivalent in this book. They are defined in Cartesian coordinates in the following way

$$\operatorname{rot} \mathbf{w} \equiv \operatorname{curl} \mathbf{w} = \left(\frac{\partial}{\partial x^k} \mathbf{e}^k\right) \times \mathbf{w} = \epsilon_{ijk} \frac{\partial w^k}{\partial x_j} \mathbf{e}^i.$$

Chapter 3

Kinematics

3.1 Basic notions

We proceed to discuss the kinematics of the continuum. The main notions are the velocity field $\mathbf{v}(\mathbf{X},t)$ and the acceleration field $\mathbf{a}(\mathbf{X},t)$. They are defined by the following relations

$$\mathbf{v}(\mathbf{X},t) = \frac{\partial \mathbf{f}}{\partial t}(\mathbf{X},t), \quad \mathbf{a}(\mathbf{X},t) = \frac{\partial \mathbf{v}}{\partial t}(\mathbf{X},t). \tag{3.1}$$

Another quantity frequently appearing in the theory of continuous bodies is the gradient of velocity $\mathbf{L}$. It is defined by the time derivative of the deformation gradient $\mathbf{F}$

$$\mathbf{L} = \frac{\partial \mathbf{F}}{\partial t}\mathbf{F}^{-1}. \tag{3.2}$$

Later we discuss this notion in some details.

In relation to the kinematics of the body, it is useful to introduce a certain transformation group which has a great influence on the construction of constitutive relations. Namely, it is assumed that material properties of bodies cannot change by changing the reference frame in such a way that *distances* of material points in the configuration space remain unchanged. If we introduce two reference systems, say, with position vectors without and with star – in Figure 4 we demonstrate such systems without a rotation of the body – then we require that in both systems the distance between two arbitrary points of the body must be the same. Vectors $\mathbf{r}_2 - \mathbf{r}_1$, $\mathbf{r}_2^* - \mathbf{r}_1^*$ may not be the same due to the rigid rotation of the body but for both reference systems (*observers*) we have

$$|\mathbf{r}_2 - \mathbf{r}_1| = |\mathbf{r}_2^* - \mathbf{r}_1^*|. \tag{3.3}$$

This is the property of the configuration space which we call *isometry*. The most general form of the transformation which leads to the relation

(3.3) is as follows

$$\mathbf{x}^* = \mathbf{O}(t)\,\mathbf{x} + \mathbf{c}(t)\,, \quad \mathbf{O}^T = \mathbf{O}^{-1}, \tag{3.4}$$

where $\mathbf{O}$ is an arbitrary time dependent orthogonal tensor and $\mathbf{c}$ an arbitrary time dependent vector. This class of transformations forms an *isometry group* and each member of this group is called an *Euclidean transformation.* We require that material properties are independent of the choice of two reference systems which differ on the isometry transformation. Incidentally, the transformation in which $\mathbf{O}$ and $\mathbf{c}$ are constant is called *Galilean.* Classical equations of motion are invariant with respect to these transformations.

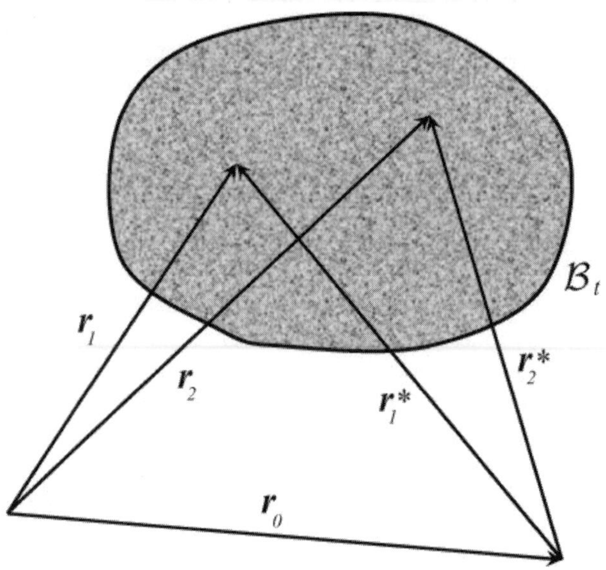

Fig. 4: *Change of a reference system in Euclidean spaces*

It is useful to check the transformation properties of objects which we were discussing up to now. After easy calculations we obtain

$$
\begin{aligned}
\mathbf{f}^*(\mathbf{X},t) &= \mathbf{O}(t)\,\mathbf{f}(\mathbf{X},t) + \mathbf{c}(t),\\
\mathbf{v}^* &= \mathbf{O}\mathbf{v} + \dot{\mathbf{O}}\mathbf{x} + \dot{\mathbf{c}}, \quad \dot{\mathbf{O}} := \frac{d\mathbf{O}}{dt}, \quad \dot{\mathbf{c}} := \frac{d\mathbf{c}}{dt},\\
\mathbf{F}^* &= \mathbf{O}\mathbf{F},\\
\mathbf{C}^* &= \mathbf{C}, \quad \mathbf{B}^* = \mathbf{O}\mathbf{B}\mathbf{O}^T,\\
\mathbf{a}^* &= \mathbf{O}\mathbf{a} + 2\dot{\mathbf{O}}\mathbf{v} + \ddot{\mathbf{O}}\mathbf{x} + \ddot{\mathbf{c}}, \quad \ddot{\mathbf{O}} := \frac{d^2\mathbf{O}}{dt^2}, \quad \ddot{\mathbf{c}} := \frac{d^2\mathbf{c}}{dt^2},\\
\mathbf{L}^* &= \mathbf{O}\mathbf{L}\mathbf{O}^T + \boldsymbol{\Omega}, \quad \boldsymbol{\Omega} := \dot{\mathbf{O}}\mathbf{O}^T.
\end{aligned}
\tag{3.5}
$$

Scalars which do not change due to the transformation (3.4): $\varphi^* = \varphi$, vectors which change according to the rule: $\mathbf{b}^* = \mathbf{O}\mathbf{b}$, and tensors which transform according to the rule: $\mathbf{T}^* = \mathbf{O}\mathbf{T}\mathbf{O}^T$ are called *objective*. Hence in the above quoted examples $\mathbf{B}$ is objective, $\mathbf{F}$ behaves like a collection of three objective vectors (objects in parentheses): $\mathbf{F} = (F_{kK}\mathbf{e}_k)\mathbf{e}_K$, $\mathbf{C}$ behaves like a collection of six scalars C_{KL}. The remaining objects are nonobjective. It is convenient to write them in the form in which the deviation from the objectivity is better exposed. For the velocity and acceleration we have

$$
\begin{aligned}
\mathbf{O}\mathbf{v} &= \mathbf{v}^* - \boldsymbol{\Omega}(\mathbf{x}^* - \mathbf{c}) - \dot{\mathbf{c}}, \quad \boldsymbol{\Omega} := \dot{\mathbf{O}}\mathbf{O}^T, \quad \boldsymbol{\Omega}^T = -\boldsymbol{\Omega},\\
\mathbf{O}\mathbf{a} &= \mathbf{a}^* - 2\boldsymbol{\Omega}(\mathbf{v}^* - \dot{\mathbf{c}}) + \boldsymbol{\Omega}^2(\mathbf{x}^* - \mathbf{c}) - \dot{\boldsymbol{\Omega}}(\mathbf{x}^* - \mathbf{c}) - \ddot{\mathbf{c}},
\end{aligned}
\tag{3.6}
$$

where the antisymmetric tensor $\boldsymbol{\Omega}$ denotes the *spin matrix* (matrix of relative angular velocities of both reference systems). The contributions to the acceleration are called: $2\boldsymbol{\Omega}(\mathbf{v}^* - \dot{\mathbf{c}})$ – Coriolis, $-\boldsymbol{\Omega}^2(\mathbf{x}^* - \mathbf{c})$ – centrifugal, $\dot{\boldsymbol{\Omega}}(\mathbf{x}^* - \mathbf{c})$ – Euler and $\ddot{\mathbf{c}}$ – relative translational accelerations, respectively. They play an important role in the description of motion with respect to the so-called *noninertial frames of reference*.

It is also convenient to separate objective and nonobjective contributions to the velocity gradient $\mathbf{L}$

$$
\begin{aligned}
\mathbf{L} &= \mathbf{D} + \mathbf{W},\\
\mathbf{D} &= \frac{1}{2}\left(\mathbf{L} + \mathbf{L}^T\right) = \mathbf{D}^T, \quad \mathbf{D}^* = \mathbf{O}\mathbf{D}\mathbf{O}^T,\\
\mathbf{W} &= \frac{1}{2}\left(\mathbf{L} - \mathbf{L}^T\right) = -\mathbf{W}^T, \quad \mathbf{W}^* = \mathbf{O}\mathbf{W}\mathbf{O}^T + \boldsymbol{\Omega}.
\end{aligned}
\tag{3.7}
$$

Hence the *stretching* tensor $\mathbf{D}$ is objective and the *spin* tensor $\mathbf{W}$ is nonobjective.

3.2 Objective time derivatives

We proceed to consider the problem of the so-called *objective time derivatives*. This problem appears in constructions of constitutive laws.

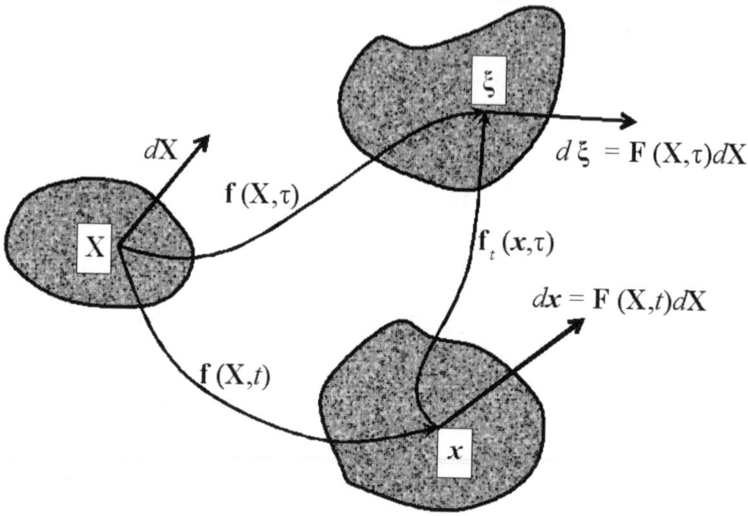

Fig. 5: *Relative deformation gradient*

Let us begin with the analysis of a change of the reference configuration. This is demonstrated in Fig. 5. The purpose is to use the current configuration at the instant of time t as the reference configuration for the motion in the vicinity of the instant t. The function of motion defined on the current configuration will be denoted by $\mathbf{f}_t(.,.)$. For an arbitrary point $\boldsymbol{\xi}$ in the configuration at the instant of time τ it is given by the following relation

$$\boldsymbol{\xi} = \mathbf{f}\left[\mathbf{f}^{-1}(\mathbf{x},t),\tau\right] = \mathbf{f}_t(\mathbf{x},\tau). \tag{3.8}$$

Corresponding deformation gradients are as follows

$$d\boldsymbol{\xi} = \mathbf{F}(\mathbf{X},\tau)\,d\mathbf{X} = \mathbf{F}\left(\mathbf{f}^{-1}(\mathbf{x},t),\tau\right)\mathbf{F}^{-1}\left(\mathbf{f}^{-1}(\mathbf{x},t),t\right)d\mathbf{x} \tag{3.9}$$
$$= \mathbf{F}_t(\mathbf{x},\tau)\,d\mathbf{x} \quad\Longrightarrow\quad \mathbf{F}_t(\boldsymbol{\xi},\tau) = \mathbf{F}_t\left(\mathbf{f}_t^{-1}(\boldsymbol{\xi},\tau),\tau\right), \quad \mathbf{F}_t(\boldsymbol{\xi},\tau)\big|_{\tau=t} = 1,$$

with an appropriate change of variables given by (3.8). The quantity $\mathbf{F}_t(\mathbf{x},\tau)$ is called the *relative deformation gradient* with respect to the current configuration.

In order to see time changes at the current configuration, we investigate a material vector $\mathbf{Q}(\mathbf{X})$. Its images in two instances of time t and τ are as follows

$$\begin{aligned} &\mathbf{q}(\mathbf{x},t) = \mathbf{F}\left(\mathbf{f}^{-1}(\mathbf{x},t),t\right)\mathbf{Q}\left(\mathbf{f}^{-1}(\mathbf{x},t)\right), \\ &\mathbf{q}(\boldsymbol{\xi},\tau) = \mathbf{F}\left(\mathbf{f}^{-1}(\boldsymbol{\xi},\tau),\tau\right)\mathbf{Q}\left(\mathbf{f}^{-1}(\boldsymbol{\xi},\tau)\right) \\ \Longrightarrow \quad &\mathbf{q}(\mathbf{x},t) = \mathbf{F}_t^{-1}(\boldsymbol{\xi},\tau)\,\mathbf{q}(\boldsymbol{\xi},\tau)\big|_{\boldsymbol{\xi}=\mathbf{f}_t^{-1}(\mathbf{x},\tau)}. \end{aligned} \tag{3.10}$$

We define the time derivative of $\mathbf{q}(\mathbf{x},t)$ as a limit $\tau \to t$ of the time derivative of $\mathbf{q}(\boldsymbol{\xi},\tau)$. In this way we account for time changes due to explicit dependence on time, due to the changes of position of the material point $\mathbf{X}$ as well as due to rotation of basis vectors along the trajectory. Such an operator is called *Lie derivative related to the field of velocity* $\mathbf{v}$ (for mathematical foundations of this notion see [298]). We have

$$\mathcal{L}_{\mathbf{v}}\mathbf{q}(\mathbf{x},t) = \left.\frac{d\left[\mathbf{F}_t^{-1}(\boldsymbol{\xi},\tau)\,\mathbf{q}(\boldsymbol{\xi},\tau)\right]}{d\tau}\right|_{\tau=t}$$

$$= \frac{\partial \mathbf{q}(\mathbf{x},t)}{\partial t} + \mathbf{v}\cdot\operatorname{grad}\mathbf{q}(\mathbf{x},t) + \left.\frac{d\left[\mathbf{F}_t^{-1}(\boldsymbol{\xi},\tau)\right]}{d\tau}\right|_{\tau=t}\mathbf{q}(\mathbf{x},t).$$

Bearing (3.9) in mind we get

$$\frac{d\mathbf{F}_t^{-1}\mathbf{F}_t}{d\tau} = 0 = \frac{d\mathbf{F}_t^{-1}}{d\tau}\mathbf{F}_t + \mathbf{F}_t^{-1}\frac{d\mathbf{F}_t}{d\tau} \quad \Longrightarrow \quad \frac{d\mathbf{F}_t^{-1}}{d\tau} = -\mathbf{F}_t^{-1}\frac{d\mathbf{F}_t}{d\tau}\mathbf{F}_t^{-1},$$

i.e.

$$\left.\frac{d\left[\mathbf{F}_t^{-1}(\boldsymbol{\xi},\tau)\right]}{d\tau}\right|_{\tau=t} = -\frac{d\mathbf{F}\left(\mathbf{f}^{-1}(\mathbf{x},t),t\right)}{dt}\mathbf{F}^{-1}\left(\mathbf{f}^{-1}(\mathbf{x},t),t\right) = -\mathbf{L}(\mathbf{x},t).$$

Hence

$$\begin{aligned} \mathcal{L}_{\mathbf{v}}\mathbf{q}(\mathbf{x},t) &= \frac{\partial \mathbf{q}(\mathbf{x},t)}{\partial t} + \mathbf{v}\cdot\operatorname{grad}\mathbf{q}(\mathbf{x},t) - \mathbf{L}(\mathbf{x},t)\,\mathbf{q}(\mathbf{x},t) \\ &\equiv \dot{\mathbf{q}} - \mathbf{L}\mathbf{q}, \quad \dot{\mathbf{q}} = \frac{\partial \mathbf{q}(\mathbf{x},t)}{\partial t} + \mathbf{v}\cdot\operatorname{grad}\mathbf{q}(\mathbf{x},t). \end{aligned} \tag{3.11}$$

It is easy to check that this derivative is objective, i.e.

$$\mathcal{L}_{\mathbf{v}}\mathbf{q}^*(\mathbf{x}^*,t) = \mathbf{O}(t)\,\mathcal{L}_{\mathbf{v}}\mathbf{q}(\mathbf{x},t)\big|_{\mathbf{x}=\mathbf{O}^T\mathbf{x}^*}. \tag{3.12}$$

In the similar way we can define the time derivatives for material tensors of the second order. For example, one can introduce the following *Rivlin-Ericksen tensors* [267] (see also [237]) describing the rate of deformation i/ as the time derivative of the right Cauchy-Green tensor $\dot{\mathbf{C}}$ is nonobjective one defines the Lie derivative of the relative Cauchy-Green tensor

$$\mathbf{C}_t(\tau) = \mathbf{F}_t^T(\tau)\,\mathbf{F}_t(\tau) \quad \Longrightarrow \quad \mathbf{A}_1(t) = \left.\frac{\partial C_t(\tau)}{\partial \tau}\right|_{\tau=t} = \mathbf{L}^T + \mathbf{L} = 2\mathbf{D}, \tag{3.13}$$

ii/ higher order Rivlin-Ericksen tensors

$$\mathbf{A}_n(t) = \mathbf{C}_t^{(n)}(t) = \left.\frac{\partial^n \mathbf{C}_t(\tau)}{\partial \tau^n}\right|_{\tau=t}, \quad n = 1, ..., \tag{3.14}$$

In the same way one can introduce objective time derivatives of nonmaterial vectors and tensors. For instance, the time derivative of a vector $\boldsymbol{\nu}$ orthogonal to a material surface (compare (2.14)) has the following form

$$\boldsymbol{\nu}(\mathbf{x},t) = \mathbf{F}^{-T}(t)\,\mathbf{F}^T(\tau)\,\boldsymbol{\nu}(\boldsymbol{\xi},\tau)\big|_{\substack{\mathbf{X}=\mathbf{f}^{-1}(\mathbf{x},t)\\ \boldsymbol{\xi}=\mathbf{f}(\mathbf{X},\tau)}} \tag{3.15}$$

$$\Longrightarrow \quad \mathcal{L}_v \boldsymbol{\nu} = \overset{\bullet}{\boldsymbol{\nu}} + \mathbf{L}^T \boldsymbol{\nu}, \quad \overset{\bullet}{\boldsymbol{\nu}} := \frac{\partial \boldsymbol{\nu}}{\partial t} + \mathbf{v}\cdot \operatorname{grad} \boldsymbol{\nu}.$$

Let us mention in passing that the time derivative $\overset{\bullet}{(...)}$ introduced above is the so-called *material time derivative*. It describes time changes along trajectories of material points and it is applied in the Eulerian description which we discuss in the next Chapter.

Jerrold E. Marsden
1942–

Thomas J. R. Hughes
1943–

Stuart S. Antman
1945–

There exist many modern textbooks on continuum mechanics which can be used as amendment to the above presentation. To quote just a few: Rather sophisticated mathematically but with many examples is the book of J. E. Marsden and T. J. R. Hughes [200]. A brief description of foundations and many examples of material laws for nonlinear elastic solids can be found in the book of R. Batra [27]. Applications of the code *Mathematica* to nonlinear problems of continuum mechanics are presented in the book of A. Romano, R. Lancellotta, A. Marasco [273]. An extensive presentation of the linear acoustics following from the theory of continuous media contains the book of L. M. Brekhovskikh and V. Goncharov [49]. Rigorous transitions from the general continuum to theories of elastic rods and plates can be found in the work of S. S. Antman [13].

3.3 Boundaries, kinematical compatibility condition for singular surfaces

3.3.1 *Introduction*

As we have already indicated, many kinds of the motion of continua cannot be described by a field theory in which the continuous function of motion $\mathbf{f}(\mathbf{X}, t)$ describes the mappings between the reference and current configurations. Fracture and welding (Fig. 6a), tear (Fig. 6b) or creation of new boundary points indicated in Fig. 6c, are examples of such ("non-topological") motions. Their characteristic feature is that, locally on the new boundary, two material points are created from a single material point. Such problems must be either solved as contact problems or they cannot be described by a continuum at all.

However, there is a very extensive class of singular motions which can be described by a continuum. Their common feature is the existence of a singular surface which we have already mentioned in Section 2.6. We proceed to describe some properties of such surfaces.

The first kind of singular surface defined within continuum mechanics was introduced by Helmholtz [142]. It was an orientable regular surface $\mathcal{S}_t$ modeling the *slip*. On such a surface, the inverse function of motion $\mathbf{X} = \mathbf{f}^{-1}(\mathbf{x}, t)$ is discontinuous. Each place in the configuration space is occupied on such a surface by two material points. In this motion, two different bodies slip past one another without penetration and the surface of contact is material for each body. A surface across which the tangential velocity suffers a discontinuity

$$[[\mathbf{v} - (\mathbf{v} \cdot \mathbf{n})\,\mathbf{n}]] \neq \mathbf{0}, \tag{3.16}$$

is called the *vortex sheet*. Slip surfaces are vortex sheets.

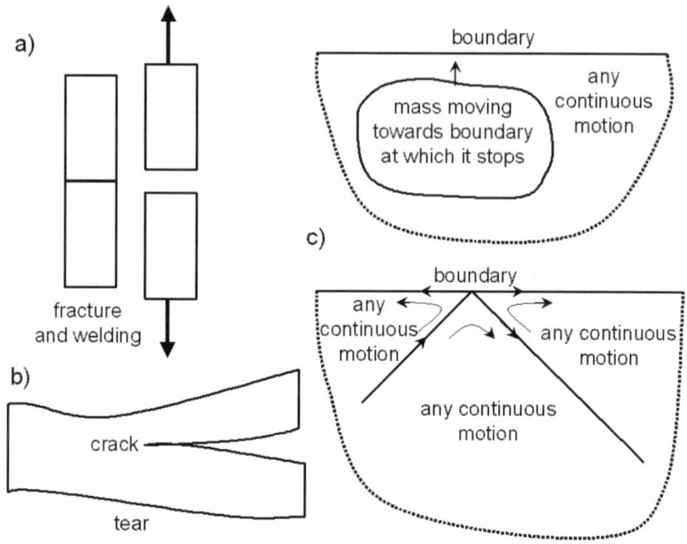

Fig. 6: *Examples of discontinuous motions in which new material points, lines and surfaces are created*

Another example of the motion in which the inverse function of motion is double-valued is the *dislocation* of Volterra [341]. Such defects are singular surfaces intended to represent deformations corresponding to removal or insertion of matter within another continuum. On such surfaces the function $\mathbf{f}^{-1}(\mathbf{x},t)$ is double-valued and surfaces are not material.

A *shock surface* is one across which the normal velocity is discontinuous

$$[[\mathbf{v}\cdot\mathbf{n}]] \neq 0. \tag{3.17}$$

Dislocations may be shock surfaces.

3.3.2 *Modicum of surface geometry*

We assume that a moving singular surface is given by a single valued function. In the Lagrangian description, the surface $\mathcal{S}$ is assumed to be given by a differentiable function Σ for each instant of time t

$$\Sigma(\mathbf{X},t) = 0, \quad \text{i.e.} \quad \sigma(\mathbf{x},t) = 0 \quad \text{where} \quad \Sigma(\mathbf{X},t) = \sigma(\mathbf{f}(\mathbf{X},t),t), \tag{3.18}$$

and, obviously, $\sigma(\mathbf{x},t)$ yields the current (Eulerian) image $\mathcal{S}_t$ of $\mathcal{S}$ at the instant of time t.

A surface imbedded in the three-dimensional Euclidean space is usually defined in a parametric form. For purposes of the continuum mechanics, we can write in the reference configuration

$$\mathbf{X} = \mathbf{S}\left(\Xi^A, t\right), \quad A = 1, 2, \tag{3.19}$$

where Ξ^A are surface coordinates for $\mathcal{S}$. Obviously, if we eliminate Ξ^A from these three relations we obtain $(3.18)_1$. Similarly, in the current configuration, $\mathcal{S}_t$, we can use the relations

$$\mathbf{x} = \mathbf{s}\left(\xi^\alpha, t\right), \quad \alpha = 1, 2, \tag{3.20}$$

where ξ^α denote surface coordinates. Their elimination yields the equation $(3.18)_2$.

The partial derivatives of functions $\mathbf{S}\left(., t\right), \mathbf{s}\left(., t\right)$ define *tangent vectors* to parametric lines on surfaces

$$\begin{aligned} \mathbf{T}_A &= \frac{\partial \mathbf{S}}{\partial \Xi^A} = T_A^K \mathbf{e}_K, \quad \text{tangent to } \Xi^B = const, \quad B \neq A, \\ \mathbf{t}_\alpha &= \frac{\partial \mathbf{s}}{\partial \xi^\alpha} = t_\alpha^k \mathbf{e}_k, \quad \text{tangent to } \xi^\beta = const, \quad \beta \neq \alpha. \end{aligned} \tag{3.21}$$

The *metric tensors* of surfaces $\mathcal{S}$ and $\mathcal{S}_t$ (the *first fundamental form of the surface*), respectively, are defined by relations

$$\begin{aligned} A_{AB} &= \mathbf{T}_A \cdot \mathbf{T}_B = T_A^K T_{KB}, \\ a_{\alpha\beta} &= \mathbf{t}_\alpha \cdot \mathbf{t}_\beta = t_\alpha^k t_{k\beta}. \end{aligned} \tag{3.22}$$

They allow to calculate distances and angles on the surface and, consequently, specify the intrinsic geometry of surfaces.

The unit normal vectors may be defined by differentiation of relations (3.18). In addition to smoothness of functions Σ and σ, it is assumed that both representations of the surface are *orientable* (compare the counterexample of the Möbius strip which is smooth but not orientable) with unit normal vectors given by relations

$$\mathbf{N} = \frac{\operatorname{Grad}\Sigma}{\left|\operatorname{Grad}\Sigma\right|}, \quad \mathbf{n} = \frac{\operatorname{grad}\sigma}{\left|\operatorname{grad}\sigma\right|}, \quad \operatorname{Grad}\Sigma = \mathbf{F}^T \operatorname{grad}\sigma. \tag{3.23}$$

These definitions are equivalent to definitions based on vector products of tangent vectors

$$\begin{aligned}
\mathbf{N} &= \frac{\mathbf{T}_1 \times \mathbf{T}_2}{|\mathbf{T}_1 \times \mathbf{T}_2|} = \frac{\epsilon_{KLM} T_1^L T_2^M}{\sqrt{A}} \mathbf{e}^K = \frac{1}{2} \frac{\epsilon_{KLM} \epsilon^{AB} T_A^L T_B^M}{\sqrt{A}} \mathbf{e}^K, \\
A &= \epsilon^{AB} A_{1A} A_{2B}, \\
\mathbf{n} &= \frac{\mathbf{t}_1 \times \mathbf{t}_2}{|\mathbf{t}_1 \times \mathbf{t}_2|} = \frac{\epsilon_{klm} t_1^l t_2^m}{\sqrt{a}} \mathbf{e}^k = \frac{1}{2} \frac{\epsilon_{klm} \epsilon^{\alpha\beta} t_\alpha^l t_\beta^m}{\sqrt{a}} \mathbf{e}^k, \\
a &= \epsilon^{\alpha\beta} a_{1\alpha} a_{2\beta},
\end{aligned} \tag{3.24}$$

where $\epsilon^{AB}, \epsilon^{\alpha\beta}$ are two-dimensional permutation symbols $\begin{bmatrix} 0 & 1 \\ -1 & 0 \end{bmatrix}$. The magnitudes of vector products yield the following form of surface elements corresponding to relations (2.10) and (2.12)

$$d\mathcal{S} = \sqrt{A} d\Xi^1 d\Xi^2, \quad d\mathcal{S}_t = \sqrt{a} d\xi^1 d\xi^2. \tag{3.25}$$

Both the tangent vectors and normal vectors change as we move along the surface. For the changes of the vectors $\mathbf{T}_A, \mathbf{t}_\alpha$ we have

$$\frac{\partial \mathbf{T}_A}{\partial \Xi^B} = \Gamma_{AB}^C \mathbf{T}_C + B_{AB} \mathbf{N}, \quad \frac{\partial \mathbf{t}_\alpha}{\partial \xi^\beta} = \Gamma_{\alpha\beta}^\gamma \mathbf{t}_\gamma + b_{\alpha\beta} \mathbf{n}, \tag{3.26}$$

where $\Gamma_{AB}^C, \Gamma_{\alpha\beta}^\gamma$ are Christoffel symbols of the second kind of surfaces (see: Appendix 1)

$$\Gamma_{AB}^C = \frac{\partial \mathbf{T}_A}{\partial \Xi^B} \cdot \mathbf{T}^C, \quad \Gamma_{\alpha\beta}^\gamma = \frac{\partial \mathbf{t}_\alpha}{\partial \xi^\beta} \cdot \mathbf{t}^\gamma, \tag{3.27}$$

and the tensors $B_{AB}, b_{\alpha\beta}$ are called *curvature tensors* (the *second fundamental form of the surface*)

$$\begin{aligned}
B_{AB} &= -\frac{\partial \mathbf{N}}{\partial \Xi^A} \cdot \mathbf{T}_B, \\
b_{\alpha\beta} &= -\frac{\partial \mathbf{n}}{\partial \xi^\alpha} \cdot \mathbf{t}_\beta.
\end{aligned} \tag{3.28}$$

The eigenvalues of these symmetric tensors are called *principal curvatures.*

The above relations yield easily the relation for changes of the normal vectors $\mathbf{N}$ and $\mathbf{n}$. As the normal vectors are unit we have

$$\begin{aligned}
&\frac{\partial \mathbf{N} \cdot \mathbf{N}}{\partial \Xi^A} = 0 \quad \Rightarrow \quad \mathbf{N} \cdot \frac{\partial \mathbf{N}}{\partial \Xi^A} = 0 \quad \Rightarrow \quad \frac{\partial \mathbf{N}}{\partial \Xi^A} = K_A^B \mathbf{T}_B, \\
&\frac{\partial \mathbf{n} \cdot \mathbf{n}}{\partial \xi^\alpha} = 0 \quad \Rightarrow \quad \mathbf{n} \cdot \frac{\partial \mathbf{n}}{\partial \xi^\alpha} = 0 \quad \Rightarrow \quad \frac{\partial \mathbf{n}}{\partial \xi^\alpha} = k_\alpha^\beta \mathbf{t}_\beta;
\end{aligned}$$

then the substitution in (3.28) yields

$$K_A^C \mathbf{T}_C \cdot \mathbf{T}_B = K_A^C A_{CB} = -B_{AB} \quad \Rightarrow \quad K_{AB} = -B_{AB},$$
$$k_\alpha^\gamma \mathbf{t}_\gamma \cdot \mathbf{t}_\beta = k_\alpha^\gamma a_{\gamma\beta} = -b_{\alpha\beta} \quad \Rightarrow \quad k_{\alpha\beta} = -b_{\alpha\beta}.$$

Hence

$$\frac{\partial \mathbf{N}}{\partial \Xi^A} = -A^{BC} B_{CA} \mathbf{T}_B, \quad \frac{\partial \mathbf{n}}{\partial \xi^\alpha} = -a^{\beta\gamma} b_{\gamma\alpha} \mathbf{t}_\beta. \tag{3.29}$$

Therefore, as expected, for a flat surface (i.e. the curvature tensor vanishes: $B_{AB} = 0$ or $b_{\alpha\beta} = 0$) the normal vector is constant on the whole surface.

In addition, for the existence of the surface within the classical geometry of Euclidean spaces, the classical Mainardi-Codazzi and Gauss conditions must be satisfied

$$B_{AB;C} = B_{AC;B}, \quad \overset{\mathbf{A}}{R}_{ABCD} = B_{AB} B_{CD} - B_{BC} B_{AD}, \tag{3.30}$$

$$b_{\alpha\beta;\gamma} = b_{\alpha\gamma;\beta}, \quad \overset{\mathbf{a}}{R}_{\alpha\beta\gamma\delta} = b_{\alpha\beta} b_{\gamma\delta} - b_{\beta\gamma} b_{\alpha\delta}, \tag{3.31}$$

where $\overset{\mathbf{A}}{R}_{ABCD}$ is the Riemann tensor based upon the metric A_{AB}, and $\overset{\mathbf{a}}{R}_{\alpha\beta\gamma\delta}$ the Riemann tensor based upon the metric $a_{\alpha\beta}$ (comp. Sec. 2.5 and [205]).

Finally, let us recall the divergence theorem for surfaces. It follows from the Kelvin transformation quoted in Appendix 1 (relation (A.82)). We consider a surface $\mathcal{S}_t$ with the boundary curve $\partial\mathcal{S}_t$ (for surfaces in Lagrangian description, the considerations are dual). We consider a vector $\boldsymbol{\varphi} = \varphi^k \mathbf{e}_k$ tangent to this surface. From Kelvin's relation it follows

$$\int_{\mathcal{S}_t} \epsilon_{kij} n_j \frac{\partial \varphi^i}{\partial x^k} dS_t = \oint_{\partial\mathcal{S}_t} \varphi^k h_k dl_t, \tag{3.32}$$

where $\mathbf{h}$ is the unit vector tangent to the curve $\partial\mathcal{S}_t$. Since both $\boldsymbol{\varphi}$ and $\mathbf{t}$ are tangent to $\mathcal{S}_t$, we can write

$$\boldsymbol{\varphi} = \varphi^\alpha \mathbf{t}_\alpha, \quad \mathbf{h} = h^\alpha \mathbf{t}_\alpha. \tag{3.33}$$

Bearing (3.24) in mind, we obtain from (3.32)

$$\oint_{\partial\mathcal{S}_t} a_{\alpha\beta} \varphi^\alpha h^\beta dl_t = \int_{\mathcal{S}_t} \frac{1}{\sqrt{a}} \frac{\partial \left(\varphi^\gamma a_{\gamma\alpha}\right)}{\partial \xi^\beta} \epsilon^{\alpha\beta} dS_t. \tag{3.34}$$

We simplify this relation by introducing the following notation

$$\psi^{\alpha} = \frac{1}{\sqrt{a}} \epsilon^{\alpha\beta} a_{\beta\gamma} \varphi^{\gamma}, \quad \upsilon^{\alpha} = \sqrt{a} a^{\alpha\beta} \epsilon_{\beta\gamma} h^{\gamma}. \tag{3.35}$$

It is easy to show that vectors $\boldsymbol{\varphi}, \boldsymbol{\psi}$ are perpendicular to each other and have the equal magnitude. The same concerns the vectors $\mathbf{h}, \boldsymbol{\upsilon}$. It means that $\boldsymbol{\upsilon}$ is the unit vector normal to $\partial \mathcal{S}_t$ and tangential to $\mathcal{S}_t$.

Finally, we obtain

$$\oint_{\partial \mathcal{S}_t} \boldsymbol{\psi} \cdot \boldsymbol{\upsilon} dl_t = \int_{\mathcal{S}_t} \psi^{\alpha}_{;\alpha} dS_t, \tag{3.36}$$

where we have used the formula

$$\frac{\partial \sqrt{a}}{\partial \xi^{\beta}} = \sqrt{a} \Gamma^{\alpha}_{\alpha\beta}. \tag{3.37}$$

The formula (3.36) is the *divergence theorem on the surface.*

3.3.3 *Time dependence of surfaces*

Two particular cases of time dependence in (3.18) are important: if σ is independent of time we say that the surface is *stationary* and if Σ is independent of time we say that the surface is *material.*

If these representations are neither stationary nor material they move – the first one in the reference configuration, the second one in the current configuration with *normal speeds* defined by the following relations

$$U = -\frac{\frac{\partial \Sigma}{\partial t}}{|\operatorname{Grad} \Sigma|}, \quad c = -\frac{\frac{\partial \sigma}{\partial t}}{|\operatorname{grad} \sigma|}. \tag{3.38}$$

Bearing (3.18) in mind, we obtain immediately

$$|\operatorname{Grad} \Sigma| \, U = |\operatorname{grad} \sigma| \left(c - \mathbf{v} \cdot \mathbf{n} \right), \quad \mathbf{v} \left(\mathbf{X}.t \right) = \frac{\partial \mathbf{f}}{\partial t}. \tag{3.39}$$

The quantity U is called the *speed of propagation* while c is the *speed of displacement* of the surface.

If the full speed of surface points of $\mathcal{S}$ and, correspondingly, $\mathcal{S}_t$ is given by the relations

$$\mathbf{U} = U\mathbf{N}, \quad \mathbf{c} = c\mathbf{n}, \tag{3.40}$$

then we say that surface points perform the *normal motion.*

Let us mention that, in contrast to (3.18), equations $(3.19)_1$, $(3.20)_2$, also yield relations for *tangential velocities* of surfaces (*gliding*). Differentiating of (3.19) and (3.20) with respect to time, we obtain

$$\frac{\partial \mathbf{S}}{\partial t} = \frac{\mathbf{T}_A}{|\mathbf{T}_A|} U_{\|}^A + U\mathbf{N}, \tag{3.41}$$

$$\frac{\partial \mathbf{s}}{\partial t} = \frac{\mathbf{t}_\alpha}{|\mathbf{t}_\alpha|} c_{\|}^\alpha + c\mathbf{n}. \tag{3.42}$$

Taking the product with $\mathbf{T}_B$ and $\mathbf{t}_\beta$ (i.e. vectors which are tangent to parametric lines on $\mathcal{S}$ and $\mathcal{S}_t$), respectively, we obtain for tangential velocities

$$U_{\|A} = \frac{\partial \mathbf{S}}{\partial t} \cdot \frac{\mathbf{T}_A}{|\mathbf{T}_A|}, \quad A = 1, 2 \quad \text{(no sum!)}, \tag{3.43}$$

$$c_{\|\alpha} = \frac{\partial \mathbf{s}}{\partial t} \cdot \frac{\mathbf{t}_\alpha}{|\mathbf{t}_\alpha|}, \quad \alpha = 1, 2 \quad \text{(no sum!)}. \tag{3.44}$$

The above introduced description of surfaces indeed represents smooth *moving* surfaces provided certain differential conditions for time changes of their two fundamental forms, A_{AB}, B_{AB} for $\mathcal{S}$ and $a_{\alpha\beta}, b_{\alpha\beta}$ for $\mathcal{S}_t$, are satisfied. These *compatibility conditions* have been derived by J. Ericksen and their derivation can be found in [335] (Sec. 178).

The compatibility conditions have the form

$$\frac{\partial A_{AB}}{\partial t} = 2\left(U_{\|(A;B)} - U B_{AB}\right), \tag{3.45}$$

$$\frac{\partial B_{AB}}{\partial t} - U_{\|}^C B_{AB;C} - B_{(A}^C U_{\| B);C} = U_{;AB} - U B_A^C B_{CB},$$

for $\mathcal{S}$ and

$$\frac{\partial a_{\alpha\beta}}{\partial t} = 2\left(c_{\|(\alpha;\beta)} - c b_{\alpha\beta}\right), \tag{3.46}$$

$$\frac{\partial b_{\alpha\beta}}{\partial t} - c_{\|}^\gamma b_{\alpha\beta;\gamma} - b_{(\alpha}^\gamma c_{\| \beta);\gamma} = c_{;\alpha\beta} - c b_\alpha^\gamma b_{\gamma\beta},$$

for $\mathcal{S}_t$.

The conditions (3.45) and (3.46) are necessary and sufficient for the existence of the moving surfaces, $\mathcal{S}$ or $\mathcal{S}_t$, respectively, when $A_{AB}, B_{AB}, U, U_{\|A}$ or $a_{\alpha\beta}, b_{\alpha\beta}, c, c_{\|\alpha}$ are given.

Differentiation of the determinants A and a yields immediately

$$\frac{\partial A}{\partial t}=\frac{\partial A}{\partial A_{AB}}\frac{\partial A_{AB}}{\partial t}=AA^{AB}\frac{\partial A_{AB}}{\partial t}=2A\left(U^{A}_{\|;A}-2K_{M0}U\right),\qquad(3.47)$$
$$\frac{\partial a}{\partial t}=\frac{\partial a}{\partial a_{\alpha\beta}}\frac{\partial a_{\alpha\beta}}{\partial t}=aa^{\alpha\beta}\frac{\partial a_{\alpha\beta}}{\partial t}=2a\left(c^{\alpha}_{\|;\alpha}-2K_{M}c\right),$$

where (3.45), (3.46) were used and

$$K_{M0}=\frac{1}{2}A^{AB}B_{AB},\quad K_{M}=\frac{1}{2}a^{\alpha\beta}b_{\alpha\beta},\qquad(3.48)$$

are the *mean curvatures* of surfaces $\mathcal{S},\mathcal{S}_t$, respectively. Consequently, bearing (3.25) in mind,

$$(d\mathcal{S})^{\bullet}=\left(U^{A}_{\|;A}-2K_{M0}U\right)d\mathcal{S},\quad (d\mathcal{S}_t)^{\bullet}=\left(c^{\alpha}_{\|;\alpha}-2K_{M}c\right)d\mathcal{S}_t.\qquad(3.49)$$

3.3.4 *Kinematical compatibility conditions*

We proceed to discuss kinematical compatibility conditions which concern the time behavior of fields in the vicinity of singular surfaces. We present only the conditions for the current configuration of the surface $\mathcal{S}_t$ but, of course, relations for $\mathcal{S}$ are dual.

First of all, we need the notion of the displacement derivative.

The moving surface $\mathcal{S}_t$ divides two regions $\mathcal{P}_t^{+}$ and $\mathcal{P}_t^{-}$. The surface is supposed to be singular for the field $\varphi(\mathbf{x},t)$ as defined by (2.87). The problem can be solved generally by means of Hadamard's Lemma (Sec. 2.6) applied to the four-dimensional space-time. Then the geometrical compatibility conditions of Section 2.6 and the kinematical compatibility conditions which we proceed to consider would follow as particular cases. However, for practical purposes it is more convenient to separate these conditions. It may happen, for instance, in continuum mechanics that geometrical compatibility conditions are satisfied and kinematical conditions are not. Such a problem appears, for example, when we split the body into two parts or when we join two parts together.

In order to specify jump conditions on a moving surface, we have to introduce a rate of change of limits of a function $\varphi(\mathbf{x},t)$ caused by the motion of the surface. This notion has been introduced by T. Y. Thomas (comp. [323], Chapter 2). We consider a tangential path on the hypersurface $\sigma(\mathbf{x},t)=0$, of the 4-dimensional x-t-space, tangent to the vector $(c\mathbf{n},1)$. Hence, for functions $\varphi^{+}(\mathbf{x},t),\varphi^{-}(\mathbf{x},t)$ defined as limits of $\varphi(\mathbf{x},t)$ at the point $\mathbf{x}=\mathbf{s}(\xi^{\alpha},t)$ on the moving surface $\mathcal{S}_t$, we specify their rates of change $\delta\varphi^{+}/\delta t,\delta\varphi^{-}/\delta t$ as apparent to an observer moving with the normal velocity c of the surface. They have the form of directional derivatives (see [322])

$$\frac{\delta\varphi^+}{\delta t} = \lim_{\Delta t \to 0} \frac{\varphi^{+\prime} - \varphi^+}{\Delta t}, \quad \frac{\delta\varphi^-}{\delta t} = \lim_{\Delta t \to 0} \frac{\varphi^{-\prime} - \varphi^-}{\Delta t}, \tag{3.50}$$

where

$$\varphi^{+\prime} = \varphi^+ \left(\mathbf{s}\left(\xi^\alpha, t\right) + c\Delta t\mathbf{n}, t\right), \quad \varphi^{-\prime} = \varphi^- \left(\mathbf{s}\left(\xi^\alpha, t\right) + c\Delta t\mathbf{n}, t\right), \tag{3.51}$$

i.e. the increments of φ^+, φ^- are caused by the displacement

$$\Delta\mathbf{s} = c\Delta t\mathbf{n}. \tag{3.52}$$

This means that we evaluate functions with prime in points of the surface $\mathcal{S}_{t+\Delta t}$ parallel to $\mathcal{S}_t$ shifted by means of the normal vector $c\Delta t\mathbf{n}$. Hence

$$\frac{\delta \left[\left[\varphi\right]\right]}{\delta t} = \lim_{\Delta t \to 0} \left[\frac{\varphi^{+\prime} - \varphi^{-\prime}}{\Delta t} - \frac{\varphi^+ - \varphi^-}{\Delta t} \right] = \lim_{\Delta t \to 0} \left[\frac{\varphi^{+\prime} - \varphi^+}{\Delta t} - \frac{\varphi^{-\prime} - \varphi^-}{\Delta t} \right],$$

and, consequently,

$$\frac{\delta \left[\left[\varphi\right]\right]}{\delta t} = \left[\left[\frac{\delta\varphi}{\delta t} \right]\right]. \tag{3.53}$$

In particular, according to (3.50) we have

$$\frac{\delta\mathbf{s}}{\delta t} = \lim_{\Delta t \to 0} \frac{\Delta\mathbf{s}}{\Delta t} = \lim_{\Delta t \to 0} \frac{\mathbf{s}\left(\xi^\alpha, t\right) + c\Delta t\mathbf{n} - \mathbf{s}\left(\xi^\alpha, t\right)}{\Delta t} = c\mathbf{n}. \tag{3.54}$$

Simultaneously, using the same properties which we have specified in the proof of Hadamard's Lemma in Section 2.6,

$$\begin{aligned} \lim_{\Delta t \to 0} \frac{\varphi^{+\prime} - \varphi^+}{\Delta t} &= \lim_{\Delta t \to 0} \frac{\Delta\mathbf{s}}{\Delta t} \cdot \left(\operatorname{grad}\varphi\right)^+ + \left(\frac{\partial\varphi}{\partial t}\right)^+, \\ \lim_{\Delta t \to 0} \frac{\varphi^{-\prime} - \varphi^-}{\Delta t} &= \lim_{\Delta t \to 0} \frac{\Delta\mathbf{s}}{\Delta t} \cdot \left(\operatorname{grad}\varphi\right)^- + \left(\frac{\partial\varphi}{\partial t}\right)^-. \end{aligned} \tag{3.55}$$

Consequently, we obtain

$$\begin{aligned} \frac{\delta\varphi^+}{\delta t} &= c\mathbf{n} \cdot \left(\operatorname{grad}\varphi\right)^+ + \left(\frac{\partial\varphi}{\partial t}\right)^+, \\ \frac{\delta\varphi^-}{\delta t} &= c\mathbf{n} \cdot \left(\operatorname{grad}\varphi\right)^- + \left(\frac{\partial\varphi}{\partial t}\right)^-. \end{aligned} \tag{3.56}$$

These relations yield immediately the following condition for the jump

$$\frac{\delta}{\delta t}\left[\left[\varphi\right]\right]=\left[\left[\frac{\partial\varphi}{\partial t}\right]\right]+c\mathbf{n}\cdot\left[\left[\operatorname{grad}\varphi\right]\right]. \tag{3.57}$$

This is the *kinematical compatibility condition* or *Hadamard condition* for the scalar field φ. It has been derived by J. Hadamard whose book on the wave propagation, published in 1903 [138] contains the first modern wave analysis within continuum mechanics based on the notion of a singular surface. We present further some results of this analysis extended by Richard Courant, Peter Lax, Fritz John and many others.

If φ is a tensor, the above relation does not possess proper invariance properties. This can be improved by defining an appropriate Lie derivative as we did in Sec. 3.2. We shall not present it in this book and refer the reader to the book of Truesdell and Toupin [335].

In the case of a spatial tensor $\boldsymbol{\psi}=\psi^{k\ldots l}_{m\ldots n}\mathbf{e}_k\otimes\ldots\otimes\mathbf{e}^m$, the relation (3.57) defines the tensor and we call it the *displacement derivative*

$$\frac{\delta_d\boldsymbol{\psi}}{\delta t}=\frac{\partial\boldsymbol{\psi}}{\partial t}+c\mathbf{n}\cdot\operatorname{grad}\boldsymbol{\psi}. \tag{3.58}$$

Herrman von Helmholtz
1821 – 1894

Jacques S. Hadamard
1865 – 1963

Richard Courant
1888 – 1972

In particular, for the normal vector $\mathbf{n}$ of the surface $\mathcal{S}_t$ we can easily prove (e.g. [323])

$$\frac{\delta_d\mathbf{n}}{\delta t}=-a^{\alpha\beta}\mathbf{t}_\alpha c_{;\beta}. \tag{3.59}$$

The geometrical interpretation of this relation is straightforward. If we take the scalar product of this relation with the tangent vector $\mathbf{t}_\beta$ we obtain

$$\mathbf{t}_\alpha \cdot \frac{\delta_d \mathbf{n}}{\delta t} = -c_{;\alpha}, \tag{3.60}$$

which means that the length of the projection of the displacement derivative of the normal onto a given direction on the surface is the negative of the speed of displacement in this direction. Obviously, $\delta_d \mathbf{n}/\delta t \cdot \mathbf{n} = 0$.

In the case of a continuous field, we have from (3.57)

$$[[\varphi]] = 0 \quad \Rightarrow \quad \left[\left[\frac{\partial \varphi}{\partial t}\right]\right] = -c\mathbf{n} \cdot [[\operatorname{grad} \varphi]] \,. \tag{3.61}$$

Bearing the geometrical compatibility condition (2.92) in mind, we obtain

$$\left[\left[\frac{\partial \varphi}{\partial t}\right]\right] = -cb. \tag{3.62}$$

This relation is expressed in terms appropriate for a scalar but it applies to arbitrary tensors.

As in the case of geometrical compatibility conditions, one can easily derive the iterated kinematical compatibility conditions. We shall not do so here and refer the reader again to the extensive analysis in the book of Truesdell and Toupin ([335], Sec. 181; see also [323]).

Apart from processes yielding surface effects such as wave propagation or interactions through ideal boundaries, continuum thermodynamics deals often with processes on two-dimensional bodies modelling, for instance, membranes. These may combine with classical boundaries and lead to essential modifications of boundary conditions (e.g. the well-known boundary conditions of the third kind for heat conduction or permeable boundaries of porous materials). In Secs. 2.6 and 3.3, we have presented some geometrical and kinematical conditions for the persistence of such surfaces. Further we amend these properties with the discussion of balance equations which provide the dynamic compatibility conditions for singular surfaces.

The geometrical interpretation of this relation is straightforward. If we take [illegible]

Chapter 4

Balance equations

4.1 Balance laws in Lagrangian description

As we have already mentioned in the first Chapter, fundamental quantities describing thermomechanical processes such as mass, momentum, angular momentum, energy and entropy satisfy balance equations. These notions are defined on a family of measurable subsets of the body $\mathcal{B}_0$. Let us choose a member of this family, say $\mathcal{P} \subset \mathcal{B}_0$. Then $\Phi(\mathcal{P}, t)$ denotes any of the above quantities prescribed to the subbody $\mathcal{P}$ at the instant of time t. It is the quantity which can be measured in laboratories. It is assumed that the set function $\Phi(., t)$ is *additive*, i.e. for two subbodies $\mathcal{P}_1$ and $\mathcal{P}_2$ which are separate $\mathcal{P}_1 \cap \mathcal{P}_2 = \emptyset$, $\Phi(\mathcal{P}_1 \cup \mathcal{P}_2, t) = \Phi(\mathcal{P}_1, t) + \Phi(\mathcal{P}_2, t)$. For instance, the energy of two subbodies which are not overlapping is the sum of energies of both subbodies. This assumption is usually slightly weaker in order to admit a concentration of energy on interfaces. We return to this problem in Section 4.3. In addition, it is assumed that this set function is *continuous* with respect to the volume measure, i.e. there exists a constant $\alpha > 0$ such that for any subbody $\mathcal{P}$

$$|\Phi(\mathcal{P}, t)| \leq \alpha \operatorname{vol} \mathcal{P}, \tag{4.1}$$

where $\operatorname{vol}\mathcal{P}$ is the volume of $\mathcal{P}$. According to the measure theory, these two assumptions, additivity and continuity, yield the existence of the *density* $\varphi(\mathbf{X}, t)$, $\mathbf{X} \in \mathcal{B}_0$ (the so-called *Radon-Nikodym derivative*) such that

$$\Phi(\mathcal{P}, t) = \int_{\mathcal{P}} \varphi(\mathbf{X}, t)\, dV, \tag{4.2}$$

where on the right-hand side we have the so-called Lebesgue integral. The above representation is the most fundamental feature of continuous models. It is obvious that the requirement of additivity eliminates long-range interactions from the model (compare remarks in Chapter 1). A contribution of such interactions would mean, for instance, that the energy of two separate subbodies would not be equal to the sum of energies of both subbodies but it would rather contain as well an energy of interaction.

The quantity Φ is assumed to satisfy the *balance equation*

$$\forall_{\mathcal{P}} \frac{d\Phi}{dt}(\mathcal{P},t) = \Psi_{\mathcal{S}}(\partial\mathcal{P},t) + \Psi_{\mathcal{P}}(\mathcal{P},t), \tag{4.3}$$

where $\Psi_{\mathcal{S}}$ describes the *flux* of the quantity Φ through the surface $\partial\mathcal{P}$ of the subbody $\mathcal{P}$ and $\Psi_{\mathcal{P}}$ is the sum of the volume *supply* of the quantity Φ from the external world and of the *production* of Φ in the subbody $\mathcal{P}$. These two functions are assumed to satisfy axioms similar to (4.1) and, consequently, it can be proved (e.g. [329], [330]) that they possess the following representations

$$\Psi_{\mathcal{S}}(\partial\mathcal{P},t) = \oint_{\partial\mathcal{P}} \psi_{\mathcal{S}}(\mathbf{X},t)\,d\mathcal{S}, \quad \Psi_{\mathcal{P}}(\mathcal{P},t) = \int_{\mathcal{P}} [\psi_V(\mathbf{X},t) + \hat{\varphi}(\mathbf{X},t)]\,dV, \tag{4.4}$$

where $\psi_{\mathcal{S}}$ is the flux density per unit surface and unit time of the field density φ, ψ_V is the density of the *volume supply* of φ and $\hat{\varphi}$ is the *production* (*source*) per unit volume and time of the field φ[1].

Additionally, it is assumed that the surface $\partial\mathcal{P}$ is orientable and the dependence of the flux $\psi_{\mathcal{S}}(\mathbf{X},t)$ on a choice of the surface reduces only to the dependence on the unit vector $\mathbf{N}(\mathbf{X},t)$ orthogonal to the surface at the point $\mathbf{X}$ and oriented outwards with respect to $\mathcal{P}$, i.e. $\psi_{\mathcal{S}}(\mathbf{X},t) = \psi(\mathbf{N},\mathbf{X},t)$. Then one can show the following [62]:

Cauchy Theorem: There exists a function $\boldsymbol{\psi}(\mathbf{X},t)$ such that

$$\psi_{\mathcal{S}}(\mathbf{X},t) = \boldsymbol{\psi}(\mathbf{X},t)\cdot\mathbf{N}(\mathbf{X},t). \tag{4.5}$$

It means that $\psi_{\mathcal{S}}$ is a *linear homogeneous* function of the unit vector $\mathbf{N}$.

We prove this property.

We show first that $\psi_{\mathcal{S}}$ changes sign when the surface changes orientation: $\mathbf{N}\to-\mathbf{N}$. We divide a subbody $\mathcal{P}$ into two subbodies $\mathcal{P}_1$ and $\mathcal{P}_2$, $\mathcal{P} = \mathcal{P}_1\cup\mathcal{P}_2$, such that they have a common part of the boundary, $\mathcal{S}$: $\mathcal{P}_1\cap\mathcal{P}_2 = \mathcal{S}$. This surface has the outward orientation $\mathbf{N}$ for $\mathcal{P}_1$ and, consequently, it has the outward orientation $-\mathbf{N}$ for $\mathcal{P}_2$. Then the balance equation has the

---

[1]The difference between the volume supply ψ_V and the production $\hat{\varphi}$ can be recognized only in relation to the constitutive definition of the material. Then the volume supply is a quantity which is controlled from the external world – it can be, for instance, switched off, and the production (source) is controlled by constitutive variables which characterize a particular material.

form

$$\frac{d}{dt}\int_{\mathcal{P}_1\cup\mathcal{P}_2}\varphi dV = \oint_{\partial\mathcal{P}}\psi_{\mathcal{S}}d\mathcal{S} + \int_{\mathcal{P}_1\cup\mathcal{P}_2}\left[\psi_V + \hat{\varphi}\right]dV.$$

Using the balance equations for $\mathcal{P}_1$ and $\mathcal{P}_2$ separately we obtain

$$\int_{\mathcal{S}}\psi_{\mathcal{S}}d\mathcal{S} = -\int_{\mathcal{S}}\psi_{-\mathcal{S}}d\mathcal{S},$$

where $-\mathcal{S}$ indicates the opposite orientation of the surface. Hence $\psi_{\mathcal{S}} = -\psi_{-\mathcal{S}}$ or, bearing the assumption on dependence on the normal vector in mind,

$$\psi(\mathbf{N},\mathbf{X},t) = -\psi(-\mathbf{N},\mathbf{X},t). \tag{4.6}$$

Now we are in the position to prove the linearity of the above function with respect to $\mathbf{N}$. As this function is defined only for unit vectors we define first the following extension on the space $\mathcal{V}^3$ of arbitrary vectors $\mathbf{W}$

$$\tilde{\psi}(\mathbf{W},\mathbf{X},t) = \begin{cases} |\mathbf{W}|\,\psi\left(\frac{\mathbf{W}}{|\mathbf{W}|},\mathbf{X},t\right) \text{ for } \mathbf{W}\neq 0, \\ 0 \text{ for } \mathbf{W} = 0. \end{cases} \tag{4.7}$$

We show that this extension is the linear function with respect to $\mathbf{W}$, i.e. for two arbitrary numbers a and b we have $\tilde{\psi}(a\mathbf{W}_1 + b\mathbf{W}_2,\mathbf{X},t) = a\tilde{\psi}(\mathbf{W}_1,\mathbf{X},t) + b\tilde{\psi}(\mathbf{W}_2,\mathbf{X},t)$. This condition can be replaced by the following two conditions:

i/ for any real number a and any vector $\mathbf{W}$: $\tilde{\psi}(a\mathbf{W},\mathbf{X},t) = a\tilde{\psi}(\mathbf{W},\mathbf{X},t)$,

ii/ for any two vectors $\mathbf{W}_1,\mathbf{W}_2$: $\tilde{\psi}(\mathbf{W}_1+\mathbf{W}_2,\mathbf{X},t) = \tilde{\psi}(\mathbf{W}_1,\mathbf{X},t) + \tilde{\psi}(\mathbf{W}_2,\mathbf{X},t)$.

It is clear that the function (4.7) satisfies the above conditions for either $a = 0$ or $\mathbf{W} = \mathbf{0}$ or $a > 0, \mathbf{W}\neq\mathbf{0}$. Hence we confine our interest to the case $a < 0$ and $\mathbf{W}\neq\mathbf{0}$. We have

$$\begin{aligned}\tilde{\psi}(a\mathbf{W},\mathbf{X},t) &= \tilde{\psi}(-|a|\,\mathbf{W},\mathbf{X},t) = |a|\,\tilde{\psi}(-\mathbf{W},\mathbf{X},t)\\ &= -|a|\,\tilde{\psi}(\mathbf{W},\mathbf{X},t) = a\tilde{\psi}(\mathbf{W},\mathbf{X},t),\end{aligned}$$

which proves i/.

In the case of linearly dependent vectors $\mathbf{W}_1,\mathbf{W}_2$, property ii/ reduces to i/. Therefore we assume that these vectors are linearly independent. Let

$$\mathbf{W}_3 = -(\mathbf{W}_1 + \mathbf{W}_2). \tag{4.8}$$

We consider a triangular block $\mathcal{P}_\delta$, containing $\mathbf{X}$, with the faces $\mathcal{S}_1, \mathcal{S}_2, \mathcal{S}_3$ normal to $\mathbf{W}_1, \mathbf{W}_2, \mathbf{W}_3$, respectively, and the two parallel end triangles $\mathcal{S}_4, \mathcal{S}_5$ apart by the distance δ (Fig. 7). Let ε be the height of the triangles $\mathcal{S}_4, \mathcal{S}_5$ and $A_i, i = 1, 2, 3$ be the areas of $\mathcal{S}_i$. From the construction of the block we have

$$\frac{A_1}{|\mathbf{W}_1|} = \frac{A_2}{|\mathbf{W}_2|} = \frac{A_3}{|\mathbf{W}_3|}. \tag{4.9}$$

The balance equation written for $\mathcal{P}_\delta$ yields

$$\frac{1}{\varepsilon}\int_{\mathcal{P}_\delta}\left(\frac{\partial\varphi}{\partial t} - \psi_V - \hat{\varphi}\right) dV - \frac{1}{\varepsilon}\int_{\mathcal{S}_4\cup\mathcal{S}_5} \psi_S\left(\mathbf{N}, \mathbf{X}, t\right) d\mathcal{S}$$

$$= \frac{1}{\varepsilon}\sum_{i=1}^{3}\int_{\mathcal{S}_i} \psi_S\left(\frac{\mathbf{W}_i}{|\mathbf{W}_i|}, \mathbf{X}, t\right) d\mathcal{S}.$$

It is easy to see that $\operatorname{vol}\mathcal{P}_\delta$ and A_4, A_5 are of the order ε^2 whereas $A_i, i = 1, 2, 3$ is of order ε. Hence, we obtain

$$\lim_{\varepsilon\to 0}\sum_{i=1}^{3}\frac{1}{\varepsilon}\int_{\mathcal{S}_i} \psi_S\left(\frac{\mathbf{W}_i}{|\mathbf{W}_i|}, \mathbf{X}, t\right) d\mathcal{S} = 0.$$

Now let us apply the mean value theorem to the above relation. We have

$$\lim_{\varepsilon\to 0}\sum_{i=1}^{3}\frac{1}{\varepsilon}A_i\psi_S\left(\frac{\mathbf{W}_i}{|\mathbf{W}_i|}, \mathbf{X}^{(i)}, t\right) = 0,$$

where $\mathbf{X}^{(i)} \in \mathcal{S}_i$. Bearing (4.9) in mind, we finally arrive at

$$\lim_{\varepsilon\to 0}\sum_{i=1}^{3}|\mathbf{W}_i|\,\psi_S\left(\frac{\mathbf{W}_i}{|\mathbf{W}_i|}, \mathbf{X}^{(i)}, t\right)$$

$$= |\mathbf{W}_1|\,\psi_S\left(\frac{\mathbf{W}_1}{|\mathbf{W}_1|}, \mathbf{X}, t\right) + |\mathbf{W}_2|\,\psi_S\left(\frac{\mathbf{W}_2}{|\mathbf{W}_2|}, \mathbf{X}, t\right) - |\mathbf{W}_3|\,\psi_S\left(\frac{\mathbf{W}_3}{|\mathbf{W}_3|}, \mathbf{X}, t\right) = 0,$$

which yields condition ii/. This completes the proof.

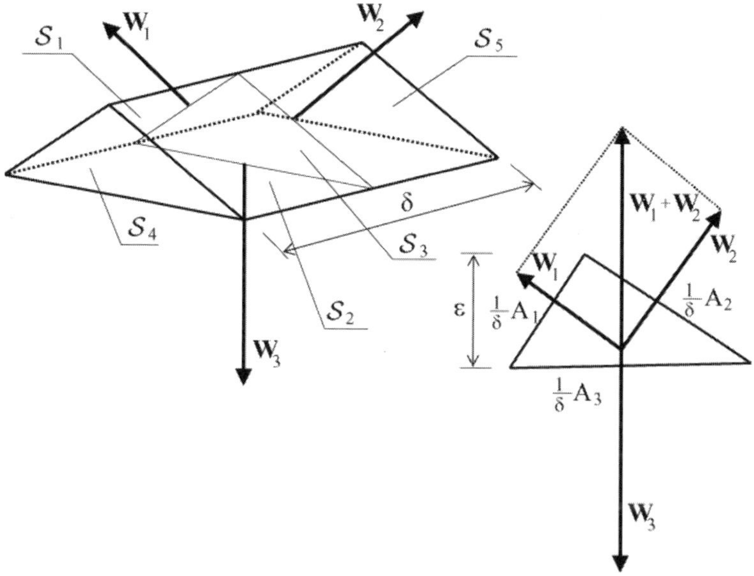

Fig. 7: *Triangular block used in the proof of Cauchy's Theorem*

Bearing the above results in mind we can write the general balance equation in the following form

$$\forall_{\mathcal{P}} \ \frac{d}{dt}\int_{\mathcal{P}} \varphi(\mathbf{X},t)\,dV = \oint_{\partial\mathcal{P}} \boldsymbol{\psi}(\mathbf{X},t)\cdot\mathbf{N}(\mathbf{X},t)\,dS + \int_{\mathcal{P}} \left[\psi_V(\mathbf{X},t)+\hat{\varphi}(\mathbf{X},t)\right]dV. \tag{4.10}$$

This result can be transformed to the *local form.* We consider two special cases of this form – one which holds in *regular points*, i.e. in points $\mathbf{X} \in \mathcal{B}_0$ in which all densities appearing in (4.10) are continuous and, secondly, in *points of a singular surface* $\mathcal{S}$ which may move through the body with a speed $U\mathbf{N}$. This is the velocity of the points on the surface in direction perpendicular to the surface. As we see further the balance equation (4.10) does not contribute anything to the description of motion which is tangential to the surface, i.e. gliding of $\mathcal{S}$ along tangential directions is immaterial for our considerations. In points of a singular surface limits of densities of the relation (4.10) may be different on both sides of the surface, i.e. they may suffer *finite jumps.*

Let us first consider the case of a regular point $\mathbf{X} \in \mathcal{B}_0$. We construct an infinite descending family of subbodies $\{\mathcal{P}_i\}_{i=1}^{\infty}$ with three properties: i/ each set of this family contains the point $\mathbf{X}$, ii/ for each i, $\mathcal{P}_{i+1} \subset \mathcal{P}_i$ and iii/ $\lim_{i\to\infty} \operatorname{vol}\mathcal{P}_i = 0$, where $\operatorname{vol}\mathcal{P}_i = \int_{\mathcal{P}_i} dV$ is the volume of $\mathcal{P}_i$. Then using the Stokes Theorem for the surface integral we obtain

$$\lim_{i\to\infty}\frac{1}{\operatorname{vol}\mathcal{P}_i}\int\left[\frac{\partial\varphi}{\partial t}-\operatorname{Div}\boldsymbol{\psi}-\psi_V-\hat{\varphi}\right]dV=0,$$

and, accounting for the mean value Theorem for integrals, the following *local balance equation* follows

$$\frac{\partial\varphi}{\partial t}=\operatorname{Div}\boldsymbol{\psi}+\psi_V+\hat{\varphi}, \tag{4.11}$$

for almost all points of $\mathcal{B}_0$.

In thermomechanics this equation is written for mass, momentum, angular momentum, energy and entropy. We list the corresponding densities in Table 2.

Table 2: *Densities of thermomechanics*

Name of density	density φ	flux $\boldsymbol{\psi}$	supply ψ_V	source $\hat{\varphi}$
mass	ρ_0	0	0	0
momentum	$\rho_0\mathbf{v}$	$\mathbf{P}$	$\rho_0\mathbf{b}$	0
angular momentum	$\rho_0\mathbf{x}\times\mathbf{v}$	$M_k^M\mathbf{e}^k\otimes\mathbf{e}_M$ $M_k^M=\epsilon_{klm}x^lP^{mM}$	$\rho_0\mathbf{x}\times\mathbf{b}$	0
energy	$\rho_0\left(\varepsilon+\frac{1}{2}v^2\right)$	$-\mathbf{Q}+\mathbf{P}^T\mathbf{v}$	$\rho_0\mathbf{v}\cdot\mathbf{b}+$ $+\rho_0 r$	0
entropy	$\rho_0\eta$	$-\mathbf{H}$	$\rho_0 s$	$\hat{\eta}$

All densities are, of course, referred to the unit volume in the undeformed (reference) configuration $\mathcal{B}_0$. $\mathbf{P}$ is the so-called Piola-Kirchhoff stress tensor, $\mathbf{b}$ is the body force per unit mass, ε denotes the specific internal energy per unit mass, $\frac{1}{2}\rho_0 v^2\equiv\frac{1}{2}\rho_0\mathbf{v}\cdot\mathbf{v}$ is the density of kinetic energy per unit reference volume, $\mathbf{Q}$ is the heat flux vector in the reference configuration, r is the density of energy radiation per unit mass, η is the specific entropy per unit mass, $\mathbf{H}$ is the entropy flux vector in the reference configuration, and $\hat{\eta}$ is the source of entropy per unit mass.

The explicit form of balance equations corresponding to the quantities listed in Table 2 is as follows

$$
\begin{aligned}
\frac{\partial \rho_0}{\partial t} &= 0, \\
\frac{\partial \rho_0 \mathbf{v}}{\partial t} &= \operatorname{Div} \mathbf{P} + \rho_0 \mathbf{b}, \\
\frac{\partial \rho_0 \mathbf{x} \times \mathbf{v}}{\partial} &= \operatorname{Div} \mathbf{x} \times \mathbf{P} + \rho_0 \mathbf{x} \times \mathbf{b}, \\
\frac{\partial \rho_0 \left(\varepsilon + \frac{1}{2} v^2\right)}{\partial t} &= \operatorname{Div} \left(-\mathbf{Q} + \mathbf{P}^T \mathbf{v}\right) + \rho_0 \left(\mathbf{v} \cdot \mathbf{b} + r\right), \\
\frac{\partial \rho_0 \eta}{\partial t} &= \operatorname{Div} \left(-\mathbf{H}\right) + \rho_0 s + \hat{\eta},
\end{aligned} \tag{4.12}
$$

where

$$
\mathbf{x} \times \mathbf{P} = \epsilon_{klm} x^l P^{mM} \mathbf{e}^k \otimes \mathbf{e}_M. \tag{4.13}
$$

The body force $\rho_0 \mathbf{b}$ appearing in the momentum balance equation contains the action of the external world on the body but it may also contain contributions stemming from *noninertial frames of reference*.

$$
\begin{aligned}
\rho_0 \mathbf{b} &= \rho_0 \mathbf{b}^0 + \rho_0 \mathbf{i}, \\
\rho_0 \mathbf{i} &= 2\rho_0 \boldsymbol{\Omega} \left(\mathbf{v}^* - \dot{\mathbf{c}}\right) - \rho_0 \boldsymbol{\Omega}^2 \left(\mathbf{x}^* - \mathbf{c}\right) + \rho_0 \dot{\boldsymbol{\Omega}} \left(\mathbf{x}^* - \mathbf{c}\right) + \rho_0 \ddot{\mathbf{c}}.
\end{aligned} \tag{4.14}
$$

Then the force $\rho_0 \mathbf{b}$ is called the *apparent body force* because it consists of the true external force $\rho_0 \mathbf{b}^0$ and of the *inertial body force* $\rho_0 \mathbf{i}$ which in turn, possesses the following contributions

$2\rho_0 \boldsymbol{\Omega} \left(\mathbf{v}^* - \dot{\mathbf{c}}\right)$ – *Coriolis force*,
$-\rho_0 \boldsymbol{\Omega}^2 \left(\mathbf{x}^* - \mathbf{c}\right)$ – *centrifugal force*,
$\rho_0 \boldsymbol{\Omega} \left(\mathbf{x}^* - \mathbf{c}\right)$ – *Euler force*,
$\rho_0 \ddot{\mathbf{c}}$ – *inertial force of relative translation*.

We return later to the detailed discussion of the definitions and the interpretations of all these quantities.

Let us note that except for entropy all other sources are zero. Such balance equations are called *conservation laws*. We see in the theory of multicomponent systems that for some field quantities of mixtures it does not have to be the case.

Now let us turn our attention to points on a singular surface $\mathcal{S}$. We construct again a descending family of subbodies $\{\mathcal{P}_i\}$ with three properties: i/ for each i, $\mathcal{P}_i \cap \mathcal{S} = \mathcal{P}_{i+1} \cap \mathcal{S}$, ii/ for each i, $\mathcal{P}_{i+1} \subset \mathcal{P}_i$ and iii/ $\lim_{i \to \infty} \operatorname{vol} \mathcal{P}_i = 0$. Such a family is demonstrated in Fig. 8.

First we estimate the derivative on the left-hand side of the balance equation. We have

$$\frac{d}{dt}\int_{\mathcal{P}_i}\varphi dV = \frac{d}{dt}\int_{\mathcal{P}_i^+}\varphi dV + \frac{d}{dt}\int_{\mathcal{P}_i^-}\varphi dV$$

$$= \int_{\mathcal{P}_i^+}\frac{\partial\varphi}{\partial t}dV - \int_{\mathcal{S}\cap\partial\mathcal{P}_i^+}\varphi^+ U d\mathcal{S} + \int_{\mathcal{P}_i^-}\frac{\partial\varphi}{\partial t}dV + \int_{\mathcal{S}\cap\partial\mathcal{P}_i^-}\varphi^- U d\mathcal{S},$$

where $\mathcal{P}_i^+, \mathcal{P}_i^-$ are the parts of $\mathcal{P}_i$ lying above and below the surface $\mathcal{S}$, respectively, φ^+, φ^- are limits of φ calculated from the positive (with respect to the orientation $\mathbf{N}$) and negative sides of $\mathcal{S}$. The difference in sign in surface integrals follows from the opposite orientation of the surfaces $\mathcal{S}\cap\partial\mathcal{P}_i^+$ and $\mathcal{S}\cap\partial\mathcal{P}_i^-$.

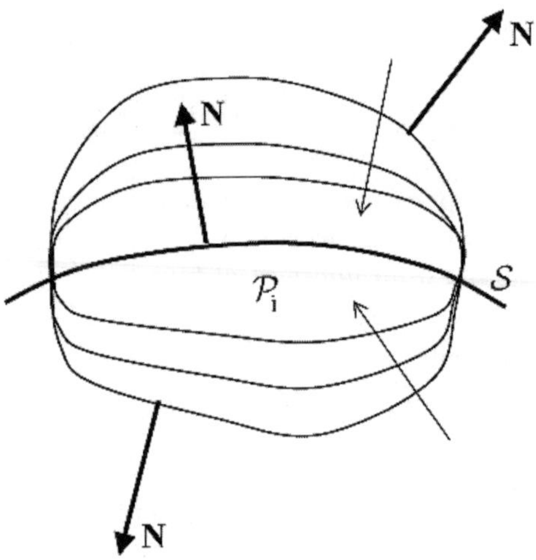

Fig. 8: *Transition to a singular surface*

In the limit $i\to\infty$ volume integrals vanish. The flux term can be written in the form

$$\oint_{\partial\mathcal{P}_i}\boldsymbol{\psi}\cdot\mathbf{N}d\mathcal{S} = \int_{\partial\mathcal{P}_i\cap\partial\mathcal{P}_i^+}\boldsymbol{\psi}\cdot\mathbf{N}d\mathcal{S} + \int_{\partial\mathcal{P}_i\cap\partial\mathcal{P}_i^-}\boldsymbol{\psi}\cdot\mathbf{N}d\mathcal{S}. \tag{4.15}$$

Taking the limit in the whole balance equation we obtain

$$\int_{\mathcal{S}\cap\mathcal{P}_i}\left[-\left(\varphi^+ - \varphi^-\right)U - \left(\boldsymbol{\psi}^+\cdot\mathbf{N} - \boldsymbol{\psi}^-\cdot\mathbf{N}\right)\right]d\mathcal{S} = 0.$$

We can localize this relation as well and for an arbitrary point of the surface $\mathcal{S}$ we obtain the following *Kotchine condition*

$$[[\varphi]]\,U + [[\boldsymbol{\psi}]]\cdot\mathbf{N} = 0, \quad [[...]] := (...)^+ - (...)^- . \tag{4.16}$$

It has been assumed that the source is volume continuous, i.e. that the surface $\mathcal{S}$ does not contribute to the production. It does not have to be the case for some surfaces such as membranes. We discuss this problem later.

For thermomechanical fields the balance equations (4.12) and their counterparts on singular surfaces are collected in Table 3 in the same order as in Table 2.

In the evaluation of the second law of thermodynamics it is convenient to work with field equations of the first order. Then neither $\mathbf{F}$ should be considered as the gradient of the function of motion $\mathbf{f}$ nor $\mathbf{v}$ should be the time derivative of this function. The function $\mathbf{f}$ does not appear in such an approach. Its existence is secured by the integrability conditions of $\mathbf{F}$ and $\mathbf{v}$

$$\frac{\partial \mathbf{F}}{\partial t} = \operatorname{Grad} \mathbf{v}, \quad \operatorname{Grad} \mathbf{F} = (\operatorname{Grad} \mathbf{F})^{T^{23}} \text{ i.e. } \frac{\partial F_{kK}}{\partial X_L} = \frac{\partial F_{kL}}{\partial X_K}. \tag{4.17}$$

Table 3: *Balance equations of thermomechanical model in Lagrangian description*
Left – regular points, right – points of a singular surface

$\frac{\partial \rho_0}{\partial t} = 0$	$[[\rho_0]]\, U = 0$
$\rho_0 \frac{\partial \mathbf{v}}{\partial t} = \operatorname{Div} \mathbf{P} + \rho_0 \mathbf{b}$	$[[\rho_0 \mathbf{v}]]\, U + [[\mathbf{P}]]\, \mathbf{N} = 0$
$\mathbf{P}\mathbf{F}^T = \mathbf{F}\mathbf{P}^T$	identity
$\rho_0 \frac{\partial}{\partial t}\left(\varepsilon + \frac{1}{2}v^2\right) + \operatorname{Div}\left(\mathbf{Q} - \mathbf{P}^T\mathbf{v}\right) = \rho_0 \mathbf{v} \cdot \mathbf{b} + \rho_0 r$	$\left[\left[\rho_0\left(\varepsilon + \frac{1}{2}v^2\right)\right]\right] U - \left[\left[\mathbf{Q} - \mathbf{P}^T\mathbf{v}\right]\right] \cdot \mathbf{N} = 0$
$\rho_0 \frac{\partial \eta}{\partial t} + \operatorname{Div} \mathbf{H} = \rho_0 s + \hat{\eta}$	$[[\rho_0 \eta]]\, U - [[\mathbf{H}]] \cdot \mathbf{N} = 0$

Clearly, these relations are identically satisfied if we introduce the function $\mathbf{f}$. Otherwise they have to be used in the model in the same way as other field equations. Usually the second condition is directly incorporated in the evaluation of thermodynamical admissibility (comp. Section 2.5 for its consequences). However, the first one remains as an additional equation.

It is convenient to write the integrability condition $(4.17)_1$ in the form of balance equation. We have

$$\frac{\partial \mathbf{F}}{\partial t} - \operatorname{Div}(\mathbf{v} \otimes \mathbf{1}) = 0 \quad \Longrightarrow \quad \frac{d}{dt}\int_{\mathcal{P}} \mathbf{F}dV - \oint_{\partial \mathcal{P}} \mathbf{v} \otimes \mathbf{N}dS = 0. \quad (4.18)$$

It means that we have an additional kinematical jump condition on singular surfaces

$$[[\mathbf{F}]]\, U + [[\mathbf{v} \otimes \mathbf{N}]] = 0. \quad (4.19)$$

This is one of the Hadamard kinematic compatibility conditions which form the basis of wave analysis in continua (see Section 3.3). It yields two important conclusions: on singular surfaces on which the velocity is continuous the deformation gradient must also be continuous and on material surfaces of contact of two bodies ($U = 0$) the velocity is continuous.

4.2 Balance laws in Eulerian description

The above form of balance equations related to the reference configuration at a chosen instant of time t_0 is often inconvenient in practical applications. For instance, the fluid mechanics never relies on such a description and it uses a current configuration as the reference. We call the above presented description *Lagrangian* and we proceed now to formulate *Eulerian description* in which the current configuration is used.

Let us begin with the proof of the Euler, Piola and Jacobi identities [250], [154] which are frequently used by the transformation of balance equations. Namely

$$\operatorname{Div}\left(J\mathbf{F}^{-T}\right) = 0, \quad \operatorname{div}\left(J^{-1}\mathbf{F}^{T}\right) = 0. \quad (4.20)$$

We prove the first one. The dual identity follows in the similar manner. We write it in Cartesian coordinates

$$\frac{\partial \left(JF^{-1}_{Kk}\right)}{\partial X_K} = \frac{\partial J}{\partial X_K} F^{-1}_{Kk} + J\frac{\partial F^{-1}_{Kk}}{\partial X_K}$$
$$= JF^{-1}_{Ll}\frac{\partial F_{lL}}{\partial X_K} F^{-1}_{Kk} - JF^{-1}_{Kl}F^{-1}_{Lk}\frac{\partial F_{lL}}{\partial X_K},$$

and (4.20) follows when we use the symmetry $\partial F_{lL}/\partial X_K = \partial F_{lK}/\partial X_L$. In the derivation we have used the identity

$$\frac{\partial\left(F_{Kk}^{-1}F_{kL}\right)}{\partial X_K}=0=\frac{\partial F_{Kk}^{-1}}{\partial X_K}F_{kL}+F_{Kk}^{-1}\frac{\partial F_{kL}}{\partial X_K}$$
$$\Longrightarrow\quad \frac{\partial F_{Kk}^{-1}}{\partial X_K}=-F_{Kl}^{-1}F_{Lk}^{-1}\frac{\partial F_{lL}}{\partial X_K}. \tag{4.21}$$

These identities follow as well directly from the Euler relation (2.8).

The transformation of Lagrangian to Eulerian description relies on the substitution of the inverse function of motion $\mathbf{X}=\mathbf{f}^{-1}(\mathbf{x},t)$. We have, for instance

$$\begin{aligned}&\mathbf{v}=\bar{\mathbf{v}}(\mathbf{x},\mathbf{t})=\mathbf{v}\left(\mathbf{f}^{-1}(\mathbf{x},t),t\right),\quad \mathbf{a}=\bar{\mathbf{a}}(\mathbf{x},\mathbf{t})=\mathbf{a}\left(\mathbf{f}^{-1}(\mathbf{x},t),t\right),\\&\mathbf{L}=\bar{\mathbf{L}}(\mathbf{x},\mathbf{t})=\mathbf{L}\left(\mathbf{f}^{-1}(\mathbf{x},t),t\right),\\&\mathbf{B}=\bar{\mathbf{B}}(\mathbf{x},\mathbf{t})=\mathbf{F}\left(\mathbf{f}^{-1}(\mathbf{x},t),t\right)\mathbf{F}^{T}\left(\mathbf{f}^{-1}(\mathbf{x},t),t\right),\end{aligned} \tag{4.22}$$

where the bar indicates a function dependent on $\mathbf{x}$ rather than $\mathbf{X}$.

Transformation of the velocity gradient $\mathbf{L}$ has a special bearing. We have

$$\mathbf{L}=\frac{\partial\mathbf{F}}{\partial t}\mathbf{F}^{-1}=(\operatorname{Grad}\mathbf{v})\,\mathbf{F}^{-1}=\operatorname{grad}\bar{\mathbf{v}}, \tag{4.23}$$

and this relation explains the name of $\mathbf{L}$.

We also have the following relations for derivatives of J

$$\begin{aligned}\frac{\partial J}{\partial t}&=J\mathbf{F}^{-T}\cdot\frac{\partial\mathbf{F}}{\partial t}=\bar{J}\operatorname{tr}\bar{\mathbf{L}}=\bar{J}\operatorname{div}\bar{\mathbf{v}},\\ \operatorname{Grad}J&=\bar{\mathbf{F}}^{T}\operatorname{grad}\bar{J}.\end{aligned} \tag{4.24}$$

Whenever it will be clear from the context that we work in spatial coordinates $(\mathbf{x},t)$ we shall skip the bar over Eulerian quantities.

Let us investigate the balance equations. The transformation $\mathbf{X}\rightarrow\mathbf{x}$ in (4.10) yields

$$\begin{aligned}\frac{d}{dt}\int_{\mathcal{P}_t}\bar{\varphi}\bar{J}^{-1}dv&=\oint_{\partial\mathcal{P}_t}\left(\bar{J}^{-1}\bar{\mathbf{F}}\bar{\psi}\right)\cdot\mathbf{n}dS_t+\int_{\mathcal{P}_t}\left[\bar{\psi}_V+\hat{\varphi}'\right]\bar{J}^{-1}dv,\\ \hat{\varphi}'&:=\hat{\varphi}\left(\mathbf{f}^{-1}(\mathbf{x},t),t\right),\end{aligned} \tag{4.25}$$

where we have used the formula for the transformation of variables $\mathbf{X}\rightarrow\mathbf{x}$ known from the classical analysis (compare (2.15) and (2.21)). The domain of integration is given by the transformation of the *material* volume $\mathcal{P}_t=\mathbf{f}(\mathcal{P},t)$, where $\partial\mathcal{P}_t$ is its boundary and $\mathbf{n}$ the unit normal vector given by (2.14). This formula explains the presence of the contribution $\bar{J}^{-1}\bar{\mathbf{F}}$ in this relation. dv denotes the infinitesimal volume element in the current configuration and dS_t the infinitesimal surface element in the current configuration. $\bar{J}^{-1} = J^{-1}(\mathbf{x}, t)$ is in this relation, of course, the Jacobian of the transformation.

We introduce the following notation which will be particularly useful in thermodynamics of multicomponent systems

$$\varphi_t = \varphi J^{-1}, \quad \boldsymbol{\psi}_t = J^{-1}\mathbf{F}\boldsymbol{\psi}, \quad \psi_{Vt} = \psi_V J^{-1}, \quad \hat{\varphi}_t = \hat{\varphi}', \tag{4.26}$$

all of them being functions of $(\mathbf{x}, t)$. Then the general balance equation in Eulerian description has the form

$$\frac{d}{dt}\int_{\mathcal{P}_t} \varphi_t dv = \oint_{\partial\mathcal{P}_t} \boldsymbol{\psi}_t \cdot \mathbf{n} dS_t + \int_{\mathcal{P}_t} [\psi_t + \hat{\varphi}_t]\, dv. \tag{4.27}$$

As in the Lagrangian description we can derive the local form of this equation in regular points and in points on a singular surface.

In the first case the left-hand side has the form

$$\frac{d}{dt}\int_{\mathcal{P}_t} \varphi_t dv = \int_{\mathcal{P}_t} \frac{\partial \varphi_t}{\partial t} dv + \oint_{\partial\mathcal{P}_t} \varphi_t \mathbf{v}\cdot\mathbf{n} dS_t = \int_{\mathcal{P}_t}\left[\frac{\partial \varphi_t}{\partial t} + \operatorname{div}(\varphi_t \mathbf{v})\right] dv, \tag{4.28}$$

which results from the fact that the domain is material.

Hence by means of Stokes Theorem and the localization procedure discussed before we obtain the local balance equation in Eulerian description

$$\frac{\partial \varphi_t}{\partial t} + \operatorname{div}(\varphi_t \mathbf{v} - \boldsymbol{\psi}_t) = \hat{\varphi}_t, \tag{4.29}$$

for almost all points of $\mathbf{f}(\mathcal{B}_0, t)$. This is the Eulerian counterpart of equation (4.11) in the Lagrangian description.

In the second case, i.e. on a singular surface $\mathcal{S}_t$ we obtain the following counterpart of the Kochine condition (4.16)

$$[[\varphi_t (c - \mathbf{v}\cdot\mathbf{n}) + \boldsymbol{\psi}_t \cdot \mathbf{n}]] = 0. \tag{4.30}$$

We illustrate the above general considerations by the mass balance. We have

$$\frac{d}{dt}\int_{\mathcal{P}} \rho_0 dV = \frac{d}{dt}\int_{\mathbf{f}(\mathcal{P},t)} \bar{\rho}_0 \bar{J}^{-1} dv = 0.$$

Consequently, it is convenient to introduce the following notion of the *current mass density*

$$\rho(\mathbf{x}, t) = \rho_0\left(\mathbf{f}^{-1}(\mathbf{x}, t)\right) J^{-1}\left(\mathbf{f}^{-1}(\mathbf{x}, t), t\right). \tag{4.31}$$

It satisfies the following balance law (conservation of mass)

$$\frac{d}{dt}\int_{\mathcal{P}_t}\rho dv = 0, \quad \mathcal{P}_t := \mathbf{f}(\mathcal{P}, t). \tag{4.32}$$

In regular points, we can transform this relation in the following way

$$\begin{aligned}\frac{d}{dt}\int_{\mathcal{P}_t}\rho dv &= \int_{\mathcal{P}_t}\frac{\partial\rho}{\partial t}dv + \oint_{\partial\mathcal{P}_t}\rho\mathbf{v}\cdot\mathbf{n}dS_t \\ &= \int_{\mathcal{P}_t}\left[\frac{\partial\rho}{\partial t} + \operatorname{div}(\rho\mathbf{v})\right]dv = 0 \quad \text{i.e.} \quad \frac{\partial\rho}{\partial t} + \operatorname{div}(\rho\mathbf{v}) = 0.\end{aligned} \tag{4.33}$$

We have used in these manipulations the fact that the surface $\partial\mathcal{P}_t$ of the material domain $\mathcal{P}_t$ (i.e. the domain whose motion in the current configurations is determined by material points forming the domain) moves with the speed $\mathbf{v}\cdot\mathbf{n}$, where $\mathbf{n}$ denotes the outward normal vector of this surface.

Making use of relations (4.24) it can be easily shown that the relation (4.31) $\rho = \rho_0 J^{-1}$ is the solution of equation (4.33) with the initial condition $\rho(\mathbf{x}, t = t_0) = \rho_0(\mathbf{f}(\mathbf{X}, t_0))$. This is the reason for the fact that, in contrast to fluid mechanics, in solid mechanics in which Lagrangian description is used the continuity equation (i.e. conservation of mass (4.33)) is not included in the set of fundamental field equations.

For a singular surface $\mathcal{S}_t$ which moves with the speed $c\mathbf{n}$ we can derive the jump condition. The procedure is similar to that used in the Lagrangian description. We have

$$\frac{d}{dt}\int_{\mathcal{P}_t}\rho dv = \int_{\mathcal{P}_t^+\cup\mathcal{P}_t^-}\frac{\partial\rho}{\partial t}dv + \oint_{\partial\mathcal{P}_t^+}\rho\mathbf{w}\cdot\mathbf{n}dS_t + \oint_{\partial\mathcal{P}_t^-}\rho\mathbf{w}\cdot\mathbf{n}dS_t, \tag{4.34}$$

where $\mathbf{w}\cdot\mathbf{n} = \mathbf{v}\cdot\mathbf{n}$ on material surfaces $\partial\mathcal{P}_t^+ \cap \partial\mathcal{P}_t$ and $\partial\mathcal{P}_t^- \cap \partial\mathcal{P}_t$ and $\mathbf{w}\cdot\mathbf{n} = \pm c$ on the singular surface $\mathcal{S}_t \cap \partial\mathcal{P}_t^{\pm}$. The difference in sign appears again due to the difference in the orientation. As before we form a descending family of subsets and taking the limit of balance equations we obtain

$$\int_{\mathcal{S}_t\cap\mathcal{P}_t}\left[\rho^+\left(\mathbf{v}^+\cdot\mathbf{n}-c\right) - \rho^-\left(\mathbf{v}^-\cdot\mathbf{n}-c\right)\right]dS_t = 0$$

$$\Longrightarrow \quad [[\rho(c - \mathbf{v}\cdot\mathbf{n})]] = 0. \tag{4.35}$$

This Eulerian jump condition (continuity of mass through the singular surface) is the counterpart of the Lagrangian relation quoted in Table 3.

In the case of momentum balance, we have

$$\frac{d}{dt}\int_{\mathcal{P}_t}\rho\mathbf{v}dv=\oint_{\partial\mathcal{P}_t}\mathbf{T}\mathbf{n}d\mathcal{S}_t+\int_{\mathcal{P}_t}\rho\mathbf{b}dv. \tag{4.36}$$

In order to find the relation between the tensor $\mathbf{T}$ and the Piola-Kirchhoff stress tensor $\mathbf{P}$ of the Lagrangian description one can use either the general relation (4.25) or transform directly the local momentum balance in a regular point

$$\begin{aligned}\frac{\partial\left(\rho_0 J^{-1}J\mathbf{v}\right)}{\partial t}&=J\frac{\partial\left(\rho\mathbf{v}\right)}{\partial t}+\rho\mathbf{v}\frac{\partial J}{\partial t}=\bar{J}\left[\frac{\partial\left(\bar{\rho}\bar{\mathbf{v}}\right)}{\partial t}+\bar{\mathbf{v}}\cdot\operatorname{grad}\bar{\mathbf{v}}\right]+\bar{\rho}\bar{\mathbf{v}}\bar{J}\operatorname{div}\bar{\mathbf{v}}\\&=\bar{J}\left[\frac{\partial\left(\bar{\rho}\bar{\mathbf{v}}\right)}{\partial t}+\operatorname{div}\left(\bar{\rho}\bar{\mathbf{v}}\otimes\bar{\mathbf{v}}\right)\right]\\&=\bar{J}\operatorname{div}\left(\bar{J}^{-1}\bar{\mathbf{P}}\bar{\mathbf{F}}^T\right)+\rho_0 J^{-1}J\mathbf{b}=\bar{J}\left(\operatorname{div}\bar{\mathbf{T}}+\bar{\rho}\bar{\mathbf{b}}\right),\end{aligned}$$

i.e. skipping the bar for Eulerian quantities

$$\frac{\partial\left(\rho\mathbf{v}\right)}{\partial t}+\operatorname{div}\left(\rho\mathbf{v}\otimes\mathbf{v}-\mathbf{T}\right)=\rho\mathbf{b},\quad \mathbf{T}=J^{-1}\mathbf{P}\mathbf{F}^T. \tag{4.37}$$

This is the local form of momentum conservation law. $\mathbf{T}$ is called the *Cauchy stress tensor* [107], [62].

Inspection of the above relations shows that the transformation from Lagrangian to Eulerian description in regular points requires the following transformation of operators

$$\frac{\partial}{\partial t}\to\frac{\partial}{\partial t}+\mathbf{v}\cdot\operatorname{grad},\quad \operatorname{Grad}\to\bar{\mathbf{F}}^T\operatorname{grad}. \tag{4.38}$$

The time derivative appearing in the above relations is called *material* and it is sometimes denoted by a dot on top of the symbol.

In Table 4 we have collected the balance equations in Eulerian description. It should be stressed that these balance equations hold in arbitrary – inertial or noninertial – frames of reference. In the latter case, the body force $\rho\mathbf{b}$ is *apparent* (comp. (4.14))

$$\begin{aligned}\rho\mathbf{b}&=\rho\mathbf{b}^0+\rho\mathbf{i},\\\rho\mathbf{i}&=2\rho\boldsymbol{\Omega}\left(\mathbf{v}^*-\dot{\mathbf{c}}\right)-\rho\boldsymbol{\Omega}^2\left(\mathbf{x}^*-\mathbf{c}\right)+\rho\dot{\boldsymbol{\Omega}}\left(\mathbf{x}^*-\mathbf{c}\right)+\rho\ddot{\mathbf{c}}.\end{aligned} \tag{4.39}$$

We complete the considerations of balance equations for subbodies $\mathcal{P}_t\subset\mathcal{B}_t$ with a few remarks concerning particular cases of jump conditions (conditions on a singular surface).

We have already mentioned that a singular surface which is *material*, i.e. a surface of contact between two bodies, does not move in the Lagrangian description ($U = 0$) and it means that $c = \mathbf{v} \cdot \mathbf{n}$ in the Eulerian description. The jump of the mass density can be in such cases arbitrary and the remaining conditions have the following form

$$[[\mathbf{T}]]\,\mathbf{n} = 0, \quad [[\mathbf{q}]] \cdot \mathbf{n} = 0, \quad [[\mathbf{h}]] \cdot \mathbf{n} = 0. \tag{4.40}$$

We have used the fact that on material surfaces not only the normal component of velocity $\mathbf{v} \cdot \mathbf{n}$ but the full velocity $\mathbf{v}$ is continuous. This is the consequence of the Hadamard condition (4.19).

The first two relations play an important role in the formulation of boundary conditions for continua. The first one – continuity of tractions, means that we may prescribe forces on the boundary and these will be transmitted into the body by the stress vector $\mathbf{Tn}$. The second one – continuity of the heat flux is used as one of the boundary conditions in the theory of heat conduction. The second two conditions have the great importance for properties of the so-called *ideal walls* which are a part of the second law of thermodynamics. We discuss them further.

Table 4: *Balance equations of thermomechanical model in Eulerian description*
Left – regular points, right – points of a singular surface

$\frac{\partial \rho}{\partial t} + \operatorname{div}(\rho \mathbf{v}) = 0, \quad \rho = \rho_0 \bar{J}^{-1}$	$[[\rho\,(c - \mathbf{v} \cdot \mathbf{n})]] = 0$
$\frac{\partial(\rho \mathbf{v})}{\partial t} + \operatorname{div}(\rho \mathbf{v} \otimes \mathbf{v} - \mathbf{T}) = \rho \mathbf{b}$, $\mathbf{T} = \bar{J}^{-1} \bar{\mathbf{P}} \bar{\mathbf{F}}^T$	$[[\rho\,(c - \mathbf{v} \cdot \mathbf{n})\,\mathbf{v}]] + [[\mathbf{T}]]\,\mathbf{n} = 0$
$\mathbf{T} = \mathbf{T}^T$	identity
$\frac{\partial}{\partial t}\left[\rho\left(\varepsilon + \frac{1}{2}v^2\right)\right] + {}$ $+ \operatorname{div}\left(\rho\left(\varepsilon + \frac{1}{2}v^2\right)\mathbf{v} + \mathbf{q} - \mathbf{T}\mathbf{v}\right)$ $= \rho \mathbf{v} \cdot \mathbf{b} + \rho r, \quad \mathbf{q} = \bar{J}^{-1} \bar{\mathbf{F}} \bar{\mathbf{Q}}$	$\left[\left[\rho\,(c - \mathbf{v} \cdot \mathbf{n})\left(\varepsilon + \frac{1}{2}v^2\right)\right]\right]$ $- [[\mathbf{q} - \mathbf{T}\mathbf{v}]] \cdot \mathbf{n} = 0$
$\frac{\partial(\rho \eta)}{\partial t} + \operatorname{div} \mathbf{h} = \rho s + \hat{\eta}_t, \quad \hat{\eta}_t = \bar{J}^{-1} \hat{\eta}$	$[[\rho\,(c - \mathbf{v} \cdot \mathbf{n})\,\eta]] - [[\mathbf{h}]] \cdot \mathbf{n} = 0$

On surfaces carrying jump of velocity, we can write the above conditions in the alternative form

$$m := \rho^+ \left(c - \mathbf{v}^+ \cdot \mathbf{n}\right) \equiv \rho^- \left(c - \mathbf{v}^- \cdot \mathbf{n}\right),$$

$$m\,[[\mathbf{v}]] + [[\mathbf{T}]] \cdot \mathbf{n} = 0, \tag{4.41}$$

$$m \left[\left[\varepsilon + \frac{1}{2}v^2\right]\right] - [[\mathbf{q} - \mathbf{T}\mathbf{v}]] \cdot \mathbf{n} = 0.$$

These equations for the stress tensor reduced to pressure $\mathbf{T} = -p\mathbf{1}$, which is characteristic for gas dynamics, are called *Rankine-Hugoniot conditions* and they form a foundation for the theory of shock waves in gases. The coefficient m – the mass transport coefficient is related to the *Mach number.*

4.3 Extension on membranes and interfaces

As already mentioned in Sec. 3.3.4, the continuum thermodynamics is able to model not only smooth processes but also processes which appear on some singular surfaces. Balance equations formulated in Sec. 4.1 admit only a certain class of such singularities. Namely, the assumption (4.1) yields Kotchine's jump condition (4.16) for the field φ whose characteristic feature is lack of any contribution of the singular surface itself. If we want to add some physical properties of the surface we have to weaken the assumption (4.1) in the following way

$$|\Phi(\mathcal{P}, t)| \leq \alpha \operatorname{vol} \mathcal{P} + \beta \operatorname{area} \mathcal{S} \cap \mathcal{P}, \tag{4.42}$$

where $\mathcal{S}$ is the singular surface, $\operatorname{area} \mathcal{S} \cap \mathcal{P}$ denotes the area of the part of singular surface common with the subbody $\mathcal{P}$ and β is a positive constant. Then under certain additional assumptions on the family of subbodies $\mathcal{P} \subset \mathcal{B}_0$ which we shall not present in this book (e.g. see [331], [354]) the set function Φ has the representation [112]

$$\Phi(\mathcal{P}, t) = \int_{\mathcal{P}} \varphi(\mathbf{X}, t)\, dV + \int_{\mathcal{S} \cap \mathcal{P}} \phi(\mathbf{X}, t)\, dS, \tag{4.43}$$

where φ and ϕ are volume and surface densities of Φ, respectively.

We investigate time changes of Φ in its Eulerian representation which follows from relations (3.25) for the transformation of surface elements.

We write the transformation of the relation (4.43) assuming for simplicity that surface coordinates are convective, i.e. $\xi^\alpha = \delta^\alpha_A \Xi^A$. Then $d\mathcal{S} = \sqrt{A} d\xi^1 d\xi^2$, $d\mathcal{S}_t = \sqrt{a} d\xi^1 d\xi^2$ and we obtain

$$\Phi_t(\mathcal{P}_t, t) = \int_{\mathcal{P}_t} \varphi_t(\mathbf{x}, t)\, dv + \int_{\mathcal{S}_t \cap \mathcal{P}_t} \phi_t(\mathbf{x}, t)\, d\mathcal{S}_t, \tag{4.44}$$

where

$$\varphi_t = \varphi J^{-1}, \quad \phi_t = \phi \sqrt{\frac{A}{a}}. \tag{4.45}$$

Obviously, A and a are determinants of the first fundamental form in reference and current configurations of the surface, respectively (see (3.22)).

As indicated in (4.3), the rate of change of Φ_t appears for three reasons: the flux to $\mathcal{P}_t$ and $\mathcal{S}_t \cap \mathcal{P}_t$, the supply to these two regions and the production within them. They are written in the form

$$\Psi_{\mathcal{S}_t}(\partial\mathcal{P}_t, t) = \oint_{\partial\mathcal{P}_t} \boldsymbol{\psi}_t(\mathbf{x}, t) \cdot \mathbf{n} d\mathcal{S}_t + \int_{\mathcal{S}_t \cap \partial\mathcal{P}_t} \overset{s}{\boldsymbol{\psi}}_t(\mathbf{x}, t) \cdot \boldsymbol{v} dl_t, \tag{4.46}$$

$$\Psi_{\mathcal{P}_t}(\mathcal{P}_t, t) = \int_{\mathcal{P}_t} \psi_t(\mathbf{x}, t)\, dv + \int_{\mathcal{S}_t \cap \partial\mathcal{P}_t} \overset{s}{\psi}_t(\mathbf{x}, t)\, d\mathcal{S}_t, \tag{4.47}$$

$$\hat{\Psi}_{\mathcal{P}_t}(\mathcal{P}_t, t) = \int_{\mathcal{P}_t} \hat{\varphi}_t(\mathbf{x}, t)\, dv + \int_{\mathcal{S}_t \cap \partial\mathcal{P}_t} \hat{\varphi}^s_t(\mathbf{x}, t)\, d\mathcal{S}_t. \tag{4.48}$$

This is similar to (4.4) but there are additional contributions of the flux density $\overset{s}{\boldsymbol{\psi}}_t$ per line element, the supply to the singular surface $\overset{s}{\psi}_t$, and the production on the singular surface $\hat{\varphi}^s_t$.

The general form of the balance equation in Eulerian description is in this more general case of the form

$$\frac{d}{dt}\left(\int_{\mathcal{P}_t} \varphi_t(\mathbf{x}, t)\, dv + \int_{\mathcal{S}_t \cap \mathcal{P}_t} \phi_t(\mathbf{x}, t)\, d\mathcal{S}_t \right)$$

$$\begin{aligned} &= \oint_{\partial\mathcal{P}_t} \boldsymbol{\psi}_t(\mathbf{x}, t) \cdot \mathbf{n} d\mathcal{S}_t + \int_{\mathcal{S}_t \cap \partial\mathcal{P}_t} \overset{s}{\boldsymbol{\psi}}_t(\mathbf{x}, t) \cdot \boldsymbol{v} dl_t \\ &\quad + \int_{\mathcal{P}_t} \psi_t(\mathbf{x}, t)\, dv + \int_{\mathcal{S}_t \cap \mathcal{P}_t} \overset{s}{\psi}_t(\mathbf{x}, t)\, d\mathcal{S}_t \\ &\quad + \int_{\mathcal{P}_t} \hat{\varphi}_t(\mathbf{x}, t)\, dv + \int_{\mathcal{S}_t \cap \mathcal{P}_t} \hat{\varphi}^s_t(\mathbf{x}, t)\, d\mathcal{S}_t. \end{aligned} \tag{4.49}$$

Obviously, the local counterpart of balance equation in regular points $\mathbf{x} \in \mathcal{B}_t \backslash \mathcal{S}_t$ is identical with (4.29). We make again the pillbox argument as demonstrated in Fig. 8. For the time derivative of the contribution of the singular surface we can write

$$\frac{d}{dt}\int_{\mathcal{S}_t\cap\mathcal{P}_t}\phi_t(\mathbf{x},t)\,d\mathcal{S}_t=\int_{\mathcal{S}\cap\mathcal{P}}\left(\phi_t(\mathbf{X},t)\sqrt{a}\right)^{\bullet}d\xi^1d\xi^2$$

$$=\int_{\mathcal{S}\cap\mathcal{P}}\left(\overset{\bullet}{\phi}_t(\mathbf{X},t)\sqrt{\frac{a}{A}}+\frac{1}{2}\frac{\overset{\bullet}{a}}{\sqrt{a}}\phi_t(\mathbf{X},t)\right)d\xi^1d\xi^2$$

$$=\int_{\mathcal{S}\cap\mathcal{P}}\left(\overset{\bullet}{\phi}_t(\mathbf{X},t)+\phi_t(\mathbf{X},t)\left(c^{\alpha}_{\|;\alpha}-2K_Mc\right)\right)\sqrt{a}d\xi^1d\xi^2 \tag{4.50}$$

$$=\int_{\mathcal{S}_t\cap\mathcal{P}_t}\left(\overset{\bullet}{\phi}_t(\mathbf{x},t)+\phi_t(\mathbf{x},t)\left(c^{\alpha}_{\|;\alpha}-2K_Mc\right)\right)d\mathcal{S}_t,$$

where the relation (3.47) has been used.

We have to transform also the line integral (4.49). We have

$$\int_{\mathcal{S}_t\cap\partial\mathcal{P}_t}\overset{s}{\boldsymbol{\psi}}_t\cdot\boldsymbol{\upsilon}dl_t=\int_{\mathcal{S}_t\cap\partial\mathcal{P}_t}\left(\overset{s}{\psi}{}^{\alpha}{}_t\mathbf{t}_\alpha+\mathbf{n}\cdot\overset{s}{\boldsymbol{\psi}}_t\mathbf{n}\right)\cdot\boldsymbol{\upsilon}dl_t$$

$$=\int_{\mathcal{S}_t\cap\partial\mathcal{P}_t}\overset{s}{\psi}{}^{\alpha}{}_t\mathbf{t}_\alpha\cdot\boldsymbol{\upsilon}dl_t=\int_{\mathcal{S}_t\cap\mathcal{P}_t}\overset{s}{\psi}{}^{\alpha}{}_{t;\alpha}d\mathcal{S}_t. \tag{4.51}$$

We have applied the relation (3.36).

Now, taking the limit indicated in Fig. 8, we easily arrive at the following local balance equation on the singular surface

$$\overset{\bullet}{\phi}_t+\phi_t\left(c^{\alpha}_{\|;\alpha}-2K_Mc\right)-\overset{s}{\psi}{}^{\alpha}{}_{t;\alpha}-\overset{s}{\psi}_t-\hat{\varphi}^s_t \tag{4.52}$$
$$=\left[\left[\varphi_t(c-\mathbf{v}\cdot\mathbf{n})+\boldsymbol{\psi}_t\cdot\mathbf{n}\right]\right],\quad \text{a.e. on } \mathcal{S}_t.$$

Certainly, without additional surface contributions collected on the left-hand side of this equation, we would obtain again the Kotchine condition (4.30) derived earlier in this Chapter.

Hence, the three basic surface fields: surface mass density ρ^s, surface momentum density $\rho^s\mathbf{v}^s$, surface energy density $\rho^s\varepsilon^s$ must satisfy the following balance equations

$$\overset{\bullet}{\rho^s}+\rho^s\left(c^{\alpha}_{\|;\alpha}-2K_Mc\right)=\left[\left[\rho(c-\mathbf{v}\cdot\mathbf{n})\right]\right], \tag{4.53}$$

$$\left(\rho^s\mathbf{v}^s\right)^{\bullet}+\rho^s\mathbf{v}^s\left(c^{\alpha}_{\|;\alpha}-2K_Mc\right)-\overset{s}{t}{}^{i\alpha}_{;\alpha}-\rho^s\overset{s}{\mathbf{b}}=\left[\left[\rho\mathbf{v}(c-\mathbf{v}\cdot\mathbf{n})+\mathbf{T}\mathbf{n}\right]\right], \tag{4.54}$$

$$\left(\rho^s\varepsilon^s\right)^\bullet + \rho^s\varepsilon^s\left(c^\alpha_{\|;\alpha} - 2K_M c\right) + \overset{s}{q}{}^\alpha_{;\alpha}$$

$$-\left\{\overset{s}{t}{}^{i\alpha}v^s_{i;\alpha} - \left[\left[\frac{1}{2}\left(\mathbf{v}^s - \mathbf{v}\right)\cdot\left(\mathbf{v}^s - \mathbf{v}\right)\rho\left(c - \mathbf{v}\cdot\mathbf{n}\right) + \mathbf{T}\cdot\left(\left(\mathbf{v}^s - \mathbf{v}\right)\otimes\mathbf{n}\right)\right]\right]\right\}$$

$$-\rho^s r^s = \left[\left[\rho\varepsilon\left(c - \mathbf{v}\cdot\mathbf{n}\right) - \mathbf{q}\cdot\mathbf{n}\right]\right]. \tag{4.55}$$

In the derivation of the last relation the balance of momentum (4.54) has been used.

In these relations, $\overset{s}{t}{}^{i\alpha}$ denotes the *full surface stress* components, $\overset{s}{\mathbf{b}}$ the *external surface forces*, $\overset{s}{q}{}^\alpha$ the *surface heat flux*, $\overset{s}{r}$ the *surface radiation*. The stress can be decomposed into two parts

$$\overset{s}{t}{}^{i\alpha}\mathbf{e}_i = S^{\alpha\beta}\mathbf{t}_\alpha + S^\alpha\mathbf{n}, \tag{4.56}$$

where $S^{\alpha\beta}$ is called the *surface stress* and S^α the *normal surface stress*. If there is no spin, i.e. if angular momentum reduces to the moment of momentum, the corresponding balance equation of angular momentum (comp. (4.12)) yields the following condition

$$\epsilon_{ijk}\overset{s}{t}{}^{j\alpha}\,\overset{s}{t}{}^k_\alpha = 0 \quad\Rightarrow\quad S^{\alpha\beta} = S^{\beta\alpha}, \quad S^\alpha = 0. \tag{4.57}$$

In addition to the above fields, if there is an additional spin on the singular surface, one needs the condition for a full angular momentum (including, apart from moment of momentum, the spin), and, in the case of coupling with electromagnetic field, the balance equation for electric charges. The latter coupling requires as well the consideration of balance on singular lines. We do not consider these problems in this book and, for this reason, we refer the reader to the considerations presented by Ingo Müller [222] or Kolumban Hutter *et al.* [152]. Thermodynamics of processes described by surface balance equations has been initiated by G. P. Moeckel in 1975 [210] but only a little progress has been made in this field up to now.

In order to see the applicability of the above conditions let us write the momentum balance for an impermeable membrane at rest. Equation (4.54) reads then

$$\left[\left[\mathbf{T}\right]\right]\mathbf{n} = -\overset{s}{t}{}^{i\alpha}_{;\alpha}, \tag{4.58}$$

when we leave out the external forces. Substitution of (4.57) yields

$$[[\mathbf{T}]]\,\mathbf{n} = -\left(S^{\alpha\beta}\mathbf{t}_\alpha\right)_{;\beta} = -S^{\alpha\beta}_{\ ;\beta}\mathbf{t}_\alpha - S^{\alpha\beta}\frac{\partial \mathbf{t}_\alpha}{\partial \xi^\beta}. \tag{4.59}$$

Consequently, the jump of the normal component of stresses has the form

$$\mathbf{n}\cdot[[\mathbf{T}]]\,\mathbf{n} = -S^{\alpha\beta}b_{\alpha\beta}. \tag{4.60}$$

This leads to one of the fundamental relations in the theory of capillarity and surface tension. Namely, in the case of fluids, when the stress tensors reduce to pressures the equilibrium of the membrane is described by the *Young-Laplace equation* following from (4.60)

$$\mathbf{T} = -p\mathbf{1}, \quad S^{\alpha\beta} = \sigma a^{\alpha\beta} \quad \Rightarrow \quad [[p]] = \sigma\left(\frac{1}{R_1} + \frac{1}{R_2}\right), \tag{4.61}$$

where σ is the material parameter called the *surface tension* and R_1, R_2 are radii of *principal curvatures* with the mean curvature given by $K_M = \frac{1}{2}\left(\frac{1}{R_1} + \frac{1}{R_2}\right)$.

Many applications of surface balance equations in the theory of cavitation, bubble dynamics, rubber balloons etc. can be found in the books of Christopher Brennen [50] and Ingo Müller, Peter Strehlow [226].

Chapter 5

Second law of thermodynamics

5.1 Entropy inequality, thermodynamical admissibility

5.1.1 *Motivation*

Since a long time it has been clear to scientists that macroscopic processes are *irreversible.* This means that there exist no macroscopic systems which, when disturbed, return spontaneously (i.e. without any influence of the external world) to their initial state. Since XIX century there exist models of physical systems in which macroscopic properties of processes are being derived from microscopic description, in most cases stemming from the classical mechanics of discrete systems. It has been observed that such a construction of models yields the paradox. Microscopic mechanical models are reversible, i.e. all processes possible for a given choice of the time variable are also possible after a reversal of time. This contradicts the macroscopic irreversibility. It can be shown rigorously, for example, that equations of dynamics of many interacting particles lead to solutions in which after a sufficiently long time the system *spontaneously* returns to an arbitrarily small neighborhood of its initial state. This time is called the *recurrence time* of Poincaré's cycle. One can estimate this time and for large systems containing, say 10^{23} particles (the order of magnitude of the Avogadro number), the recurrence time exceeds the time of existence of the Universe by many orders of magnitude.

The above described properties of large systems led to vehement discussions among physicists of the end of XIX century and the beginning of XX century. L. Boltzmann proposed a model of gases – the so-called *kinetic theory,* in which the microscopic model was reversible (noninteracting particles flying free in space and exchanging momentum and energy in elastic collisions) and the macroscopic result described by the so-called H-Theorem, was irreversible [42], [43]. This result has been opposed by many physicists who were using, for instance, the argument based on the

Poincaré cycle, that the model must contain some flaws. On Zermelo's criticism pointing out the existence of the recurrence time Boltzmann supposedly replied: "You should wait that long!" However, for Boltzmann the result of this discussion had a tragic end. He committed suicide.

Before we present the modern version of the principle of macroscopic irreversibility we discuss briefly two simple models motivating this principle on the basis of microscopic considerations.

The first example has been constructed by P. and T. Ehrenfest in 1907 [98], [99] and it is called "Dog and flea model" or "urn model". We present the idea of this model in Fig. 9.

For urn model we consider the dynamics of two urns containing N balls labelled from 1 to N. For simplicity, let us assume that initially all balls are in urn (1) and urn (2) is empty. The motion of balls is given by random drawing a number between 1 and N and then moves a ball with this number from one urn to the other. In the dog-flea model the motion occurs because fleas are jumping from one dog to the other looking for better breeding conditions. It is obvious that the "flux" must start from urn (1) in direction of urn (2). It may happen that we draw the same number again and the ball then returns to urn (1). However, for a very large N, say $N = 10^{23}$, it is much more probable that we get a different number and another ball moves from urn (1) to urn (2). Hence, despite small fluctuations, an average flux of balls has a definite direction until we reach the macrostate in which both urns have almost equal number of balls $N/2$. Although it cannot be excluded that we draw the sequence of numbers which would make urn (2) empty again, it is clear that such a process is extremely exceptional. This is connected to the fact that the state with all balls in one urn is only one while the state of equally distributed balls may appear in $W = 2^N$ variations if we ignore the numbering of balls. It is said that the equal distribution between two urns yields the *maximum of the entropy* $S = k \ln W$ (k – Boltzmann constant, see Section 6.1). We shall return to this definition later.

Let us note that the above analysis assumes a very large number of possible states of the system. It is still a matter of dispute how small a system may be in order to admit a macroscopic modelling. For instance, the diffusion process of DNA particles within cells (translation through nanopores) is described by a variation between a few hundred states. Still some macroscopic thermodynamical arguments are applied to such systems.

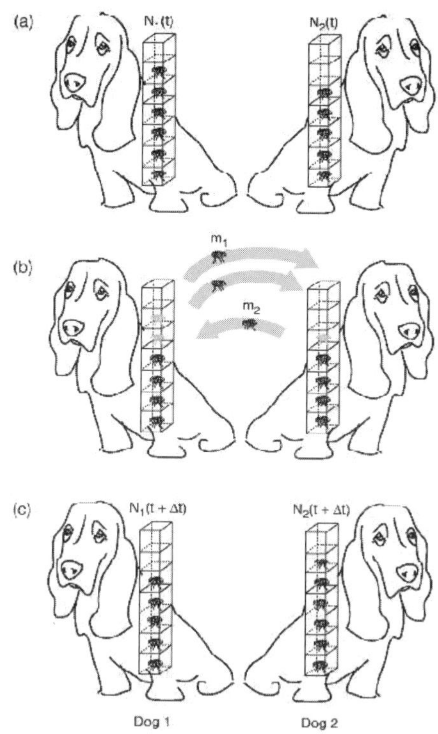

Fig. 9: *Ehrenfest's dog-flea (urn) model (1907)*
a/ state of the system at time t; b/ a particular microtrajectory to which two fleas jump from the dog on the left and one flea jumps from the dog on the right; c/ occupancies of the dogs at time $t + \Delta t$ (K. Gosh et al., Am. J. Phys. 74(2), 2006)

We proceed to present an ingenious model of Marc Kac [161] which illustrates the problem of irreversibility of Boltzmann's kinetic theory. Simple calculations presented below demonstrate that the irreversibility results from a certain assumption on randomness of the system ("*Stoßzahlansatz*") and this is, of course, related to the large number properties of the above presented dog-flea model.

The system consists of N white or black circles distributed in the equal distance Δt on the circumference of a large circle (Fig. 10). Dynamics of the system is introduced by the rotation of the wheel with the set, S, of M spokes distributed at random in different middle positions between small circles. Whenever the wheel rotates on the angle Δt, the small circle through which the spoke passes changes the color. We assume that both N and M are very large but simultaneously $M \ll N$. The instantaneous number of white and black circles is described by the following "equation of motion" with the discrete time

$$N_w(t+\Delta t) = N_w(t) + N_b(S,t) - N_w(S,t), \tag{5.1}$$

with

$$\forall_t\, N_w(t) + N_b(t) = N, \quad N_w(S,t) + N_b(S,t) = M. \tag{5.2}$$

Certainly, $N_b(S,t)$ denotes the number of black circles which change color in the step $t \to t + \Delta t$, and $N_w(S,t)$ the number of white circles which change color in this step.

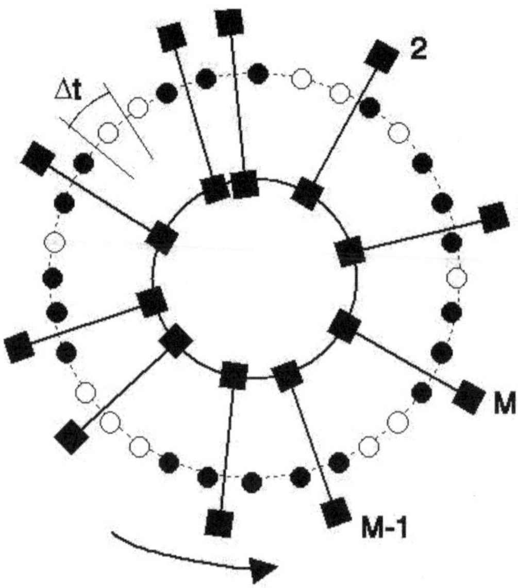

Fig. 10: *Irreversibility and Stoßzahlansatz – model of M. Kac (1956)* [161].

Equation (5.1) constitutes the counterpart of the *Liouville equation* for N particles and describes reversible processes on the microlevel: after two full rotations of the wheel the system returns to its initial state. This means that the recurrence time of Poincaré's cycle is equal for this model to $t_P = 2N\Delta t$.

The change in the surplus of white circles at the instant of time $t + \Delta t$ is, according to the equation of motion, given by the following equation

$$[N_w(t+\Delta t) - N_b(t+\Delta t)] = [N_w(t) - N_b(t)] - 2[N_w(S,t) - N_b(S,t)]. \tag{5.3}$$

Now, we make the *reasonable assumption* that, owing to the randomness of the set S, after sufficiently many time steps the black and white circles will be regularly distributed, i.e.

$$N_w(S,t) = \frac{M}{N} N_w(t), \quad N_b(S,t) = \frac{M}{N} N_b(t). \tag{5.4}$$

Substitution of this assumption in (5.3) yields the following difference equation

$$[N_w(t+\Delta t) - N_b(t+\Delta t)] = \left(1 - 2\frac{M}{N}\right)[N_w(t) - N_b(t)]. \tag{5.5}$$

This equation can be solved immediately and we obtain

$$[N_w(t) - N_b(t)] = N\left(1 - 2\frac{M}{N}\right)^t. \tag{5.6}$$

Let us assume that the initial state of the system was $N_w(t=0) = N$, $N_b(t=0) = 0$. Then it follows

$$\frac{N_w(t)}{N} = \frac{1}{2} + \frac{1}{2}\left(1 - 2\frac{M}{N}\right)^t. \tag{5.7}$$

Consequently

$$\frac{N_w(t=0)}{N} = 1, \quad \lim_{t\to\infty} \frac{N_w(t)}{N} = \frac{1}{2}. \tag{5.8}$$

The most important property of the solution (5.7) is its irreversibility. It can be easily checked that, independently of the initial state, we always obtain the same asymptotic end state with equal number of white and black circles. There is no trace left of the microscopical periodicity of processes. The reason for this behavior is hidden in the only assumption which we made – (5.4) – which corresponds to the Stoßzahlansatz of Boltzmann's theory. We see that conditions under which the solution (5.7) makes sense are the randomness of the set S and long, but not too long times of observations: $1 \ll t \ll t_P$.

5.1.2 *Entropy inequality*

5.1.2.1 *Entropy principle*

The above examples motivate the following formulation of macroscopic models. Constitutive equations which define a particular class of materials should have such a form that solutions of field equations will not be invariant with respect to the time reversal (*irreversibility*) and, secondly, that disturbances of finite time duration should produce solutions which relax (i.e. possess a time limit in infinity) to an *equilibrium state* characteristic for a given class of materials and for given boundary conditions.

Lars Onsager
1903 – 1976

Bernard D. Coleman
1930–

Ingo Müller
1936–

This program is usually realized by means of an additional scalar inequality which constraints the class of solutions. Its form has been varying since XIX century. Clausius [74], [75] and Duhem [91] (he added the surface integral to the entropy inequality yielding the entropy flux $\mathbf{H}$) proposed the first continuous version of such an inequality and this inequality, for the first time clearly formulated by Eckart in 1940 [93], has been very intensively investigated in 60ies by C. Truesdell, B. Coleman, W. Noll any many others (see [332] for the detailed discussion of the evolution of this approach). It has been found that results of classical thermostatics obtained by Gibbs [120], Carathéodory [57] and many others follow as a particular case (equilibrium properties) from this inequality. It has also been proved that the linear nonequilibrium thermodynamics proposed by Onsager [239], [240], [85] is a particular case of this inequality for single component systems. Simultaneously, it has been shown that the Clausius-Duhem inequality is inadequate in description of multicomponent systems as well as for some nonmechanical (e.g. electromagnetic [152]) fields. The extension of this inequality has been proposed in 1971 by I. Müller [219], [218]. It is usually called the *entropy inequality* and this seems to be the most general formulation of the continuous second law of thermodynamics which has been proposed up to now.

There is an alternative approach to the formulation of the second law of thermodynamics which stems from the classical XIX century works of S. Carnot [58] (English translation in [206]) and F. Reech on *cyclic processes.* This notion has been extended and applied to nonequilibrium thermodynamics by J. Serrin [292] and M. Šilhavý [297] and then summarized by C. Truesdell and S. Bharatha [336].

Further in this book, we discuss various applications of the entropy inequality. We begin with a formulation appropriate for all models which we present later and, for the purpose of this Chapter we assume that the notion of temperature is known. However, we return in the next Chapter to the discussion of notions of empirical and absolute temperatures and various problems which they may create, particularly for multicomponent systems.

The strategy of continuum thermodynamics in construction of macroscopic models relies on the assumption that solutions of the field equations identically satisfy the *second law of thermodynamics* also called the *entropy principle.* This law consists of four parts. We formulate them in Lagrangian description and then, when needed, change variables to Eulerian description.

Entropy principle:

i/ There exist a nontrivial entropy density function η and the entropy flux $\mathbf{H}$ which are both dependent on the same constitutive variables as other constitutive functions of the model.

ii/ The entropy density satisfies the balance equation whose form in regular points is as follows

$$\rho_0 \frac{\partial \eta}{\partial t} + \operatorname{Div} \mathbf{H} = \hat{\eta}, \tag{5.9}$$

where $\hat{\eta}$ denotes the entropy source.

iii/ The entropy source is nonnegative for all *solutions* of field equations, i.e. for all *thermodynamic processes.* Consequently, the following inequality holds

$$\forall \text{ thermodynamic processes } \rho_0 \frac{\partial \eta}{\partial t} + \operatorname{Div} \mathbf{H} \geq 0. \tag{5.10}$$

iv/ There exist *ideal walls* on which there is no entropy production, i.e.

$$[[\mathbf{H} \cdot \mathbf{N}]] + [[\eta]] \, \rho_0 U = 0, \tag{5.11}$$

these walls are assumed to be material, i.e. $U = 0$.

The last condition is modified for multicomponent systems and concerns the existence of semipermeable membranes. We discuss this notion further in this book.

This formulation has been modified by H. W. Alt [9] in order to include some nonequilibrium phase transformations whose description requires a weaker mathematical approach (see also the work of H. W. Alt and I.

Pawlow: [10]). We shall not present these considerations in this book as results for all cases which we discuss are in both approaches identical.

I-Shih Liu has proposed in 1972 [187] a method of exploitation of the inequality (5.9) which reminds the classical method of Lagrange multipliers in mechanics[1]. Namely, instead of this inequality restricting the *solutions of field equations* we consider solutions of an extended inequality which should hold for *all fields.* This can be done if we consider field equations as constraints on solutions of the entropy inequality.

5.1.2.2 *Thermoelasticity*

Before we formulate Liu's Theorem we consider a simple example of the thermodynamical construction of the field model. It describes the so-called *thermoelastic material.* *Thermoelasticity* is the theory which describes changes of two fields: the function of motion $\mathbf{f}$ describing time dependent large deformations of the material and the temperature T responsible for the energy transfer in the material in the form of heat conduction. Further we discuss in details the notion of temperature. For the purpose of this example we do not go into any details concerning this field. In the Lagrangian description we do not need to consider the mass density ρ_0 because, according to the mass conservation, it does not change in time. In the case of homogeneous materials it is even constant.

For the fields $\{\mathbf{f}, T\}$ as functions of variables $(\mathbf{X}, t) \in \mathcal{B}_0 \times \mathcal{T}$ we must formulate *field equations.* As we require from the model that it satisfies the conservation laws of momentum, moment of momentum and energy, these laws are chosen as the foundation for the construction of field equations

$$\rho_0 \frac{\partial^2 \mathbf{f}}{\partial t^2} = \operatorname{Div} \mathbf{P} + \rho_0 \mathbf{b}, \quad \rho_0 \frac{\partial \varepsilon}{\partial t} + \operatorname{Div} \mathbf{Q} = \mathbf{P} \cdot \operatorname{Grad} \frac{\partial \mathbf{f}}{\partial t} + \rho r. \tag{5.12}$$

In addition we have the restriction following from the conservation of moment of momentum: $\mathbf{P}\,(\operatorname{Grad} \mathbf{f})^T = (\operatorname{Grad} \mathbf{f})\,\mathbf{P}^T$. The energy conservation law was reduced by means of the momentum conservation. Consequently, we obtain the balance equation for the internal energy ε which does not have the form of the conservation law (the so-called divergent form). There appears a source term which describes the power of stresses $\mathbf{P} \cdot \operatorname{Grad}(\partial \mathbf{f}/\partial t)$.

Equations (5.12) are not yet field equations. We must perform the so-called *closure* which defines the Piola-Kirchhoff stress tensor $\mathbf{P}$, the internal energy ε, and the heat flux $\mathbf{Q}$ in terms of the fields $\mathbf{f}$, T. This is done in the form of *constitutive relations* which limit the applicability of the model

[1]It has been pointed out by R. A. Hauser and N. P. Kirchner [139] that the result of I-Shih Liu is a particular case of Farkas-Minkowski Theorem known in the duality theory of linear programming.

to a particular class of materials. For thermoelastic materials it is *assumed* that constitutive relations have the following form

$$\mathbf{P} = \mathbf{P}(\mathbf{v}, \mathbf{F}, T, \mathbf{G}), \quad \varepsilon = \varepsilon(\mathbf{v}, \mathbf{F}, T, \mathbf{G}), \qquad (5.13)$$
$$\mathbf{Q} = \mathbf{Q}(\mathbf{v}, \mathbf{F}, T, \mathbf{G}), \quad \mathbf{G} := \operatorname{Grad} T.$$

These are the simplest possible relations which do not yield a triviality of the model. They possess a few features characteristic for such a construction

i/ among variables we have the first gradients of the fields $\mathbf{F} = \operatorname{Grad}\mathbf{f}$, $\mathbf{G} = \operatorname{Grad} T$ which account for the influence of a neighborhood of a point $\mathbf{X} \in \mathcal{B}_0$ on the properties of the material at this point,

ii/ they do not contain a dependence on the function of motion $\mathbf{f}$. This is related to the *principle of material objectivity* which we discuss further; as a matter of fact the same principle eliminates a dependence on the velocity $\mathbf{v}$ as well,

iii/ the constitutive relations are *functions* and not functionals which would be able to account for the dependence on the past history of processes. Such functionals would appear, for instance, in cases in which at least one of the constitutive quantities $\mathbf{P}$, ε, $\mathbf{Q}$ would be given by an *evolution equation.* We discuss such classes of materials (e.g. viscoelastic solids) further in this book.

The constitutive relations must be further restricted by, for example, a condition of thermodynamical admissibility. However, if we are lucky we may formulate the above relations on the basis of experiments and then no further restrictions would be needed. This does not seem to be the case ever. At least some hints from a general model on how to conduct experiments are always needed and this is the motivation for the thermodynamical construction of models.

We close this example with an alternative formulation of the thermoelastic model which is more convenient for thermodynamical considerations. It has been mentioned already that the field $\mathbf{f}$ can be replaced by two fields $\mathbf{F}, \mathbf{v}$ and then we have to require certain integrability conditions in order to be able to integrate $\mathbf{F}$ and $\mathbf{v}$ *a posteriori* in order to find the motion $\mathbf{f}$. The model in this setting has the following form

i/ fields $\{\mathbf{v}, \mathbf{F}, T\}$

ii/ conservation laws

$$\rho_0 \frac{\partial \mathbf{v}}{\partial t} - \operatorname{Div}\mathbf{P} = \mathbf{0}, \quad \frac{\partial \mathbf{F}}{\partial t} - \operatorname{Grad}\mathbf{v} = 0, \qquad (5.14)$$
$$\rho_0 \frac{\partial \varepsilon}{\partial t} + \operatorname{Div}\mathbf{Q} = \mathbf{P} \cdot \operatorname{Grad}\mathbf{v},$$

iii/ constitutive relations

$$\mathbf{P}=\mathbf{P}\left(\mathbf{F},T,\mathbf{G}\right),\quad \varepsilon=\varepsilon\left(\mathbf{F},T,\mathbf{G}\right),\quad \mathbf{Q}=\mathbf{Q}\left(\mathbf{F},T,\mathbf{G}\right). \tag{5.15}$$

This problem still contains a dependence on the temperature gradient $\mathbf{G}$ which yields the set of the second order equations. This can be changed as well by exchanging the role of $\mathbf{G}$ and $\mathbf{Q}$. We demonstrate further this way of constructing field equations which is known as the *extended thermodynamics*.

We have left out the constitutive dependence on the velocity $\mathbf{v}$. Also body forces $\mathbf{b}$ and radiation r have been assumed to be zero. This assumption follows from the fact that the entropy inequality should hold for arbitrary external sources. It means that constitutive restrictions which follow from the second law should be independent of such sources. Consequently, restrictions which we obtain without sources must be identical as these when those sources are present. This argument is different from the argument used, for example, in a series of papers of B. D. Coleman [80], [77] and his followers.

The above problem should satisfy the entropy inequality of the following form

$$\forall \text{ thermodynamic processes } \rho_0\frac{\partial \eta}{\partial t}+\operatorname{Div}\mathbf{H}\geq 0, \tag{5.16}$$

$$\eta=\eta\left(\mathbf{F},T,\mathbf{G}\right),\quad \mathbf{H}=\mathbf{H}\left(\mathbf{F},T,\mathbf{G}\right).$$

Using the chain rule of differentiation, we can write this inequality in the explicit form

$$\forall \text{ fields } \rho_0\left[\frac{\partial \eta}{\partial F_{kK}}\frac{\partial F_{kK}}{\partial t}+\frac{\partial \eta}{\partial T}\frac{\partial T}{\partial t}+\frac{\partial \eta}{\partial G_K}\frac{\partial G_K}{\partial t}\right]$$

$$+\frac{\partial H_K}{\partial F_{lL}}\frac{\partial F_{lL}}{\partial X_K}+\frac{\partial H_K}{\partial T}G_K+\frac{\partial H_K}{\partial G_L}\frac{\partial G_L}{\partial X_K}\geq 0 \tag{5.17}$$

where

$$\frac{\partial F_{kK}}{\partial t}=\frac{\partial v_k}{\partial X_K},$$

$$\frac{\partial T}{\partial t}=\left(\frac{\partial \varepsilon}{\partial T}\right)^{-1}\left[-\frac{\partial \varepsilon}{\partial F_{kK}}\frac{\partial F_{kK}}{\partial t}-\frac{\partial \varepsilon}{\partial G_K}\frac{\partial G_K}{\partial t}\right.$$

$$- \frac{1}{\rho_0} \left(\frac{\partial Q_K}{\partial F_{lL}} \frac{\partial F_{lL}}{\partial X_K} + \frac{\partial Q_K}{\partial T} G_K + \frac{\partial Q_K}{\partial G_L} \frac{\partial G_L}{\partial X_K} - P_{kK} \frac{\partial v_k}{\partial X_K} \right) \Bigg], \tag{5.18}$$

and these derivatives are eliminated from (5.17).

Clearly, this inequality is linear with respect to the following derivatives

$$\left\{ \frac{\partial F_{kK}}{\partial t}, \frac{\partial G_K}{\partial t}, \frac{\partial F_{lL}}{\partial X_K}, \frac{\partial G_L}{\partial X_K}, \frac{\partial v_k}{\partial X_K} \right\}. \tag{5.19}$$

As the inequality must hold for all fields – we have eliminated constraints imposed by field equations (note that linearity of the momentum equation with respect to the acceleration $\partial v_k / \partial t$ does not impose any restrictions because the acceleration does not appear in the entropy inequality) coefficients of the above derivatives have to vanish identically. Otherwise we could choose the fields in such a way that one negative term would dominate all others and the inequality would be violated. Hence, we obtain

$$\begin{aligned} \frac{\partial \eta}{\partial F_{kK}} - \Lambda^\varepsilon \frac{\partial \varepsilon}{\partial F_{kK}} + \frac{1}{\rho_0} P_{kK} = 0, \quad \Lambda^\varepsilon := \frac{\partial \eta}{\partial T} \left(\frac{\partial \varepsilon}{\partial T} \right)^{-1} \\ \frac{\partial \eta}{\partial G_K} - \Lambda^\varepsilon \frac{\partial \varepsilon}{\partial G_K} = 0, \\ \frac{\partial H_{(K}}{\partial F_{lL)}} - \Lambda^\varepsilon \frac{\partial Q_{(K}}{\partial F_{lL)}} = 0, \\ \frac{\partial H_{(K}}{\partial G_{L)}} - \Lambda^\varepsilon \frac{\partial Q_{(K}}{\partial G_{L)}} = 0, \end{aligned} \tag{5.20}$$

where the parentheses denote the symmetric part, e.g. $\partial Q_{(K} / \partial G_{L)} = \frac{1}{2} \left(\partial Q_K / \partial G_L + \partial Q_L / \partial G_K \right)$. We have made the assumption $\partial \varepsilon / \partial T \neq 0$ which is physically justified as this derivative defines the specific heat in thermodynamical equilibrium (see Chapter 6).

There remains the *residual inequality* which does not contain linear contributions anymore

$$\mathcal{D} := \left(\frac{\partial H_K}{\partial T} - \Lambda^\varepsilon \frac{\partial Q_K}{\partial T} \right) G_K \geq 0. \tag{5.21}$$

$\mathcal{D}$ is called the *dissipation* function for thermoelastic materials.

The exploitation of identities (5.20) requires certain additional techniques which we present in the next Section. They are either based on a simplifying assumption that the material is isotropic or, in more general cases, one has to use a Proportionality Theorem (Section 5.5) proved by I-Shih Liu in 1996 [190] on the relation between the entropy flux and heat flux. Under rather general conditions this Theorem states that in our case identities $(5.20)_{3,4}$ yield

$$H_K = \Lambda^\varepsilon Q_K. \tag{5.22}$$

Then the substitution of the above relation in $(5.20)_{3,4}$ leads to

$$\Lambda^\varepsilon = \Lambda^\varepsilon (T). \tag{5.23}$$

Let us consider a contact surface between a thermoelastic material and an ideal gas. For the latter we have the relation $\mathbf{H} = (1/T)\,\mathbf{Q}$ which we motivate further in this book (Chapter 7) by microscopic (kinetic) considerations. Bearing the continuity of fluxes on the surface of contact in mind (jump conditions and the last part of the second law of thermodynamics) we obtain

$$[[\mathbf{H}]] \cdot \mathbf{N} = 0, \quad [[\mathbf{Q}]] \cdot \mathbf{N} = 0 \quad \Longrightarrow \quad \Lambda^\varepsilon = \frac{1}{T}. \tag{5.24}$$

Consequently, we obtain the classical Fourier result for fluxes

$$\mathbf{H} = \frac{1}{T}\mathbf{Q}. \tag{5.25}$$

Substitution in the definition of Λ^ε $(5.20)_1$ leads to the classical thermodynamical identity

$$\frac{\partial \varepsilon}{\partial T} - T\frac{\partial \eta}{\partial T} = 0. \tag{5.26}$$

It is useful to define the following function

$$\psi = \varepsilon - T\eta. \tag{5.27}$$

Then differentiation with respect to the temperature T yields immediately

$$\eta = -\frac{\partial \psi}{\partial T}, \quad \varepsilon = \psi - T\frac{\partial \psi}{\partial T}, \quad \psi = \psi\,(T, \mathbf{F}), \tag{5.28}$$

where the identity $(5.20)_2$ has been exploited. The function ψ is called the *Helmholtz free energy*. If we use it in $(5.20)_1$ we obtain

$$\mathbf{P} = \rho_0 \frac{\partial \psi}{\partial \mathbf{F}}. \tag{5.29}$$

Hence, if the constitutive relation for the Helmholtz free energy $\psi = \psi\,(T, \mathbf{F})$ is known it determines constitutive relations for $\varepsilon, \eta, \mathbf{P}$. This is one of the main results of the second law of thermodynamics: it yields the existence of *thermodynamical potentials* which reduce the number of required constitutive relations.

By means of the above results we can immediately show that the following differential relation holds

$$d\eta = \frac{1}{T}\left(d\varepsilon - \frac{1}{\rho_0}\mathbf{P}\cdot\mathbf{dF}\right). \tag{5.30}$$

In classical thermodynamics this relation is called the *Gibbs equation* and it is often mistakenly called the second law of thermodynamics. Obviously, it yields relations (5.26) and (5.29). However, it does not contain the residual inequality (5.21) which also follows from the second law. In other words, Gibbs equation cannot determine the *dissipation* $\mathcal{D}$ in the system. Bearing the above results in mind we can write the dissipation in the following form

$$\mathcal{D} = -\frac{1}{T^2}\mathbf{Q}\cdot\mathbf{G} \geq 0. \tag{5.31}$$

This is, of course, the requirement that heat cannot flow from colder to hotter regions. If we assume the linear Fourier law for heat conduction, we obtain the classical result for the heat conductivity, K,

$$\mathbf{Q} = -K\mathbf{G} \quad \Longrightarrow \quad K \geq 0. \tag{5.32}$$

This is related to the so-called stability of the thermodynamical equilibrium. Namely, we define the *thermodynamical equilibrium* as a state in which the dissipation $\mathcal{D}$ vanishes. In our case it means that the temperature gradient in equilibrium must be zero. Hence, all isothermal processes $T = const$ of arbitrary large deformations contain only equilibrium states, i.e. these are *equilibrium processes*. Simultaneously, the inequality (5.31) shows that equilibrium states are states of *minimum* of the dissipation function. Consequently the following condition must hold

$$\forall\, \mathbf{Z} \quad \mathbf{Z}\cdot\frac{\partial^2\mathcal{D}}{\partial\mathbf{G}\partial\mathbf{G}}\mathbf{Z} > 0, \tag{5.33}$$

i.e. the Hessian matrix of $\mathcal{D}$ is positive definite. This is the so-called *stability condition of equilibrium*.

5.1.3 *I-Shih Liu Theorem*

The above results show how powerful the entropy inequality is. However, we can essentially improve the method of derivation of these results. Instead of directly eliminating some derivatives by means of field equations we can use the I-Shih Liu Theorem on the existence of Lagrange multipliers which help to eliminate the constraints imposed by field equations on the entropy inequality.

Inspection of the above derivation shows that we are solving an *algebraic* problem. Constitutive relations and the chain rule of differentiation yield the entropy inequality whose part is a linear function of some derivatives, in our example listed in (5.19) and a nonlinear part which leads to the residual inequality. If we eliminate restrictions imposed by field equations we have to solve an inequality for *arbitrary independent* linear contributions.

We formulate the general problem in the following abstract way.

We construct a continuous model for fields which form an n-dimensional vector $\mathbf{w} \equiv \{w_\gamma\}_{\gamma=1}^n \in \mathcal{V}^n$. The set of field equations for $\mathbf{w}$ is of the following form

$$\frac{\partial F_{0\gamma}}{\partial t} + \frac{\partial F_{K\gamma}}{\partial X_K} = \hat{f}_\gamma, \tag{5.34}$$

$$\{F_{0\gamma}\}_{\gamma=1}^n \in \mathcal{V}^n, \quad \{F_{K\gamma}\}_{\gamma=1}^n \in \mathcal{V}^n, K = 1,2,3, \quad \left\{\hat{f}_\gamma\right\}_{\gamma=1}^n \in \mathcal{V}^n,$$

where the following differentiable constitutive relations are fulfilled

$$F_{0\gamma} = F_{0\gamma}(\mathfrak{w}), \quad F_{K\gamma} = F_{K\gamma}(\mathfrak{w}), \quad \hat{f}_\gamma = \hat{f}_\gamma(\mathfrak{w}), \quad \mathfrak{w} \in \mathcal{V}^N. \tag{5.35}$$

The choice of constitutive variables $\mathfrak{w}$ defines the class of materials to be described. The constitutive relations (5.35) must be such that solutions of field equations identically satisfy the entropy inequality

$$\rho_0 \frac{\partial \eta}{\partial t} + \frac{\partial H_K}{\partial X_K} \geq 0, \quad \eta = \eta(\mathfrak{w}), \quad H_K = H_K(\mathfrak{w}). \tag{5.36}$$

Obviously the constitutive variables $\mathfrak{w}$ may contain both fields $\mathbf{w}$ as well as their various temporal and spatial derivatives.

Let us introduce the following notation

$$\mathfrak{a} = \left\{\rho_0 \frac{\partial \eta}{\partial \mathfrak{w}_1}, ..., \rho_0 \frac{\partial \eta}{\partial \mathfrak{w}_N}, \frac{\partial H_1}{\partial \mathfrak{w}_1}, ..., \frac{\partial H_3}{\partial \mathfrak{w}_N}\right\} \in \Re^{4N}, \tag{5.37}$$

$$\mathfrak{X} = \left\{\frac{\partial \mathfrak{w}_1}{\partial t}, ..., \frac{\partial \mathfrak{w}_N}{\partial t}, \frac{\partial \mathfrak{w}_1}{\partial X_1}, ..., \frac{\partial \mathfrak{w}_N}{\partial X_3}\right\}^T \in \Re^{4N}, \tag{5.38}$$

where $\mathfrak{w}_a, a = 1, ..., N$ are components of the vector $\mathfrak{w}$. Then the field equations can be written in the form

$$\mathsf{A}\mathfrak{X} - \hat{\mathbf{f}} = 0, \quad \hat{\mathbf{f}} \equiv \left\{\hat{f}_\gamma\right\}_{\gamma=1}^n \tag{5.39}$$

where

$$\mathsf{A} = \begin{pmatrix} \frac{\partial F_{01}}{\partial \mathfrak{w}_1} & \frac{\partial F_{01}}{\partial \mathfrak{w}_2} & \cdots & \frac{\partial F_{31}}{\partial \mathfrak{w}_N} \\ \frac{\partial F_{02}}{\partial \mathfrak{w}_1} & \frac{\partial F_{02}}{\partial \mathfrak{w}_2} & \cdots & \frac{\partial F_{32}}{\partial \mathfrak{w}_N} \\ \vdots & \vdots & & \vdots \\ \frac{\partial F_{0n}}{\partial \mathfrak{w}_1} & \frac{\partial F_{0n}}{\partial \mathfrak{w}_2} & \cdots & \frac{\partial F_{3n}}{\partial \mathfrak{w}_N} \end{pmatrix} \in \Re^n \times \Re^{4N}. \tag{5.40}$$

Now we are in the position to formulate the following

I-Shih Liu Theorem (*on the existence of Lagrange multipliers*) [187], [191], [364]. Let A be given by (5.40) and $\mathfrak{X}$ by (5.38), and

$$\mathsf{S} = \left\{ \mathfrak{X} \in \Re^{4N} \middle| \mathsf{A}\mathfrak{X} - \hat{\mathbf{f}} = 0 \right\} \neq \emptyset. \tag{5.41}$$

Then, the following conditions are equivalent:

$$\text{i/} \quad \forall \mathfrak{X} \in \mathsf{S}: \quad \mathfrak{a} \cdot \mathfrak{X} \geq 0, \tag{5.42}$$

where $\mathfrak{a}$ is given by (5.37),

$$\text{ii/} \quad \exists \mathbf{\Lambda} \in \Re^n, \mathbf{\Lambda} \neq \mathbf{0} \quad \forall \mathfrak{X} \in \Re^{4N}: \quad \mathfrak{a} \cdot \mathfrak{X} - \mathbf{\Lambda} \cdot \left(\mathsf{A}\mathfrak{X} - \hat{\mathbf{f}} \right) \geq 0, \tag{5.43}$$

$$\text{iii/} \quad \exists \mathbf{\Lambda} \in \Re^n, \mathbf{\Lambda} \neq \mathbf{0}: \quad \mathfrak{a} - \mathsf{A}^T \mathbf{\Lambda} = \mathbf{0}, \quad \mathbf{\Lambda} \cdot \hat{\mathbf{f}} \geq 0. \tag{5.44}$$

Proof: We prove the following implications

$$\text{i/} \Longleftarrow \text{ii/} \Longleftrightarrow \text{iii/} \Longleftarrow \text{i/}. \tag{5.45}$$

1. The implication ii/$\Longrightarrow$i/ is immediate.

2. We shall prove the equivalence of ii/ and iii/. The relation (5.43) can be written in the form

$$\forall \mathfrak{X} \in \Re^{4N}: \quad \left(\mathfrak{a} - \mathsf{A}^T \cdot \mathbf{\Lambda} \right) \cdot \mathfrak{X} + \mathbf{\Lambda} \cdot \hat{\mathbf{f}} \geq 0. \tag{5.46}$$

Since this inequality holds for arbitrary $\mathfrak{X}$, it follows necessarily that

$$\left(\mathfrak{a} - \mathsf{A}^T \cdot \mathbf{\Lambda} \right) = 0 \quad \Longrightarrow \quad \mathbf{\Lambda} \cdot \hat{\mathbf{f}} \geq 0. \tag{5.47}$$

3. It remains to prove the implication i/$\Longrightarrow$iii/. With this aim, we

define the following sets:

$$
\begin{aligned}
\mathsf{H} &\equiv \left\{ \mathfrak{X} \in \Re^{4N} \middle| \mathfrak{a} \cdot \mathfrak{X} \geq 0 \right\}, \\
\mathsf{H}_0 &\equiv \left\{ \mathfrak{X} \in \Re^{4N} \middle| \mathfrak{a} \cdot \mathfrak{X} = 0 \right\}, \quad \mathsf{S}_0 \equiv \left\{ \mathfrak{X} \in \Re^{4N} \middle| \mathsf{A}\mathfrak{X} = 0 \right\}, \\
\mathsf{H}_0^{\perp} &\equiv \left\{ \mathfrak{Y} \in \Re^{4N} \middle| \forall \mathfrak{X} \in \mathsf{H}_0 : \mathfrak{Y} \cdot \mathfrak{X} = 0 \right\}, \\
\mathsf{S}_0^{\perp} &\equiv \left\{ \mathfrak{Y} \in \Re^{4N} \middle| \forall \mathfrak{X} \in \mathsf{S}_0 : \mathfrak{Y} \cdot \mathfrak{X} = 0 \right\}.
\end{aligned}
\tag{5.48}
$$

To prove the assertion we show first that

$$\mathsf{H}_0^{\perp} \subset \mathsf{S}_0^{\perp}. \tag{5.49}$$

Let us first motivate the purpose of this relation. It is easy to see from definitions (5.48) that the set of derivatives of the entropy and entropy flux with respect to the constitutive variables $\mathfrak{w}$, i.e. the vector $\mathfrak{a}$ belongs to the subspace $\mathsf{H}_0^{\perp}$ because it is orthogonal to all vectors belonging to H_0. Consequently, if we prove (5.49) we can claim that this vector belongs to $\mathsf{S}_0^{\perp}$. Simultaneously, we can introduce a special representation of such vectors based on the matrix A which has a physical bearing. Components of $\mathfrak{a}$ in this representation form the vector of Lagrange multipliers $\mathbf{\Lambda}$.

It is easy to notice that $\mathsf{H}_0, \mathsf{H}_0^{\perp}, \mathsf{S}_0^{\perp}$ are subspaces of $\Re^{4N}$. Simultaneously, i/ implies that $\mathsf{S} \subset \mathsf{H}$; we obtain $\mathsf{S}_0 \subset \mathsf{H}_0$.

Suppose that the above relation does not hold. Then

$$\exists \mathfrak{Y} \in \mathsf{S}_0 : \quad \mathfrak{a} \cdot \mathfrak{Y} \neq 0.$$

However, S_0 is the linear subspace of $\Re^{4N}$. Therefore

$$\forall a \in \Re : \quad a\mathfrak{Y} \in \mathsf{S}_0 \quad \Longrightarrow \quad \forall \mathfrak{Z} \in \mathsf{S} : \quad \mathfrak{Z} + a\mathfrak{Y} \in \mathsf{S}.$$

On the other hand

$$\mathfrak{a} \cdot (\mathfrak{Z} + a\mathfrak{Y}) = a\mathfrak{a} \cdot \mathfrak{Y} + \mathfrak{a} \cdot \mathfrak{Z}.$$

Hence

$$\exists a \in \Re : \quad a\mathfrak{a} \cdot \mathfrak{Y} \leq -\mathfrak{a} \cdot \mathfrak{Z} \quad \Longrightarrow \quad \mathfrak{Z} + a\mathfrak{Y} \notin \mathsf{H},$$

which is a contradiction. Hence, $\mathsf{S}_0 \subset \mathsf{H}_0$.

We proceed to prove the implication i/$\Longrightarrow$iii/. By definition, we have $\mathfrak{a} \in \mathsf{H}_0^{\perp}$ and consequently, $\mathfrak{a} \in \mathsf{S}_0^{\perp}$. Let us construct, from the matrix A, the sequence of vectors $\{\mathsf{A}_1, ..., \mathsf{A}_n\}$ whose coordinates coincide with the rows of A. Then

$$\forall 1 \leq \gamma \leq n: \quad \mathsf{A}_\gamma \in \mathsf{S}_0^\perp. \tag{5.50}$$

On the other hand, we have

$$\dim \mathsf{S}_0^\perp = \operatorname{rank} \mathsf{A}, \tag{5.51}$$

and, consequently, the vectors $\{\mathsf{A}_\gamma\}_{\gamma=1}^n$ span the space $\mathsf{S}_0^\perp$. It follows that

$$\exists \mathbf{\Lambda} \in \Re^n, \mathbf{\Lambda} \neq \mathbf{0}: \quad \mathfrak{a} = \sum_{\gamma=1}^n \mathsf{A}_\gamma \Lambda^\gamma = \mathsf{A}^T \mathbf{\Lambda}. \tag{5.52}$$

Finally,

$$\forall \mathfrak{X} \in \mathsf{S}: \quad \mathfrak{a} \cdot \mathfrak{X} = \left(\mathsf{A}^T \mathbf{\Lambda}\right) \cdot \mathfrak{X} = \mathbf{\Lambda} \cdot (\mathsf{A}\mathfrak{X}) = \mathbf{\Lambda} \cdot \hat{\mathbf{f}}; \tag{5.53}$$

since $\mathsf{S} \subset \mathsf{H}$, i.e. $\mathfrak{a} \cdot \mathfrak{X} \geq 0$, we obtain

$$\mathbf{\Lambda} \cdot \hat{\mathbf{f}} \geq 0, \tag{5.54}$$

which completes the proof.

For the example of thermoelastic materials (5.14), (5.15) the objects appearing in the above theorem have the following form

$$\begin{gathered}
\mathbf{w} = \{\mathbf{v}, \mathbf{F}, T\} \in \mathcal{V}^{13}, \quad n = 13, \\
\mathfrak{w} := \{\mathbf{v}, \mathbf{F}, T, \mathbf{G}\} \in \mathcal{V}^{16}, \quad N = 16, \\
\{F_{0\gamma}\}_{\gamma=1}^{13} = \{\rho_0 v_k, F_{kK}, \rho_0 \varepsilon\} \in \mathcal{V}^{13}, \\
\{F_{K\gamma}\}_{\gamma=1}^{13} = \{P_{kK}, v_k \delta_{LK}, Q_K\}, \quad K = 1, 2, 3, \\
\left\{\hat{f}_\gamma\right\}_{\gamma=1}^{13} = \left\{\mathbf{0}, \mathbf{0}, P_{kK} \frac{\partial v_k}{\partial X_K}\right\}.
\end{gathered} \tag{5.55}$$

The Lagrange multipliers have been shown to be

$$\{\Lambda_\gamma\}_{\gamma=1}^n = \left\{\mathbf{0}, \Lambda_{kK}, \frac{1}{T}\right\}, \tag{5.56}$$

where Λ_{kK} has not been derived in the explicit form because the constraint due to the condition $(5.14)_2$ has been eliminated by substitution. The multipliers of momentum equations are zero because these equations contain a linear contribution of the acceleration. Further we show various generalizations of this model and demonstrate the full exploitation of the Liu Theorem.

5.2 Isotropy, material objectivity

5.2.1 *Example - rigid heat conductor*

We begin the presentation of technical tools for the exploitation of thermodynamical identities with a brief discussion of a very simple classical problem of the heat conduction in a rigid heat conductor. This example demonstrates all basic features of the exploitation of the second law of thermodynamics without many technical details.

We consider the problem defined by a single scalar field of the absolute temperature T on the domain $\mathcal{B}_0$ of the rigid heat conductor (undeformable body). The function $T(\mathbf{X}, t)$ is assumed to follow from the energy conservation law

$$\rho_0 \frac{\partial \varepsilon}{\partial t} + \operatorname{Div} \mathbf{Q} = 0, \tag{5.57}$$

where ρ_0 is the constant mass density, and we have left out the energy radiation r. The latter contribution is immaterial for the exploitation of the second law. We assume that the internal energy density ε and the heat flux $\mathbf{Q}$ satisfy the constitutive relations

$$\varepsilon = \varepsilon(T, \mathbf{G}), \quad \mathbf{Q} = \mathbf{Q}(T, \mathbf{G}), \quad \mathbf{G} = \operatorname{Grad} T, \tag{5.58}$$

which are sufficiently smooth functions of both variables.

The second law of thermodynamics is given in the form of entropy inequality

$$\rho_0 \frac{\partial \eta}{\partial t} + \operatorname{Div} \mathbf{H} \geq 0, \quad \eta = \eta(T, \mathbf{G}), \quad \mathbf{H} = \mathbf{H}(T, \mathbf{G}), \tag{5.59}$$

which must hold for all thermodynamical processes, i.e. for all solutions of the field equation which follows from (5.57) by substitution of (5.58).

According to Liu's Theorem there exists a Lagrange multiplier $\Lambda^{\varepsilon}(T, \mathbf{G})$ which allows to write the above inequality in the following modified form

$$\rho_0 \frac{\partial \eta}{\partial t} + \operatorname{Div} \mathbf{H} - \mathbf{\Lambda}^{\varepsilon}\left(\rho_0 \frac{\partial \varepsilon}{\partial t} + \operatorname{Div} \mathbf{Q}\right) \geq 0, \tag{5.60}$$

and this inequality must hold for all fields T and not only for solutions of field equations. Chain rule of differentiation yields the explicit form of this inequality

$$\rho_0 \left(\frac{\partial \eta}{\partial T} - \Lambda^\varepsilon \frac{\partial \varepsilon}{\partial T} \right) \frac{\partial T}{\partial t} + \rho_0 \left(\frac{\partial \eta}{\partial \mathbf{G}} - \Lambda^\varepsilon \frac{\partial \varepsilon}{\partial \mathbf{G}} \right) \cdot \frac{\partial \mathbf{G}}{\partial t}$$

$$+ \left(\frac{\partial \mathbf{H}}{\partial T} - \Lambda^\varepsilon \frac{\partial \mathbf{Q}}{\partial T} \right) \cdot \mathbf{G} + \left(\frac{\partial H_K}{\partial G_L} + \Lambda^\varepsilon \frac{\partial Q_K}{\partial G_L} \right) \frac{\partial^2 T}{\partial X_K X_L} \geq 0. \tag{5.61}$$

The last contribution has been written in coordinates in order to reveal the symmetry with respect to indices K and L. The above inequality is linear with respect to the following derivatives

$$\mathfrak{X} := \left\{ \frac{\partial T}{\partial t}, \frac{\partial \mathbf{G}}{\partial t}, \frac{\partial^2 T}{\partial X_K X_L} \right\}. \tag{5.62}$$

Obviously, the fourth contribution proportional to $\mathbf{G}$ is nonlinear. Hence, Liu's Theorem implies

$$\frac{\partial \eta}{\partial T} - \Lambda^\varepsilon \frac{\partial \varepsilon}{\partial T} = 0, \quad \frac{\partial \eta}{\partial \mathbf{G}} - \Lambda^\varepsilon \frac{\partial \varepsilon}{\partial \mathbf{G}} = 0,$$

$$\frac{\partial H_{(K}}{\partial G_{L)}} + \Lambda^\varepsilon \frac{\partial Q_{(K}}{\partial G_{L)}} = 0, \tag{5.63}$$

where the parentheses denote the symmetrical part of the tensor. There remains the *residual inequality*

$$\mathcal{D} = \left(\frac{\partial \mathbf{H}}{\partial T} - \Lambda^\varepsilon \frac{\partial \mathbf{Q}}{\partial T} \right) \cdot \mathbf{G} \geq 0, \tag{5.64}$$

which defines the *dissipation* $\mathcal{D}(T, \mathbf{G})$ in the rigid heat conductor.

We begin the exploitation of the above identities from $(5.63)_3$. In the case of *isotropic heat conductors* the vector functions $\mathbf{Q}, \mathbf{H}$ must have the following form

$$\mathbf{Q} = Q\left(T, G^2\right)\mathbf{G}, \quad \mathbf{H} = H\left(T, G^2\right)\mathbf{G}, \tag{5.65}$$

where Q, H are arbitrary scalar functions of T and of the *invariant* of $\mathbf{G}$, i.e. its length $G = \sqrt{\mathbf{G} \cdot \mathbf{G}}$. For convenience we use the square of this variable. Then the identity (5.63) has the form

$$2\left(\frac{\partial H}{\partial G^2} - \Lambda^\varepsilon \frac{\partial Q}{\partial G^2} \right) G_K G_L + \left(H - \Lambda^\varepsilon Q \right) \delta_{KL} = 0. \tag{5.66}$$

Taking the deviatoric part of this identity

$$2\left(\frac{\partial H}{\partial G^2} - \Lambda^\varepsilon \frac{\partial Q}{\partial G^2}\right)\left(G_K G_L - \frac{1}{3}G^2\delta_{KL}\right) = 0,$$

we obtain that the first contribution should be zero, and, consequently, the second contribution must vanish as well

$$\frac{\partial H}{\partial G^2} - \Lambda^\varepsilon \frac{\partial Q}{\partial G^2} = 0, \quad H - \Lambda^\varepsilon Q = 0. \tag{5.67}$$

Hence the substitution of the second identity in the first one gives the following result

$$\mathbf{H} = \Lambda^\varepsilon \mathbf{Q}, \quad \Lambda^\varepsilon = \Lambda^\varepsilon (T). \tag{5.68}$$

There are only few results of this type for anisotropic materials. In the case of the rigid heat conductor which is *linear* with respect to the temperature gradient we can prove it easily. Namely, if we assume

$$\begin{aligned} H_K &= \mathrm{H}_{KL}\frac{\partial T}{\partial X_L}, \quad Q_K = \mathrm{Q}_{KL}\frac{\partial T}{\partial X_L}, \\ \mathrm{H}_{KL} &= \mathrm{H}_{KL}(T), \quad \mathrm{Q}_{KL} = \mathrm{Q}_{KL}(T), \end{aligned} \tag{5.69}$$

where $\mathrm{H}_{KL}, \mathrm{Q}_{KL}$ are *conductivity matrices*, then the identity $(5.63)_3$ has the form

$$\mathrm{H}_{(KL)} - \Lambda \mathrm{Q}_{(KL)} = 0. \tag{5.70}$$

The skew symmetric parts of matrices $\mathrm{H}_{KL}, \mathrm{Q}_{KL}$ are not restricted. If we assume that they are identically zero then we obtain again the result (5.68).

The question is if the result (5.68) is characteristic only for materials of a single vectorial constitutive variable. We shall see that differential identities of the form $(5.63)_3$ give rise to the similar result also for much more general cases of materials. This is the subject of Liu's Proportionality Theorem presented in Section 5.5.

We use now the assumption on the existence of ideal walls. According to jump conditions on such walls we have

$$[[\mathbf{H}]] \cdot \mathbf{N} = 0, \quad [[\mathbf{Q}]] \cdot \mathbf{N} = 0 \quad \Longrightarrow \quad [[\Lambda^\varepsilon (T)]] = 0. \tag{5.71}$$

There exists one system for which we can calculate the relation between the heat flux $\mathbf{Q}$ and the entropy flux $\mathbf{H}$ from a microscopic model (kinetic theory; see the remark for thermoelastic materials in Section 5.1.2). This is the ideal gas. For this material the relation has the form $\mathbf{H} = \mathbf{Q}/T$, where T

is the absolute temperature[2]. We return frequently to this relation further in this book. Consequently, if we bring the rigid heat conductor to the contact with the ideal gas through the ideal wall we obtain

$$\Lambda^{\varepsilon}(T) = \frac{1}{T}. \tag{5.72}$$

This relation holds true in all points of $\mathcal{B}_0$ and not only on the contact surface because the multiplier Λ^{ε} does not depend on $\mathbf{X}$ in the explicit manner. Hence, for isotropic rigid heat conductors, the relation between fluxes has the classical form

$$\mathbf{H} = \frac{1}{T}\mathbf{Q}. \tag{5.73}$$

It is appropriate to make a comment on this relation. In many works on continuum thermodynamics this relation is *assumed* from the beginning and then the bulk entropy inequality has the following form

$$\frac{d}{dt}\int_{\mathcal{P}} \rho\eta dV + \oint_{\partial\mathcal{P}} \frac{\mathbf{Q}\cdot\mathbf{N}}{T} dS - \int_{\mathcal{P}} \rho s dV \geq 0. \tag{5.74}$$

This is the so-called *Clausius-Duhem inequality*. With the entropy radiation $s = r/T$, it has been very extensively used in works of Coleman, Eringen, Noll, Serrin, Šilhavý, Williams, Truesdell and many others (e.g. [332], [297], [292], [104]). This form of the entropy inequality follows indeed in many cases of single component materials from the more general inequality (5.10) used in this book. However, multicomponent systems which we consider further are an example that the Calusius-Duhem inequality is too restrictive [220], [222].

We return to the remaining identities $(5.63)_{1,2}$. They can be written in the form

$$\eta = -\frac{1}{T}\frac{\partial\psi}{\partial T}, \quad \varepsilon = \psi - T\frac{\partial\psi}{\partial T}, \quad \psi := \varepsilon - T\eta, \quad \psi = \psi(T), \tag{5.75}$$

where ψ is the trivial example of the Helmholtz free energy. Consequently, neither the free energy ψ nor the internal energy ε, nor the entropy η can depend on the temperature gradient.

Finally, bearing the relation (5.73) in mind we obtain the residual inequality

$$\mathbf{Q}\cdot\mathbf{G} \leq 0. \tag{5.76}$$

[2]This relation follows also from the macroscopic thermodynamics of homogeneous systems based on the so-called Caratheodory principle. We show some results of this approach in the next Chapter.

This is the classical statement of the second law of thermodynamics which says that the heat flux has the orientation opposite to the temperature gradient, i.e. the energy flows from hot to cold regions. This inequality can be written for isotropic heat conductors as follows

$$K \geq 0, \quad K \equiv -Q. \tag{5.77}$$

Substitution of all the above results in the energy conservation equation (5.57) yields the classical heat conduction equation [232]

$$\rho_0 c_v \frac{\partial T}{\partial t} = \operatorname{Div}\left(K \operatorname{Grad} T\right), \quad c_v := \frac{\partial \varepsilon}{\partial T}, \tag{5.78}$$

where c_v is the *specific heat* under the constant volume. This quantity is positive and this follows from the stability condition of thermodynamical equilibrium. We discuss this problem in the next Chapter. The *thermodynamical equilibrium* is defined as a state in which the dissipation vanishes. For the rigid heat conductors it is the case when the distribution of temperature is homogeneous in space: $\operatorname{Grad} T = 0$.

5.2.2 *Isotropy*

We return now to certain general properties of constitutive relations which are helpful in the evaluation of identities following from the second law of thermodynamics.

As indicated above in the majority of cases we have to assume the *isotropy of materials*. For anisotropic materials results are scarce and usually extensions from isotropic to anisotropic properties are made *ad hoc.*

Isotropy of materials is one of the properties related to the behavior of models under the transformation of frames. However, we can perform such a transformation either on the reference configuration or in the current configuration. In the first case invariance properties specify the *symmetry* of the material and the corresponding group of transformations is the so-called *symmetry group of the material* [334]. In the particular case when the group is the *full orthogonal group* the material is called *isotropic.* In the second case the transformation is specified by the isometry of the space of motions which we have discussed before. This yields the *principle of frame indifference* or *the principle of material objectivity.* It plays an important role in the continuum mechanics and thermodynamics by delivering general restrictions on the formulation of constitutive functions of material bodies. It reflects the idea that material properties should be independent of observations made by different observers. Obviously, different observers, as we pointed out in Chapter 1, are related by time-dependent Euclidean transformation of frames. For this reason material frame-indifference is sometimes interpreted as invariance under superimposed *rigid body motions.*

We present further many examples of invariance properties related to both transformations. In order to investigate such invariance properties, we have to introduce the mathematical description of isotropic functions. We proceed to do so.

We limit our attention to the full orthogonal group which consists of orthogonal tensors $\mathbf{O}$, $\det \mathbf{O} = \pm 1$. We consider a scalar φ, a vector $\mathbf{h}$, and a symmetric tensor $\mathbf{T}$ which transform in the following (objective) way

$$\varphi^* = \varphi, \quad \mathbf{h}^* = \mathbf{O}\mathbf{h}, \quad \mathbf{T}^* = \mathbf{O}\mathbf{T}\mathbf{O}^T. \tag{5.79}$$

In the particular case of the single vector variable which transforms according to the rule $\mathbf{w}^* = \mathbf{O}\mathbf{w}$, where $\mathbf{O}$ is orthogonal, $\mathbf{w}$ is an arbitrary vector and $\mathbf{w}^*$ is its transformation, arbitrary scalar function $\varphi(\mathbf{w})$, vector function $\mathbf{h}(\mathbf{w})$, second order symmetric tensor function $\mathbf{T}(\mathbf{w})$ of a single vector variable $\mathbf{w}$ are said to be *isotropic* if for an arbitrary orthogonal tensor $\mathbf{O}$ the following conditions are satisfied

$$\varphi(\mathbf{w}^*) = \varphi(\mathbf{w}), \quad \mathbf{h}(\mathbf{w}^*) = \mathbf{O}\mathbf{h}(\mathbf{w}), \quad \mathbf{T}(\mathbf{w}^*) = \mathbf{O}\mathbf{T}(\mathbf{w})\mathbf{O}. \tag{5.80}$$

It is essential in these relations that functions $\varphi(...), \mathbf{h}(...), \mathbf{T}(...)$ remain the same and only their coordinates change in the new reference.

For the scalar function the representation result is immediate. As the orthogonal transformation changes only the direction but not the length of an arbitrary vector, a scalar function may satisfy the condition $(5.80)_1$ if it depends only on the length of the vector, i.e.

$$\varphi(\mathbf{w}) = \varphi(w), \quad w \equiv |\mathbf{w}| = \sqrt{\mathbf{w}\cdot\mathbf{w}}, \quad w^* = w. \tag{5.81}$$

The representation for the vector function has the following form

$$\mathbf{h}(\mathbf{w}) = h(w)\,\mathbf{w}. \tag{5.82}$$

This relation has been used in previous considerations. We proceed to prove it.

For $\mathbf{w} = 0$ we have

$$\forall \mathbf{O}: \quad \mathbf{h}(\mathbf{0}) = \mathbf{O}\mathbf{h}(\mathbf{0}) \quad \Longrightarrow \quad \mathbf{h}(\mathbf{0}) = \mathbf{0}, \tag{5.83}$$

and (5.82) is satisfied. Let us assume that $\mathbf{w} \neq \mathbf{0}$. Then

$$\mathbf{h}(\mathbf{w}) = \left[\frac{1}{w^2}\mathbf{w}\cdot\mathbf{h}(\mathbf{w})\right]\mathbf{w} + \left[\mathbf{h}(\mathbf{w}) - \left(\frac{1}{w^2}\mathbf{w}\cdot\mathbf{h}(\mathbf{w})\right)\mathbf{w}\right]. \tag{5.84}$$

Obviously, the second vector on the right-hand side is orthogonal to $\mathbf{w}$. After the transformation this relation has the following form

$$\begin{aligned}\mathbf{h}(\mathbf{w}^*) &= \left[\frac{1}{w^2}\mathbf{Ow}\cdot(\mathbf{h}(\mathbf{Ow}))\right]\mathbf{Ow}+\left[\mathbf{h}(\mathbf{Ow})-\left(\frac{1}{w^2}\mathbf{Ow}\cdot(\mathbf{h}(\mathbf{Ow}))\right)\mathbf{Ow}\right]\\ &= \left[\frac{1}{w^2}\mathbf{Ow}\cdot\mathbf{Oh}(\mathbf{w})\right]\mathbf{Ow}+\left[\mathbf{Oh}(\mathbf{w})-\left(\frac{1}{w^2}\mathbf{Ow}\cdot\mathbf{Oh}(\mathbf{w})\right)\mathbf{Ow}\right]\\ &= \left[\frac{1}{w^2}\mathbf{w}\cdot\mathbf{h}(\mathbf{w})\right]\mathbf{Ow}+\mathbf{O}\left[\mathbf{h}(\mathbf{w})-\left(\frac{1}{w^2}\mathbf{w}\cdot\mathbf{h}(\mathbf{w})\right)\mathbf{w}\right].\end{aligned}$$

Let us consider a particular case of the transformation which is the rotation by 180° about the vector $\mathbf{w}$, i.e. $\mathbf{w}=\mathbf{Ow}$. In this case the second contribution perpendicular to $\mathbf{w}$ changes the sign

$$\mathbf{h}(\mathbf{w}) = \left[\frac{1}{w^2}\mathbf{w}\cdot\mathbf{h}(\mathbf{w})\right]\mathbf{w}-\left[\mathbf{h}(\mathbf{w})-\left(\frac{1}{w^2}\mathbf{w}\cdot\mathbf{h}(\mathbf{w})\right)\mathbf{w}\right]. \quad (5.85)$$

Comparison of (5.84) and (5.85) shows that the second term must vanish, i.e.

$$\mathbf{h}(\mathbf{w}) = \left[\frac{1}{w^2}\mathbf{w}\cdot\mathbf{h}(\mathbf{w})\right]\mathbf{w}. \quad (5.86)$$

The coefficient is obviously an objective scalar. This completes the proof.

For the symmetric tensor we find the representation by investigation of the following auxiliary function

$$\mathbf{h}(\mathbf{w}) = \mathbf{T}(\mathbf{w})\,\mathbf{w}. \quad (5.87)$$

This function is isotropic. Namely

$$\mathbf{h}(\mathbf{Ow}) = \mathbf{T}(\mathbf{Ow})\,\mathbf{Ow} = \mathbf{OT}(\mathbf{w})\,\mathbf{O}^T\mathbf{Ow} = \mathbf{Oh}(\mathbf{w}).$$

Therefore, there exists a scalar function $h(w)$ such that

$$\mathbf{h}(\mathbf{w}) = h(w)\,\mathbf{w} \quad \Longrightarrow \quad [\mathbf{T}(\mathbf{w}) - h(w)\,\mathbf{1}]\,\mathbf{w} = \mathbf{0}. \quad (5.88)$$

Hence, $\mathbf{w}$ is the eigenvector of $\mathbf{T}$. Since the tensor $\mathbf{T}$ is symmetric we have the following spectral representation

$$\mathbf{T}(\mathbf{w}) = \alpha^{(1)}(\mathbf{w})\,\mathbf{w}\otimes\mathbf{w}+\alpha^{(2)}(\mathbf{w})\,\mathbf{u}^{(1)}\otimes\mathbf{u}^{(1)}+\alpha^{(3)}(\mathbf{w})\,\mathbf{u}^{(2)}\otimes\mathbf{u}^{(2)}, \quad (5.89)$$

where $\mathbf{u}^{(1)},\mathbf{u}^{(2)}$ are the remaining unit eigenvectors of $\mathbf{T}$. Now let us choose a rotation about the vector $\mathbf{w}$ which satisfies the conditions[3]

$$\mathbf{Ow} = \mathbf{w}, \quad \mathbf{Ou}^{(1)} = \mathbf{u}^{(2)}, \quad \mathbf{Ou}^{(2)} = \mathbf{u}^{(1)}. \tag{5.90}$$

Then it follows from the spectral decomposition

$$\mathbf{T}(\mathbf{w}) = \alpha^{(1)}(\mathbf{w})\,\mathbf{w}\otimes\mathbf{w} + \alpha^{(3)}(\mathbf{w})\,\mathbf{u}^{(1)}\otimes\mathbf{u}^{(1)} + \alpha^{(2)}(\mathbf{w})\,\mathbf{u}^{(2)}\otimes\mathbf{u}^{(2)},$$

which yields

$$\alpha^{(2)}(\mathbf{w}) = \alpha^{(3)}(\mathbf{w}).$$

Therefore the spectral representation of $\mathbf{T}$ reduces to the following form

$$\begin{aligned} \mathbf{T}(\mathbf{w}) &= \tau_0(w)\,\mathbf{1} + \tau_1(w)\,\mathbf{w}\otimes\mathbf{w}, \\ \tau_1(w) &\equiv \alpha^{(1)}(\mathbf{w}) - \frac{1}{w^2}\alpha^{(2)}(\mathbf{w}). \end{aligned} \tag{5.91}$$

This relation specifies the representation of the isotropic symmetric tensor function of one vector variable.

Due to its practical importance we present in some details another special case of isotropic functions of a single symmetric second rank tensors. Then we have

Representation Theorem (Rivlin, Ericksen) [267].

Let $\varphi, \mathbf{h}, \mathbf{T}$ be isotropic scalar-, vector-, and symmetric tensor-valued functions of a symmetric tensor variable $\mathbf{A}$. Then it is necessary and sufficient that they have the following representations

$$\begin{aligned} &1.\quad \varphi(\mathbf{A}) = \varphi\left(a^{(1)}, a^{(2)}, a^{(3)}\right), \\ &2.\quad \mathbf{h}(\mathbf{A}) = 0, \\ &3.\quad \mathbf{T}(\mathbf{A}) = \tau_0\mathbf{1} + \tau_1\mathbf{A} + \tau_2\mathbf{A}^2, \end{aligned} \tag{5.92}$$

where $\varphi, \tau_0, \tau_1, \tau_2$ are scalar functions of the three eigenvalues $a^{(1)}, a^{(2)}, a^{(3)}$ of $\mathbf{A}$.

Proof. The sufficiency is trivial. We prove the necessity. For the vector function we have

$$\mathbf{h}\left(\mathbf{OAO}^T\right) = \mathbf{Oh}(\mathbf{A}),$$

and, choosing $\mathbf{O} = -\mathbf{1}$, we obtain $\mathbf{h}(\mathbf{A}) = -\mathbf{h}(\mathbf{A})$, which proves the Theorem.

---

[3]i.e. in the basis $\{\mathbf{u}^{(1)}, \mathbf{u}^{(2)}, \mathbf{w}\}$ $\mathbf{O}$ ($\det\mathbf{O} = -1$) is given by the matrix $\begin{pmatrix} 0&1&0 \\ 1&0&0 \\ 0&0&1 \end{pmatrix}$.

For the scalar function, we have to show that whenever two tensors $\mathbf{A}$ and $\mathbf{B}$ have the same eigenvalues, say $\left\{a^{(i)}\right\}_{i=1}^{3}$ then $\varphi(\mathbf{A}) = \varphi(\mathbf{B})$. Using the spectral representation we obtain

$$\mathbf{A} = \sum_{i=1}^{3} a^{(i)} \mathbf{u}^{(i)} \otimes \mathbf{u}^{(i)}, \quad \mathbf{B} = \sum_{i=1}^{3} a^{(i)} \mathbf{v}^{(i)} \otimes \mathbf{v}^{(i)}, \tag{5.93}$$

where $\mathbf{u}^{(i)}, \mathbf{v}^{(i)}$ are unit eigenvectors of $\mathbf{A}$ and $\mathbf{B}$, respectively. Let us choose the transformation $\mathbf{O}$ in such a way that

$$\mathbf{u}^{(i)} = \mathbf{O}\mathbf{v}^{(i)}, \tag{5.94}$$

then the spectral representations (5.93) yield

$$\mathbf{A} = \mathbf{O}\mathbf{B}\mathbf{O}^T. \tag{5.95}$$

Hence

$$\varphi(\mathbf{A}) = \varphi\left(\mathbf{O}\mathbf{B}\mathbf{O}^T\right), \tag{5.96}$$

and the definition of isotropy: $\varphi\left(\mathbf{O}\mathbf{B}\mathbf{O}^T\right) = \varphi(\mathbf{B})$ yields the Theorem.

It remains to prove the representation for the tensor function $\mathbf{T}$. We prove it in two steps. First of all, let us show that any eigenvector of $\mathbf{A}$ is also an eigenvector of $\mathbf{T}(\mathbf{A})$. Let us choose $\mathbf{O}$ to be the rotation by 180° about the eigenvector $\mathbf{u}^{(1)}$ of $\mathbf{A}$. Then

$$\begin{aligned} \mathbf{O}\mathbf{u}^{(1)} &= \mathbf{u}^{(1)}, \quad \mathbf{O}\mathbf{u}^{(2)} = -\mathbf{u}^{(2)}, \quad \mathbf{O}\mathbf{u}^{(3)} = -\mathbf{u}^{(3)} \\ &\Longrightarrow \quad \mathbf{O}\mathbf{A}\mathbf{O}^T = \mathbf{A}. \end{aligned} \tag{5.97}$$

Since $\mathbf{T}(\mathbf{A})$ is isotropic, we have for this $\mathbf{O}$

$$\mathbf{O}\mathbf{T}(\mathbf{A})\mathbf{O}^T = \mathbf{T}\left(\mathbf{O}\mathbf{A}\mathbf{O}^T\right) \quad \Longrightarrow \quad \mathbf{O}\mathbf{T}(\mathbf{A}) = \mathbf{T}(\mathbf{A})\mathbf{O}. \tag{5.98}$$

Hence

$$\mathbf{O}\mathbf{T}(\mathbf{A})\mathbf{u}^{(1)} = \mathbf{T}(\mathbf{A})\mathbf{O}\mathbf{u}^{(1)} = \mathbf{T}(\mathbf{A})\mathbf{u}^{(1)}, \tag{5.99}$$

and, for our choice of $\mathbf{O}$ it means that $\mathbf{T}(\mathbf{A})\mathbf{u}^{(1)}$ must be parallel to $\mathbf{u}^{(1)}$. Therefore $\mathbf{u}^{(1)}$ is the eigenvector of $\mathbf{T}$. For the remaining eigenvectors the proof is identical. Consequently, we can write the tensor $\mathbf{T}$ in the following form

$$\mathbf{T}(\mathbf{A}) = \sum_{i=1}^{3} b^{(i)} \mathbf{u}^{(i)} \otimes \mathbf{u}^{(i)}, \tag{5.100}$$

where $b^{(i)}$ are functions of $\mathbf{A}$.

In the second step we show that this result implies the representation for tensors. Let us first consider the case of distinct eigenvalues $a^{(i)}$ of the tensor $\mathbf{A}$. We consider the set of the following three equations for τ_0, τ_1, τ_2

$$\tau_0 + a^{(i)}\tau_1 + \left(a^{(i)}\right)^2 \tau_2 = b^{(i)}, \quad i = 1, 2, 3. \tag{5.101}$$

Since the determinant

$$\begin{vmatrix} 1 \; a^{(1)} \; \left(a^{(1)}\right)^2 \\ 1 \; a^{(2)} \; \left(a^{(2)}\right)^2 \\ 1 \; a^{(3)} \; \left(a^{(3)}\right)^2 \end{vmatrix} = \left(a^{(1)} - a^{(2)}\right)\left(a^{(2)} - a^{(3)}\right)\left(a^{(3)} - a^{(1)}\right) \tag{5.102}$$

does not vanish we can solve (5.101) with respect to τ_0, τ_1, τ_2. Substitution of (5.101) in (5.100) yields

$$\begin{aligned} \mathbf{T}(\mathbf{A}) &= \tau_0 \sum_{i=1}^{3} \mathbf{u}^{(i)} \otimes \mathbf{u}^{(i)} + \tau_1 \sum_{i=1}^{3} a^{(i)} \mathbf{u}^{(i)} \otimes \mathbf{u}^{(i)} + \tau_2 \sum_{i=1}^{3} \left(a^{(i)}\right)^2 \mathbf{u}^{(i)} \otimes \mathbf{u}^{(i)} \\ &= \tau_0(\mathbf{A})\,\mathbf{1} + \tau_1(\mathbf{A})\,\mathbf{A} + \tau_2(\mathbf{A})\,\mathbf{A}^2. \end{aligned} \tag{5.103}$$

According to the *Cayley-Hamilton Theorem* for tensors generated by a three-dimensional vector space we have

$$\mathbf{A}^3 - I\mathbf{A}^2 + II\mathbf{A} - III\mathbf{1} = \mathbf{0}, \tag{5.104}$$

$$\begin{aligned} I &= \operatorname{tr} \sum_{i=1}^{3} a^{(i)}, \\ II &= \frac{1}{2}\operatorname{tr}\left(I^2 - \operatorname{tr}\mathbf{A}^2\right) = a^{(1)}a^{(2)} + a^{(2)}a^{(3)} + a^{(1)}a^{(3)}, \\ III &= \det \mathbf{A} = a^{(1)}a^{(2)}a^{(3)}, \end{aligned} \tag{5.105}$$

where I, II, III are *the principal invariants* of $\mathbf{A}$. It means that the set $\left\{\mathbf{1}, \mathbf{A}, \mathbf{A}^2\right\}$ is the *basis* of three linearly independent second order tensors for the space of all symmetric second order tensors. It means that (5.103) is the representation of $\mathbf{T}$ in this basis. Consequently, isotropy of $\mathbf{T}$ implies the isotropy of coefficients τ_0, τ_1, τ_2. This proves the Theorem for three distinct eigenvalues.

For two distinct eigenvalues the proof is similar and it yields $\tau_2 = 0$. Finally, for three identical eigenvalues every vector is the eigenvector of $\mathbf{A}$ and, according to (5.100), this yields $\tau_1 = \tau_2 = 0$. This completes the proof.

The above considerations demonstrate problems which arise in proofs of theorems on the representation of isotropic functions. We shall not go into any details of this painstaking tedious analysis (see original works of R. Rivlin, J. M. Spencer [309], [307], [308], [310], [305], [306], G. F. Smith [300], [299], C.-C. Wang [345], etc.) and present Tables 5 to 7 with final results. We skip here the problem of a maximum set of invariants needed for the representation of a given scalar-, vector- or tensor-valued function. This is related to the question of functional *independence* of invariants which we do not discuss in this book. We refer to the original literature and to the book of I-Shih Liu where some of these questions, also for anisotropic materials, are presented in details [191].

Let us demonstrate on a simple example how to use the Tables. Let us say that the symmetric tensor $\mathbf{T}$ is an isotropic function of two symmetric tensors $\mathbf{A}, \mathbf{B}$. Then, according to Table 6, it must have the form

$$\mathbf{T} = \alpha_0 \mathbf{1} + \alpha_1 \mathbf{A} + \alpha_2 \mathbf{A}^2 + \beta_1 \mathbf{B} + \beta_2 \mathbf{B}^2 + \gamma_1 (\mathbf{AB} + \mathbf{BA}) + \gamma_2 \mathbf{ABA} + \gamma_3 \mathbf{BAB}, \tag{5.106}$$

where scalar coefficients $\alpha_0, \alpha_1, \alpha_2, \beta_1, \beta_2, \gamma_1, \gamma_2, \gamma_3$ are functions of the scalar invariants

$$\begin{aligned} &\operatorname{tr}\mathbf{A}, \quad \operatorname{tr}\mathbf{A}^2, \quad \operatorname{tr}\mathbf{A}^3, \quad \operatorname{tr}\mathbf{B}, \quad \operatorname{tr}\mathbf{B}^2, \quad \operatorname{tr}\mathbf{B}^2, \\ &\operatorname{tr}\mathbf{AB}, \quad \operatorname{tr}\mathbf{AB}^2, \quad \operatorname{tr}\mathbf{A}^2\mathbf{B}, \quad \operatorname{tr}\mathbf{A}^2\mathbf{B}^2, \end{aligned} \tag{5.107}$$

which follow from Table 5.1.

5.2.3 *Material objectivity, example: Monney-Rivlin materials*

We can now investigate the invariance properties of the example of thermoelastic model which we have discussed in Section 5.1.2. The constitutive relations (5.13) were assumed to have the following form

$$\begin{aligned} &\mathbf{P} = \mathbf{P}(\mathbf{v}, \mathbf{F}, T, \mathbf{G}), \quad \varepsilon = \varepsilon(\mathbf{v}, \mathbf{F}, T, \mathbf{G}), \\ &\mathbf{Q} = \mathbf{Q}(\mathbf{v}, \mathbf{F}, T, \mathbf{G}), \quad \mathbf{G} := \operatorname{Grad} T. \end{aligned} \tag{5.108}$$

Quantities appearing in these relations transform under the Euclidean time dependent transformation of frames in the configuration space in the following way

$$\begin{aligned} &\mathbf{P}^* = \mathbf{OP}, \quad \varepsilon^* = \varepsilon, \quad \mathbf{Q}^* = \mathbf{Q}, \quad \mathbf{v}^* = \mathbf{Ov} + \dot{\mathbf{O}}\mathbf{x} + \dot{\mathbf{c}}, \\ &\mathbf{F}^* = \mathbf{OF}, \quad T^* = T, \quad \mathbf{G}^* = \mathbf{G}. \end{aligned} \tag{5.109}$$

The principle of material objectivity (material frame-indifference) requires that constitutive functions in (5.108) remain unaffected by this transformation, i.e. in the new frame

$$\mathbf{P}^* = \mathbf{P}(\mathbf{v}^*, \mathbf{F}^*, T^*, \mathbf{G}^*), \quad \varepsilon^* = \varepsilon(\mathbf{v}^*, \mathbf{F}^*, T^*, \mathbf{G}^*), \qquad (5.110)$$
$$\mathbf{Q}^* = \mathbf{Q}(\mathbf{v}^*, \mathbf{F}^*, T^*, \mathbf{G}^*),$$

Table 5.1: *Isotropic scalar invariants*

	Invariant elements
One variable:	
$\mathbf{v}$	$\mathbf{v}\cdot\mathbf{v}$
$\mathbf{A}=\mathbf{A}^T$:	$\operatorname{tr}\mathbf{A}, \operatorname{tr}\mathbf{A}^2, \operatorname{tr}\mathbf{A}^3$
$\mathbf{W}=-\mathbf{W}^T$:	$\operatorname{tr}\mathbf{W}^2$
two variables:	
$\mathbf{v}_1, \mathbf{v}_2$:	$\mathbf{v}_1\cdot\mathbf{v}_2$
$\mathbf{v}, \mathbf{A}$:	$\mathbf{v}\cdot\mathbf{A}\mathbf{v}, \mathbf{v}\cdot\mathbf{A}^2\mathbf{v}$
$\mathbf{v}, \mathbf{W}$:	$\mathbf{v}\cdot\mathbf{W}^2\mathbf{v}$
$\mathbf{A}_1, \mathbf{A}_2$:	$\operatorname{tr}\mathbf{A}_1\mathbf{A}_2, \operatorname{tr}\mathbf{A}_1\mathbf{A}_2^2, \operatorname{tr}\mathbf{A}_1^2\mathbf{A}_2, \operatorname{tr}\mathbf{A}_1^2\mathbf{A}_2^2,$
$\mathbf{W}_1, \mathbf{W}_2$:	$\operatorname{tr}\mathbf{W}_1\mathbf{W}_2$
three variables:	
$\mathbf{v}_1, \mathbf{v}_2, \mathbf{A}$:	$\mathbf{v}_1\cdot\mathbf{A}\mathbf{v}_2, \mathbf{v}_1\cdot\mathbf{A}^2\mathbf{v}_2$
$\mathbf{v}_1, \mathbf{v}_2, \mathbf{W}$:	$\mathbf{v}_1\cdot\mathbf{W}\mathbf{v}_2, \mathbf{v}_1\cdot\mathbf{W}^2\mathbf{v}_2$
$\mathbf{v}, \mathbf{A}_1, \mathbf{A}_2$:	$\mathbf{v}\cdot\mathbf{A}_1\mathbf{A}_2\mathbf{v}_2$
$\mathbf{v}, \mathbf{W}_1, \mathbf{W}_2$:	$\mathbf{v}\cdot\mathbf{W}_1\mathbf{W}_2\mathbf{v}, \mathbf{v}\cdot\mathbf{W}_1\mathbf{W}_2^2\mathbf{v},$ $\mathbf{v}\cdot\mathbf{W}_2\mathbf{W}_1^2\mathbf{v}$
$\mathbf{v}, \mathbf{A}, \mathbf{W}$:	$\mathbf{v}\cdot\mathbf{W}\mathbf{A}\mathbf{v}, \mathbf{v}\cdot\mathbf{W}\mathbf{A}^2\mathbf{v}, \mathbf{v}\cdot\mathbf{W}\mathbf{A}\mathbf{W}^2\mathbf{v}$
$\mathbf{A}_1, \mathbf{A}_2, \mathbf{A}_3$:	$\operatorname{tr}\mathbf{A}_1\mathbf{A}_2\mathbf{A}_3$
$\mathbf{W}_1, \mathbf{W}_2, \mathbf{W}_3$:	$\operatorname{tr}\mathbf{W}_1\mathbf{W}_2\mathbf{W}_3$
$\mathbf{A}_1, \mathbf{A}_2, \mathbf{W}$:	$\operatorname{tr}\mathbf{A}_1\mathbf{A}_2\mathbf{W}, \operatorname{tr}\mathbf{A}_1\mathbf{A}_2^2\mathbf{W},$ $\operatorname{tr}\mathbf{A}_2\mathbf{A}_1^2\mathbf{W}, \operatorname{tr}\mathbf{A}_1\mathbf{W}\mathbf{A}_2\mathbf{W}^2$
$\mathbf{A}, \mathbf{W}_1, \mathbf{W}_2$:	$\operatorname{tr}\mathbf{A}\mathbf{W}_1\mathbf{W}_2, \operatorname{tr}\mathbf{A}\mathbf{W}_1\mathbf{W}_2^2,$ $\operatorname{tr}\mathbf{A}\mathbf{W}_2\mathbf{W}_1^2$
four variables:	
$\mathbf{v}_1, \mathbf{v}_2, \mathbf{A}_1, \mathbf{A}_2$:	$\mathbf{v}_1\cdot\mathbf{A}_1\mathbf{A}_2\mathbf{v}_2, \mathbf{v}_1\cdot\mathbf{A}_2\mathbf{A}_1\mathbf{v}_2$
$\mathbf{v}_1, \mathbf{v}_2, \mathbf{W}_1, \mathbf{W}_2$:	$\mathbf{v}_1\cdot\mathbf{W}_1\mathbf{W}_2\mathbf{v}_2, \mathbf{v}_1\cdot\mathbf{W}_2\mathbf{W}_1\mathbf{v}_2$
$\mathbf{v}_1, \mathbf{v}_2, \mathbf{A}, \mathbf{W}$:	$\mathbf{v}_1\cdot\mathbf{A}\mathbf{W}\mathbf{v}_2, \mathbf{v}_1\cdot\mathbf{W}\mathbf{A}\mathbf{v}_2$

Table 5.2: *Isotropic vector invariants*

Generator elements	
one variable:	
$\mathbf{v}$:	$\mathbf{v}$
$\mathbf{A}$ or $\mathbf{W}$:	$\mathbf{0}$
two variables:	
$\mathbf{v}, \mathbf{A}$:	$\mathbf{A}\mathbf{v}, \mathbf{A}^2\mathbf{v}$
$\mathbf{v}, \mathbf{W}$:	$\mathbf{W}\mathbf{v}, \mathbf{W}^2\mathbf{v}$
three variables:	
$\mathbf{v}, \mathbf{A}_1, \mathbf{A}_2$:	$\mathbf{A}_1\mathbf{A}_2\mathbf{v}$, $\mathbf{A}_2\mathbf{A}_1\mathbf{v}$
$\mathbf{v}, \mathbf{W}_1, \mathbf{W}_2$:	$\mathbf{W}_1\mathbf{W}_2\mathbf{v}$, $\mathbf{W}_2\mathbf{W}_1\mathbf{v}$
$\mathbf{v}, \mathbf{A}, \mathbf{W}$:	$\mathbf{A}\mathbf{W}\mathbf{v}, \mathbf{W}\mathbf{A}\mathbf{v}$

Table 6: *Isotropic symmetric tensor invariants*

Generator elements	
no variable:	
0:	$\mathbf{1}$
one variable:	
$\mathbf{v}$:	$\mathbf{v} \otimes \mathbf{v}$
$\mathbf{A}$:	$\mathbf{A}, \mathbf{A}^2$
$\mathbf{W}$:	$\mathbf{W}^2$
two variables:	
$\mathbf{v}_1, \mathbf{v}_2$:	$\mathbf{v}_1 \otimes \mathbf{v} + \mathbf{v}_2 \otimes \mathbf{v}_1$
$\mathbf{v}, \mathbf{A}$:	$\mathbf{v} \otimes \mathbf{A}\mathbf{v} + \mathbf{A}\mathbf{v} \otimes \mathbf{v}, \mathbf{A}\mathbf{v} \otimes \mathbf{A}\mathbf{v}$
$\mathbf{v}, \mathbf{W}$:	$\mathbf{v} \otimes \mathbf{W}\mathbf{v} + \mathbf{W}\mathbf{v} \otimes \mathbf{v}, \mathbf{W}\mathbf{v} \otimes \mathbf{W}\mathbf{v}$, $\mathbf{W}\mathbf{v} \otimes \mathbf{W}^2\mathbf{v} + \mathbf{W}^2\mathbf{v} \otimes \mathbf{W}\mathbf{v}$
$\mathbf{A}_1, \mathbf{A}_2$:	$\mathbf{A}_1\mathbf{A}_2 + \mathbf{A}_2\mathbf{A}_1, \mathbf{A}_1\mathbf{A}_2\mathbf{A}_1, \mathbf{A}_2\mathbf{A}_1\mathbf{A}_2$
$\mathbf{W}_1, \mathbf{W}_2$:	$\mathbf{W}_1\mathbf{W}_2 + \mathbf{W}_2\mathbf{W}_1, \mathbf{W}_1\mathbf{W}_2^2 - \mathbf{W}_2^2\mathbf{W}_1$, $\mathbf{W}_1^2\mathbf{W}_2 - \mathbf{W}_2\mathbf{W}_1^2$
$\mathbf{A}, \mathbf{W}$:	$\mathbf{A}\mathbf{W} - \mathbf{W}\mathbf{A}, \mathbf{W}\mathbf{A}\mathbf{W}, \mathbf{A}^2\mathbf{W} - \mathbf{W}\mathbf{A}^2$, $\mathbf{W}\mathbf{A}\mathbf{W}^2 - \mathbf{W}^2\mathbf{A}\mathbf{W}$

Table 7: *Isotropic skew symmetric tensor invariants*

	Generator elements
one variable:	
$\mathbf{v}$ or $\mathbf{A}$:	$\mathbf{0}$
$\mathbf{W}$:	$\mathbf{W}$
two variables:	
$\mathbf{v}_1, \mathbf{v}_2$:	$\mathbf{v}_1 \otimes \mathbf{v}_2 - \mathbf{v}_2 \otimes \mathbf{v}_1$
$\mathbf{v}, \mathbf{A}$:	$\mathbf{v} \otimes \mathbf{A}\mathbf{v} - \mathbf{A}\mathbf{v} \otimes \mathbf{v}, \mathbf{v} \otimes \mathbf{A}^2\mathbf{v} - \mathbf{A}^2\mathbf{v} \otimes \mathbf{v},$ $\mathbf{A}\mathbf{v} \otimes \mathbf{A}^2\mathbf{v} - \mathbf{A}^2\mathbf{v} \otimes \mathbf{A}\mathbf{v}$
$\mathbf{v}, \mathbf{W}$:	$\mathbf{v} \otimes \mathbf{W}\mathbf{v} - \mathbf{W}\mathbf{v} \otimes \mathbf{v}, \mathbf{v} \otimes \mathbf{W}^2\mathbf{v} - \mathbf{W}^2\mathbf{v} \otimes \mathbf{v}$
$\mathbf{A}_1, \mathbf{A}_2$:	$\mathbf{A}_1\mathbf{A}_2 - \mathbf{A}_2\mathbf{A}_1,, \mathbf{A}_1\mathbf{A}_2^2 - \mathbf{A}_2^2\mathbf{A}_1, \mathbf{A}_1^2\mathbf{A}_2 - \mathbf{A}_2\mathbf{A}_1^2,$ $\mathbf{A}_1\mathbf{A}_2^2\mathbf{A}_1 - \mathbf{A}_1^2\mathbf{A}_2\mathbf{A}_1, \mathbf{A}_2\mathbf{A}_1\mathbf{A}_2^2 - \mathbf{A}_2^2\mathbf{A}_1\mathbf{A}_2$
$\mathbf{W}_1, \mathbf{W}_2$:	$\mathbf{W}_1\mathbf{W}_2 - \mathbf{W}_2\mathbf{W}_1$
$\mathbf{A}, \mathbf{W}$:	$\mathbf{A}\mathbf{W} + \mathbf{W}\mathbf{A}, \mathbf{A}\mathbf{W}^2 - \mathbf{W}^2\mathbf{A}$
three variables:	
$\mathbf{v}_1, \mathbf{v}_2, \mathbf{A}$:	$\mathbf{v}_1 \otimes \mathbf{A}\mathbf{v}_2 - \mathbf{A}\mathbf{v}_1 \otimes \mathbf{v}_2, \mathbf{v}_2 \otimes \mathbf{A}\mathbf{v}_1 - \mathbf{A}\mathbf{v}_1 \otimes \mathbf{v}_2$
$\mathbf{v}_1, \mathbf{v}_2, \mathbf{W}$:	$\mathbf{v}_1 \otimes \mathbf{W}\mathbf{v}_2 - \mathbf{W}\mathbf{v}_1 \otimes \mathbf{v}_2, \mathbf{v}_2 \otimes \mathbf{W}\mathbf{v}_1 - \mathbf{W}\mathbf{v}_1 \otimes \mathbf{v}_2$
$\mathbf{v}, \mathbf{A}_1, \mathbf{A}_2$:	$\mathbf{A}_1\mathbf{v} \otimes \mathbf{A}_2\mathbf{v} - \mathbf{A}_2\mathbf{v} \otimes \mathbf{A}_1\mathbf{v}, \mathbf{A}_1\mathbf{A}_2\mathbf{v} \otimes \mathbf{v} - \mathbf{v} \otimes \mathbf{A}_1\mathbf{A}_2\mathbf{v},$ $\mathbf{A}_2\mathbf{A}_1\mathbf{v} \otimes \mathbf{v} - \mathbf{v} \otimes \mathbf{A}_2\mathbf{A}_1\mathbf{v}$
$\mathbf{A}_1, \mathbf{A}_2, \mathbf{A}_3$:	$\mathbf{A}_1\mathbf{A}_2\mathbf{A}_3 - \mathbf{A}_3\mathbf{A}_2\mathbf{A}_1, \mathbf{A}_2\mathbf{A}_3\mathbf{A}_1 - \mathbf{A}_1\mathbf{A}_3\mathbf{A}_2,$ $\mathbf{A}_3\mathbf{A}_1\mathbf{A}_2 - \mathbf{A}_2\mathbf{A}_1\mathbf{A}_3$

where functions $\mathbf{P}(...), \varepsilon(...), \mathbf{Q}(...)$ are *the same* in both frames. Consequently, the combination of these relations yields

$$\begin{aligned} \mathbf{OP}(\mathbf{v}, \mathbf{F}, T, \mathbf{G}) &= \mathbf{P}\left(\mathbf{O}\mathbf{v} + \dot{\mathbf{O}}\mathbf{x} + \dot{\mathbf{c}}, \mathbf{OF}, T, \mathbf{G}\right), \\ \varepsilon(\mathbf{v}, \mathbf{F}, T, \mathbf{G}) &= \varepsilon\left(\mathbf{O}\mathbf{v} + \dot{\mathbf{O}}\mathbf{x} + \dot{\mathbf{c}}, \mathbf{OF}, T, \mathbf{G}\right), \\ \mathbf{Q}(\mathbf{v}, \mathbf{F}, T, \mathbf{G}) &= \mathbf{Q}\left(\mathbf{O}\mathbf{v} + \dot{\mathbf{O}}\mathbf{x} + \dot{\mathbf{c}}, \mathbf{OF}, T, \mathbf{G}\right), \end{aligned} \tag{5.111}$$

for all orthogonal $\mathbf{O}(t)$ and for all vectors $\mathbf{c}(t)$.

Clearly, for $\mathbf{O} = \mathbf{1}$ the above relations indicate that constitutive functions cannot be dependent on the velocity $\mathbf{v}$ as we have indicated in Section 5.1.2.

It remains to investigate the relations

$$\begin{aligned} &\mathbf{OP}(\mathbf{F}, T, \mathbf{G}) = \mathbf{P}(\mathbf{OF}, T, \mathbf{G}), \quad \varepsilon(\mathbf{F}, T, \mathbf{G}) = \varepsilon(\mathbf{OF}, T, \mathbf{G}), \\ &\mathbf{Q}(\mathbf{F}, T, \mathbf{G}) = \mathbf{Q}(\mathbf{OF}, T, \mathbf{G}). \end{aligned} \tag{5.112}$$

The polar decomposition yields the relation $\mathbf{F} = \mathbf{RU}$, where $\mathbf{R}$ is orthogonal. Consequently, if we choose $\mathbf{O} = \mathbf{R}^T$, we obtain for the scalar function

$$\varepsilon(\mathbf{F}, T, \mathbf{G}) = \varepsilon(\mathbf{U}, T, \mathbf{G}), \tag{5.113}$$

which shows that constitutive functions cannot contain a dependence on $\mathbf{R}$ and they may be dependent only on $\mathbf{U}$ or, equivalently, on $\mathbf{C} = \mathbf{U}^2$. Hence, bearing relations (5.28), (5.29) in mind, the result of the material objectivity is as follows

$$\begin{aligned} &\varepsilon = \varepsilon(\mathbf{C}, T, \mathbf{G}), \quad \psi = \psi(\mathbf{C}, T, \mathbf{G}), \quad \eta = \eta(\mathbf{C}, T, \mathbf{G}), \\ &\mathbf{P} = 2\rho_0 \mathbf{F} \frac{\partial \psi}{\partial \mathbf{C}}(\mathbf{C}, T, \mathbf{G}), \quad \mathbf{Q} = \mathbf{Q}(\mathbf{C}, T, \mathbf{G}). \end{aligned} \tag{5.114}$$

Hence the dependence on $\mathbf{R}$ appears in the explicit manner in the relation for Piola-Kirchhoff stress tensor and this is the result of the rule of transformation for this tensor which behaves from the left as a vector in the current configuration (Eulerian).

In general, we can formulate the principle of material objectivity as follows

The constitutive function of an objective quantity must be independent of the frame in the space of configurations.

On the other hand, the orthogonal transformation of frames in the reference configuration requires

$$\begin{aligned} \mathbf{P}(\mathbf{C}, T, \mathbf{G})\,\mathbf{O}^T &= \mathbf{P}\left(\mathbf{OCO}^T, T, \mathbf{OG}\right), \\ \varepsilon(\mathbf{C}, T, \mathbf{G}) &= \varepsilon\left(\mathbf{OCO}^T, T, \mathbf{OG}\right), \\ \mathbf{OQ}(\mathbf{C}, T, \mathbf{G}) &= \mathbf{Q}\left(\mathbf{OCO}^T, T, \mathbf{OG}\right), \end{aligned} \tag{5.115}$$

the first transformation characteristic for three vectors $P_{kK}\mathbf{e}_K$, $k = 1, 2, 3$, the second one for the scalar, and the last one again for the vector $Q_K\mathbf{e}_K$. These isotropic functions depend on one vector and one symmetric tensor. The dependence on the scalar T is immaterial for transformation properties. Bearing the results presented in Tables 5.1 and 5.2 in mind we obtain for the internal energy ε

$$\varepsilon = \varepsilon(T, I, II, III, |\mathbf{G}|, IV, V), \tag{5.116}$$

where the scalar invariants are defined as follows

$$I = \operatorname{tr}\mathbf{C}, \quad II = \frac{1}{2}\left(I^2 + \operatorname{tr}\mathbf{C}^2\right), \quad III = \det\mathbf{C}, \qquad (5.117)$$
$$IV = \mathbf{G}\cdot\mathbf{C}\mathbf{G}, \quad V = \mathbf{G}\cdot\mathbf{C}^2\mathbf{G}.$$

This general result simplifies considerably when we account for the thermodynamical admissibility which we investigated before (see (5.28)). Then the dependence on $|\mathbf{G}|, IV, V$ cannot appear. Simultaneously, relations between the Helmholtz free energy, internal energy and the Piola-Kirchhoff stress tensor combined together yield

$$\mathbf{P} = \mathbf{P}\left(T, I, II, III\right). \qquad (5.118)$$

Hence, according to (5.114) we obtain the following relation for the Piola-Kirchhoff stress tensor

$$\mathbf{P} = 2\rho_0\mathbf{F}\left[\frac{\partial\psi}{\partial I}\mathbf{1} + \frac{\partial\psi}{\partial II}\left(I\mathbf{1} - \mathbf{C}\right) + \frac{\partial\psi}{\partial III}III\mathbf{C}^{-1}\right] \qquad (5.119)$$
$$= \sqrt{III}\left[\beth_0\mathbf{1} + \beth_1\mathbf{B} + \beth_{-1}\mathbf{B}^{-1}\right]\mathbf{F}^{-T},$$

where

$$\beth_0 = 2\rho\left(II\frac{\partial\psi}{\partial II} + III\frac{\partial\psi}{\partial III}\right), \qquad (5.120)$$
$$\beth_1 = 2\rho\frac{\partial\psi}{\partial I}, \quad \beth_{-1} = -2\rho III\frac{\partial\psi}{\partial II}, \quad \rho = \rho_0\sqrt{III},$$

are the so-called *elasticities* or *response coefficients* depending on invariants I, II, III and the temperature T, and the constitutive relation (5.119) describes the *compressible Mooney-Rivlin material*. In this derivation we have used the mass balance (4.31), the Caley-Hamilton Theorem and the fact that invariants for the Cauchy-Green deformation tensors $\mathbf{C}$ and $\mathbf{B}$ are identical. Mooney-Rivlin material and its various simplifications are the most frequently used models of the nonlinear elastic behavior of materials such as rubber (comp. [131], [334], [346], [234], [226]).

The relation for the Cauchy stress $\mathbf{T}$ follows in the form

$$\mathbf{T} = \beth_0\mathbf{1} + \beth_1\mathbf{B} + \beth_{-1}\mathbf{B}^{-1}. \qquad (5.121)$$

Obviously, due to isotropy of the material, the spectral representation of this stress tensor is as follows

$$\mathbf{T} = \sum_{\alpha=1}^{3}\sigma^{(\alpha)}\mathbf{k}_B^{(\alpha)}\otimes\mathbf{k}_B^{(\alpha)}, \quad \sigma^{(\alpha)} = \beth_0 + \beth_1\lambda^{(\alpha)} + \beth_{-1}\frac{1}{\lambda^{(\alpha)}}, \qquad (5.122)$$

where $\lambda^{(\alpha)}$, $\mathbf{k}_B^{\alpha}$ are eigenvalues and, respectively, eigenvectors of the left Cauchy-Green deformation tensor $\mathbf{B}$.

In the particular case

$$\beth_1 = const, \quad \beth_{-1} = 0, \tag{5.123}$$

the material is called *neo-Hookean*.

For the heat flux, we can use the result reported in Table 5.2 and obtain

$$\mathbf{Q} = Q_1 \mathbf{G} + Q_2 \mathbf{C} \mathbf{G} + Q_3 \mathbf{C}^2 \mathbf{G}, \tag{5.124}$$

where Q_1, Q_2, Q_3 are functions of $T, I, II, III, |\mathbf{G}|, IV, V$.

5.3 Materials with constraints

In many cases of practical interest modeling of materials relies on a simplifying assumption that fields describing processes are not only fulfilling field equations but also some additional restrictions. This is the case in which one assumes the *incompressibility* (no volume changes under hydrostatic pressure) or *inextensibility* (no extension under tensile load in a particular direction) limiting the class of admissible motions or *degenerate thermal conductivity* properties when the material does not conduct heat in a particular direction. These restrictions may be *bilateral* (*two-sided*), i.e. symmetric with respect to the change of sign of the force or *unilateral* (*one-sided*), e.g. cables carrying only tensile loading (compare: tensegrity structures [351], [241], strings [339]) or masonry structures carrying only compressible loading (compare [338]). Conditions may be additionally time dependent.

For a modern mathematical theory of continua with constraints one can consult the book of Wozniak [382] (in Polish!). Some thermodynamical considerations for constrained thermoelastic materials can be found in works of Casey [59], [60], [21]. A brief review of the subject was recently published by Rajagopal and Saccomandi [258].

In this Section, we consider a simple class of such additional conditions which limit the class of motions of the material. We shall not consider thermal restrictions and we skip the time dependence of these conditions, i.e. constraints are *scleronomic* [334], [12], [14]. Consequently, in constitutive relations (5.13) we skip the arguments $\mathbf{v}, T, \mathbf{G}$ and assume additional conditions in the form [364]

$$\forall \mathbf{F} : \quad \mathcal{J}^{(\alpha)}(\mathbf{F}) = 0, \quad \alpha = 1, ..., a, \tag{5.125}$$

where each function $\mathcal{J}^{(\alpha)}$ is assumed to be continuously differentiable. For each α, this condition means that in the nine-dimensional space of $\mathbf{F}$ the motion of the medium is restricted to a $(9-a)$-dimensional hypersurface. Consequently, the maximum value of the number a is nine. As a matter of fact, it is lower if we assume that conditions (5.125) should satisfy the principle of material objectivity. Then, for arbitrary time dependent orthogonal tensors $\mathbf{O}(t)$

$$\mathcal{J}^{(\alpha)}(\mathbf{OF}) = \mathcal{J}^{(\alpha)}(\mathbf{F}) \quad \Rightarrow \quad \mathcal{J}^{(\alpha)}(\mathbf{C}) = 0. \tag{5.126}$$

This means that the hypersurface of constraints lies in the six-dimensional space of tensors $\mathbf{C}$ and its dimension: $a \leq 6$.

The set of conditions (5.125) means that the set of field equations for the function of motion $\mathbf{f}$ is ill posed. It is overdetermined. Consequently, we may expect the existence of solutions only if we are able to introduce additional fields. These fields define the so-called *reaction forces on constraints* and we proceed to investigate their structure by means of the second law of thermodynamics.

Certain caution is required in this analysis in relation to the temperature dependence. It is known (see: Chapter 6) that for gases volume changes of the material are coupled to changes of temperature and this yields changes in the pressure. The same effect is observed by the creation of thermal stresses in solids due to the thermal expansion. These effects may not be fully admissible if we introduce constraints on motions. The requirement of *stability of thermodynamical equilibrium* imposes restrictions on material parameters which may be involved in constraint conditions. For instance, we show in Section 6.1 that the upper bound of the parameter of thermal expansion is proportional to the compressibility modulus (see (6.48)). Both are nonnegative and the latter goes to zero for incompressible materials. It means that in such materials thermal expansion is not admissible either. This restriction has been ignored in some works on constraint materials (e.g. [11]). For this reason, the constitutive temperature dependence should be restricted simultaneously with mechanical restriction. Little has been done in this respect. Even the most recent works on this subject (e.g. [258]) avoid this problem. We return to some aspects of incompressible materials in Section 9.4. For the further presentation in this Section we introduce *a posteriori* some consequences of thermodynamical stability.

As the condition (5.126) must hold for all $\mathbf{C}$ we can write it as the local condition

$$\forall d\mathbf{F} \text{ compatible with field equations :} \tag{5.127}$$

$$\operatorname{tr}\left(\frac{\partial \mathcal{J}^{(\alpha)}}{\partial \mathbf{F}} d\mathbf{F}^T\right) = 0, \quad d\mathbf{F} = \frac{\partial \mathbf{F}}{\partial t} dt + \operatorname{Grad} \mathbf{F} d\mathbf{X}.$$

Hence the condition (5.126) can be replaced by the following two differential conditions

$$\operatorname{tr}\left[\frac{\partial \mathcal{J}^{(\alpha)}}{\partial \mathbf{F}}\left(\frac{\partial \mathbf{F}}{\partial t}\right)^{T}\right]=0, \quad (\operatorname{Grad}\mathbf{F})^{\overset{13}{T}}\cdot\left(\frac{\partial \mathcal{J}^{(\alpha)}}{\partial \mathbf{F}}\right)^{T}=0, \tag{5.128}$$

$$\frac{\partial \mathcal{J}^{(\alpha)}}{\partial \mathbf{F}}=2\mathbf{F}\frac{\partial \mathcal{J}^{(\alpha)}}{\partial \mathbf{C}},$$

where

$$(\operatorname{Grad}\mathbf{F})^{\overset{13}{T}}=\frac{\partial F^k_K}{\partial X^L}\mathbf{e}^L\otimes\mathbf{e}^K\otimes\mathbf{e}_k, \tag{5.129}$$

$$(\operatorname{Grad}\mathbf{F})^{\overset{13}{T}}\cdot\left(\frac{\partial \mathcal{J}^{(\alpha)}}{\partial \mathbf{F}}\right)^{T}=\frac{\partial F^k_K}{\partial X^L}\frac{\partial \mathcal{J}^{(\alpha)}}{\partial F^k_L}\mathbf{e}^K.$$

Consequently, restricted changes of the deformation gradient, $\mathbf{F}$, must be such that they are tangent to the hypersurface (5.126) as the derivatives $\partial\mathcal{J}^{(\alpha)}/\partial\mathbf{F}$ define nine-dimensional normal hypervector of this surface. Consequently, if we perform an arbitrary variation of $\mathbf{F}$ which does not satisfy this orthogonality condition, i.e. it is not perpendicular to the vectors $\partial\mathcal{J}^{(\alpha)}/\partial\mathbf{F}$, the components of this variation violating this condition must be eliminated by reaction forces preventing such motions. Hence, the reaction forces must be perpendicular to the hypersurface of constraints. It means that they do not work on changes of the deformation gradient which are compatible with constraints. Such constraints are called *ideal.*

Let us investigate the restrictions imposed by the second law of thermodynamics. We consider processes in thermoelastic materials described by the fields $\{\mathbf{v},\mathbf{F},T,\mathbf{Q}\}$ satisfying equations (5.14), (5.15), (5.25) and (5.32) and restricted by constraints (5.128). According to Liu's Theorem arbitrary fields $\{\mathbf{v},\mathbf{F},T,\mathbf{Q}\}$ must fulfil the inequality

$$\rho_0\frac{\partial \eta}{\partial t}+\operatorname{Div}\left(\frac{\mathbf{Q}}{T}\right)-\mathbf{\Lambda}^v\cdot\left(\rho_0\frac{\partial \mathbf{v}}{\partial t}-\operatorname{Div}\mathbf{P}\right)$$

$$-\Lambda^\varepsilon\left(\rho_0\frac{\partial \varepsilon}{\partial t}+\operatorname{Div}\mathbf{Q}-\operatorname{tr}\mathbf{P}\left(\frac{\partial \mathbf{F}}{\partial t}\right)\right)$$

$$-\operatorname{tr}\mathbf{\Lambda}^F\left(\frac{\partial \mathbf{F}}{\partial t}-\operatorname{Grad}\mathbf{v}\right)-\mathbf{\Lambda}^Q\cdot(\mathbf{Q}+K\operatorname{Grad}T)$$

$$-\sum_{\alpha=1}^{a}\left[\boldsymbol{\Lambda}^{\mathcal{J}^{(\alpha)}}\cdot\left((\operatorname{Grad}\mathbf{F})^{\overset{13}{T}}\cdot\left(\frac{\partial\mathcal{J}^{(\alpha)}}{\partial\mathbf{F}}\right)^{T}\right)+\lambda^{\mathcal{J}^{(\alpha)}}\left(\frac{\partial\mathcal{J}^{(\alpha)}}{\partial\mathbf{F}}\cdot\frac{\partial\mathbf{F}}{\partial t}\right)\right]\geq 0. \tag{5.130}$$

Obviously, the multipliers $\boldsymbol{\Lambda}^{v}, \Lambda^{\varepsilon}, \boldsymbol{\Lambda}^{F}, \boldsymbol{\Lambda}^{Q}, \Lambda^{\mathcal{J}}, \lambda^{\mathcal{J}}$ must be functions of the fields $\{\mathbf{v}, \mathbf{F}, T, \mathbf{Q}\}$.

Linearity of this inequality with respect to derivatives $\partial\mathbf{v}/\partial t$, $\operatorname{Grad}\mathbf{v}$ and $\operatorname{Grad}\mathbf{F}$ yields

$$\boldsymbol{\Lambda}^{v}=0,\quad \boldsymbol{\Lambda}^{F}=0,\quad \boldsymbol{\Lambda}^{\mathcal{J}^{(\alpha)}}=0. \tag{5.131}$$

The last condition follows from the fact that $\operatorname{Grad}\mathbf{F}$ possesses 18 (3×6 – due to the symmetry of the second gradient, $\operatorname{Grad}\mathbf{F}$) independent components and the number of components of all multipliers $\boldsymbol{\Lambda}^{\mathcal{J}^{(\alpha)}}$ does not exceed 18 ($3\times a$, $a\leq 6$).

Now the chain rule of differentiation yields the following form of the inequality

$$\rho_0\left(\frac{\partial\eta}{\partial T}-\Lambda^{\varepsilon}\frac{\partial\varepsilon}{\partial T}\right)\frac{\partial T}{\partial t}$$

$$+\operatorname{tr}\left(2\rho_0\mathbf{F}\frac{\partial\eta}{\partial\mathbf{C}}-\Lambda^{\varepsilon}2\rho_0\mathbf{F}\frac{\partial\varepsilon}{\partial\mathbf{C}}+\Lambda^{\varepsilon}\mathbf{P}-\sum_{\alpha=1}^{a}\lambda^{\mathcal{J}^{(\alpha)}}2\mathbf{F}\frac{\partial\mathcal{J}^{(\alpha)}}{\partial\mathbf{C}}\right)\left(\frac{\partial\mathbf{F}}{\partial t}\right)^{T}$$

$$+\left(\frac{1}{T}-\Lambda^{\varepsilon}\right)\operatorname{Div}\mathbf{Q}-\left(\frac{\mathbf{Q}}{T^2}+K\boldsymbol{\Lambda}^{Q}\right)\cdot\operatorname{Grad}T-\boldsymbol{\Lambda}^{Q}\cdot\mathbf{Q}\geq 0. \tag{5.132}$$

Hence, the linearity with respect to $\partial T/\partial t$ $\partial\mathbf{F}/\partial t$, $\operatorname{Div}\mathbf{Q}$, $\operatorname{Grad}T$ yields

$$\Lambda^{\varepsilon}=\frac{1}{T},\quad \boldsymbol{\Lambda}^{Q}=\frac{\mathbf{Q}}{T^2K},$$

$$\eta=-\frac{\partial\psi}{\partial T},\quad \varepsilon=\psi-T\frac{\partial\psi}{\partial T},\quad \psi=\varepsilon-T\eta=\psi(\mathbf{C},T), \tag{5.133}$$

$$\mathbf{P}=2\rho_0\mathbf{F}\frac{\partial\psi}{\partial\mathbf{C}}+\sum_{\alpha=1}^{a}2T\lambda^{\mathcal{J}^{(\alpha)}}\mathbf{F}\frac{\partial\mathcal{J}^{(\alpha)}}{\partial\mathbf{C}}, \tag{5.134}$$

$$K\geq 0.$$

Relation (5.134) shows that the Helmholtz free energy, ψ, is not a thermodynamic potential for stresses if the motion is constrained. The stress tensor consists of two parts: the first one is determined by ψ while the second one is given in terms of the functions $\mathcal{J}^{(\alpha)}$ defining constraints and unspecified multipliers $\lambda^{\mathcal{J}^{(\alpha)}}$. These multipliers are unknown fields extending the set $\{\mathbf{v}, \mathbf{F}, T, \mathbf{Q}\}$ which makes the set of field equations and constraint conditions $\mathcal{J}^{(\alpha)} = 0$ well posed.

If we introduce the notation

$$\mathbf{N} = -\sum_{\alpha=1}^{a} p^{(\alpha)} \mathbf{F} \frac{\partial \mathcal{J}^{(\alpha)}}{\partial \mathbf{C}} \mathbf{F}^T, \quad p^{(\alpha)} = -2\frac{T}{J}\lambda^{\mathcal{J}^{(\alpha)}}, \quad \mathbf{N}^T = \mathbf{N}, \tag{5.135}$$

then the Cauchy stress tensor $\mathbf{T} = J^{-1}\mathbf{P}\mathbf{F}^T$ becomes

$$\mathbf{T} - \mathbf{N} = 2\rho \mathbf{F} \frac{\partial \psi}{\partial \mathbf{C}} \mathbf{F}^T. \tag{5.136}$$

The symmetric tensor, $\mathbf{N}$, is called the *reaction stress* and clearly its form follows neither from the constitutive laws nor from the constraint conditions. Consequently, it must be found from the field equations.

Let us consider again the condition $(5.128)_1$. It implies

$$\operatorname{tr}\left[\mathbf{F}\frac{\partial \mathcal{J}^{(\alpha)}}{\partial \mathbf{C}} \mathbf{F}^T \mathbf{F}^{-T} \left(\frac{\partial \mathbf{F}}{\partial t}\right)^T\right] = 0 \quad \Rightarrow \quad \operatorname{tr} \mathbf{N} \mathbf{L}^T = 0, \tag{5.137}$$

where $\mathbf{L}$ is the velocity gradient defined by (3.2).

This means that reaction stresses $\mathbf{N}$ do not work on the real deformations in the case of ideal constraints. This is, of course, the result of the orthogonality of $\mathbf{N}$ to the hypersurface of constraints $\mathcal{J}^{(\alpha)} = 0$.

We demonstrate the consequences of the above results in a few particular cases of practical importance.

Incompressible materials

We consider the constraint condition of the following form ($\alpha = 1$)

$$\mathcal{J}(\mathbf{F}) = (\det \mathbf{F})^2 - 1 \equiv \det \mathbf{C} - 1 = 0. \tag{5.138}$$

Such a material, in contrast to compressible thermoelastic materials in isochoric processes, cannot change the volume in any real process of deformation: $J = 1$. Therefore, it is called *incompressible*. We have

$$\begin{aligned} \frac{\partial \mathcal{J}}{\partial \mathbf{C}} &= \mathbf{C}^{-1} \det \mathbf{C} \qquad (5.139) \\ \Rightarrow \quad \mathbf{N} &= -p\mathbf{F}\left(\mathbf{C}^{-1}\det \mathbf{C}\right)\mathbf{F}^T = -p\mathbf{F}\left(\mathbf{F}^T\mathbf{F}\right)^{-1}\mathbf{F}^T = -p\mathbf{1}. \end{aligned}$$

The above result of Poincaré shows that only the deviatoric part of Cauchy's stress can be described by the constitutive law

$$\mathbf{T} + p\mathbf{1} = 2\rho\mathbf{F}\frac{\partial\psi}{\partial\mathbf{C}}\mathbf{F}^T, \quad \psi = \psi(\mathbf{C}, T). \tag{5.140}$$

Simultaneously, the dependence on temperature, T, should be such that the thermal expansion coefficient, i.e. changes of volume, J, by changing temperature should be zero. This is the consequence of the stability of thermodynamical equilibrium. We investigate this problem in some details for incompressible Maxwellian fluids in Section 9.4. For solids, this problem is still open. In most cases of practical importance, models are constructed for isothermal processes.

In the case of isotropic materials the Helmholtz free energy, ψ, depends only on invariants I, II because $III = 1$. Consequently,

$$\frac{\partial\psi}{\partial\mathbf{C}} = \frac{\partial\psi}{\partial I}\mathbf{1} + \frac{\partial\psi}{\partial II}(I\mathbf{1} - \mathbf{C})$$

$$\Rightarrow \quad \mathbf{T} = -p\mathbf{1} + \daleth_1\mathbf{B} + \daleth_{-1}\mathbf{B}^{-1}, \quad \daleth_1 = 2\rho\frac{\partial\psi}{\partial I}, \quad \daleth_{-1} = -2\rho\frac{\partial\psi}{\partial II}, \tag{5.141}$$

and the field of pressure was redefined

$$p \to p - 2\rho\frac{\partial\psi}{\partial II}II. \tag{5.142}$$

The equation (5.141) describes the *incompressible Mooney-Rivlin material.*

Inextensible materials

Some composite materials are modelled as a deformable matrix with a reinforcement by inextensible fibers (cords). Let us consider a material with a single family of such cords whose orientation in the reference configuration is given by the field of unit vectors $\mathbf{K}(\mathbf{X})$. Then the constraint condition (5.126) has the form

$$\mathcal{J}(\mathbf{F}) = (\mathbf{FK})\cdot(\mathbf{FK}) - 1 = \mathbf{K}\cdot\mathbf{CK} - 1 = 0. \tag{5.143}$$

Hence

$$\frac{\partial\mathcal{J}}{\partial\mathbf{C}} = \mathbf{K}\otimes\mathbf{K} \quad \Rightarrow \quad \mathbf{N} = -p\mathbf{F}(\mathbf{K}\otimes\mathbf{K})\mathbf{F}^T, \tag{5.144}$$

i.e.

$$\boldsymbol{N} = N\frac{\mathbf{FK}}{|\mathbf{FK}|} \otimes \frac{\mathbf{FK}}{|\mathbf{FK}|}, \quad N = -p\,|\mathbf{FK}|^2 . \tag{5.145}$$

This result of Adkins and Rivlin [2] shows that the traction in the direction **FK** cannot be described by the constitutive relation and this result reflects the reaction force on the inextensibility in this direction. Obviously, in the case of higher number of such families the reaction force will be the sum of reaction forces for all local directions of cords.

Again, the problem of the thermal expansion is not properly incorporated in these models and it is assumed that the models describe isothermal processes.

Rigid bodies

If we assume that the material is inextensible in an arbitrary direction, the whole stress tensor, $\mathbf{T}$, is not described by the constitutive law. It becomes the reaction force on the constraints which do not allow any deformation of the body. Hence, we deal with the *rigid body*. Certainly, for rigid bodies the thermal expansion is also assumed to be zero.

5.4 Constitutive relations for various thermoelastic materials

In this Section we collect definitions of some nonlinear models of elastic incompressible materials. We follow the presentation of R. Batra [27] (see also [128]).

5.4.1 *Neo-Hookean material*

For this incompressible material the Helmholtz free energy function has the form

$$\rho\psi = \frac{\beth_1}{2}\left(I - 3\right), \tag{5.146}$$

and the Cauchy stress tensor follows as

$$\mathbf{T} = -p\mathbf{1} + \beth_1 \mathbf{B}.$$

This corresponds to the relation (5.123) for compressible materials. In the linear model (small deformation) the parameter $\beth_1$ becomes identical with the shear modulus μ.

One can also introduce the so-called *Varga material* which is defined by the analogous relation with the invariant I of $\mathbf{C}$ replaced by the invariant i of the stretch $\mathbf{U}$:

$$\rho\psi = \frac{\beth_1}{2}(i-3), \quad i = \operatorname{tr}\mathbf{U}. \tag{5.147}$$

5.4.2 *Mooney-Rivlin material*

The *incompressible Mooney-Rivlin material* is described by the Helmholtz free energy [214]

$$\rho\psi = \frac{\beth_1}{2}(I-3) + \frac{\beth_{-1}}{2}(II-3), \tag{5.148}$$

and, in the classical case, material parameters may only be the functions of temperature. For the Cauchy stress we obtain (see (5.141))

$$\mathbf{T} = -p\mathbf{1} + \beth_1\mathbf{B} + \beth_{-1}\mathbf{B}^{-1}, \tag{5.149}$$

where the sum $(\beth_1 + \beth_{-1})$ is the shear modulus, μ, for the linear model. As an example, in the room temperature the values of material parameters for rubber are [226] $\beth_1 = 300$ kPa, $\beth_{-1} = -30$ kPa. Both are almost linearly dependent on temperature which follows from the property of rubber model in which the elastic properties are specified by the entropic contribution to the Helmholtz free energy (the so-called *entropic elasticity*, comp. [222]).

Again one can introduce the model in which invariants of $\mathbf{C}$ are replaced by invariants of $\mathbf{U}$ [235]:

$$\rho\psi = \frac{\beth_1}{2}(i-3) + \frac{\beth_{-1}}{2}(ii-3), \quad ii = \frac{1}{2}\left(i^2 - \operatorname{tr}\mathbf{U}^2\right). \tag{5.150}$$

5.4.3 *Biological tissue*

Biological tissues [31] are often modelled by a nonlinear incompressible elastic material whose Helmholtz free energy has the form

$$\rho\psi = \frac{\mu}{2\gamma}\left[e^{\gamma(I-3)} - 1\right], \tag{5.151}$$

where μ is the shear modulus of the linear model and $\gamma \geq 0$ is another constant. The Cauchy stress has the form

$$\mathbf{T} = -p\mathbf{1} + \mu e^{\gamma(I-3)}\mathbf{B}. \tag{5.152}$$

5.4.4 *Ogden's materials*

R. W. Ogden [234] has introduced a class of isotropic nonlinear elastic materials whose Helmholtz free energy function depends directly on eigenvalues rather than invariants of the deformation tensor **C** (equivalently, **B**). Namely

$$\rho\psi = \sum_n \frac{\mu_n}{\alpha_n}\left(\lambda^{(1)\alpha_n} + \lambda^{(2)\alpha_n} + \lambda^{(3)\alpha_n} - 3\right), \tag{5.153}$$

where $\lambda^{(3)} = 1/\left(\lambda^{(1)}\lambda^{(2)}\right)$.

This relation reduces to the neo-Hookean material for $n = 1$ and $\alpha_1 = 1$, and to the Monney-Rivlin material for $n = 2, \alpha_1 = 1, \alpha_2 = -1$. Ogden has shown that the following values of parameters

$$\begin{aligned} &\alpha_1 = 0.65, \quad \alpha_2 = 2.5, \quad \alpha_3 = -1.0, \\ &\mu_1 = 618 \text{ kPa}, \quad \mu_2 = 1.2 \text{ kPa}, \quad \mu_3 = -9.81 \text{ kPa}, \end{aligned} \tag{5.154}$$

give the good fit for experimental data of the vulcanized rubber.

5.4.5 *Gent's material*

Rubber exhibits a strain-hardening effect at large strains. This is not well encompassed by the Mooney-Rivlin model. The following improvement was proposed by Gent [118]

$$\rho\psi = \daleth_1 J_m \ln\left(1 - \frac{(I-3)}{J_m}\right) + \daleth_{-1}\ln\frac{II}{3}, \tag{5.155}$$

where $\daleth_1, \daleth_{-1}, J_m$ are constants. Thus

$$\mathbf{T} = -p\mathbf{1} - \frac{2\daleth_1 J_m}{(J_m - I + 3)}\mathbf{B} - \frac{6\daleth_{-11}}{II}\mathbf{B}^{-1}.$$

Physically, the constant J_m is the maximum stretch ratio for a typical rubber.

5.4.6 *Generalized Blatz-Ko material*

This model, motivated by modelling of a class of foamed polyurethane elastomers, describes compressible nonlinear elastic materials. The free energy is as follows

$$\rho\psi = \frac{\mu f}{2}\left[(I-3) - \frac{2}{q}\left(III^{q/2}-1\right)\right] \tag{5.156}$$
$$+\frac{\mu(1-f)}{2}\left[\left(\frac{II}{III}-3\right) - \frac{2}{q}(III^{-q/2}-1\right],$$

where

$$q = -\frac{2\nu}{1-2\nu} = -\frac{\lambda}{\mu}, \tag{5.157}$$

f is a material parameter between 0 and 1, ν is Poisson's ratio at zero strain, and λ, μ are Lamé constants at zero strain. Indeed, the relation (5.156) describes well foamed polyurethane elastomers and it can also be used for solid polyurethane rubbers for which $f = 1$ gives the good fit with Blatz-Ko experiments. The Cauchy stress for this model has the form

$$\mathbf{T} = \mu\left[\left(-fIII^{(q-1)/2} + (1-f)\,III^{-(q-1)/2}\right)\mathbf{1} + \frac{f}{III^{1/2}}\mathbf{B} - \frac{1-f}{III^{1/2}}\mathbf{B}^{-1}\right]. \tag{5.158}$$

5.4.7 *Signiorini's material*

The following two models are closest to the classical linear elasticity. Both describe compressible nonlinear elastic materials.

The model of Signorini [296] is exactly of the second order in dependence of the Cauchy stress $\mathbf{T}$ on the Almansi-Hamel deformation tensor $\mathbf{e}$. The Helmholtz free energy has the following form

$$\rho_0\psi(\mathbf{e}) = \frac{\left(\mu + \frac{1}{2}c\right)(1-I_e) + \frac{1}{2}\left(\lambda + \mu - \frac{1}{2}c\right)I_e^2 + cII_e}{\sqrt{1-2I_e+4II_e-8III_e}}, \tag{5.159}$$

where

$$I_e = \operatorname{tr}\mathbf{e}, \quad II_e = \frac{1}{2}\left(I_e^2 - \operatorname{tr}\mathbf{e}^2\right), \quad III_e = \det\mathbf{e}. \tag{5.160}$$

Hence, the relation for the stress tensor is as follows

$$\mathbf{T} = \left[\lambda I_e + \frac{1}{2}\left(\lambda+\mu-\frac{1}{2}c\right)I_e^2 + cII_e\right]\mathbf{1} + 2\left[\mu - \left(\lambda+\mu+\frac{1}{2}c\right)I_e\right]\mathbf{e} - 2c\mathbf{e}^2. \tag{5.161}$$

Even though frequently used, the model with $c = 0$ seems to have some flaws (see [326]).

5.4.8 *St. Venant-Kirchoff material*

This model [311], [163] is similar to the classical linear elasticity

$$\rho\psi = \frac{1}{2}(\lambda + 2\mu) I_E^2 - 2\mu II_E, \quad I_E = \operatorname{tr}\mathbf{E}, \quad II_E = \frac{1}{2}\left(I_E^2 - \operatorname{tr}\mathbf{E}^2\right), \tag{5.162}$$

where $\mathbf{E}$ is Green-St.Venant deformation tensor, and λ, μ are Lamé constants at zero strain. The Cauchy stress in terms of $\mathbf{B}$ and its invariants I, II, III is as follows

$$\mathbf{T} = \frac{\mu}{III^{1/2}}\left[-II\mathbf{1} + \left(I - 1 - \frac{q}{2}(I-3)\right)\mathbf{B} + III\mathbf{B}^{-1}\right], \tag{5.163}$$
$$q = -\frac{\lambda}{\mu}.$$

5.5 Proportionality Theorem

We close this Chapter with a Theorem which enables the exploitation of some identities following from the entropy inequality.

Thermodynamic identities relate derivatives of the entropy flux $\mathbf{H}$ and the heat flux $\mathbf{Q}$ or, in Eulerian description, $\mathbf{h}$ and $\mathbf{q}$. It is of primary importance to check if these identities yield the relation $\mathbf{H} = \mathbf{Q}/T$, or $\mathbf{h} = \mathbf{q}/T$, respectively. Then the entropy inequality (5.9) reduces to the Clausius-Duhem inequality (5.74).

As we have seen in Section 5.2, the relation (5.63) between derivatives of $\mathbf{H}$ and $\mathbf{Q}$ indeed yields for isotropic functions the relation (5.68) between fluxes $\mathbf{H}$ and $\mathbf{Q}$. Such a relation was assumed without proof in derivations of Section 5.3.

Hence the question arises if the following statement is true:

For arbitrary isotropic vector functions $\mathbf{H}$ and $\mathbf{Q}$ of an arbitrary number N of vector and some symmetric tensor variables there holds the following implication

$$\frac{\partial H_{(K}}{\partial v^n_{L)}} - \Lambda\frac{\partial Q_{(K}}{\partial v^n_{L)}} = 0 \quad \text{for} \quad 1 \le n \le N \quad \Rightarrow \quad H_K - \Lambda Q_K = 0, \tag{5.164}$$

where Λ is a scalar function.

Such implications are sometimes assumed in the literature (e.g. [150], [152]) in the exploitation of the second law of thermodynamics. I-Shih Liu [190] has shown that the implication may be wrong and simultaneously, he has proved the Theorem in which conditions for the validity of (5.164) were demonstrated. We quote first the counter example to the above statement constructed by I-Shih Liu.

Let $\mathbf{H}$ and $\mathbf{Q}$ be vector functions and Λ be a scalar function of three vectors $(\mathbf{u}, \mathbf{v}, \mathbf{w})$ defined as follows

$$\mathbf{H} = \gamma \mathbf{u} \times \mathbf{v}, \quad \mathbf{Q} = -\frac{1}{\gamma} \mathbf{u} \times \mathbf{v}, \quad \Lambda = \gamma^2, \quad \gamma = (\mathbf{u} \times \mathbf{v}) \cdot \mathbf{w} \neq 0. \quad (5.165)$$

These functions transform under an orthogonal transformation $\mathbf{O}$ according to relations

$$\begin{aligned} \mathbf{Ou} \times \mathbf{Ov} &= \det(\mathbf{O})\, \mathbf{O}(\mathbf{u} \times \mathbf{v}), \\ (\mathbf{Ou} \times \mathbf{Ov}) \cdot \mathbf{Ow} &= \det(\mathbf{O})\, ((\mathbf{u} \times \mathbf{v}) \cdot \mathbf{w}) \\ \Rightarrow \quad \gamma^* = \det(\mathbf{O})\, \gamma, \quad \mathbf{H}^* &= \mathbf{OH}, \quad \mathbf{Q}^* = \mathbf{OQ}, \quad \Lambda^* = \Lambda, \end{aligned} \quad (5.166)$$

because $[\det(\mathbf{O})]^2 = 1$. Consequently functions (5.165) satisfy the conditions for objective functions (5.79).

Simultaneously, they satisfy the identities

$$\frac{\partial H_{(K}}{\partial u_{L)}} - \Lambda \frac{\partial Q_{(K}}{\partial u_{L)}} = 0, \quad \frac{\partial H_{(K}}{\partial v_{L)}} - \Lambda \frac{\partial Q_{(K}}{\partial v_{L)}} = 0, \quad \frac{\partial H_{(K}}{\partial w_{L)}} - \Lambda \frac{\partial Q_{(K}}{\partial w_{L)}} = 0. \quad (5.167)$$

However

$$\mathbf{H} - \Lambda \mathbf{Q} = \gamma \mathbf{u} \times \mathbf{v}, \quad (5.168)$$

which means that the implication (5.164) does not hold. Consequently, the relations for derivatives of $\mathbf{H}$ and $\mathbf{Q}$ are, in general, not sufficient for such a conclusion.

We proceed to present I-Shih Liu's Theorem in which sufficient conditions are formulated. We begin with the quotation of two Lemmas.

Lemma 1: Let $\mathcal{F}$ be a function (scalar-, vector-, or tensor-valued) of the second order tensors such that $\mathcal{F}(\mathbf{O}) = 0$ for any orthogonal tensor $\mathbf{O}$. Then the gradient of this function at the identity tensor is symmetric, i.e. for any skew-symmetric tensor $\mathbf{W}$, $\partial_A \mathcal{F}(\mathbf{1})[\mathbf{W}] = 0$ (notation in components: $\partial \mathcal{F}/\partial A_{KL}(\mathbf{A})\, W_{KL} = \partial_A \mathcal{F}(\mathbf{A})[\mathbf{W}]$)[4].

Proof: For $\mathbf{W} = -\mathbf{W}^T$ and $0 < \varepsilon \ll 1$, the tensor $\mathbf{1} + \varepsilon \mathbf{W}$ is orthogonal to within the second order term $o(2)$ in ε (i.e. for ε: $o(2) = o(\varepsilon)$ such that $\lim_{\varepsilon \to 0} [o(\varepsilon)/\varepsilon] = 0$). Namely: $(\mathbf{1} + \varepsilon \mathbf{W})(\mathbf{1} + \varepsilon \mathbf{W})^T = \mathbf{1} - \varepsilon^2 \mathbf{W}^2 = \mathbf{1} + o(\varepsilon)$. Hence, there exists an orthogonal tensor $\mathbf{O}_W = \mathbf{1} + \varepsilon \mathbf{W} + o(\varepsilon)$.

[4]The definition of the gradient can be introduced, for instance, by the relation

$$\partial_A \mathcal{F}(\mathbf{A})[\mathbf{B}] = \frac{d}{dt} \mathcal{F}(\mathbf{A} + t\mathbf{B})\big|_{t=0}.$$

By the assumption, we have $\mathcal{F}(\mathbf{O}_W) = 0$ and $\mathcal{F}(\mathbf{1}) = 0$. Consequently, the gradient satisfies the condition

$$\begin{aligned}\mathcal{F}(\mathbf{1} + \varepsilon \mathbf{W}) - \mathcal{F}(\mathbf{1}) &= \partial_A \mathcal{F}(\mathbf{1})[\varepsilon \mathbf{W}] + o(\varepsilon) \\ &= \mathcal{F}(\mathbf{O}_W + o(\varepsilon)) - \mathcal{F}(\mathbf{O}_W) \\ &= \partial_A \mathcal{F}(\mathbf{O}_W)[o(\varepsilon)] + o(\varepsilon) = o(\varepsilon),\end{aligned}$$

which implies that the first order term $\varepsilon \partial_A \mathcal{F}(\mathbf{1})[\mathbf{W}]$ must vanish.

Lemma 2: Let $\mathbf{H}(\mathbf{A}, \mathbf{v})$ be an isotropic function. Then it satisfies the following condition

$$(\delta_{KL} H_M - \delta_{KM} H_L) = \left(\frac{\partial H_K}{\partial v_L} v_M - \frac{\partial H_K}{\partial v_M} v_L \right) \tag{5.169}$$

$$+ \left(\frac{\partial H_K}{\partial A_{LN}} A_{MN} + \frac{\partial H_K}{\partial A_{NL}} A_{NM} - \frac{\partial H_K}{\partial A_{MN}} A_{LN} - \frac{\partial H_K}{\partial A_{NM}} A_{NL} \right).$$

Proof: Let us define

$$\mathcal{F}(\mathbf{G}) = \mathbf{H}\left(\mathbf{G}\mathbf{A}\mathbf{G}^T, \mathbf{G}\mathbf{v}\right) - \mathbf{G}\mathbf{H}(\mathbf{A}, \mathbf{v}). \tag{5.170}$$

Then

$$\partial_G \mathcal{F}(\mathbf{1})[\mathbf{W}] = \partial_A \mathbf{H}\left[\mathbf{W}\mathbf{A} + \mathbf{A}\mathbf{W}^T\right] + \partial_v \mathbf{H}[\mathbf{W}\mathbf{v}] - \mathbf{W}\mathbf{h}.$$

Due to isotropy of $\mathbf{H}$ we have for any $\mathbf{G} = \mathbf{O}$, $\mathbf{O}$ – orthogonal: $\mathcal{F}(\mathbf{O}) = 0$. Consequently, we can apply Lemma 1 and obtain

$$\left(\frac{\partial H_K}{\partial A_{LN}} A_{MN} + \frac{\partial H_K}{\partial A_{NL}} A_{NM} - \frac{\partial H_K}{\partial v_L} A_M - \delta_{KL} H_M \right) W_{LM} = 0,$$

for all skew symmetric tensors $\mathbf{W}$. This proves the Lemma.

The relation (5.169) can be immediately generalized to isotropic vector functions of an arbitrary number of vector and tensor variables

$$(\delta_{KL} H_M - \delta_{KM} H_L) = \sum_{n=1}^{N} \left(\frac{\partial H_K}{\partial v_L^n} v_M^n - \frac{\partial H_K}{\partial v_M^n} v_L^n \right) \tag{5.171}$$

$$+ \sum_{b=1}^{M} \left(\frac{\partial H_K}{\partial A_{LN}^b} A_{MN}^b + \frac{\partial H_K}{\partial A_{NL}^b} A_{NM}^b - \frac{\partial H_K}{\partial A_{MN}^b} A_{LN}^b - \frac{\partial H_K}{\partial A_{NM}^b} A_{NL}^b \right).$$

In order to simplify manipulations we introduce the following notation

$$K_K = H_K - \Lambda Q_K, \quad K^v_{KL} = \frac{\partial H_K}{\partial v_L} - \Lambda \frac{\partial Q_K}{\partial v_L}, \tag{5.172}$$
$$K^A_{KLM} = \frac{\partial H_K}{\partial A_{LM}} - \Lambda \frac{\partial Q_K}{\partial A_{LM}}.$$

The Theorem proved by I-Shih Liu in 1996 has the following form:

Liu's Proportionality Theorem [190]: Let **H** and **Q** be isotropic vector functions, and Λ be an isotropic scalar function, of an arbitrary number of vector and tensor variables. Assume that

i/ for N vector variables $\mathbf{v}^n, n = 1, ..., N$,

$$\left(\frac{\partial H_K}{\partial v^n_L} + \frac{\partial H_L}{\partial v^n_K}\right) - \Lambda \left(\frac{\partial Q_K}{\partial v^n_L} + \frac{\partial Q_L}{\partial v^n_K}\right) = 0, \tag{5.173}$$

ii/ for every other vector variable **u**,

$$\frac{\partial H_K}{\partial u_L} - \Lambda \frac{\partial Q_K}{\partial u_L} = 0, \tag{5.174}$$

iii/ for every tensor variable **A**,

$$\frac{\partial H_K}{\partial A_{LM}} - \Lambda \frac{\partial Q_K}{\partial A_{LM}} = 0. \tag{5.175}$$

Then Λ is constant and

$$\mathbf{H} = \Lambda \mathbf{Q} \tag{5.176}$$

holds, for $N = 1$ and $N = 2$ with the assumption that **Q** and $\mathbf{v}^1 \times \mathbf{v}^2$ be functionally independent (i.e. **Q** does not contain linear contributions proportional to $\mathbf{v}^1 \times \mathbf{v}^2$).

Proof: The assumption i/ means that $\mathbf{K}^{v^n}$ is skew symmetric

$$K^{v^n}_{KL} + K^{v^n}_{LK} = 0, \quad n = 1, ..., N, \tag{5.177}$$

and the assumptions ii/ and iii/ yield

$$K^u_{KL} = 0, \quad K^A_{KLM} = 0. \tag{5.178}$$

According to Lemma 2 for vectors **H** and **Q** we have

$$(\delta_{KL}H_M - \delta_{KM}H_L) = \sum_v \left(\frac{\partial H_K}{\partial v_L}v_M - \frac{\partial H_K}{\partial v_M}v_L\right)$$
$$+\sum_A \left(\frac{\partial H_K}{\partial A_{LN}}A_{MN} + \frac{\partial H_K}{\partial A_{NL}}A_{NM} - \frac{\partial H_K}{\partial A_{MN}}A_{LN} - \frac{\partial H_K}{\partial A_{NM}}A_{NL}\right),$$

$$(\delta_{KL}Q_M - \delta_{KM}Q_L) = \sum_v \left(\frac{\partial Q_K}{\partial v_L}v_M - \frac{\partial Q_K}{\partial v_M}v_L\right)$$
$$+\sum_A \left(\frac{\partial Q_K}{\partial A_{LN}}A_{MN} + \frac{\partial Q_K}{\partial A_{NL}}A_{NM} - \frac{\partial Q_K}{\partial A_{MN}}A_{LN} - \frac{\partial Q_K}{\partial A_{NM}}A_{NL}\right),$$

where the summation is taken over all vectors $\mathbf{v}^1, ..., \mathbf{v}^N$ and $\mathbf{u}$ as well as all tensors $\mathbf{A}$. Using the abbreviations (5.172) and identities (5.177) and (5.178) we obtain

$$(\delta_{KL}K_M - \delta_{KM}K_L) = \sum_{n=1}^{N} \left(K^{v^n}_{KL}v^n_M - K^{v^n}_{KM}v^n_L\right). \tag{5.179}$$

Hence, the relations

$$K_1 = \sum_{n=1}^{N} K^{v^n}_{12}v^n_2, \quad K_2 = \sum_{n=1}^{N} K^{v^n}_{23}v^n_3, \quad K_3 = \sum_{n=1}^{N} K^{v^n}_{31}v^n_1, \tag{5.180}$$

as well as the following set of equations

$$\begin{aligned}
&\sum_{n=1}^{N}\left(K^{v^n}_{12}v^n_2 + K^{v^n}_{31}v^n_3\right) = 0, \quad \sum_{n=1}^{N}\left(K^{v^n}_{12}v^n_3 + K^{v^n}_{31}v^n_2\right) = 0,\\
&\sum_{n=1}^{N}\left(K^{v^n}_{23}v^n_3 + K^{v^n}_{12}v^n_1\right) = 0, \quad \sum_{n=1}^{N}\left(K^{v^n}_{23}v^n_1 + K^{v^n}_{12}v^n_3\right) = 0,\\
&\sum_{n=1}^{N}\left(K^{v^n}_{13}v^n_1 + K^{v^n}_{23}v^n_2\right) = 0, \quad \sum_{n=1}^{N}\left(K^{v^n}_{31}v^n_2 + K^{v^n}_{23}v^n_1\right) = 0.
\end{aligned} \tag{5.181}$$

The three equations on the right can be further reduced to

$$\sum_{n=1}^{N} K^{v^n}_{12}v^n_3 = 0, \quad \sum_{n=1}^{N} K^{v^n}_{23}v^n_1 = 0, \quad \sum_{n=1}^{N} K^{v^n}_{31}v^n_2 = 0. \tag{5.182}$$

The proof will now be made for $N = 1$ and $N = 2$ separately.

In the first case, relations (5.182) substituted in (5.180) imply

$$K_1 = K_2 = K_3 = 0 \quad \Rightarrow \quad \mathbf{H} = \Lambda \mathbf{Q}. \tag{5.183}$$

Then conditions of the Theorem imply that Λ is independent of all vectorial and tensorial arguments. Consequently, the Theorem is proved.

For $N = 2$, the set (5.179), (5.180) reduces to

$$\begin{aligned}
K_{12}^{v^1} v_2^1 + K_{31}^{v^1} v_3^1 + K_{12}^{v^2} v_2^2 + K_{31}^{v^2} v_3^2 &= 0, \\
K_{23}^{v^1} v_3^1 + K_{12}^{v^1} v_1^1 + K_{23}^{v^2} v_3^2 + K_{12}^{v^2} v_1^2 &= 0, \\
K_{31}^{v^1} v_1^1 + K_{23}^{v^1} v_2^1 + K_{31}^{v^2} v_1^2 + K_{23}^{v^2} v_2^2 &= 0, \\
K_{12}^{v^1} v_3^1 + K_{12}^{v^2} v_3^2 &= 0, \\
K_{23}^{v^1} v_1^1 + K_{23}^{v^2} v_1^2 &= 0, \\
K_{31}^{v^1} v_2^1 + K_{31}^{v^2} v_2^2 &= 0.
\end{aligned} \tag{5.184}$$

This is the homogeneous set of equations and, consequently, it yields the parametric solution

$$\frac{K_{23}^{v^1}}{v_1^2} = \frac{K_{31}^{v^1}}{v_2^2} = \frac{K_{12}^{v^1}}{v_3^2} = -\frac{K_{23}^{v^2}}{v_1^1} = -\frac{K_{31}^{v^2}}{v_2^1} = -\frac{K_{12}^{v^2}}{v_3^1} = \gamma, \tag{5.185}$$

which imply from (5.180) that

$$\mathbf{K} = \mathbf{H} - \Lambda \mathbf{Q} = \gamma \left(\mathbf{v}^1 \times \mathbf{v}^2 \right), \tag{5.186}$$

where γ is a scalar function of vector and tensor variables.

Substitution in conditions ii/ and iii/ yields

$$\frac{\partial \Lambda}{\partial X} \mathbf{Q} + \frac{\partial \gamma}{\partial X} \left(\mathbf{v}^1 \times \mathbf{v}^2 \right) = 0, \tag{5.187}$$

where X stands for any vector $\mathbf{u}$ and any tensor $\mathbf{A}$. As $\mathbf{Q}$ and $\mathbf{v}^1 \times \mathbf{v}^2$ are functionally independent, we obtain

$$\begin{aligned}
&\frac{\partial \Lambda}{\partial X} = 0, \quad \frac{\partial \gamma}{\partial X} = 0 \quad \Rightarrow \quad \Lambda = \Lambda \left(\mathbf{v}^1, \mathbf{v}^2 \right), \quad \gamma = \gamma \left(\mathbf{v}^1, \mathbf{v}^2 \right) \\
&\Rightarrow \quad \mathbf{K} = \mathbf{K} \left(\mathbf{v}^1, \mathbf{v}^2 \right) = K_{(1)} \mathbf{v}^1 + K_{(2)} \mathbf{v}^2,
\end{aligned} \tag{5.188}$$

where the last relation follows from the isotropy. $K_{(1)}, K_{(2)}$ are isotropic scalar functions of $\mathbf{v}^1, \mathbf{v}^2$. Taking the scalar product of this relation with the vector $\left(\mathbf{v}^1 \times \mathbf{v}^2 \right)$ we see that $\gamma = 0$. Consequently $\mathbf{K} = 0$ and it means $\mathbf{H} - \Lambda \mathbf{Q} = 0$. Consequently, conditions ii/ and iii/ of the Theorem yield that Λ is independent of $\mathbf{v}^1$, $\mathbf{v}^2$ and, hence, of any vector and tensor variables. This completes the proof of the theorem.

Corollary.

The above Theorem remains valid

i/ if for any symmetric tensor variable $\mathbf{A}$, the condition (5.175) is replaced by

$$\left(\frac{\partial H_K}{\partial A_{LM}} + \frac{\partial H_M}{\partial A_{LK}}\right) - \Lambda\left(\frac{\partial Q_K}{\partial A_{LM}} + \frac{\partial Q_M}{\partial A_{LK}}\right) = 0; \qquad (5.189)$$

ii/ if for any skew symmetric tensor variable $\mathbf{W}$, the condition (5.175) is replaced by

$$\left(\frac{\partial H_K}{\partial W_{LM}} - \frac{\partial H_M}{\partial W_{LK}}\right) - \Lambda\left(\frac{\partial Q_K}{\partial W_{LM}} - \frac{\partial Q_M}{\partial W_{LK}}\right) = 0. \qquad (5.190)$$

These results can be generalized to cases in which the right-hand side of differential identities is different from zero.

The proof similar to the proof of the Theorem as well as some further results can be found in the I-Shih Liu paper [190].

We use these results further in the exploitation of the entropy inequality for systems more complex than thermoelastic materials.

Chapter 6

Equilibrium Gibbs thermodynamics

6.1 Thermostatics of gases

Classical works on thermodynamics were done primarily in relation to engineering applications. One of the first who worked on the second law of thermodynamics was Sadi Carnot [Nicolas Léonard Sadi (a name given after a medieval Persian poet and philosopher Sadi of Shiraz) Carnot, 1796-1832; he died during the cholera epidemic in Paris][18]. The description of his work on heat engines (Pierre R. Roberge) indicates what kind of thermodynamics was developed at this time: "Carnot devised an ideal engine in which a gas is allowed to expand to do work, absorbing heat in the process, and is expanded again without transfer of heat but with a temperature drop. The gas is then compressed, heat being given off, and finally it is returned to its original condition by another compression, accompanied by a rise of temperature. This series of operations, known as *Carnot's cycle*, shows that even under ideal conditions a heat engine cannot convert into mechanical energy all the heat energy supplied to it; some of the heat energy must be rejected."

The above description indicates that thermodynamics of these times was concerned with collections of homogeneous systems for which processes were describing transfers between different equilibrium states. Neither temporal nor spatial variables were appearing in these considerations. In this Chapter, we present some basic notions which arise in such a formulation. We limit the attention to the simplest substance modelled by thermodynamics of equilibria (thermostatics) – an ideal gas.

Extensive presentation of equilibrium thermodynamics can be found in numerous books, for instance by M. Planck [252], D. Elwell and A. J. Pointon [100], H. B. Callen [56], A. Sommerfeld [302], G. Emch, Chuang Liu [101] or A. Shavit, C. Gutfinger [293].

Nicolas Carnot
1796 – 1832

Juliuss Robert Mayer
1814 – 1878

Josiah Willard Gibbs
1839 – 1903

Rudolf Clausius
1822 – 1888

Pierre Maurice Marie Duhem
1861 – 1916

Constantin Carathéodory
1873 – 1950

We assume that a one-component system undergoes homogeneous and quasistatic processes which are described by changes of the *volume* V (units: $\left[\mathrm{m}^3\right]$), and of the *energy* E (units: joul $[\mathrm{J}] = \left[\mathrm{kg}\cdot\mathrm{m}^2/\mathrm{s}^2\right]$). The latter is identical with the potential energy because the kinetic energy is negligible in quasistatic processes. Each combination of these two variables (E, V) is called the *state* of the system and *processes* in this model are identified by their initial and final states and, additionally, by some quantities not belonging to the space of states.

The volume V is frequently replaced by some other equivalent variables describing the configuration of the system. One of them is the number of *moles*, n, of the substance contained in the system. One mole is defined as the mass of the system consisting of $A = 6.0237 \cdot 10^{23}$ molecules of the gas. The number A is called the *Avogadro number*. Below we quote masses corresponding to one mole of different substances. We use in these relations the notion of the relative molecular mass $M_r = \mu/\mu_0$, where μ is the molecular mass of a gas, and μ_0 is either the atomic mass of the hydrogen $\mu_H = 1.67329 \cdot 10^{-24}$ g, or, in more recent formulations, 1/12 of the atomic mass of carbon $\mu_0 = 1.66011 \cdot 10^{-24}$ g. In practical applications, the difference between these two definitions is usually ignored.

molecular hydrogen H_2	$M_r = 2$	$1 \text{ mol} = A \cdot M_r \cdot \mu_H = 2.015879394$ g
molecular oxygen O_2	$M_r = 32$	$1 \text{ mol} = A \cdot M_r \cdot \mu_H = 32.25407030$ g
molecular nitrogen N_2	$M_r = 28$	$1 \text{ mol} = A \cdot M_r \cdot \mu_H = 28.22231152$ g
carbon C	$M_r = 12$	$1 \text{ mol} = A \cdot M_r \cdot \mu_H = 12.09527636$ g
argon Ar	$M_r = 40$	$1 \text{ mol} = A \cdot M_r \cdot \mu_H = 40.31758788$ g
chlor Cl	$M_r = 35$	$1 \text{ mol} = A \cdot M_r \cdot \mu_H = 40.31758788$ g
natrium Na	$M_r = 23$	$1 \text{ mol} = A \cdot M_r \cdot \mu_H = 23.18261303$ g

Then the following combinations of variables are frequently used

$m = n\mu = const$ [kg]	– total mass of the system,
$\rho = m/V$ [kg/m^3]	– mass density,
$v = 1/\rho$ [m^3/kg]	– specific volume,
$\varepsilon = E/m$ [J/kg]	– specific energy.

In contrast to densities which we were using before, these are not fields. They are not related to the space variable $\mathbf{x}$. As systems are homogeneous they may be considered to be fields constant in space.

The number of moles is a very useful variable when we consider mixtures of many gases. Then the number of moles of a chosen component of the mixture gives a contribution of this component to the whole mixture. We discuss such mixtures in the second part of the Chapter.

Let us first consider balance equations which we have investigated in previous Chapters in the particular case of homogeneous systems.

As the motion is not considered, the momentum balance equation is trivially satisfied. Only boundary conditions for the body $\mathcal{B}$ (mechanical equilibrium with the external world) must be verified.

We consider first the conservation of energy for processes without exchange of heat. Then the energy balance equation reduces to the form

$$\frac{dE}{dt} = -\oint_{\partial\mathcal{B}} p\mathbf{v} \cdot \mathbf{n} dS + \int_{\mathcal{B}} \rho \mathbf{b} \cdot \mathbf{v} dV, \quad \mathbf{T} = -p\mathbf{1}. \tag{6.1}$$

The time does not appear in classical thermostatics and, consequently, the above relations must be written in the form of increments

$$dE = d'W, \tag{6.2}$$

where

$$d'W = \left(\int_{\mathcal{B}} \rho \mathbf{b} \cdot \mathbf{v} dV - \oint_{\partial \mathcal{B}} p \mathbf{v} \cdot \mathbf{n} dS \right) dt. \tag{6.3}$$

This relation indicates that both approaches to thermodynamics – the field model and the model of homogeneous systems may coincide only if relaxation processes to thermodynamical equilibrium are sufficiently fast (i.e. the characteristic relaxation times are much shorter than characteristic times of macroscopic observations) and, simultaneously, sufficiently slow in order to consider them to be quasistatic.

Bearing homogeneity assumption in mind ($p = const$), according to the mass balance we obtain

$$\oint_{\partial \mathcal{B}} \mathbf{v} \cdot \mathbf{n} dS = \int_{\mathcal{B}} \operatorname{div} \mathbf{v} dV \approx \frac{dV}{dt}, \quad V = \operatorname{vol} \mathcal{B}. \tag{6.4}$$

Hence, when we neglect the influence of body forces,

$$dE + pdV = 0, \quad \text{where} \quad p = p(E, V). \tag{6.5}$$

The contribution $d'W = -pdV$ is called the *mechanical power*. The notation $d'W$ means that, in contrast to dE the mechanical power is usually not a full differential of any function.

The relation for p is the constitutive law, characteristic for the ideal gas. We transform it later to the form which is called the caloric state equation. The above form of the energy conservation law is characteristic for system in *adiabatic isolation*.

It is easy to construct examples of systems in which the volume remains constant but the energy of the system changes: $dV = 0, dE \neq 0$. In such cases, relation (6.2) cannot be satisfied and we have to add the contribution describing the exchange of energy with the external world which undergoes in constant volume. We denote this contribution by $d'Q$ and call it the *heat exchange*. Such systems are said to be in *diathermal isolation*. Comparison with the energy conservation which we were discussing for the field model yields

$$d'Q = \left(\int_{\mathcal{B}} \rho r dV - \oint_{\partial \mathcal{B}} \mathbf{q} \cdot \mathbf{n} dS \right) dt. \tag{6.6}$$

If the system is neither in adiabatic nor in diathermal isolation both changes are possible and the energy conservation law has the form

$$dE = d'W + d'Q. \tag{6.7}$$

In the classical thermostatics this relation is called the *first law of thermodynamics* [204].

Now we consider two systems $\mathcal{A}$ and $\mathcal{B}$ in the adiabatic isolation. Their states are given by (E_A, V_A) and (E_B, V_B), respectively. The corresponding pressures are $p_A = p(E_A, V_A)$ and $p_B = p(E_B, V_B)$. We create the following process. Without violating the adiabatic isolation from the external world, we bring the two systems to a contact through a diathermal wall. There appears an exchange of energy between the systems and, according to the first law of thermodynamics we have

$$dE_A = d'Q_A, \quad dE_B = d'Q_B. \tag{6.8}$$

As we did already in the field theory we assume that the energy is an *additive function.* Then, for the system which consists of $\mathcal{A}$ and $\mathcal{B}$

$$dE = dE_A + dE_B = 0 \quad \Longrightarrow \quad d'Q_A + d'Q_B = 0, \tag{6.9}$$

due to the adiabatic isolation of both systems from the external world.

The assumption on additivity implies, as before that there are no long-range interactions between systems and that the energy does not concentrate on the wall of contact between both systems. Such walls or *interfaces* are called *ideal.*

The process created in the above described manner yields new states of equilibrium $(E'_A, V_A), (E'_B, V_B)$ in which the exchange of energy between both systems ceases. As the full energy of the system does not change only one of the quantities E'_A, E'_B is independent. Consequently, for these two systems one can introduce a function θ_{AB} which determines the states of equilibrium

$$\theta_{AB}(E'_A, V_A) = \theta_{AB}(E'_B, V_B). \tag{6.10}$$

Properties of this function are determined by the so-called *zeroth law of thermodynamics.* It says that the equilibrium states of systems are *transitive*, i.e. for three arbitrary systems $\mathcal{A}$, $\mathcal{B}$, $\mathcal{C}$, the equilibrium between two arbitrary pairs, say, ($\mathcal{A}$,$\mathcal{B}$) and ($\mathcal{A}$,$\mathcal{C}$) implies the equilibrium of the remaining pair ($\mathcal{C}$,$\mathcal{B}$). In terms of the function θ this condition has the form

$$\begin{aligned} &\theta_{AB}(E_A, V_A) = \theta_{AB}(E_B, V_B) \text{ and} \\ &\theta_{AC}(E_A, V_A) = \theta_{AC}(E_C, V_C) \quad \Longrightarrow \quad \theta_{BC}(E_B, V_B) = \theta_{BC}(E_C, V_C). \end{aligned} \tag{6.11}$$

It means that there exists a function $\theta(E, V)$ which determines whether two systems in contact through a diathermal wall are in equilibrium or not. This function is called the *empirical temperature.*

We assume additionally that the empirical temperature is invertible with respect to the first argument, i.e. we can write $E(\theta, V)$ and the state of the system is described by the pair (θ, V).

We proceed to formulate the second law of thermodynamics for homogeneous systems considered in this Chapter. It consists of two parts. First of all, let us consider a state (θ, V) and some vicinity of this state. Some of the states of this vicinity are attainable from (θ, V) by purely mechanical (adiabatic) processes, some other not. For latter, some heat must be exchanged with the external world. This *assumption* has already been made by Sadi Carnot in his work on efficiency of heat engines [58]. As a mathematical statement it has been introduced to thermodynamics by C. Carathéodory in 1909 [57]: *in any neighborhood of an arbitrary state there exist states which are not attainable on an adiabatic path (a process in which $d'Q = 0$).*

In the simple case which we consider in this Chapter it means that, for such states $dE - d'W \neq 0$. Carathéodory proved that his assumption yields the *integrability* of this 1-form, i.e. the existence of functions S and Λ such that

$$dS = \Lambda \left(dE - d'W \right). \tag{6.12}$$

This statement is trivial in the case of two variables θ, V – the 1-form of two variables is always integrable. However the statement is not trivial in cases of systems described by more variables. We consider some of them further in this Chapter.

The state function S is called the *entropy* of the system. We proceed to prove some properties of this function. However, it should be stressed that, in contrast to claims in some books on the classical thermostatics, the above relation is only a part of the second law. We return later to this point.

Let us begin with the investigation of the multiplier Λ. We consider two systems $\mathcal{A}$, $\mathcal{B}$ in thermodynamical equilibrium and possessing states $(\theta, V_A), (\theta, V_B)$. Bearing (6.12) in mind we have

$$dS_A = \Lambda_A \left(dE_A + p_A dV_A \right), \quad dS_B = \Lambda_B \left(dE_B + p_B dV_B \right), \tag{6.13}$$

for an infinitesimal change of these states. Simultaneously, the state of the system which consists of $\mathcal{A}$ and $\mathcal{B}$ in the diathermal contact is described by the parameters (θ, V_A, V_B) and its changes of energy E and entropy S are as follows

$$\begin{aligned} dE = dE_A + dE_B &= -\left(p_A dV_A + p_B dV_B \right) \\ \Rightarrow \quad dS &= \Lambda \left(dE_A + dE_B + p_A dV_A + p_B dV_B \right), \end{aligned} \tag{6.14}$$

where the additivity of the energy was used.

Let us transform the variables

$$(\theta, V_A) \to (\theta, S_A), \quad (\theta, V_B) \to (\theta, S_B). \tag{6.15}$$

Bearing (6.13) in mind we obtain

$$dS = \frac{\partial S}{\partial \theta} d\theta + \frac{\partial S}{\partial S_A} dS_A + \frac{\partial S}{\partial S_B} dS_B = \Lambda \left(\frac{dS_A}{\Lambda_A} + \frac{dS_B}{\Lambda_B} \right).$$

Hence

$$\frac{\partial S}{\partial \theta} = 0, \quad \frac{\partial S}{\partial S_A} = \frac{\Lambda}{\Lambda_A}, \quad \frac{\partial S}{\partial S_B} = \frac{\Lambda}{\Lambda_B}. \tag{6.16}$$

These relations imply the following *integrability conditions*

$$\begin{aligned} \frac{\partial^2 S}{\partial \theta \partial S_A} &= \frac{\partial^2 S}{\partial S_A \partial \theta} \quad \Rightarrow \quad \frac{\partial}{\partial \theta} \left(\frac{\Lambda}{\Lambda_A} \right) = 0, \\ \frac{\partial^2 S}{\partial \theta \partial S_B} &= \frac{\partial^2 S}{\partial S_B \partial \theta} \quad \Rightarrow \quad \frac{\partial}{\partial \theta} \left(\frac{\Lambda}{\Lambda_B} \right) = 0. \end{aligned} \tag{6.17}$$

They can be written in the form

$$\frac{1}{\Lambda} \frac{\partial \Lambda}{\partial \theta} = \frac{1}{\Lambda_A} \frac{\partial \Lambda_A}{\partial \theta} = \frac{1}{\Lambda_B} \frac{\partial \Lambda_B}{\partial \theta} = \mu(\theta), \tag{6.18}$$

where $\mu(\theta)$ is an arbitrary function of the empirical temperature θ. Integration of these relations yields the following *multiplicative* representation of multipliers

$$\begin{aligned} \Lambda_A(\theta, S_A) &= \nu(S_A) \exp\left(\int \mu(\theta) \, d\theta \right), \\ \Lambda_B(\theta, S_B) &= \nu(S_B) \exp\left(\int \mu(\theta) \, d\theta \right), \\ \Lambda(\theta, S_A, S_B) &= \nu(S_A, S_B) \exp\left(\int \mu(\theta) \, d\theta \right). \end{aligned} \tag{6.19}$$

Let us introduce the notation

$$T(\theta) = \left[C \exp\left(\int \mu(\theta) \, d\theta \right) \right]^{-1}, \tag{6.20}$$

where C is an arbitrary positive constant. The quantity T is called the *absolute temperature*. It is clear from the above construction that it is a *universal* function, i.e. it is the same for all systems described by the above model.

Substitution of results (6.19), (6.20) in (6.13) yields

$$dS'_A = \frac{1}{T(\theta)}\left(dE_A + p_A dV_A\right), \quad S'_A = \frac{1}{C}\int \frac{dS_A}{\nu_A} + \text{const.} \tag{6.21}$$

This new function S' is also called the entropy. The above relation is quite general and we use it in the form

$$dS = \frac{1}{T}\left(dE - d'W\right). \tag{6.22}$$

It is called the *Gibbs equation.*

Gibbs equation describes only reversible changes of the entropy. We have seen that the second law of thermodynamics in the field approach defines as well the so-called *dissipation function.* In order to introduce a similar notion within the frame of thermostatics, we have to make a rather artificial extension of the notion of changes of entropy. Namely, we have to assume that the increment of entropy $d'S$ consists of two contributions: a reversible part which appears in the Gibbs equation and is related to the heat exchange $dS_{rev} = d'Q/T$ and from the irreversible part $d'S_{irr}$ which is nonnegative[1]. For changes of the total entropy we have then $d'S = dS_{rev} + d'S_{irr} = d'S_{irr} + \frac{d'Q}{T}$ and the following inequality is assumed to hold

$$d'S - \frac{d'Q}{T} \equiv d'S - \frac{1}{T}\left(dE + pdV\right) \geq 0, \tag{6.23}$$

where we have used the first law of thermodynamics. Gibbs equation and the above inequality form together the second law of thermodynamics in the form presented in this Chapter.

The above inequality can be written as follows

$$d\left(E - TS\right) \leq -SdT - pdV, \tag{6.24}$$

provided we assume the existence of the entropy function beyond reversible processes. This inequality forms the basis for analysis of *stability of equilibrium states.* It says that the *Helmholtz free energy* $\Psi = E - TS$ possesses a *minimum* in the state of equilibrium reached on the isothermal ($dT = 0$) and isochoric ($dV = 0$) ways. Further we show such an analysis for systems in adiabatic isolation.

Now we present a few simple conclusions from Gibbs equation. Let us assume that states are described by variables: absolute temperature T and

[1] In the general case, one cannot even assume that $d'S_{irr}$, and consequently $d'S$, are full differentials of some functions. In this sense, the nonequilibrium entropy S_{irr} and the total entropy S may not exist within this approach.

volume V. Then we have

$$dS = \frac{\partial S}{\partial T}dT + \frac{\partial S}{\partial V}dV = \frac{1}{T}\left(\frac{\partial E}{\partial T}dT + \left(\frac{\partial E}{\partial V} + p\right)dV\right). \qquad (6.25)$$

Since dT and dV are arbitrary it follows immediately

$$\frac{\partial E}{\partial T} = T\frac{\partial S}{\partial T}, \quad p = -\left(\frac{\partial E}{\partial V} - T\frac{\partial S}{\partial V}\right). \qquad (6.26)$$

These important relations couple *constitutive relations* for E, S and p. They can be written in a simpler form by means of the *Helmholtz free energy*, Ψ,

$$\Psi = E - TS = \Psi(T, V) \qquad (6.27)$$

$$\Rightarrow \quad S = -\frac{\partial \Psi}{\partial T}, \quad E = \Psi - T\frac{\partial \Psi}{\partial T}, \quad p = -\frac{\partial \Psi}{\partial V}.$$

Hence, it follows from the Gibbs equation that we have to know only one constitutive function for the Helmholtz free energy, the remaining relations for internal energy, entropy and pressure following by differentiation. We say that the Helmholtz free energy is a *thermodynamical potential* for this choice of state variables.

Let us transform the state variables in the following way

$$(T, V) \to (T, p). \qquad (6.28)$$

Then the Gibbs equation can be written in the form

$$dG = -SdT + Vdp, \quad G = E - TS + pV = G(T, p). \qquad (6.29)$$

Similarly as before we obtain by differentiation

$$S = -\frac{\partial G}{\partial T}, \quad V = \frac{\partial G}{\partial p}, \quad E = G - T\frac{\partial G}{\partial T} - p\frac{\partial G}{\partial p}. \qquad (6.30)$$

Therefore for this choice of state variables the function G is the thermodynamical potential. It is called the *Gibbs free energy*.

Finally for the transformation of variables

$$(T, V) \to (S.p), \qquad (6.31)$$

we have

$$dH = TdS + Vdp, \quad H = E + pV = H(S, p), \qquad (6.32)$$

and

$$T = \frac{\partial H}{\partial S}, \quad V = \frac{\partial H}{\partial p}, \quad E = H - p\frac{\partial H}{\partial p}. \qquad (6.33)$$

The potential H is called the *enthalpy*.

state variables	**thermodynamical potential**
(E, V)	– entropy S
(T, V)	– Helmholtz free energy $\Psi = E - TS$
(T, p)	– Gibbs free energy (free enthalpy) $G = E - TS + pV$
(S, p)	– enthalpy $H = E + pV$

The above presented transformations of variables illustrate the so-called *Legendre transformation*, in which the transformation of variables yields a corresponding transformation of the potential. A particular choice of variables depends only on a particular application of the model especially related to possibilities of control of variables in experiments. For instance, in a simple mechanical experiment of extension of the rod we can control either the force applied to the rod (*soft loading device*), or its extension (*hard loading device*). In thermodynamics it is sometimes easier to control the temperature and sometimes (e.g. for shock waves) the entropy.

Apart from the above direct consequences of the Gibbs equation constitutive relations lead to the so-called integrability conditions. We show here only one example.

Relation (6.25) contains on the left-hand side the full differential. This yields the symmetry of the second mixed derivative with respect to T, V. Hence, we have

$$\frac{\partial}{\partial V}\left(\frac{1}{T}\frac{\partial E}{\partial T}\right) = \frac{\partial}{\partial T}\left[\frac{1}{T}\left(\frac{\partial E}{\partial V} + p\right)\right], \tag{6.34}$$

or, after simplifications,

$$\frac{\partial E}{\partial V} = T\frac{\partial p}{\partial T} - p. \tag{6.35}$$

This is an example of the *Maxwell relation*. The above relation is called *Clausius-Clapeyron equation*. Such relations appear in thermodynamics of more complex systems and form the basis of chemical thermodynamics. They play, for example, a very important role in experimental verification of thermodynamical potentials. Relation (6.35) means that the derivative of internal energy does not have to be estimated from difficult calorimetric experiments but it follows from the so-called *thermal state equation*

$$p = p(T, V). \tag{6.36}$$

However, some calorimetric experiments are necessary for the estimation of internal energy. Namely, from the first law of thermodynamics we have

$$d'Q = mc_v dT + \left(\frac{\partial E}{\partial V} + p\right) dV, \quad c_v = \frac{1}{m}\frac{\partial E}{\partial T}, \tag{6.37}$$

where c_v is *specific heat under constant volume*. Bearing (6.35) in mind, we obtain

$$m\frac{\partial c_v}{\partial V} = T\frac{\partial^2 p}{\partial T^2} \quad \Rightarrow \quad mc_v = mc_v^0 + T\frac{\partial^2}{\partial T^2}\int_{V_0}^{V} pdV. \tag{6.38}$$

This relation shows that the specific heat follows from the thermal state equation (6.36) up to a function of temperature $c_v^0(T)$. This function must be found in a single calorimetric experiment for which $V = V_0$.

Relations (6.35) and (6.38) determine both derivatives of the energy and their integration yields up to a constant the following *caloric state equation*

$$E = E(T, V). \tag{6.39}$$

In many cases it is more convenient to control pressure rather than volume. From the first law of thermodynamics it follows for the state variables (T, p)

$$d'Q = d(E + pV) - V dp = \frac{\partial (E + pV)}{\partial T} dT + \left[\frac{\partial (E + pV)}{\partial T} - V\right] dp. \tag{6.40}$$

It should be stressed that $E + pV$ is here not the enthalpy H which is the potential for variables (S, p). The quantity

$$c_p = \frac{1}{m}\frac{\partial (E + pV)}{\partial T}, \quad E = E(T, p), \quad V = V(T, p),$$

is called the *specific heat under constant pressure*.

The above transformation of variables leads also to the following relation

$$c_p - c_v = \frac{1}{m}\left(\frac{\partial E}{\partial V} + p\right)\frac{\partial V}{\partial T}, \tag{6.41}$$

i.e.

$$\gamma = \frac{c_p}{c_v} = 1 + \left[\left.\left(\frac{\partial E}{\partial V}\right)\right|_{T=\text{const.}} + p\right]\frac{\partial V}{\partial T}\left.\left(\frac{\partial E}{\partial T}\right)\right|_{V=\text{const.}}^{-1}, \tag{6.42}$$

where the coefficient γ is the so-called *adiabatic exponent* frequently appearing in gas dynamics.

Apart from the above discussed laws (principles) of thermodynamics it is assumed that the state of thermodynamical equilibrium is *stable*. Let us investigate this assumption. The condition of stability for systems in adiabatic isolation (i.e. for $d'Q = 0, dS = 0$) requires that entropy should have a maximum in this state. In some practical applications this condition may not be satisfied. It concerns, in particular, the so-called metastable equilibrium states which appear in the description of phase transformations. They are sometimes called *frozen equilibria* because systems may appear in these states only for a finite time duration. We shall not discuss such problems in this book.

Let us consider two identical thermodynamical systems each of them having the energy E and the volume V. We connect these systems and assume that the joined system is in adiabatic isolation. Let us assume that the initial equilibrium state has been disturbed in such a way that the initial state of the first system becomes $(E + \delta E, V + \delta V)$ while the initial state of the second system becomes $(E - \delta E, V - \delta V)$. The system develops to a new equilibrium state in which the entropy reaches its maximum. As the entropy is an additive function we have

$$S(2E, 2V) > S(E + \delta E, V + \delta V) + S(E - \delta E, V - \delta V). \tag{6.43}$$

Let us expand this relation into the Taylor series. We obtain ($\partial S/\partial E = 0$, $\partial S/\partial V = 0$)

$$\frac{\partial^2 S}{\partial E^2}(\delta E)^2 + 2\frac{\partial^2 S}{\partial E \partial V}\delta E \delta V + \frac{\partial^2 S}{\partial V^2}(\delta V)^2 < 0. \tag{6.44}$$

It means that Hessian (the matrix of second derivatives) has to be negative definite. For chosen variables Gibbs equation indicates as well

$$\frac{\partial S}{\partial E} = \frac{1}{T}, \quad \frac{\partial S}{\partial V} = \frac{p}{T}. \tag{6.45}$$

Changing the variables $(E, V) \rightarrow (T, V)$ we obtain immediately

$$\frac{mc_v}{T^2}(\delta T)^2 - \frac{1}{T}\frac{\partial p}{\partial V}(\delta V)^2 > 0. \tag{6.46}$$

This inequality indicates the following *stability conditions* of the thermodynamical equilibrium

$$c_v > 0, \quad \kappa_T > 0, \quad \kappa_T = -V\frac{\partial p}{\partial V}, \tag{6.47}$$

where the coefficient κ_T is called *isothermal compressibility modulus*.

If we introduce pressure as the variable in (6.46) we obtain the following inequalities

$$c_p \geq c_v, \quad \kappa_T > 0, \quad \alpha^2 \leq \frac{1}{VT}c_p\kappa_T, \quad \alpha = \frac{1}{V}\frac{\partial V}{\partial T}\bigg|_{p=\text{const.}}, \tag{6.48}$$

where α is the *thermal expansion coefficient*. Hence, for the vanishing compressibility $\kappa_T \to 0$ there is no thermal expansion: $\alpha = 0$. This result has been anticipated in Section 5.3, where we have indicated that incompressible materials cannot undergo the thermal expansion.

We complete this review of classical thermostatics of single component systems with the presentation of the simplest example of the thermodynamical model of *ideal gases*. Real gases of small densities behave in high temperatures almost like ideal gases, i.e. substances whose thermal state equation $p = p(\rho, T)$ has the form

$$p = \rho \frac{R}{M_r} T \quad \text{where} \quad R = 8.3153 \cdot 10^3 \, \frac{\mathrm{J}}{\mathrm{kg} \cdot \mathrm{K}}, \quad M_r = \frac{\mu}{\mu_0}. \tag{6.49}$$

Depending on applications the thermal state equation is written in different form. Most frequently used relations are collected below.

$$\begin{aligned}
pV &= m \frac{R}{M_r} T, \qquad\qquad (6.50)\\
p\nu &= \frac{R}{M_r} T \quad \text{where} \quad \nu = \frac{1}{\rho} = \frac{V}{m}, \quad \text{or applying} \quad m = N\mu, \quad M_r = \frac{\mu}{\mu_0}\\
pV &= NkT \quad \text{where} \quad k = R\mu_0 = 1.38044 \cdot 10^{-23} \tfrac{\mathrm{J}}{\mathrm{K}} \text{ – Boltzmann constant,}\\
p &= \frac{kT}{v\mu} = \frac{\rho k T}{\mu}.
\end{aligned}$$

R is the *universal gas constant*. Thermal state equation implies for normal conditions: $p = 1$ atm., $T = 273.15$ K (0° C), $N = A$ (Avogadro number)

$$V_{mol} = \frac{NkT}{p} = 22.4207 \text{ liter}.$$

The air is the mixture of gases

78.08% nitrogen N_2, 20.95% oxygen O_2, 0.94% argon Ar, 0.03% carbon dioxide CO_2.

Hence the molecular mass of the air is as follows

$$\begin{aligned}
\mu_{air} &= 0.7808\mu_{N_2} + 0.2095\mu_{O_2} + 0.009\mu_{Ar} + 0.0003\mu_{CO_2} \quad \Longrightarrow\\
&\Rightarrow \quad M_r^{air} = \frac{\mu_L}{\mu_0} = 28.96.
\end{aligned}$$

In contrast to the thermal state equation, specifying the constitutive relation for pressure, *caloric state equation* specifies the *constitutive relation* for the specific *internal energy* $\varepsilon(\rho, T)$. For ideal gases

$$\varepsilon = z \frac{R}{M_r} T + \alpha, \tag{6.51}$$

where z is a constant:

$$z = \begin{cases} 3/2 \text{ for monatomic gases,} \\ 5/2 \text{ for two-atomic gases,} \\ 3 \text{ for polyatomic gases.} \end{cases} \tag{6.52}$$

The constant α is different for different gases and it has a bearing in description of chemical reactions. It is important to notice that the internal energy of ideal gases is independent of pressure.

6.2 Thermostatic theory of homogeneous mixtures

We proceed to present some elements of the thermostatic mixture theory. We consider a thermodynamical system which is the homogeneous mixture of A distinguishable components. In thermostatics we are not interested in a relative motion of these components (diffusion processes). This will be the subject of further Chapters. Thermostatics of homogeneous multicomponent systems (mixtures) was constructed by Gibbs [120], [119].

The following notions are characteristic for a mixture of A components

m_a	– mass of the component a, $a = 1, ..., A$
$\rho_a = m_a/V$	– partial mass density of the component a
$v_a = V/m_a$	– partial specific volume (V - volume under pressure p and temperature T)
$c_a = m_a/m$	– concentration of the component a ($m = \sum_{b=1}^{A} m_a$ - total mass)
N_a	– particle number of the component a
$\nu_a = N_a/A$	– mol number (A – Avogadro number: $6.0221367 \times 10^{23}\,\text{mol}^{-1}$)
$n_a = N_a/V$	– particle density
$X_a = \nu_a/\nu$	– mol fraction ($\nu = \sum_{a=1}^{A} \nu_a$ – total number of moles)
V_a/V	– volume fraction (V_a – volume of the pure substance a under p and T)
p_a/p	– pressure fracture

where p_a is the so-called partial pressure, and the total (bulk) pressure p, the total mass density ρ, the total (bulk) specific internal energy ε and the total (bulk) specific entropy η are given by the relations

$$p = \sum_{a=1}^{A} p_a, \quad \rho = \sum_{a=1}^{A} \rho_a, \quad \rho\varepsilon = \sum_{a=1}^{A} \rho_a \varepsilon_a, \quad \rho\eta = \sum_{a=1}^{A} \rho_a \eta_a. \tag{6.53}$$

As the system is homogeneous the total energy, entropy, etc. follow by the multiplication with the total mass m, e.g. $E = m\varepsilon, S = m\eta$, etc.

The most important thermodynamical potential for an arbitrary component a of the mixture is the so-called *chemical potential* μ_a. Its importance follows from the fact that, in contrast to the partial pressure, p_a, it is continuous across semipermeable membranes. *Semipermeable membranes* are walls (interfaces) between systems which are permeable only for some components of these systems. We proceed to present some details.

Let us consider a system at a given constant temperature T schematically shown in Fig. 11. On both sides of the membrane B permeable for the component a (i.e. the membrane is material with respect to all components except for a which may flow through the membrane) there is a mixture in which the schematically indicated pistons sustain constant total pressures p^I and p^{II}. According to the stability condition (6.24) we have for this system

$$d(E - TS) \leq -SdT - p^I dV^I - p^{II} dV^{II} \tag{6.54}$$

$$\Longrightarrow \quad d\left(E - TS + p^I V^I + p^{II} V^{II}\right) \leq -SdT + V^I dp^I + V^{II} dp^{II} = 0.$$

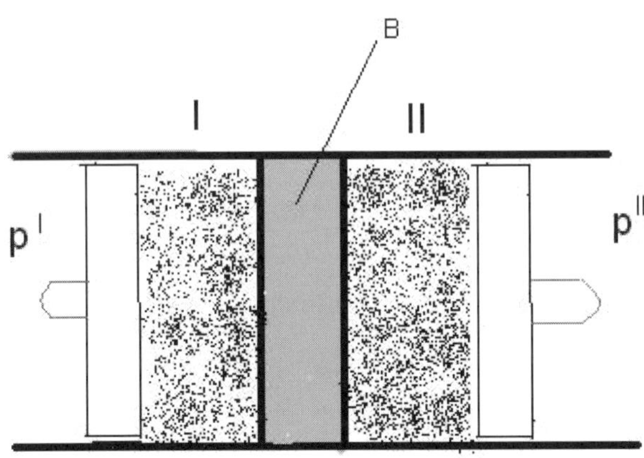

Fig. 11: *Semipermeable membrane* B *between two mixtures containing the component* a.

Hence the quantity $G = \left(E^I - TS^I + p^I V^I\right) + \left(E^{II} - TS^{II} + p^{II} V^{II}\right)$ reaches minimum in the equilibrium state (in an arbitrary process approaching equilibrium it must decay). This quantity is called the *free enthalpy* of the whole system and it is the sum of free enthalpies of both subsystems. Due to the fact that mixtures on both sides of the membrane may be different, these free enthalpies may be functions of different variables. Of course, they must be functions of temperature and total pressure but they may depend on mass contributions of components as well. For the total free enthalpy G, we have then

$$G = G^I\left(T, p^I, m_1^I, ..., m_a^I, ..., m_{A^I}^I\right) + G^{II}\left(T, p^{II}, m_1^{II}, ..., m_a^{II}, ..., m_{A^{II}}^{II}\right). \tag{6.55}$$

Due to the fact that the membrane B is permeable only for the component a, all variables except of m_a^I, m_a^{II} are in this relation constant and, additionally, the sum $m_a = m_a^I + m_a^{II}$ must be constant as well. Consequently, the minimum condition has the form

$$\frac{\partial G}{\partial m_a^I} = 0 \quad \Rightarrow \quad \frac{\partial G^I}{\partial m_a^I} = \frac{\partial G^{II}}{\partial m_a^{II}}. \tag{6.56}$$

It means that the derivative of the free enthalpy with respect to the mass contribution of the component a is continuous across the semipermeable membrane. For this reason, we define the *chemical potential* in the following way

$$\mu_a = \frac{\partial G}{\partial m_a}. \tag{6.57}$$

Then the equilibrium condition has the form

$$\mu_a^I\left(T, p^I, m_1^I, ..., m_{A^I}^I\right) = \mu_a^{II}\left(T, p^{II}, m_1^{II}, ..., m_{A^{II}}^{II}\right). \tag{6.58}$$

Definition (6.57) yields immediately the following integrability condition

$$\frac{\partial \mu_a}{\partial m_b} = \frac{\partial \mu_b}{\partial m_a}. \tag{6.59}$$

Additionally, additivity of the free enthalpy G leads to the relation

$$G = \sum_{a=1}^{A} \mu_a m_a. \tag{6.60}$$

Namely, let us consider a z-tuple enlargement of the system. We have then

$$G\left(T, p, zm_1, ..., zm_A\right) = zG\left(T, p, m_1, ..., m_A\right).$$

Differentiating with respect to z we obtain

$$\sum_{a=1}^{A} \frac{\partial G\left(T,p,zm_1,...,zm_A\right)}{\partial\left(zm_a\right)} m_a = \sum_{a=1}^{A} \frac{\partial G\left(T,p,m_1,...,m_A\right)}{\partial m_a} m_a$$
$$= G\left(T,p,m_1,...,m_A\right).$$

Bearing (6.57) in mind we obtain (6.60).

This means as well that additivity of G yields the invariance of the chemical potential with respect to the enlargement of the system, i.e.

$$\mu_a\left(T,p,zm_1,...,zm_A\right) = \mu_a\left(T,p,m_1,...,m_A\right).$$

This is possible only if the chemical potential μ_a does not depend directly on the mass of components but on their fractions such as concentrations c_b, or mol fractions X_b, $b = 1,...,A-1$ $(X_A = -\sum_{a=1}^{A-1} X_a!)$,

$$\mu_a = \mu_a\left(T,p,X_1,...,X_{A-1}\right). \tag{6.61}$$

This property of the chemical potential shows that for a one-component system the chemical potential and the density of the free enthalpy $g = G/m$ are identical.

Differentiation of the relation (6.60) with respect to m_b yields

$$\frac{\partial G}{\partial m_b} = \mu_b + \sum_{a=1}^{A} \frac{\partial \mu_a}{\partial m_b} m_a \quad \Rightarrow \quad \sum_{a=1}^{A} \frac{\partial \mu_a}{\partial m_b} m_a = 0,$$

and, accounting for the integrability condition (6.59),

$$\sum_{a=1}^{A} \frac{\partial \mu_b}{\partial m_a} m_a = 0. \tag{6.62}$$

This relation is called the *Gibbs-Duhem relation*.

It remains to formulate the Gibbs equation for mixtures. According to the definition of the chemical potential (6.57) we obtain

$$dG = -SdT + Vdp + \sum_{a=1}^{A} \mu_a dm_a. \tag{6.63}$$

The following *Maxwell relations* (integrability conditions) result from this Gibbs equation

$$\frac{\partial S}{\partial m_a} = -\frac{\partial \mu_a}{\partial T} \quad \text{and} \quad \frac{\partial V}{\partial m_a} = \frac{\partial \mu_a}{\partial p}. \tag{6.64}$$

This completes the review of thermostatics.

Chapter 7

Kinetic theories

7.1 Liouville equation

Apart from the experience with macroscopic systems which yields the phenomenological formulation of macroscopic continuum thermodynamical models, it is advisable to justify such models by their derivation from more fundamental microscopic models. Such derivations are based on the so-called *kinetic theories*. These, in turn, could be successfully constructed for classical systems of interacting particles. Macroscopic models of some fluids follow from the *kinetic theory of ideal monatomic gases*. Attempts to proceed in a similar way in cases of nonideal gases in which particles interact with each other by complex forces, mixtures of charged particles or granular and molecular substances (e.g. solutions of high polymers) have not been fully successful and only partial results are available. In this Chapter, we present some fundamental features of transition from mechanics of interacting particles to a kinetic theory of ideal rarefied gases and, subsequently, to the macroscopic description. The last transition motivates a macroscopic thermodynamical strategy of constructing models which is called the *extended thermodynamics*.

It should be kept in mind that these constructions motivate macroscopic models but usually do not replace them. The reason is not only their mathematical complexity but also the fact that their physical superiority over purely phenomenological macroscopic models is illusory. At any level of description we have to make strong and artificial assumptions (e.g. on structure of molecules, on their distribution in space – the course-graining in space, the molecular chaos in derivation of collision operators, etc.). It means that micro-macro transitions give a better physical insight and, in this sense, models on different levels are complementary rather than competing.

In the diagram below we demonstrate a scheme of the above indicated transitions. For the purpose of this book we do not need to go into all details of models named in this diagram. We show solely the structure of Liouville's equation, the construction of the BBGKY-hierarchy and the position of Boltzmann's equation in this hierarchy.

Let us consider a system of N identical (point-) particles whose state is described by a point in $6N$-dimensional (Euclidean) Γ-space

$$\left\{\boldsymbol{\xi}_1, \dot{\boldsymbol{\xi}}_1, ..., \boldsymbol{\xi}_N, \dot{\boldsymbol{\xi}}_N\right\} \in \Gamma, \tag{7.1}$$

where $\boldsymbol{\xi}_i, \quad i = 1, ..., N$ is the position of i-particle, and $\dot{\boldsymbol{\xi}}_i$ its velocity. Then the notion of momentum in the case of identical particles is defined by the relation

$$\mathbf{p}_i = m\dot{\boldsymbol{\xi}}_i, \quad i = 1, ..., N, \tag{7.2}$$

where m is the mass of particles.

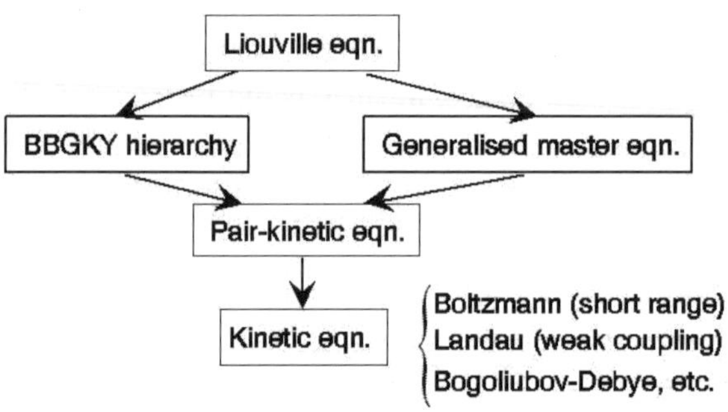

Fig. 12: *Transition from systems of material points to kinetic models*

The motion of the system is described by the set of Newton's equations

$$\dot{\boldsymbol{\xi}}_i = \frac{1}{m}\mathbf{p}_i, \quad \dot{\mathbf{p}}_i = \mathbf{F}_i, \quad i = 1, ..., N. \tag{7.3}$$

In these equations the dot denotes the time derivative and $\mathbf{F}_i$ the total force acting on the i-particle. In general, this force splits into two parts, external force $\mathbf{F}_i^{ext}$ and the force of interaction of the i-particle with the remaining $N-1$ particles $\mathbf{F}_i^{int}$. In theories of charged particles, plasmas, electrolytes, etc., the external force is of primary importance. It reflects not only the action of external fields but, in many so-called one-component models, it describes the interaction of, for instance, electrons with the ionic background which is assumed to be immobile. The interaction force $\mathbf{F}_i^{int}$ in such systems is assumed to be Lorentz force between charged particles.

In mechanical systems which we consider in this book the interaction force of neutral particles is assumed to be given by the potential of two-particle (binary) interactions. This means that we assume the relations

$$\mathbf{F}_i^{int} = - \sum_{j=1,\, j\neq i}^{N} \frac{\partial \Phi\left(\left|\boldsymbol{\xi}_i - \boldsymbol{\xi}_j\right|\right)}{\partial \boldsymbol{\xi}_i}. \tag{7.4}$$

In order to construct a solution of the set of ordinary differential equations (7.3) - (7.4) we need initial conditions which are assumed to have the form

$$\boldsymbol{\xi}_i\left(t=0\right) = \boldsymbol{\xi}_i^0, \quad \dot{\boldsymbol{\xi}}_i\left(t=0\right) = \dot{\boldsymbol{\xi}}_i^0. \tag{7.5}$$

It is convenient to write the set of ordinary differential equations (7.3) in the form of a single partial differential equation for a function, D, defined in the whole space $\Gamma \times (-\infty, \infty)$. To this aim, let us introduce the coordinates $(\mathbf{x}_i, \varsigma_i)\,, \quad i = 1, ..., N$ in Γ-space. We define

$$D\left(\mathbf{x}_1, \varsigma_1, ..., \mathbf{x}_N, \varsigma_n, t\right) = \prod_{i=1}^{N} \delta\left(\mathbf{x}_i - \boldsymbol{\xi}_i\left(t\right)\right) \delta\left(\varsigma_i - \dot{\boldsymbol{\xi}}_i\left(t\right)\right). \tag{7.6}$$

Then

$$D\left(\mathbf{x}_1, \varsigma_1, ..., \mathbf{x}_N, \varsigma_n, t=0\right) = \prod_{i=1}^{N} \delta\left(\mathbf{x}_i - \boldsymbol{\xi}_i^0\right) \delta\left(\varsigma_i - \dot{\boldsymbol{\xi}}_i^0\right). \tag{7.7}$$

By differentiation with respect to time and the use of (7.3) it is easy to show that D satisfies the following equation

$$\frac{\partial D}{\partial t} + \sum_{i=1}^{N} \left(\varsigma_i \cdot \frac{\partial D}{\partial \mathbf{x}_i} + \frac{1}{m}\mathbf{F}_i \cdot \frac{\partial D}{\partial \varsigma_i}\right) = 0, \tag{7.8}$$

with the initial condition (7.7). Namely, we can construct the solution of the above equation by the method of characteristics. These are given by the following set of equations

$$\frac{dt}{ds} = 1, \quad \frac{d\mathbf{x}_i}{ds} = \varsigma_i, \quad \frac{d\varsigma_i}{ds} = \frac{1}{m}\mathbf{F}_i, \quad \frac{dD}{ds} = 0. \tag{7.9}$$

This means, of course, that D is constant along characteristics.

The initial value problem of this characteristic problem has the form

$$\begin{aligned} &t\left(s=0\right) = 0, \quad \mathbf{x}_i\left(s=0\right) = \mathbf{x}_i^0, \quad \varsigma_i\left(s=0\right) = \varsigma_i^0, \\ &D\left(s=0\right) = \prod_{i=1}^{N} \delta\left(\mathbf{x}_i\left(s=0\right) - \boldsymbol{\xi}_i^0\right) \delta\left(\varsigma_i\left(s=0\right) - \dot{\boldsymbol{\xi}}_i^0\right), \end{aligned} \tag{7.10}$$

where $(\mathbf{x}_i^0, \varsigma_i^0)$ is an arbitrary point of Γ and δ is the Dirac delta.

It is immediately seen that $D = 0$ for the characteristic which goes through the point $(0, \mathbf{x}_i^0, \varsigma_i^0) \neq \left(0, \boldsymbol{\xi}_i^0, \dot{\boldsymbol{\xi}}_i^0\right)$. Otherwise, the function D is given by (7.6) with $\boldsymbol{\xi}_i$ and $\dot{\boldsymbol{\xi}}_i$ satisfying equations (7.3).

It is appropriate to make at this place two important observations.

First of all, it is common in literature to claim that the set of Newton's equations of motion (7.3) with initial conditions (7.5) is equivalent to equation (7.8) with the initial condition (7.7). This statement is obviously not true in general. It is known from the theory of partial differential equations that the Cauchy problem (7.7-7.8) possesses classical solutions only locally. For instance, if we allow for collisions of particles in a sense used for the Boltzmann equation (see further in this Chapter) then, at the instant of collision, velocities of both particles suffer a jump. This corresponds to the discontinuity of characteristics (7.9). Hence, in the case of such a model of encounters the solution of the above Cauchy problem does not exist globally in the classical sense. On the other hand, the solution (7.3-7.5) exists globally even if it is only piece-wise smooth. Therefore, the two problems may be equivalent only locally in time. This non-uniqueness and even non-existence problem follows, certainly, from wild motions of particles, induced by interactions within the system. Simultaneously, one expects that the solutions of (7.8) could describe also some asymptotic properties (e.g. for $t \to \infty$) of the system of N particles. This is not the case if the equivalence can be established only in the local sense and no smooth solution can be expected for equation (7.8).

The above problem is solved in kinetic theories by different *coarse-graining procedures* on the Γ-space. This means that, instead of considering trajectories of a single system of particles one follows a family of trajectories within a certain small volume of Γ-space. We return later to this point.

Second observation concerns the symmetry properties of the function D. This point, vaguely accounted for by the founders of the BBGKY-hierarchy (see: [385], [41], [44], [45], [46], [164], [165]) has been discussed by T. Koga ([169]) and Struminskii ([317], [316], [318], [319]). It is obvious that, in spite of identity of particles (point structureless particles of the identical mass m), the motion of each particle is different from the motion of any other particle. It means that in our deterministic problem the particles can be identified by their individual trajectories. Hence

$$D(\mathbf{x}_1, \varsigma_1, ..., \mathbf{x}_i, \varsigma_i, ..., \mathbf{x}_j, \varsigma_j, ..., \mathbf{x}_N, \varsigma_N, t)$$

$$\neq D(\mathbf{x}_1, \varsigma_1, ..., \mathbf{x}_j, \varsigma_j, ..., \mathbf{x}_i, \varsigma_i, ..., \mathbf{x}_N, \varsigma_N, t).$$

This lack of symmetry complicates immensely the structure of all possible approximate solution procedures.

At the same time, for large N, we do not expect the initial conditions (7.5) to be given exactly. This means that for real physical systems we have to consider a family of initial conditions

$$\left(\boldsymbol{\xi}_1^{0\alpha}, \dot{\boldsymbol{\xi}}_1^{0\alpha}, ..., \boldsymbol{\xi}_N^{0\alpha}, \dot{\boldsymbol{\xi}}_N^{0\alpha}\right) \in \Gamma, \quad \alpha \in A, \tag{7.11}$$

where A is the set of indices numerating different initial value problems.

Now, we are in the position to assume the statistical meaning of equation (7.8). Namely, we assume that the appearance of one of the initial conditions (7.11) is given with a certain, so-called *a priori probability*. Hence, we assume the existence of a function f_0^N on the *ensemble* (7.11) such that

$$\underset{\alpha\in A}{\forall} 0 \leq f_0^N\left(\boldsymbol{\xi}_1^{0\alpha}, \dot{\boldsymbol{\xi}}_1^{0\alpha}, ..., \boldsymbol{\xi}_N^{0\alpha}, \dot{\boldsymbol{\xi}}_N^{0\alpha}\right) \leq 1,$$

$$\sum_{\alpha\in A} f_0^N\left(\boldsymbol{\xi}_1^{0\alpha}, \dot{\boldsymbol{\xi}}_1^{0\alpha}, ..., \boldsymbol{\xi}_N^{0\alpha}, \dot{\boldsymbol{\xi}}_N^{0\alpha}\right) = 1. \tag{7.12}$$

If we also assume that the probability of appearance of any trajectory of the system does not change in time then

$$\frac{df^N}{ds^\alpha} = 0, \tag{7.13}$$

where s^α is the parametrization of the trajectory, corresponding to the solution of (7.3) with initial conditions

$$\boldsymbol{\xi}_i\,(t=0) = \boldsymbol{\xi}_i^{0\alpha}, \quad \dot{\boldsymbol{\xi}}_i\,(t=0) = \dot{\boldsymbol{\xi}}_i^{0\alpha}. \tag{7.14}$$

The problem of the ensemble distribution (7.11) can now be transformed into the problem of Γ-space distribution by assuming that the family (7.11) covers the whole Γ-space and f^N is also defined on Γ with condition $(7.12)_2$ replaced by a proper volume integration. Bearing this assumption in mind, we return to the partial differential equation from the characteristic form (7.13). We obtain then the following *Liouville equation* for f^N

$$\frac{\partial f^N}{\partial t} + \sum_{i=1}^{N}\left(\frac{\partial f^N}{\partial \mathbf{x}_i}\cdot\varsigma_i + \frac{1}{m}\mathbf{F}_i\cdot\frac{\partial f^N}{\partial \varsigma_i}\right) = 0, \tag{7.15}$$

$$\int_\Gamma f^N d\mathbf{x}_1 d\varsigma_1 ... d\mathbf{x}_N d\varsigma_N = 1, \tag{7.16}$$

$$f^N\left(\mathbf{x}_1, \varsigma_1, ..., \mathbf{x}_N, \varsigma_N, t=0\right) = f_0^N\left(\mathbf{x}_1, \varsigma_1, ..., \mathbf{x}_N, \varsigma_N\right).$$

In addition, due to the statistical meaning of f^N, we can now make an additional assumption that f_0^N and, consequently, also f^N is symmetric with respect to all Γ-variables.

It should be stressed that the above formulation of the Liouville equation is not equivalent any more, even locally, to the dynamic problem of motion of N particles. However, in spite of its statistical character, it is still reversible, i.e. invariant with respect to the transformation $t \to -t$ (see the discussion of this problem in the book of Wu [383]).

7.2 BBGKY-hierarchy

Let us assume that N interacting particles are enclosed in a box of volume V. For the potential of interactions (7.4) we introduce the following dimensionless notation

$$\Phi\left(|\mathbf{x}_i - \mathbf{x}_j|\right) = \Phi_0 U_{ij}, \quad \Phi_0 = \text{const.}, \tag{7.17}$$

where Φ_0 is called a *strength of potential* Φ. Simultaneously, we define the collection of integrated distribution functions (*s-point distributions*)

$$f^s = \frac{1}{V^{N-s}} \int f^N d\mathbf{x}_{s+1} d\varsigma_{s+1} ... d\mathbf{x}_N d\varsigma_N$$

$$= f^S\left(\mathbf{x}_1, \varsigma_1, ..., \mathbf{x}_s, \varsigma_s, t\right), \quad s = 1, ..., N. \tag{7.18}$$

This definition implies, of course, that f^s has the dimension

$$\left(m^{-3N}\right)\left(m/s\right)^{-3s},$$

and, simultaneously,

$$f^s = \frac{1}{V} \int f^{s+1} d\mathbf{x}_{s+1} d\varsigma_{s+1}, \quad \int f^s d\mathbf{x}_1 d\varsigma_1 ... d\mathbf{x}_s d\varsigma_s = \frac{1}{V^{N-s}}. \tag{7.19}$$

Additionally we introduce the following notation: r_0 – the *range* of potential Φ, and v_0 – the *thermal velocity* (i.e. $\theta_0 = r_0/v_0$ – *time of duration* of a single "collision"). Then the integration of equation (7.15) yields the following BBGKY-*hierarchy*

$$\frac{\partial f^s}{\partial t} + \sum_{i=1}^{s} \varsigma_i \cdot \frac{\partial f^s}{\partial \mathbf{x}_i} - \left(\frac{\Phi_0}{m v_0^2}\right) \sum_{1 \le i < j \le s} \left(\frac{\partial U_{ij}}{\partial \mathbf{x}_i} \cdot \frac{\partial f^s}{\partial \varsigma_i} + \frac{\partial U_{ij}}{\partial \mathbf{x}_j} \cdot \frac{\partial f^s}{\partial \varsigma_j}\right)$$

$$= \frac{N-s}{N}\left(nr_0^3\right)\left(\frac{\Phi_0}{mv_0^2}\right)\sum_{i=1}^{s}\int\frac{\partial U_{is+1}}{\partial \mathbf{x}_i}\cdot\frac{\partial f^{s+1}}{\partial \varsigma_i}d\mathbf{x}_{s+1}d\varsigma_{s+1}, \quad n := \frac{N}{V}, \tag{7.20}$$

where, in order to expose the coefficients, functions f^s have been renormalized: $f^s \to f^s \cdot v_0^{3s}$ and the variables $t, \mathbf{x}_i, \varsigma_i$ are dimensionless: $t \to t/\theta_0, \mathbf{x}_i \to \mathbf{x}_i/r_0, \varsigma_i \to \varsigma_i/v_0$. According to the definition (7.18), new distribution functions $f^1, ..., f^N$ are not dimensionless – f^s has the dimension r_0^{-3N} for arbitrary s. This has, of course, no influence on the order of magnitude of terms in the set (7.20).

The dimensionless form of the BBGKY-hierarchy (7.20) has been derived by G. Sandri ([282], [283]). This form exposes the following parameters

Φ_0/mv_0^2 – *strength of interactions,*
nr_0^3 – *number of particles in the range of interaction,*

which are used to classify different regimes of the gas theory. They have also been used in papers, preceding the work of Sandri without, however, a clear explanation of their origin and their rule in time coarse-graining procedures.

Kinetic theories which have been constructed on the basis of hierarchy (7.20) can be classified into the following categories

1° weakly coupled gas

$$nr_0^3 \sim 1, \quad \varepsilon := \frac{\Phi_0}{mv_0^2} \ll 1, \tag{7.21}$$

2° dilute gas with short-range forces

$$\varepsilon := nr_0^3 \ll 1, \quad \frac{\Phi_0}{mv_0^2} \sim 1, \tag{7.22}$$

3° gas with Coulomb forces in the Debye regime (one-component plasma)

$$nr_0^3 = \frac{1}{\varepsilon}, \quad \varepsilon := \frac{\Phi_0}{mv_0^2} \ll 1, \tag{7.23}$$

4° dilute weakly-coupled gas (DWC)

$$nr_0^3 \sim \varepsilon := \frac{\Phi_0}{mv_0^2} \ll 1. \tag{7.24}$$

These four regimes are indicated in the following diagram.

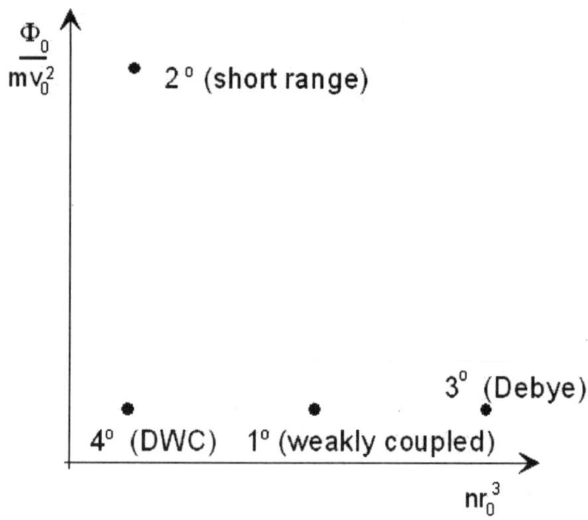

Fig. 13: *Regimes of parameters of various kinetic models.*

Obviously, all regimes lie close to the axes. This indicates the existence of *small parameters* which are used to design proper expansions. These expansions correspond to the *coarse-graining in time* which leads, in turn, to the notion of a so-called *kinetic regime*, particularly important for the construction of macroscopic models.

In contrast to the hierarchy (7.20), coarse-grained equations are *irreversible. Irreversibility of hierarchy equations requires also a number of particles* N to be sufficiently large. It is known that the period t_P of Poincaré cycle grows with growing N. Hence, we can expect the irreversibility to appear due to time coarse-graining if the characteristic time of coarse-graining procedure is small when compared with t_P. If we assume N to be large enough and consider only a few equations of hierarchy, then $s \ll N$ and we can put $\frac{N-s}{N} \approx 1$ in (7.20). It is worth mentioning that a rather common assumption on the existence of thermodynamical limit ($N \to \infty, V \to \infty, N/V \to n < \infty$) which simultaneously takes care of the length of Poincaré cycle ($t_P \to \infty$) and of the interaction with walls

($V \to \infty$) cannot be made on the level of Liouville's equation – we must first construct the BBGKY-hierarchy.

One of the possible ways of a further reduction of the hierarchy was proposed by Bogoliubov, who uses successively simpler distribution functions – from $f^N(\mathbf{x}_1, \varsigma_1, ..., \mathbf{x}_N, \varsigma_N, t)$ to $f^1(\mathbf{x}_1, \varsigma_1, t)$. This is possible due to the existence of widely different characteristic times in the system. These are

$\theta_0 = r_0/v_0$ – time of "collision",
$t_f = l_f/v_0$ – mean free time (l_f – mean free path),
$t_0 = L/c_s$ – macroscopic time (L – macroscopic length, c_s – sound speed)
t_P – Poincaré period.

Since $r_0 \sim 10^{-8}$cm, $l_f \sim 10^{-4}$cm, $L \sim 1$cm, $N \sim \infty$, we have

$$\theta_0 \ll t_f \ll t_0 \ll t_P \sim \infty. \tag{7.25}$$

Let the system, at an instant of time $t = 0$, be in an arbitrary state. In a time of order θ_0, the system rapidly changes its state on account of the intermolecular interactions Φ. Namely, the first equation of hierarchy (7.20)

$$\frac{\partial f^1}{\partial t} + \varsigma_1 \cdot \frac{\partial f^1}{\partial \mathbf{x}_1} = \left(n r_0^3\right) \left(\frac{\Phi_0}{m v_0^2}\right) \int \frac{\partial U_{12}}{\partial \mathbf{x}_1} \cdot \frac{\partial f^2}{\partial \varsigma_1} d\mathbf{x}_2 d\varsigma_2 \tag{7.26}$$

contains Φ only in the "collision" term on the right-hand side while, for $s > 1$, the potential Φ appears also on the left-hand side. Due to the presence of the coefficient $\left(n r_0^3\right)$, the influence of rapid collision changes can be made small in the case f^1. Thus, in time intervals of the order θ_0, f^1 changes very little while all the f^s for $s \geq 2$ change a great deal. The time scale set by these fast changes is called the *collision time scale*. Within this Bogoliubov-like argumentation, one studies only the averaged-out behavior of $f^s, s \geq 2$, in a time scale in which f^1 changes. That is, we change the time scale from the fine-grained to the coarse-grained. This may be expressed mathematically by the basic **Bogoliubov Ansatz:**

After a time t such that $\theta_0 \lesssim t$, a time which is sometimes called the *synchronization time*, all $f^s, s \geq 2$, are regarded as being time-dependent only through their functional dependence on f^1

$$f^s(\mathbf{x}_1, \varsigma_1, ..., \mathbf{x}_s, \varsigma_s, t) = \mathcal{F}\left(\mathbf{x}_1, \varsigma_1, ..., \mathbf{x}_s, \varsigma_s \middle| f^1(t)\right). \tag{7.27}$$

The equation for f^1

$$\frac{\partial f^1}{\partial t} = \mathcal{A}\left(\mathbf{x}_1, \varsigma_1 \middle| f^1\right), \tag{7.28}$$

is called the *kinetic equation, while the time scale* t_f on which the gas approaches some sort of local thermodynamical equilibrium, is called the *kinetic time scale.*

For times $t > t_f$, when the collisions have brought about some degree of local equilibrium, we can introduce macroscopic, *hydrodynamic* fields, varying slower than $f^1(\mathbf{x}_1, \varsigma_1, t)$.

The systematic derivation of various approximations of the BBGKY-hierarchy is based on the expansions with respect to small parameters. For instance, in the case of weakly coupled gases after appropriate change of variables the hierarchy can be written in the form

$$\frac{\partial f^s}{\partial t} + \mathcal{K}^s f^s = \varepsilon \mathcal{I}^s f^s + \varepsilon \mathcal{L}^s f^{s+1}, \quad s = 1, 2, ..., \varepsilon = \frac{\Phi_0}{m v_0^2}, n r_0^3 = 1, \quad (7.29)$$

where

$$\mathcal{K}^s = \sum_{i=1}^{s} \varsigma_i \cdot \frac{\partial}{\partial \mathbf{x}_i}, \quad \mathcal{I}^s = \sum_{1 \le i < j \le s} \left(\frac{\partial U_{ij}}{\partial \mathbf{x}_i} \cdot \frac{\partial}{\partial \mathbf{x}_i} + \frac{\partial U_{ij}}{\partial \mathbf{x}_j} \cdot \frac{\partial}{\partial \mathbf{x}_j} \right), (7.30)$$

$$\mathcal{L}^s = \sum_{i=1}^{s} \int \frac{\partial U_{is+1}}{\partial \mathbf{x}_i} \cdot \frac{\partial}{\partial \varsigma_i} d\mathbf{x}_{s+1} d\varsigma_{s+1}.$$

The direct application of the perturbation method

$$\begin{aligned} f^1 &= \overset{(0)}{f^1} + \varepsilon \overset{(1)}{f^1} + \varepsilon^2 \overset{(2)}{f^2} + O\left(\varepsilon^3\right), \\ f^2 &= \overset{(0)}{f^2} + \varepsilon \overset{(1)}{f^2} + O\left(\varepsilon^2\right), \\ f^3 &= \overset{(0)}{f^3} + O\left(\varepsilon^3\right), \end{aligned} \quad (7.31)$$

where the expansions have been written to the lowest order needed to obtain – in most cases – the kinetic equation, yields the so-called *secular behavior*, i.e. the higher order terms in expansions (7.31) become dominant for large enough times. This non-uniformity of series of regular perturbation method led in classical celestial mechanics to the Poincaré-Lighthill corrections. G. Sandri ([282],[283]) has developed a method based on the mathematical singular perturbation procedure (with rescaling of the time variable) which eliminates the above flaw. One obtains the kinetic equation for the function f^1 which coincides with this following from Boboliubov's method of synchronization and, additionally, certain conditions which prevent the secular behavior. Rather than the higher equations of the hierarchy they determine distribution functions for $s \geq 2$. As this subject goes far beyond the main topic of this book we shall not present these results any further.

7.3 Boltzmann equation

As the method of synchronization is commonly applied in the derivation of the kinetic equation for ideal gases we present it here instead of a rather tedious method of singular perturbation.

The most elementary simplification of this sort appears in the case of description by a single function $f^1(t, \boldsymbol{\xi}, \boldsymbol{\varsigma})$ which, when not normalized, is called the *phase density*. The equation for this function which is the first equation in the BBGKY-hierarchy still contains the function $f^2(t, \boldsymbol{\xi}_1, \boldsymbol{\varsigma}_1, \boldsymbol{\xi}_2, \boldsymbol{\varsigma}_2)$. However, under the additional assumption that this function factorizes $f^2(t, \boldsymbol{\xi}_1, \boldsymbol{\varsigma}_1, \boldsymbol{\xi}_2, \boldsymbol{\varsigma}_2) = f^1(t, \boldsymbol{\xi}_1, \boldsymbol{\varsigma}_1) f^1(t, \boldsymbol{\xi}_2, \boldsymbol{\varsigma}_2)$ (i.e. the probabilities for various points become independent of each other) we obtain the *closure* of the equation. This is the so-called *molecular chaos* assumption and it has been introduced by L. Boltzmann in 1868 [42].

Joseph Liouville
1809 – 1882

Ludwig Boltzmann
1844 – 1906

James Clerk Maxwell
1831 – 1879

If, in addition, interactions of points are included only through their direct elastic encounters (elastic collisions in which mass, momentum and energy are conserved), one obtains the following Boltzmann equation

$$\frac{\partial f}{\partial t} + \boldsymbol{\varsigma} \cdot \frac{\partial f}{\partial \boldsymbol{\xi}} + \left(\mathbf{b}^0 + \mathbf{i}\right) \cdot \frac{\partial f}{\partial \boldsymbol{\varsigma}} = \mathcal{C}[f, f], \tag{7.32}$$

where $\mathbf{b}^0$ is the external body force independent of $\boldsymbol{\xi}$ and $\boldsymbol{\varsigma}$ and

$$\mathbf{i} = 2\boldsymbol{\Omega}(\boldsymbol{\varsigma} - \dot{\mathbf{c}}) - \boldsymbol{\Omega}^2(\boldsymbol{\xi} - \mathbf{c}) + \dot{\boldsymbol{\Omega}}(\boldsymbol{\xi} - \mathbf{c}) + \ddot{\mathbf{c}}, \tag{7.33}$$

is the inertial acceleration following from the relative motion of the frame with respect to the inertial frame (comp. relation (4.14)). The noninertial reference frame was chosen because it demonstrates in the explicit manner the influence of Euclidean transformation on the distribution function. This has the paramount importance for the material objectivity of macroscopic quantities derived from the kinetic theory. We discuss this problem further in this Chapter. In equation (7.32), we have skipped for simplicity the index by the phase density f. This function specifies the number density of atoms. Namely, $f(t, \boldsymbol{\xi}, \boldsymbol{\varsigma})\, d\boldsymbol{\varsigma}$ is the number of particles per unit volume at the point $\boldsymbol{\xi}$ and at time t such that their velocities lie between $\boldsymbol{\varsigma}$ and $\boldsymbol{\varsigma}+d\boldsymbol{\varsigma}$, i.e. velocities whose endpoints lie in the sphere with the center $\boldsymbol{\varsigma}$ and the radius $|d\boldsymbol{\varsigma}|$. $\mathcal{C}[f, f]$ is the so-called *collision operator*. Its structure follows from the assumption on interactions of particles and from the factorization of two-point distribution functions. Boltzmann's assumptions can be satisfied only for *rarefied gases*. This yields a structure of the collision operator which contains the so-called Knutsen number

$$\mathrm{Kn} = \frac{\text{mean time of free flight}}{\text{observation time scale}} = \frac{\text{mean free path}}{\text{observation length scale}}. \tag{7.34}$$

This number forms the foundation for approximate solutions of Boltzmann's equation.

The Boltzmann equation follows from the BBGKY-hierarchy cut at the first equation containing only two-point distribution functions. Consequently, only binary collisions are accounted for in this model. Two colliding particles have the velocities $\boldsymbol{\varsigma}, \boldsymbol{\varsigma}^1$ before the collision and $\boldsymbol{\varsigma}', \boldsymbol{\varsigma}^{1\prime}$ after the collision. If they have the equal radius r then in the explicit form the collision operator is as follows

$$\mathcal{C}[f, f] = \int \left(f' f^{1\prime} - f f'\right) r g \sin\theta d\theta d\epsilon d\boldsymbol{\varsigma}^1, \tag{7.35}$$

where θ, ϵ are angles characterizing the encounter, $rg\sin\theta d\theta d\epsilon$ is the volume element in the three-dimensional space of positions $\boldsymbol{\xi}_1$ (see Fig. 14). We need further only some invariance properties of this operator.

The transition from kinetic equations to the macroscopic level of description is the subject of a very extensive research. The classical presentations of this subject can be found, for instance, in the books of S. Chapman and T. G. Cowling [69], P. Résibois, M. de Leener [261], C. A. Truesdell, R. G. Muncaster [333].

On the basis of the kinetic theory described by the Boltzmann equation we construct a series of macroscopic models which are supposed to yield *approximate solutions* for the phase density f. To this aim, we construct macroscopic fields and macroscopic field equations and this construction is based on expansions using the Knudsen number as a small parameter.

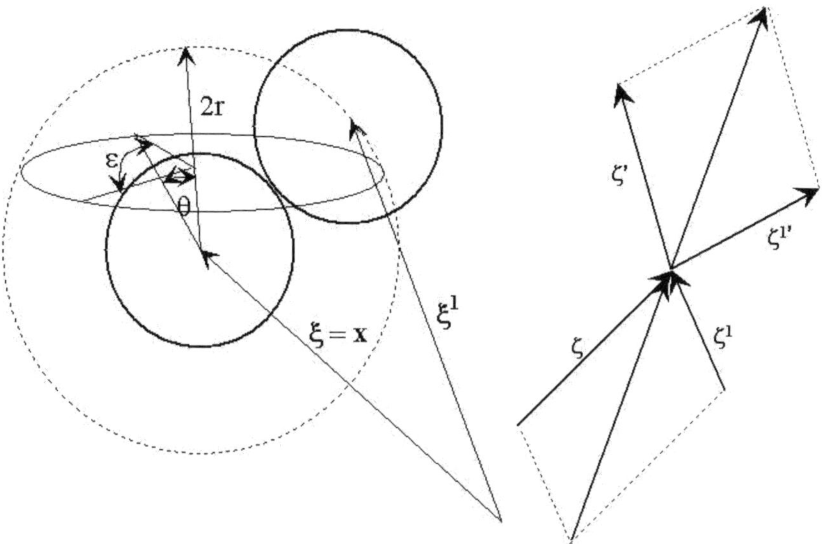

Fig. 14: *Geometry and kinematics of the encounter – construction of the collision operator.*

Most macroscopic thermodynamic fields are formed by taking *moments* of the phase density. The generic moment has the form

$$F_{i_1 i_2 \ldots i_N} = \int m\varsigma_{i_1}\varsigma_{i_1}\ldots\varsigma_{i_N} f d\varsigma, \tag{7.36}$$

where $\varsigma_{i_\alpha}, i_\alpha = 1, 2, 3$ are Cartesian coordinates of the velocity vector ς. Thus we have in particular

$$\text{mass density} \qquad F = \int m f d\varsigma = \rho, \tag{7.37}$$

$$\text{momentum density} \quad F_i = \int m\varsigma_i f d\varsigma = \rho v_i.$$

It is convenient to use in definitions of moments the so-called *peculiar velocity* $\mathbf{C} = \varsigma - \mathbf{v}$ for which

$$\int mC_i f d\varsigma = 0.$$

We use as well the following *internal moments*

$$\rho_{i_1 i_2 \ldots i_N} = \int mC_{i_1}C_{i_2}\ldots C_{i_N} f d\varsigma. \tag{7.38}$$

The first few moments and internal moments have their canonical notation and names that suggest their interpretation:

$\frac{1}{2}F_{ii} = \rho\left(\varepsilon + \frac{1}{2}v^2\right)$ -energy density	$\frac{1}{2}\rho_{ii} = \rho\varepsilon$ - internal energy density
F_{ij} - momentum flux	$-\rho_{ij} = t_{ij}$ - Cauchy stress tensor
$\frac{1}{2}F_{ijj}$ - energy flux	$\frac{1}{2}\rho_{ijj} = q_i$ - heat flux

The definition of the peculiar velocity yields simple relations between the moments and the internal moments

$$\begin{aligned} F &= \rho, \\ F_i &= \rho v_i, \\ F_{ij} &= \rho_{ij} + \rho v_i v_j, \\ F_{ijk} &= \rho_{ijk} + 3\rho_{(ij}v_{k)} + \rho v_i v_j v_k, \\ F_{ijkl} &= \rho_{ijkl} + 4\rho_{(ijk}v_{l)} + 6\rho_{(ij}v_k v_{l)} + \rho v_i v_j v_k v_l. \end{aligned} \tag{7.39}$$

Round brackets indicate symmetrization, i.e. the sum of the permutations of all N indices divided by $N!$. It is clearly seen that fluxes in the above relations have the same structure as fluxes in phenomenological local balance equations

$$\begin{aligned} F_{ij} &= \rho v_i v_j - t_{ij}, \\ \tfrac{1}{2}F_{ijj} &= \rho\left(\varepsilon + \tfrac{1}{2}v^2\right)v_i + q_i - t_{ij}v_j. \end{aligned} \tag{7.40}$$

In a monatomic ideal gas $t_{ii} = -3p$ and, consequently,

$$p = \frac{2}{3}\rho\varepsilon. \tag{7.41}$$

This is the relation between the thermal and caloric equations of state in a monatomic gas which we were discussing in the previous Chapter. In monatomic ideal gases we have as well the following thermal state equation $\varepsilon = 3kT/2m$, and this yields

$$\frac{3}{2}\frac{kT}{m} = \frac{1}{\rho}\int \frac{m}{2}C^2 f d\varsigma. \tag{7.42}$$

This relation may be used as the definition of the temperature in the kinetic theory of ideal monatomic gases.

Multiplication of the Boltzmann equation by a generic function $\psi(t, \mathbf{x}, \varsigma)$ and integration over all microscopic velocities ς yields the balance equation for the mean value $\bar{\psi}(t, \mathbf{x})$ defined by the relation

$$\rho\bar{\psi}(t, \mathbf{x}) = \int \psi(t, \mathbf{x}, \varsigma) f(t, \mathbf{x}, \varsigma) d\varsigma. \tag{7.43}$$

The balance equation for $\bar{\psi}$ is derived under the assumption that the phase density $f(t, \mathbf{x}, \varsigma)$ falls rapidly for large values of $|\varsigma|$ which means that the probability of appearance of particles with large velocities decays sufficiently rapidly as $|\varsigma| \to \infty$. Then we obtain

$$\frac{\partial \rho\bar{\psi}}{\partial t} + \frac{\partial \rho\overline{(\psi\varsigma_i)}}{\partial x_i} - \rho\overline{\left(b_i^0 + i_i\right)\frac{\partial \psi}{\partial \varsigma_i}}$$

$$= \rho\overline{\left(\frac{\partial \psi}{\partial t} + \varsigma_i \frac{\partial \psi}{\partial x_i}\right)} + \frac{1}{4}\int \left(\psi + \psi^1 - \psi' - \psi^{1\prime}\right)\left(f' f^{1\prime} - f f^1\right) rg \sin\theta d\theta d\varsigma^1 d\varsigma. \tag{7.44}$$

Of course, the third term represents the supply due to external and inertial forces, while the right-hand side is the production (source).

In derivation of macroscopic equations for ideal gases we need balance equations of moments of microscopic velocities. We set

$$\psi = m\varsigma_{i_1}\varsigma_{i_2}...\varsigma_{i_N}. \tag{7.45}$$

It follows

$$\frac{\partial F_{i_1...i_N}}{\partial t} + \frac{\partial F_{i_1...i_N j}}{\partial x_j} - NF_{(i_1...i_{N-1}}\left(f_{i_N)} + i^0_{i_N)}\right) \tag{7.46}$$

$$-NF_{k(i_1...i_{N-1}}2\Omega_{i_N)k} = S_{i_1...i_N},$$

where i^0_j is the part of the inertial acceleration i_j that is independent of the velocity ς. The right-hand side follows from the production by collisions. The symmetry of the collision operator clearly seen in (7.44) yields

$$S = 0, \quad S_i = 0, \quad S_{ii} = 0. \tag{7.47}$$

These are, of course, consequences of the mass, momentum and energy conservations during the collision.

We write equations (7.46) in the explicit form for the first thirteen moments: $F = \rho$, $F_i = \rho v_i$, F_{ij} and F_{ijj}. We have

$$\frac{\partial \rho}{\partial t} + \frac{\partial(\rho v_i)}{\partial x_i} = 0,$$

$$\frac{\partial(\rho v_i)}{\partial t}+\frac{\partial F_{ij}}{\partial x_j}-\rho\left(b_i^0+i_i^0\right)-2\rho v_k\Omega_{ik}=0,$$

$$\frac{\partial F_{ij}}{\partial t}+\frac{\partial F_{ijk}}{\partial x_k}-2\rho v_{(i}\left(b_{j)}^0+i_{j)}^0\right)-4F_{k(i}\Omega_{j)k}=S_{\langle ij\rangle},\tag{7.48}$$

$$\frac{\partial F_{ijj}}{\partial t}+\frac{\partial F_{ijjk}}{\partial x_k}-3F_{(ij}\left(b_{j)}^0+i_{j)}^0\right)-6F_{k(ij}\Omega_{j)k}=S_{ijj},$$

where the traceless (deviatoric) part of the symmetric tensor S_{ij} is defined as follows

$$S_{\langle ij\rangle}=S_{ij}-\frac{1}{3}S_{kk}\delta_{ij}.\tag{7.49}$$

In this explicit form, it becomes clear that the first two equations and the trace of the third equation represent the macroscopic conservation laws of mass, momentum and energy. Contributions of inertial forces have the familiar form known in continuum thermodynamics.

The remaining equations are balance laws for the traceless (deviatoric) part of the momentum flux F_{ij} and the energy flux F_{ijj}. These equations are not conservation laws because they contain productions due to collisions. Simultaneously, they contain explicit contributions of inertial forces. This is a very important property because it yields the answer to the question if the requirement of material objectivity follows from the above construction of the macroscopic thermodynamical model.

The set (7.48) forms the foundation of the model based on the fields $\rho, v_i, F_{ij}, F_{ijj}$. However, it is not yet the set of field equations as the fluxes $F_{\langle ijk\rangle}$ and F_{ijjk} as well as productions $S_{\langle ij\rangle}, S_{ijj}$ are not specified. Even in the simplest case of a linear isotropic representation for productions

$$\begin{aligned}S_{\langle ij\rangle}&=-\frac{1}{\tau}t_{\langle ij\rangle}=-\frac{1}{\tau}\left(F_{\langle ij\rangle}-\rho v_{\langle i}v_{j\rangle}\right),\\ S_{ijj}&=-\frac{4}{3\tau}q_i=-\frac{2}{3\tau}\left(F_{ijj}-3F_{(ij}v_{j)}-\rho v_i v^2\right),\end{aligned}\tag{7.50}$$

which appears for the simplest case of the so-called Maxwellian molecules, we have to define constitutive relations for the above mentioned fluxes. τ in the above relations denotes the mean time of the free flight and it is the same characteristic time which appears in the definition of Knudsen number. It follows from a simplified collision operator which is assumed to be proportional to the deviation of the distribution function from the Maxwell distribution [34]. This constitutes the *closure* problem of the kinetic theory which we proceed to discuss.

Let us begin with a definition proposed by L. Boltzmann for the entropy and the entropy flux. If we insert[1]

$$\psi = -k\left(\ln\frac{f}{y} - 1\right), \tag{7.51}$$

into the general balance equation (7.44), we obtain

$$\frac{\partial\left(-k\int f\ln f d\varsigma\right)}{\partial t} + \frac{\partial\left(-k\int \varsigma_i f\ln f d\varsigma\right)}{\partial x_i} \tag{7.52}$$

$$= \frac{m}{4}\int \ln\frac{f'f^{1\prime}}{ff^1}\left(f'f^{1\prime} - ff^1\right) rg\sin\theta d\theta d\varepsilon d\varsigma^1 d\varsigma.$$

It is easy to observe that the right-hand side in the above equation is always nonnegative. This property of the collision operator is called the *H-Theorem* of L. Boltzmann and it motivates the second law of thermodynamics. We have already mentioned that this Theorem was vehemently criticized by many physicists contemporary to Boltzmann as contradicting Poincare's recurrence Theorem (see Chapter 5). It is now clear that this property is the consequence of the probabilistic element of Boltzmann's derivation which appears in the assumption on the molecular chaos (compare [63], [64]).

Bearing the above property in mind, we can introduce the following definitions

$\eta = -\frac{k}{\rho}\int f\left(\ln\frac{f}{y} - 1\right)d\varsigma$	– as the specific entropy,
$\varphi_i = -k\int \varsigma_i f\left(\ln\frac{f}{y} - 1\right)d\varsigma$	– as the entropy flux,
$\hat{\eta} = \frac{k}{4}\int \ln\frac{f'f^{1\prime}}{ff^1}\left(f'f^{1\prime} - ff^1\right) \times rg\sin\theta d\theta d\varepsilon d\varsigma^1 d\varsigma$	– as the entropy production.

(7.53)

[1]The constant $1/y$ is introduced to eliminate the so-called Gibbs paradox of classical statistical mechanics. It has the interpretation of the smallest element in $(\mathbf{x},\varsigma)$ - space that can accommodate a position $\mathbf{x}$ and a velocity ς. It is equal to $h^3/m^3/(2s+1)$, where h is Planck's constant and $sh/2\pi$ is the spin of an atom. The occurrence of y and of the other constant in the definition (7.51) is unimportant for our considerations and it plays a role only in problems where the entropy constant is material (e.g. chemical reactions). For relations discussed in this Section one could use as well the definition $\psi = -k\ln f$.

The flux contains in this formulation both a convective part with an explicit dependence on the velocity as well as the intrinsic velocity-independent part. This is characteristic for the Eulerian description which follows from the derivation of macroscopic equations by means of the kinetic theory of gases. The entropy inequality follows from (7.52) in the form: *for all processes*

$$\frac{\partial \rho\eta}{\partial t} + \frac{\partial \varphi_i}{\partial x_i} \geq 0. \tag{7.54}$$

The separation of convective and nonconvective parts follows from the transformation to peculiar velocities which we present further.

The above inequality suggests the definition of the *thermodynamical equilibrium* in the kinetic approach (compare $(7.53)_3$)

$$\ln f_E + \ln f_E^1 = \ln f_E' + \ln f_E^{1\prime}, \tag{7.55}$$

for which the entropy production vanishes identically. Hence it is the *summation invariant* of the collision operator. However, we know already that there are only five such independent invariants: the atomic mass, momenta and energy. Therefore $\ln f_E$ must be their linear combination

$$\ln f_E = \alpha + \beta_i \varsigma_i + \gamma \varsigma^2, \tag{7.56}$$

where the coefficients α, β_i, γ must be determined from the conditions

$$\rho = \int m f_E d\varsigma, \quad \rho v_i = \int m\varsigma_i f_E d\varsigma, \quad \rho \frac{3}{2}\frac{kT}{m} = \int \frac{m}{2} C^2 f_E d\varsigma. \tag{7.57}$$

After simple calculations we obtain

$$f_E = \frac{\rho}{m} \sqrt[3]{\frac{m}{2\pi kT}} e^{-\frac{m(\varsigma - \mathbf{v})^2}{2kT}}. \tag{7.58}$$

This is the so-called *Maxwell distribution.*

The Maxwell distribution yields the vanishing entropy production for a large variety of fields $\rho(t, \mathbf{x}), v_i(t, \mathbf{x}), T(t, \mathbf{x})$. However, in order to satisfy the Boltzmann equation they must additionally satisfy the following field equations

$$
\begin{aligned}
\frac{\partial \rho}{\partial t} + \frac{\partial \rho v_i}{\partial x_i} &= 0, \\
\rho \left(\frac{\partial v_i}{\partial t} + v_j \frac{\partial v_i}{\partial x_j} \right) + \frac{\partial \rho \frac{k}{m} T}{\partial x_i} &= \rho \left(b_i^0 + i_i \right), \quad i_i = i_i^0 + 2\Omega_{ik} v_k, \\
\frac{\partial T}{\partial t} \delta_{ij} + 2T \frac{\partial v_{(i}}{\partial x_{j)}} &= 0, \\
\frac{\partial T}{\partial x_i} &= 0.
\end{aligned} \tag{7.59}
$$

Hence, in a monatomic gas equilibrium processes contain rigid rotations, isotropic expansions, a variety of density gradients, and a time dependent temperature whose gradient vanishes.

The extension of this equilibrium distribution to nonequilibrium situation of the thirteen moments: $F = \rho$, $F_i = \rho v_i$, F_{ij} and F_{ijj} has been proposed by H. Grad in 1949. He proposed to use the following expansion for the phase density f

$$
f = \left(a + a_i \frac{\partial}{\partial \varsigma_i} + a_{ij} \frac{\partial^2}{\partial \varsigma_i \partial \varsigma_j} + a_{ijk} \frac{\partial^3}{\partial \varsigma_i \partial \varsigma_j \partial \varsigma_k} + ... \right) f_E. \tag{7.60}
$$

Bearing relation (7.58) in mind, we obtain immediately

$$
\begin{aligned}
f = f_E \Bigg(a - a_i \frac{m}{kT} C_i + a_{ij} \left(\frac{m}{kT} \right)^2 \left(C_i C_j - \frac{kT}{m} \delta_{ij} \right) \\
- a_{ijk} \left(\frac{m}{kT} \right)^3 \left(C_i C_j C_k - \frac{kT}{m} \left(\delta_{ij} C_k + \delta_{kj} C_i + \delta_{ik} C_j \right) \right) + ... \Bigg).
\end{aligned} \tag{7.61}
$$

This is the expansion of f in terms of Hermite polynomials. Its coefficients $a, a_i, a_{ij}, a_{ijk}, ...$ follow by insertion in (7.36). For thirteen moments, this yields the so-called *Grad distribution function*

$$
f_G = f_E \left(1 - \frac{1}{2p} t_{\langle ij \rangle} \left(\frac{m}{kT} C_i C_j - \delta_{ij} \right) - \frac{1}{p} \frac{m}{kT} q_i C_i \left(1 - \frac{1}{5} \frac{m}{kT} C^2 \right) \right). \tag{7.62}
$$

Substitution of this function in relations (7.53) yields

$$
\eta = \frac{k}{m} \ln \frac{T^{3/2}}{\rho} - \underline{\frac{t_{\langle ij \rangle} t_{\langle ij \rangle}}{4 p \rho T}} - \underline{\frac{q_i q_i}{5 p^2 T}}, \tag{7.63}
$$

$$
\varphi_i = \rho \eta v_i + h_i, \quad h_i = \frac{q_i}{T} + \underline{\frac{2}{5pT} t_{\langle ij \rangle} q_j}.
$$

Hence both the entropy and the entropy flux contain a dependence on $t_{\langle ij \rangle}$ and q_i which measure the deviation from the thermodynamical equilibrium. In addition, the nonconvective entropy flux h_i is not proportional to q_i/T but it contains a term proportional to $t_{\langle ij \rangle} q_j$ as well. This is the challenge to the phenomenological thermodynamics and it is resolved, as we see in the next Chapter, by the *extended thermodynamics.*

We complete the presentation of the construction of the macroscopic thirteen fields model based on the kinetic theory by the derivation of constitutive relations. As already mentioned, we need them in order to transform equation (7.48) into field equations for the following thirteen fields $\rho \equiv F$, $v_i \equiv \frac{1}{\rho} F_i$, $T \equiv \frac{m}{3\rho k} \left(F_{ii} - \rho v^2 \right)$, $t_{\langle ij \rangle} \equiv - \left(F_{\langle ij \rangle} - \rho v_{\langle i} v_{j \rangle} \right)$, $q_i \equiv \frac{1}{2} \left(F_{ijj} - 3F_{ij} v_j + 2 v_i v^2 \right)$. We accept relations (7.50) as constitutive relations for productions, i.e.

$$S_{\langle ij \rangle} = -\frac{1}{\tau} t_{\langle ij \rangle}, \quad S_{ijj} = -\frac{4}{3\tau} q_i. \tag{7.64}$$

Consequently, we have to add relations for $\rho_{\langle ij \rangle k} \equiv F_{\langle ij \rangle k} - 3\rho_{(\langle ij \rangle} v_{k)} - \rho v_{\langle i} v_{j \rangle} v_k$ and $\rho_{ijjk} \equiv F_{ijjk} - 4\rho_{(ijj} v_{k)} - 6\rho_{(ij} v_j v_{k)} - \rho v_i v_k v^2$. In terms of peculiar velocities they have the following form

$$\rho_{\langle ij \rangle k} = \int m C_{\langle i} C_{j \rangle} C_k f d\varsigma, \quad \rho_{ijjk} = \int m C_i C^2 C_k f d\varsigma. \tag{7.65}$$

We substitute now the Grad distribution (7.62) in these relations. It follows

$$\begin{aligned} \rho_{\langle ij \rangle k} &= \frac{2}{5} \left(q_i \delta_{jk} + q_j \delta_{ik} - \frac{2}{3} q_k \delta_{ij} \right), \\ \rho_{ijjk} &= 5p \frac{kT}{m} \delta_{ij} - \frac{7}{2} t_{\langle pq \rangle} \frac{kT}{m} \left(\delta_{ip} \delta_{jq} + \delta_{iq} \delta_{jp} \right). \end{aligned} \tag{7.66}$$

Substitution of these relations in $(7.48)_{3,4}$ yields for the deviatoric part of stresses

$$t_{\langle ij \rangle} = 2p\tau \left(\underline{\frac{\partial v_{\langle i}}{\partial x_{j \rangle}}} + \frac{2}{5p} \frac{\partial q_{\langle i}}{\partial x_{j \rangle}} \right)$$

$$-\tau \left(\underbrace{\dot{t}_{\langle ij \rangle} + 2 t_{\langle n \langle i \rangle} \frac{\partial v_{j \rangle}}{\partial x_n} - 4 t_{\langle n \langle i \rangle} \Omega_{j \rangle n}} \right) - \boxed{\tau t_{\langle ij \rangle} \frac{\partial v_n}{\partial x_n}}, \tag{7.67}$$

and for the heat flux

$$q_i = -\frac{15}{4}\frac{kp\tau}{m}\left(\underline{\frac{\partial T}{\partial x_i}} + \frac{2}{5}\frac{T}{p}\frac{\partial t_{\langle in\rangle}}{\partial x_n}\right)$$

$$-\frac{3}{2}\tau\left(\underbrace{\dot{q}_i + q_n\frac{\partial v_i}{\partial x_n} - 2q_n\Omega_{in}}\right) \tag{7.68}$$

$$-\boxed{\frac{3}{2}\tau\left(\frac{7}{5}q_i\frac{\partial v_n}{\partial x_n} + \frac{4}{5}q_n\frac{\partial v_{(i}}{\partial x_{n)}} - \frac{7}{2}\frac{k}{m}t_{\langle in\rangle}\frac{\partial T}{\partial x_n} - \frac{1}{\rho}t_{\langle ik\rangle}\frac{\partial t_{\langle kn\rangle}}{\partial x_n} + \frac{1}{\rho}t_{\langle in\rangle}\frac{\partial p}{\partial x_n}\right)}.$$

First of all, the terms in boxes are nonlinear. As we neglected already nonlinear contributions to productions their contribution to field equations is not reliable. Therefore, we neglect them as well.

Secondly, the braced terms contain the combinations of terms which are altogether objective provided we tolerate some nonlinear contributions. However, they are different from those which would follow if we were using Lie derivatives discussed in Chapter 3:

$$\mathcal{L}_{\mathrm{v}}t_{\langle ij\rangle} = \dot{t}_{\langle ij\rangle} - 2t_{\langle n\,\langle i\rangle}\frac{\partial v_{j\rangle}}{\partial x_n}, \quad \mathcal{L}_{\mathrm{v}}q_i = \dot{q}_i - q_n\frac{\partial v_i}{\partial x_n}. \tag{7.69}$$

Combinations $4t_{\langle n\,\langle i\rangle}\left(\frac{\partial v_{j\rangle}}{\partial x_n} - \Omega_{j\rangle n}\right)$, $2q_n\left(\frac{\partial v_i}{\partial x_n} - \Omega_{in}\right)$ are objective (compare (3.5)), but the form of constitutive relations following from the kinetic theory is different from this which we would obtain extending phenomenological equations by addition of evolutionary contributions. This makes the extensions proposed, for example, by Cattaneo for the heat flux equation not compatible with the kinetic theory. We return to this problem in the next Chapter.

If we introduce the notation

$$\mu = p\tau \equiv \rho\frac{kT}{m}\tau, \quad K = \frac{15}{4}\frac{k}{m}p\tau \equiv \frac{15}{4}\frac{k}{m}\left(\rho\frac{kT}{m}\tau\right), \tag{7.70}$$

then (7.67) and (7.68) lead to the following constitutive relations

$$t_{\langle ij\rangle} = 2\mu\left(\underline{\frac{\partial v_{\langle i}}{\partial x_{j\rangle}}} + \frac{2}{5p}\frac{\partial q_{\langle i}}{\partial x_{j\rangle}}\right) - \tau\left(\mathcal{L}_{\mathrm{v}}t_{\langle ij\rangle} + 4t_{\langle n\,\langle i\rangle}\left(\frac{\partial v_{j\rangle}}{\partial x_n} - \Omega_{j\rangle n}\right)\right),$$

$$q_i = K\left(\underline{\frac{\partial T}{\partial x_i}} + \frac{2}{5}\frac{T}{p}\frac{\partial t_{\langle in\rangle}}{\partial x_n}\right) - \frac{3}{2}\tau\left(\mathcal{L}_{\mathrm{v}}q_i + 2q_n\left(\frac{\partial v_i}{\partial x_n} - \Omega_{in}\right)\right). \tag{7.71}$$

Clearly, if we make the assumption that both μ and K remain finite in the limit $\tau \to 0$, the above constitutive relations are identical with the classical Navier-Stokes and Fourier laws if we leave out couplings not underlined in the first terms on the right-hand sides. Otherwise, we have to deal with evolution equations for deviatoric stresses $t_{\langle ij\rangle}$ and for heat flux q_i. In the first case the speeds of shear and thermal pulses propagate with an infinite speed, while in the evolutionary case both speeds are finite. This will also be the subject of discussion of the next Chapter.

Summing up, let us note that the kinetic theory of rarefied gases yields, by means of the Grad distribution function and in the simplest linear approximation, the first order field equations for thirteen fields ρ, v_i, ε, $t_{\langle ij\rangle}$, q_i. In addition, it produces the entropy inequality with the entropy density η and the entropy flux h_i dependent on these fields. Macroscopic equations obtained in this way coincide with phenomenological equations of the Navier-Stokes-Fourier viscous fluid and, simultaneously, provide an additional physical insight by revealing the structure of the nonequilibrium entropy. The structure of this kinetic model motivates a phenomenological construction of the so-called *extended thermodynamics of ideal gases* which is the subject of the next Chapter.

For the further use we collect the results of the kinetic theory of thirteen moments:

1. Balance equations

$$\frac{\partial \rho}{\partial t} + \frac{\partial (\rho v_i)}{\partial x_i} = 0,$$

$$\frac{\partial}{\partial t}(\rho v_i) + \frac{\partial}{\partial x_j}(\rho v_i v_j + \rho_{ij}) - \rho\left(b_i^0 + i_i^0\right) - 2\rho v_k \Omega_{ik} = 0,$$

$$\frac{\partial}{\partial t}(\rho v_i v_j + \rho_{ij}) + \frac{\partial}{\partial x_k}\left(3\rho_{(ij}v_{k)} + \rho v_i v_j v_k + \rho_{ijk}\right) \tag{7.72}$$

$$-2\rho v_{(i}\left(b_{j)}^0 + i_{j)}^0\right) - 4F_{k(i}\Omega_{j)k} = S_{\langle ij\rangle},$$

$$\frac{\partial}{\partial t}\left(\rho v_i v^2 + 3\rho_{ij}v_j + \rho_{ijj}\right) + \frac{\partial}{\partial x_k}\left(\rho v_i v_k v^2 + 6\rho_{ij}v_j v_k + 4\rho_{ijj}v_k + \rho_{ijjk}\right)$$

$$-3F_{(ij}\left(b_{j)}^0 + i_{j)}^0\right) - 6F_{k(ij}\Omega_{j)k} = S_{ijj},$$

where $\rho, v_i, \frac{1}{3}\rho_{ii} = p, \rho_{\langle ij\rangle} = -t_{\langle ij\rangle}, \frac{1}{2}\rho_{jji} = q_i$ are *thirteen fields*, and $\rho_{\langle ij\rangle k}$, ρ_{ijjk}, $S_{\langle ij\rangle}$, S_{ijj} are constitutive functions.

2. Grad distribution function

$$f_G = f_E\left(1 - \frac{1}{2p}t_{\langle ij\rangle}\left(\frac{m}{kT}C_iC_j - \delta_{ij}\right) - \frac{1}{p}\frac{m}{kT}q_iC_i\left(1 - \frac{1}{5}\frac{m}{kT}C^2\right)\right), \tag{7.73}$$

where the Maxwell distribution f_E is given by the relation

$$f_E = \frac{\rho}{m}\sqrt[3]{\frac{m}{2\pi kT}}e^{-\frac{m(\varsigma - \mathbf{v})^2}{2kT}}. \tag{7.74}$$

3. Constitutive relations for fluxes $\rho_{\langle ij\rangle k}, \rho_{ijjk}$

$$\begin{aligned} \rho_{\langle ij\rangle k} &= \frac{2}{5}\left(q_i\delta_{jk} + q_j\delta_{ik} - \frac{2}{3}q_k\delta_{ij}\right), \\ \rho_{ijjk} &= 5p\frac{kT}{m}\delta_{ij} - \frac{7}{2}t_{\langle pq\rangle}\frac{kT}{m}\left(\delta_{ip}\delta_{jq} + \delta_{iq}\delta_{jp}\right). \end{aligned} \tag{7.75}$$

4. Constitutive relations for sources (productions)

$$S_{\langle ij\rangle} = \frac{1}{\tau}\rho_{\langle ij\rangle} \equiv -\frac{1}{\tau}t_{\langle ij\rangle}, \quad S_{ijj} = -\frac{2}{3\tau}\rho_{jji} \equiv -\frac{4}{3\tau}q_i. \tag{7.76}$$

5. Entropy inequality and its constitutive relations for the entropy and the entropy flux

$$\frac{\partial\rho\eta}{\partial t} + \frac{\partial\varphi_i}{\partial x_i} \geq 0, \quad \eta = \frac{k}{m}\ln\frac{T^{3/2}}{\rho} - \frac{t_{\langle ij\rangle}t_{\langle ij\rangle}}{4p\rho T} - \frac{q_iq_i}{5p^2T}, \tag{7.77}$$

$$\varphi_i = \rho\eta v_i + h_i, \quad h_i = \frac{q_i}{T} + \frac{2}{5pT}t_{\langle ij\rangle}q_j.$$

It should be mentioned as well that neither the model of thirteen moments nor Grad's method of construction of the nonequilibrium phase density f_G exhaust the results based on the Boltzmann equation. There exist numerous attempts to extend the hierarchy of equations to higher moments on the basis of the assumption that such an extension improves the approximation of a true solution f of the Boltzmann equation. Some of these extensions seem to be physically meaningful and confirmed by, for instance, light scattering experiments. However, there exists no proof of convergence of such a hierarchy. Even worse, some of such models, for instance the so-called Burnett model, corresponding to the Chapman-Enskog expansion of the Boltzmann equation with respect to the Knudsen number, Kn, as a small parameter (to the order $(Kn)^2$), possess instabilities which do not correspond to any real phenomena (see [314]).

Chapter 8

Extended thermodynamics

8.1 General structure

As indicated in the previous Chapter, we intend to construct a continuum thermodynamical model of a system whose set of fields goes beyond the set of ordinary continuum thermodynamics i.e. beyond the set of densities of mass, momentum and energy. The purpose of these additional fields is to describe deviations from the local thermodynamical equilibrium. We have demonstrated an example of such a model arising from the kinetic theory of ideal gases and this example shows that typical extensions would be fluxes of momentum and energy.

Ingo Müller
1936–

Tommaso Ruggeri
1947–

I – Shih Liu
1943–

Thermodynamical modelling for an extended set of fields should possess the following essential ingredients

– fundamental equations are of balance type; this yields mathematically a possibility of weak solutions and shock waves,

– constitutive relations (closure problem for balance equations) are local in space and time which yields sets of quasilinear field equations of the first order,
– second law of thermodynamics consists of the entropy inequality and conditions on ideal walls in which the entropy density is concave and this, in turn, yields symmetric hyperbolic field equations (well-posed Cauchy initial value problems).

One can find in the literature another extended version of thermodynamics in which the requirement of the local constitutive relation does not appear (e.g. D. Jou, J. Casas-Vázquez, G. Lebon [160], D. Jou [159], or S. Sieniutycz, P. Salamon [294]). In this approach, additional fields are introduced but the structure of constitutive laws remains classical – with contributions of gradients and time derivatives of fields. We do not present this version of thermodynamics in this book.

In this Chapter we first demonstrate a general structure of such a modeling and then we illustrate this procedure with an example of ideal gas model. This model motivates the general procedure and it can be easily compared with results of the kinetic theory. The full presentation of all aspects of extended thermodynamics can be found in the book of I. Müller and T. Ruggeri [225].

Let us begin with an analysis of a thermodynamical model which describes n-dimensional vector field $\mathbf{w} \in \mathcal{V}^n$ as a function of spatial and temporal variables $(\mathbf{x}, t)$. This field is assumed to satisfy balance equations

$$\frac{\partial \mathbf{F}_0}{\partial t} + \frac{\partial \mathbf{F}_k}{\partial x_k} = \hat{\mathbf{f}}, \quad \mathbf{F}_0, \mathbf{F}_1, \mathbf{F}_2, \mathbf{F}_3, \hat{\mathbf{f}} \in \mathcal{V}^n, \tag{8.1}$$

where *densities* $\mathbf{F}_0$, *fluxes* $\mathbf{F}_k, k = 1, 2, 3$, and *productions* $\hat{\mathbf{f}}$ are assumed to be only functions of fields $\mathbf{w}$. This is a particular case of the structure investigated in Chapter 5, where constitutive variables $\mathfrak{w} \in \mathcal{V}^N$ were assumed to contain not only fields $\mathbf{w}$ but also some derivatives of fields. The constitutive dependence on fields alone is the first characteristic feature of the extended thermodynamics and this feature was already discovered in kinetic models which we discussed in the previous Chapter. Hence, *field equations of extended thermodynamics* are provided by constitutive relations of the form

$$\mathbf{F}_0 = \mathbf{F}_0(\mathbf{w}), \quad \mathbf{F}_k = \mathbf{F}_k(\mathbf{w}), \quad k = 1, 2, 3, \quad \hat{\mathbf{f}} = \hat{\mathbf{f}}(\mathbf{w}). \tag{8.2}$$

For this reason, we say that constitutive relations of extended thermodynamics are *local* in space and time: they depend on fields in the same space-time point as the constitutive quantities without any constitutive influence of gradients or time derivatives of fields.

All solutions of field equations which follow by the above *closure* are called *thermodynamical processes.*

According to procedures which we were already discussing in this book there are some principles which restrict the class of thermodynamical processes. The main of them are as follows

- the entropy inequality,
- the convexity,
- the principle of relativity.

The combination of the first two principles is called the *entropy principle* and it is the most essential part of the *second law of thermodynamics.*

As we already know thermodynamical processes are *admissible* if the following inequality is fulfilled

$$\frac{\partial \rho\eta}{\partial t}+\frac{\partial \varphi_k}{\partial x_k}\geq 0, \quad \rho\eta=\rho\eta\left(\mathbf{w}\right), \quad \varphi_k=\varphi_k\left(\mathbf{w}\right). \tag{8.3}$$

The principle of convexity then has the form

$$\forall \mathbf{z}\in\mathcal{V}^n:\mathbf{z}\cdot\frac{\partial^2\rho\eta}{\partial\mathbf{w}\partial\mathbf{w}}\mathbf{z}<0. \tag{8.4}$$

Finally, the principle of relativity within the frame of nonrelativistic models means that field equations and the entropy inequality have the same form in all Galilean frames.

Let us investigate the condition of thermodynamical admissibility. According to Liu's Theorem we can introduce the Lagrange multipliers $\mathbf{\Lambda}\left(\mathbf{w}\right)\in\mathcal{V}^n$ such that the inequality

$$\frac{\partial \rho\eta}{\partial t}+\frac{\partial \varphi_k}{\partial x_k}-\mathbf{\Lambda}\cdot\left(\frac{\partial \mathbf{F}_0}{\partial t}+\frac{\partial \mathbf{F}_k}{\partial x_k}-\hat{\mathbf{f}}\right)\geq 0, \tag{8.5}$$

holds for all fields and not only for solutions of field equations. The solution of this inequality is as follows

$$\frac{\partial \rho\eta}{\partial \mathbf{w}}=\left(\frac{\partial \mathbf{F}_0}{\partial \mathbf{w}}\right)^T\mathbf{\Lambda}, \quad \frac{\partial \varphi_k}{\partial \mathbf{w}}=\left(\frac{\partial \mathbf{F}_k}{\partial \mathbf{w}}\right)^T\mathbf{\Lambda}, \quad \mathbf{\Lambda}\cdot\hat{\mathbf{f}}\geq 0. \tag{8.6}$$

The first two conditions can also be written in the form

$$d\left(\rho\eta\right)=\mathbf{\Lambda}\cdot d\mathbf{F}_0, \quad d\varphi_k=\mathbf{\Lambda}\cdot d\mathbf{F}_k. \tag{8.7}$$

Hence the multipliers are independent of the choice of the fields $\mathbf{w}$. In particular, assuming that $\mathbf{F}_0$ is not degenerate we can choose

$$\mathbf{w}=\mathbf{F}_0\Longrightarrow\frac{\partial \rho\eta}{\partial \mathbf{w}}=\mathbf{\Lambda}\quad\Longrightarrow\quad\frac{\partial \mathbf{\Lambda}}{\partial \mathbf{w}}=\frac{\partial^2\rho\eta}{\partial\mathbf{w}\partial\mathbf{w}}. \tag{8.8}$$

Consequently, the derivatives $\frac{\partial \mathbf{\Lambda}}{\partial \mathbf{w}}$ form the symmetric and negative definite (principle of convexity!) matrix. This means that the map $\mathbf{w} \to \mathbf{\Lambda}$ is globally invertible. By doing such a change of variables we obtain the following transformed constitutive relations

$$\begin{aligned} &\mathbf{F}_0 = \mathbf{F}_0(\mathbf{\Lambda}), \quad \mathbf{F}_k = \mathbf{F}_k(\mathbf{\Lambda}), \quad k = 1,2,3, \quad \hat{\mathbf{f}} = \hat{\mathbf{f}}(\mathbf{\Lambda}). \\ &\rho\eta = \rho\eta(\mathbf{\Lambda}), \quad \varphi_k = \varphi_k(\mathbf{\Lambda}). \end{aligned} \tag{8.9}$$

Then relations (8.7) can be written in the following form

$$d(\rho\eta - \mathbf{\Lambda} \cdot \mathbf{F}_0) = -\mathbf{F}_0 \cdot d\mathbf{\Lambda}, \quad d(\varphi_k - \mathbf{\Lambda} \cdot \mathbf{F}_k) = -\mathbf{F}_k \cdot d\mathbf{\Lambda}. \tag{8.10}$$

Hence, one can introduce the *four-potential*

$$\mathbf{h}'(\mathbf{\Lambda}) \equiv \{h'_0, h'_k\} := \{\rho\eta - \mathbf{\Lambda} \cdot \mathbf{F}_0, \varphi_k - \mathbf{\Lambda} \cdot \mathbf{F}_k\}, \tag{8.11}$$

which specifies the constitutive relations of contributions to the left-hand side of field equations

$$\mathbf{F}_0 = -\frac{\partial h'_0}{\partial \mathbf{\Lambda}}, \quad \mathbf{F}_k = -\frac{\partial h'_k}{\partial \mathbf{\Lambda}}. \tag{8.12}$$

Simultaneously

$$\rho\eta(\mathbf{\Lambda}) = h'_0 - \mathbf{\Lambda} \cdot \frac{\partial h'_0}{\partial \mathbf{\Lambda}}, \quad \varphi_k(\mathbf{\Lambda}) = h'_k - \mathbf{\Lambda} \cdot \frac{\partial h'_k}{\partial \mathbf{\Lambda}}. \tag{8.13}$$

The residual inequality has the form

$$\mathbf{\Lambda} \cdot \hat{\mathbf{f}}(\mathbf{\Lambda}) \geq 0. \tag{8.14}$$

The existence of the four-potential (8.11), (8.12) whose components are also known as *generators* and the residual inequality (8.14) exhaust the consequences of the entropy inequality.

Integrability conditions for $\mathbf{h}'(\mathbf{\Lambda})$ yield the symmetry of the matrices

$$\frac{\partial \mathbf{F}_0}{\partial \mathbf{\Lambda}}, \quad \frac{\partial \mathbf{F}_k}{\partial \mathbf{\Lambda}}. \tag{8.15}$$

These conditions, extensions of Maxwell relations of ordinary thermodynamics, provide severe restrictions on the functions $\mathbf{F}_0, \mathbf{F}_k$.

Substitution of (8.12) in field equations (8.1) yields

$$-\frac{\partial^2 h'_0}{\partial \mathbf{\Lambda} \partial \mathbf{\Lambda}} \frac{\partial \mathbf{\Lambda}}{\partial t} - \frac{\partial^2 h'_k}{\partial \mathbf{\Lambda} \partial \mathbf{\Lambda}} \frac{\partial \mathbf{\Lambda}}{\partial x_k} = \hat{\mathbf{f}}, \tag{8.16}$$

which means that the system is *symmetric* as the coefficients are Hessian matrices. This symmetry is visible if we choose Lagrange multipliers $\mathbf{\Lambda}$ as fields. For this reason these fields are called *main fields.* Together with the convexity assumption this yields the conclusion that the system is *symmetric hyperbolic.*

Hyperbolicity of the set of field equations guarantees finite speeds of propagation of disturbances. Moreover, it yields a desirable mathematical property of the well-posedness of Cauchy problems which means local existence, uniqueness, and continuous dependence of solutions on the data (stability).

In order to see that disturbances propagate with finite speeds we investigate a wave front of an acceleration wave. This is the surface on which fields, say $\mathbf{\Lambda}$, are continuous and which moves with the speed c in the normal direction and carries a discontinuity of the first derivatives of fields. Hadamard compatibility conditions (see (4.19)) yield then

$$\left[\left[\frac{\partial \mathbf{\Lambda}}{\partial x_k}\right]\right] = n_k \delta \mathbf{\Lambda}, \quad \left[\left[\frac{\partial \mathbf{\Lambda}}{\partial t}\right]\right] = -c\delta \mathbf{\Lambda}, \tag{8.17}$$

where n_k are components of the unit normal vector to the wave front, $\delta\mathbf{\Lambda}$ is the so-called amplitude of the discontinuity. Calculating the limits of equation (8.16) on both sides of the front and subtracting we obtain

$$\left(\frac{\partial^2 h_0'}{\partial \mathbf{\Lambda} \partial \mathbf{\Lambda}} c - \frac{\partial^2 h_k'}{\partial \mathbf{\Lambda} \partial \mathbf{\Lambda}} n_k\right) \delta \mathbf{\Lambda} = 0. \tag{8.18}$$

Eigenvalues of this eigenvalue problem which follow from the *dispersion relation*

$$\det\left(\frac{\partial^2 h_0'}{\partial \mathbf{\Lambda} \partial \mathbf{\Lambda}} c - \frac{\partial^2 h_k'}{\partial \mathbf{\Lambda} \partial \mathbf{\Lambda}} n_k\right) = 0, \tag{8.19}$$

are real and finite due to the symmetry of the real Hessian matrices and nonsingularity of the matrix $\frac{\partial^2 h_0'}{\partial \mathbf{\Lambda} \partial \mathbf{\Lambda}}$. They determine speeds of propagation of *weak* disturbances, i.e. discontinuities of field derivatives with a simultaneous continuity of fields. Due to the hyperbolicity the corresponding eigenvectors which specify the amplitudes are linearly independent.

In order to explain the principle of relativity we have to recall some transformations we have discussed in Chapter 3 (compare formulae (3.4), (3.5), (3.6)). The most restrictive transformation is the Euclidean transformation (3.4) which follows from the isometry of the space of motions of classical Newtonian mechanics. In Table 8 we present three transformations in the representation in Cartesian coordinates which are of interest in this Chapter.

Table 8: *Transformations of Newtonian continuum mechanics*

transfor- mation:	rotation of coordinates	Galilean transformation	Euclidean transformation
	$x_i^* = O_{ij}x_j$, $t^* = t$, O_{ij} – const	$x_i^* = O_{ij}x_j$ $+d_it$, $t^* = t$, O_{ij}, d_i – const	$x_i^* = O_{ij}(t)\,x_j$ $+c_i(t)$, $t^* = t$,
name of the invariant object	tensor	Galilean tensor	objective tensor
velocity	$v_i^* = O_{ij}v_j$,	$v_i^* = O_{ij}v_j$ $+d_j$,	$v_i^* = O_{ij}v_j$ $+\dot{O}_{ij}x_j + \dot{c}_j$,
acceleration	$a_i^* = O_{ij}a_j$,	$a_i^* = O_{ij}a_j$,	$a_i^* = O_{ij}a_j$ $+\dot{O}_{ij}v_j + \ddot{O}_{ij}x_j$ $+\ddot{c}_j$,

An invariant object with respect to any of these transformations has the following transformation rule

$$T^*_{i_1 \dots i_A} = O_{i_1 j_1} \dots O_{i_A j_A} T_{j_1 \dots j_A}. \tag{8.20}$$

Inspection of the classical momentum balance equation which yields in continuum mechanics the equation of motion shows that this equation is invariant with respect to the Galilean transformation but not with respect to the Euclidean transformation. The invariance follows from the property of acceleration which is the Galilean vector. Euclidean transformation which yields the noninertial frames of reference produces noninertial forces with which we were already dealing a few times in this book.

As extended thermodynamics contains additional balance equations, replacing some classical constitutive relations (for instance, for the heat flux) we expect that these equations will also be Galilean invariant. This is the essence of the principle of relativity for such models.

The Eulerian description yields the natural velocity-dependent contributions to balance equations. Namely, the fluxes have the following form

$$\mathbf{F}_k = \mathbf{F}_0 v_k + \mathbf{G}_k, \quad \varphi_k = \rho\eta v_k + h_k, \tag{8.21}$$

where $\mathbf{G}_k, h_k$ are *nonconvective fluxes*.

However, this does not mean that $\mathbf{F}_0, \mathbf{G}_k, h_k$ are independent of the velocity. Let us write constitutive relations in the form

$$\begin{aligned} \mathbf{F}_0 &= \mathbf{F}_0(\mathbf{v},\mathbf{u}), \quad \mathbf{G}_k = \mathbf{G}_k(\mathbf{v},\mathbf{u}), \quad \hat{\mathbf{f}} = \hat{\mathbf{f}}(\mathbf{v},\mathbf{u}), \\ \rho\eta &= \rho\eta(\mathbf{v},\mathbf{u}), \quad h_k = h_k(\mathbf{v},\mathbf{u}), \quad \hat{\eta} = \hat{\eta}(\mathbf{v},\mathbf{u}), \end{aligned} \tag{8.22}$$

where the field vector $\mathbf{w}$ has been split into $(\mathbf{v},\mathbf{u})$. Galilean transformation to $(\mathbf{v}^*,\mathbf{u}^*)$ yields the following form of these constitutive relations

$$\begin{aligned} \mathbf{F}_0^* &= \mathbf{F}_0(\mathbf{v}^*,\mathbf{u}^*), \quad \mathbf{G}_k^* = \mathbf{G}_k(\mathbf{v}^*,\mathbf{u}^*), \quad \hat{\mathbf{f}}^* = \hat{\mathbf{f}}(\mathbf{v}^*,\mathbf{u}^*), \\ \rho\eta^* &= \rho\eta(\mathbf{v}^*,\mathbf{u}^*), \quad h_k^* = h_k(\mathbf{v}^*,\mathbf{u}^*), \quad \hat{\eta}^* = \hat{\eta}(\mathbf{v}^*,\mathbf{u}^*), \end{aligned} \tag{8.23}$$

which is similar to material objectivity except that Galilean transformation instead of Euclidean transformation is used. The principle of relativity requires in addition that the field equations and the entropy balance equation in the two frames are equivalent, i.e.

$$\begin{aligned} \frac{\partial \mathbf{F}_0(\mathbf{v},\mathbf{u})}{\partial t} + \frac{\partial\left[\mathbf{F}_0(\mathbf{v},\mathbf{u})v_k + \mathbf{G}_k(\mathbf{v},\mathbf{u})\right]}{\partial x_k} &= \hat{\mathbf{f}}(\mathbf{v},\mathbf{u}), \\ \frac{\partial \rho\eta(\mathbf{v},\mathbf{u})}{\partial t} + \frac{\partial\left[\rho\eta(\mathbf{v},\mathbf{u})v_k + h_k(\mathbf{v},\mathbf{u})\right]}{\partial x_k} &= \hat{\eta}(\mathbf{v},\mathbf{u}), \end{aligned} \tag{8.24}$$

and

$$\begin{aligned} \frac{\partial \mathbf{F}_0(\mathbf{v}^*,\mathbf{u}^*)}{\partial t} + \frac{\partial\left[\mathbf{F}_0(\mathbf{v}^*,\mathbf{u}^*)v_k^* + \mathbf{G}_k(\mathbf{v}^*,\mathbf{u}^*)\right]}{\partial x_k} &= \hat{\mathbf{f}}(\mathbf{v}^*,\mathbf{u}^*), \\ \frac{\partial \rho\eta(\mathbf{v}^*,\mathbf{u}^*)}{\partial t} + \frac{\partial\left[\rho\eta(\mathbf{v}^*,\mathbf{u}^*)v_k^* + h_k(\mathbf{v}^*,\mathbf{u}^*)\right]}{\partial x_k} &= \hat{\eta}(\mathbf{v}^*,\mathbf{u}^*). \end{aligned} \tag{8.25}$$

The exploitation of this condition yields the existence of an $n \times n$ nonsingular matrix $\mathcal{X}(\mathbf{v})$ such that we have the representations

$$\begin{aligned} \mathbf{F}_0(\mathbf{v},\mathbf{u}) &= \mathcal{X}(\mathbf{v})\boldsymbol{\rho}_0(\mathbf{u}), \quad \mathbf{G}_k(\mathbf{v},\mathbf{u}) = \mathcal{X}(\mathbf{v})\boldsymbol{\rho}_k(\mathbf{u}), \\ \hat{\mathbf{f}}(\mathbf{v},\mathbf{u}) &= \mathcal{X}(\mathbf{v})\mathbf{S}(\mathbf{u}), \end{aligned} \tag{8.26}$$

where $\boldsymbol{\rho}_0(\mathbf{u}), \boldsymbol{\rho}_k(\mathbf{u}), \mathbf{S}(\mathbf{u})$ are functions in the rest frame (i.e. for $\mathbf{v} \equiv \mathbf{0}$). These representations as well as the form of the matrix $\mathcal{X}(\mathbf{v})$ have been found by T. Ruggeri in 1989 [279]. The latter is as follows

$$\mathcal{X}(\mathbf{v}) = e^{\mathbf{A}^r v_r} = \mathbf{1} + \mathbf{A}^r v_r + \frac{1}{2}\mathbf{A}^r\mathbf{A}^s v_r v_s + \ldots, \tag{8.27}$$

where $\mathbf{A}^r$ are constant matrices dependent on the tensorial character of the corresponding quantity and otherwise independent of the system.

The above representation following from the Galilean invariance is called **Ruggeri Theorem**.

The above decomposition for a symmetric tensor density of arbitrary rank A into velocity-dependent and internal parts has the following form in Cartesian coordinates

$$F_{i_1\dots i_A} = \rho_{i_1\dots i_A} + \binom{A}{1}\rho_{(i_1\dots i_{A-1}}v_{i_A)} + \binom{A}{2}\rho_{(i_1\dots i_{A-2}}v_{i_{A-1}}v_{i_A)} \qquad (8.28)$$
$$+\dots+\binom{A}{A-1}\rho_{(i_1}v_{i_2}\dots v_{i_{A-1}}v_{i_A)} + \rho v_{i_1}v_{i_{A-1}}v_{i_A},$$

and similarly for $\boldsymbol{\rho}_k$ and $\mathbf{S}$.

This completes the presentation of general properties of extended thermodynamics which we need in this book.

8.2 Nonrelativistic ideal gases

We proceed to discuss the most important example motivating the strategy of constructing models by extended thermodynamics. We show the model of ideal nonrelativistic gases [194].

In order to be able to place the model among models of extended thermodynamics, we formulate the fundamental equations for ideal gases in the notation which we presented in the first part of this Chapter. Consequently, we choose the following fields describing the system

F – the mass density,
F_i – the momentum density,
F_{ij} – the momentum flux density, $F_{ij} = F_{ji}$,
$\frac{1}{2}F_{ijj}$ – the energy flux density.

We assume that these fields satisfy in an arbitrary noninertial frame the following balance equations

$$\begin{aligned}
\frac{\partial F}{\partial t} + \frac{\partial F_k}{\partial x_k} &= 0,\\
\frac{\partial F_i}{\partial t} + \frac{\partial F_{ik}}{\partial x_k} &= F\left(b_i^0 + \overset{\cdot}{i}{}_i^0\right) + 2F_k\Omega_{ik},\\
\frac{\partial F_{ij}}{\partial t} + \frac{\partial F_{ijk}}{\partial x_k} &= \hat{f}_{\langle ij\rangle} + 2F_{(i}\left(b_{j)}^0 + \overset{\cdot}{i}{}_{j)}^0\right) + 4F_{k(i}\Omega_{j)k},\\
\frac{\partial F_{ijj}}{\partial t} + \frac{\partial F_{ijjk}}{\partial x_k} &= \hat{f}_{ijj} + 3F_{(ij}\left(b_{j)}^0 + \overset{\cdot}{i}{}_{j)}^0\right) + 6F_{kjj}\Omega_{ik}.
\end{aligned} \qquad (8.29)$$

The vector $\mathbf{i}^0$ of the inertial acceleration is given by the relation

$$i_i^0 = \left(\dot{\Omega}_{ik} - \Omega_{ij}\Omega_{jk}\right)(x_k - c_k) - 2\Omega_{ik}\dot{c}_k + \ddot{c}_k. \qquad (8.30)$$

The above set of balance equations has very peculiar symmetry properties. Namely, the flux in the nth tensorial equation acts as the density in the (n+1)th tensorial equation. We shall see that this symmetry limits the applicability to monatomic ideal gases.

The mass balance equation $(8.29)_1$, the momentum balance equation $(8.29)_2$ and the trace of $(8.29)_3$ which is the energy balance equation are conservation laws. It means the productions in these equations are zero. Simultaneously, productions $\hat{f}_{\langle ij\rangle}$ and $\hat{f}_{ijj}$ indicate that remaining balance equations describe quantities which are not conserved.

In order to construct field equations we have to solve the closure problem. To this aim we postulate the following constitutive equations

$$\begin{aligned} F_{\langle ij\rangle k} &= F_{\langle ij\rangle k}\left(F, F_i, F_{ij}, F_{ijj}\right), \\ F_{ijjk} &= F_{ijjk}\left(F, F_i, F_{ij}, F_{ijj}\right), \\ \hat{f}_{\langle ij\rangle} &= \hat{f}_{\langle ij\rangle}\left(F, F_i, F_{ij}, F_{ijj}\right), \\ \hat{f}_{ijj} &= \hat{f}_{ijj}\left(F, F_i, F_{ij}, F_{ijj}\right). \end{aligned} \qquad (8.31)$$

The set (8.29) and (8.31) forms the full model for the above chosen fields. They must still be restricted by the following principles

- the Galilean invariance of the field equations,
- the Galilean invariance of the constitutive relations for velocity-independent quantities,
- the thermodynamical admissibility.

The first condition is satisfied if the quantities appearing in the above equations have the form presented in the relation (8.28). Hence, we have (compare the identical relations (7.39) in the kinetic approach)

$$\begin{aligned} F &= \rho, \\ F_i &= \rho v_i, \\ F_{ij} &= \rho_{ij} + \rho v_i v_j, \\ F_{ijk} &= \rho_{ijk} + 3\rho_{(ij}v_{k)} + \rho v_i v_j v_k, \\ F_{ijkl} &= \rho_{ijkl} + 4\rho_{(ijk}v_{l)} + 6\rho_{(ij}v_k v_{l)} + \rho v_i v_j v_k v_k, \end{aligned} \qquad (8.32)$$

and

$$\hat{f}_{\langle ij\rangle} = S_{\langle ij\rangle}, \quad \hat{f}_{ijj} = S_{ijj} + 2S_{\langle ij\rangle}v_j. \qquad (8.33)$$

The ρ-quantities as well as S-quantities must satisfy constitutive relations following from (8.31) and satisfying the principle of material frame indifference.

In Table 9, we present these quantities in juxtaposition with the standard notation.

Table 9: *Juxtaposition of notation: extended thermodynamics vs. classical thermodynamics*

extended thermodynamics	ρ	ρ_{ij}	$\frac{1}{2}\rho_{ii}$	$\frac{1}{2}\rho_{ijj}$
classical thermodynamics	ρ	$-t_{ij}$	$\rho\varepsilon = \frac{3}{2}p$	q_i

The relation between pressure p and specific energy ε is, as we have shown in the previous Chapter, the characteristic feature of the monatomic gases.

Now, we can reformulate constitutive relations and write them in the following equivalent form

$$\begin{aligned} \rho_{\langle ijk\rangle} &= \rho_{\langle ijk\rangle}\left(\rho, \rho_{ij}, \rho_{ijj}\right), \\ \rho_{ijjk} &= \rho_{ijjk}\left(\rho, \rho_{ij}, \rho_{ijj}\right), \\ S_{\langle ij\rangle} &= S_{\langle ij\rangle}\left(\rho, \rho_{ij}, \rho_{ijj}\right), \\ S_{ijj} &= S_{ijj}\left(\rho, \rho_{ij}, \rho_{ijj}\right), \\ \rho\eta &= \rho\eta\left(\rho, \rho_{ij}, \rho_{ijj}\right), \\ h_i &= h_i\left(\rho, \rho_{ij}, \rho_{ijj}\right), \end{aligned} \tag{8.34}$$

where the last two relations for the entropy density $\rho\eta$ and the nonconvective entropy flux h_i belong to the second law of thermodynamics. We have used here the decomposition: $\rho_{\langle ij\rangle k} = \rho_{\langle ijk\rangle} - 2/15\rho_{kll}\delta_{ij} + 3/15\rho_{jll}\delta_{ik} + 3/15\rho_{ill}\delta_{jk}$ which means that we need only the constitutive relation for the fully deviatoric part $\rho_{\langle ijk\rangle}$.

The constitutive dependence on the velocity, or in the above notation on F_i, has been omitted because of the Galilean invariance. There remains a dependence on one scalar variable, one vector variable, and one variable which is the symmetric tensor of the second rank. Simultaneously, the assumption on the material objectivity of these constitutive relations yields in Cartesian coordinates

$$\begin{aligned}
O_{ir}O_{js}O_{kt}\rho_{\langle rst\rangle}\left(\rho,\rho_{ij},\rho_{ijj}\right) &= \rho_{\langle ijk\rangle}\left(\rho,O_{ir}O_{js}\rho_{rs},O_{ir}\rho_{rjj}\right),\\
O_{ir}O_{ks}\rho_{rjjs}\left(\rho,\rho_{ij},\rho_{ijj}\right) &= \rho_{ijjk}\left(\rho,O_{ir}O_{js}\rho_{rs},O_{ir}\rho_{rjj}\right),\\
O_{ir}O_{js}S_{\langle rs\rangle}\left(\rho,\rho_{ij},\rho_{ijj}\right) &= S_{\langle ij\rangle}\left(\rho,O_{ir}O_{js}\rho_{rs},O_{ir}\rho_{rjj}\right),\\
O_{ir}S_{rjj}\left(\rho,\rho_{ij},\rho_{ijj}\right) &= S_{ijj}\left(\rho,O_{ir}O_{js}\rho_{rs},O_{ir}\rho_{rjj}\right),\\
\rho\eta\left(\rho,\rho_{ij},\rho_{ijj}\right) &= \rho\eta\left(\rho,O_{ir}O_{js}\rho_{rs},O_{ir}\rho_{rjj}\right),\\
O_{ir}h_r\left(\rho,\rho_{ij},\rho_{ijj}\right) &= h_i\left(\rho,O_{ir}O_{js}\rho_{rs},O_{ir}\rho_{rjj}\right),
\end{aligned} \tag{8.35}$$

where O_{ij} are components of an arbitrary constant orthogonal tensor. Hence, the constitutive quantities are isotropic functions of one vector and one tensor variable. For these functions we have representation theorems which were quoted in Chapter 5. However, we shall not use them here and simplify the model by making an additional *assumption* that the dependence on $\rho_{\langle ij\rangle}$ (i.e. on the stress deviator $t_{\langle ij\rangle}$) and on $\frac{1}{2}\rho_{ijj}$ (i.e. on the heat flux q_i) is *linear*. Then the isotropy property yields

$$\begin{aligned}
\rho_{\langle ijk\rangle} &= 0,\\
\rho_{ijjk} &= \alpha\delta_{ik} + \beta\rho_{\langle ik\rangle},\\
S_{\langle ij\rangle} &= -\frac{1}{\tau_t}\rho_{\langle ij\rangle},\\
S_{ijj} &= -\frac{1}{\tau_q}\rho_{ijj},
\end{aligned} \tag{8.36}$$

where the coefficients $\alpha, \beta, \tau_t, \tau_q$ may be functions of ρ and ρ_{ll}.

It should be stressed that the isotropy results are the same for the requirement of the Galilean invariance and for the material frame indifference with the matrix $\mathbf{O}$ dependent on time. This is different from the ordinary thermodynamics in which the material frame indifference requiring the invariance with respect to the Euclidean transformation is more restrictive than the Galilean invariance. In this sense, the material frame indifference is not required in the extended thermodynamics and this explains as well the appearance of nonobjective contributions to constitutive relations of extended thermodynamics.

It remains to exploit the entropy principle. We begin with the direct evaluation of the entropy inequality restricted by the field equations (8.29), (8.31). In the inertial frame and without external forces we obtain the following inequality

$$\frac{\partial\rho\eta}{\partial t} + \frac{\partial\varphi_k}{\partial x_k} - \Lambda\left(\frac{\partial F}{\partial t} + \frac{\partial F_k}{\partial x_k}\right) - \Lambda_i\left(\frac{\partial F_i}{\partial t} + \frac{\partial F_{ik}}{\partial x_k}\right)$$

$$-\Lambda_{ij}\left(\frac{\partial F_{ij}}{\partial t}+\frac{\partial F_{ijk}}{\partial x_k}-\hat{f}_{\langle ij\rangle}\right)-\Lambda_{lli}\left(\frac{\partial F_{ijj}}{\partial t}+\frac{\partial F_{ijjk}}{\partial x_k}-\hat{f}_{ijj}\right)\geq 0, \quad (8.37)$$

which must hold for all fields $\left\{F, F_i, F_{ij}, \frac{1}{2}F_{ijj}\right\}$. If we insert the constitutive relations this inequality becomes linear with respect to the following derivatives

$$\frac{\partial F}{\partial t}, \frac{\partial F_i}{\partial t}, \frac{\partial F_{ij}}{\partial t}, \frac{\partial F_{ijj}}{\partial t}, \frac{\partial F}{\partial x_k}, \frac{\partial F_i}{\partial x_k}, \frac{\partial F_{ij}}{\partial x_k}, \frac{\partial F_{ijj}}{\partial x_k}. \quad (8.38)$$

This yields the thermodynamical identities which can be written in the following compact form

$$d(\rho\eta)=\Lambda dF+\Lambda_i dF_i+\Lambda_{ij}dF_{ij}+\Lambda_{lli}dF_{ijj}, \quad (8.39)$$

$$d\varphi_k=\Lambda dF_k+\Lambda_i dF_{ik}+\Lambda_{ij}dF_{ijk}+\Lambda_{lli}dF_{ijjk}. \quad (8.40)$$

Roughly speaking, the relation (8.39) determines the multipliers and the relation (8.40) gives rise to thermodynamical identities restricting constitutive relations.

There remains the residual inequality

$$\mathcal{D}=\Lambda_{ij}\hat{f}_{\langle ij\rangle}+\Lambda_{lli}\hat{f}_{ijj}\geq 0, \quad (8.41)$$

which represents the dissipation (nonnegative entropy production).

As we have shown above the Galilean invariance yields an explicit dependence on the velocity field. This also concerns Lagrange multipliers. Bearing this property in mind, after rather tedious calculations we can write the above results of the entropy principle in the following form

$$\begin{aligned} d(\rho\eta)&=\lambda d\rho+\lambda_{ij}d\rho_{ij}+\lambda_{lli}d\rho_{ijj}, \\ dh_k&=\lambda_i d\rho_{ik}+\lambda_{ij}d\rho_{ijk}+\lambda_{lli}d\rho_{ijjk}, \end{aligned} \quad (8.42)$$

$$\begin{aligned} \lambda_k\rho+\lambda_{jji}\left(2\rho_{ik}+\rho_{ll}\delta_{ik}\right)&=0, \\ 2\lambda_{kj}\rho_{ji}+\lambda_{llm}\left(2\rho_{mki}+\rho_{ijj}\delta_{mk}\right)&=\left[\rho\eta-\lambda\rho-\lambda_{jl}\rho_{jl}-\lambda_{llm}\rho_{mjj}\right]\delta_{ki}, \end{aligned} \quad (8.43)$$

$$\mathcal{D}=\lambda_{ij}S_{\langle ij\rangle}+\lambda_{lli}S_{ijj}\geq 0. \quad (8.44)$$

In these relations, λ-multipliers are velocity independent *internal* parts of Λ-multipliers and the two are related in the following way

$$\begin{aligned}
\lambda &= \Lambda + \Lambda_{ij} v_i v_j + \Lambda_{lli} v_i v^2, \\
\lambda_i &= \Lambda_i + 2\Lambda_{ij} v_j + \Lambda_{llj}\left(v^2 \delta_{ji} + 2 v_j v_i\right), \\
\lambda_{ij} &= \Lambda_{ij} + \Lambda_{llk} 3 v_{(k}\delta_{ij)}, \\
\lambda_{lli} &= \Lambda_{lli}.
\end{aligned} \tag{8.45}$$

Clearly, relations (8.42) indicate that the internal parts of multipliers are functions of constitutive variables $\{\rho, \rho_{ij}, \rho_{ijj}\}$.

Principally, we are done with the entropy principle. However, there remains a technically difficult part of evaluation of the above results. For details, we have to refer to the literature [194], [225]. We present here only the most important conclusions.

First of all, we investigate properties of the thermodynamical equilibrium, i.e. processes in which the dissipation $\mathcal{D}$ is zero. This corresponds to the vanishing sources $S_{\langle ij\rangle}$ and S_{ijj}. Such an analysis is easier if we transform the fields $\{\rho, \rho_{ij}, \rho_{ijj}\} \to \{\lambda, \lambda_i, \lambda_{lli}\}$. We have used a similar transformation at the beginning of this Chapter. Then we have to transform the four-potential in the following manner

$$\begin{aligned}
h' &= -\rho\eta + \lambda\rho + \lambda_{ij}\rho_{ij} + \lambda_{jji}\rho_{ill}, \\
h'_k &= -h_k + \lambda_{ij}\rho_{ijk} + \lambda_{jji}\rho_{illk},
\end{aligned} \tag{8.46}$$

and rewrite (8.39), (8.40) in the alternative form

$$\begin{aligned}
dh' &= \rho d\lambda + \rho_{ij} d\lambda_{ij} + \rho_{lli} d\lambda_{jji}, \\
dh'_k &= \left[\lambda_{jji} a_{im} \frac{\partial \rho_{mk}}{\partial \lambda}\right] d\lambda + \left[\rho_{ijk} + \lambda_{llm} a_{mn} \frac{\partial \rho_{nk}}{\partial \lambda_{ij}}\right] d\lambda_{ij} \\
&\quad + \left[\rho_{ijjk} + \lambda_{llm} a_{mn} \frac{\partial \rho_{nk}}{\partial \lambda_{ppi}}\right] d\lambda_{qqi}, \\
-h'\delta_{ik} &= 2\lambda_{ij}\rho_{jk} + \lambda_{jjl}\left(2\rho_{lik} + \rho_{kpp}\delta_{li}\right).
\end{aligned} \tag{8.47}$$

The multiplier λ_i has been eliminated and has brought the tensor a_{ij} defined by the relation

$$a_{ij} = \frac{1}{\rho}\left(2\rho_{ij} + \rho_{ll}\delta_{ij}\right). \tag{8.48}$$

We return now to the definition of equilibrium. In new variables, the condition $\mathcal{D} = \lambda_{\langle ij\rangle}\big|_E S_{\langle ij\rangle}\big|_E + \lambda_{lli}\big|_E S_{ijj}\big|_E = 0$ implies that this is a minimum of this function which yields

$$\lambda_{\langle ij\rangle}\big|_E = 0, \quad \lambda_{lli}\big|_E = 0. \tag{8.49}$$

Hence, with $\rho_{ii} = 2\rho\varepsilon$ we conclude (see (8.42))

$$d(\rho\eta)|_E = \frac{2}{3}\,\lambda_{ii}|_E\, d(\rho\varepsilon) + \lambda_E d\rho. \tag{8.50}$$

Comparison with the Gibbs equation yields

$$\lambda_{ii}|_E = \frac{2}{3}\frac{1}{T}, \quad \lambda_E = -\frac{g}{T}, \quad g = \varepsilon - T\eta_E + \frac{p_E}{\rho}, \tag{8.51}$$

where T is the absolute temperature and g is the Gibbs free energy (specific free enthalpy).

The above results conclude the identification of multipliers in equilibrium.

Before we apply this knowledge, we investigate the Gibbs equation which we can now write in the form

$$d\eta_E = \frac{1}{T}\left(d\varepsilon - \frac{p_E}{\rho^2}d\rho\right). \tag{8.52}$$

This equation implies the integrability condition

$$\left(\frac{\partial\varepsilon}{\partial\rho}\right)_T = \frac{1}{\rho^2}\left(p_E - T\left(\frac{\partial p_E}{\partial T}\right)_\rho\right). \tag{8.53}$$

For a monatomic gas we have $p_E = \frac{2}{3}\rho\varepsilon$ and then we can integrate the above equation. The result is

$$p_E = T^{5/2}\Phi(z), \quad \eta_E = \frac{3}{2}\int\left(\frac{\Phi'(z)}{z} - \frac{5}{3}\frac{\Phi(z)}{z^2}\right)dz + C_1, \quad z := \frac{\rho}{T^{3/2}}, \tag{8.54}$$

where $\Phi(z)$ is an arbitrary function, $\Phi' = d\Phi/dz$, and C_1 is an arbitrary constant. Consequently, the thermal equation of state is determined by a function of a single variable. It follows as well

$$\frac{g}{T} = \frac{\Phi}{z} + \int\frac{\Phi}{z^2}dz - C_1. \tag{8.55}$$

We limit our attention to nonequilibrium processes described up to terms of the first order with respect to the deviation from equilibrium. Calculations of this approximation are based on the so-called Maxwell iteration procedure. The final results are

$$
\begin{aligned}
\rho_{\langle ijk\rangle} &= O(2)\,, \\
\rho_{\langle ijjk\rangle} &= \left[5T^{7/2}\left(\int \frac{\Phi\Phi'}{z}dz + C_1\right) + C_2\right]\delta_{ik} \\
&\quad +7\frac{T}{\Phi}\left[\int \frac{\Phi\Phi'}{z}dz + C_1\right]\rho_{\langle ij\rangle} + O(2)\,, \\
S_{\langle ij\rangle} &= -\frac{\varsigma_t}{2T^{7/2}\Phi}\rho_{\langle ij\rangle} + O(2)\,, \\
S_{ijj} &= -\frac{\varsigma_q}{5T^{9/2}\left[7\left(\int \frac{\Phi\Phi'}{z}dz + C_1\right) - 5\frac{\Phi^2}{z}\right]}\rho_{ill} + O(2)\,,
\end{aligned}
\tag{8.56}
$$

where $O(2)$ denotes contribution of the second and higher orders with respect to variables $\rho_{\langle ij\rangle}$ and ρ_{ill}, and ς_t, ς_q are proportionality coefficients related to characteristic times τ_t, τ_q appearing in the representations (8.36) of the production terms

$$
\tau_t = 2T^{7/2}\Phi\frac{1}{\varsigma_t}, \quad \tau_q = 5T^{9/2}\left[7\left(\int \frac{\Phi\Phi'}{z}dz + C_1\right) - 5\frac{\Phi^2}{z}\right]\frac{1}{\varsigma_q}. \tag{8.57}
$$

The above results yield the following relations for the entropy and the entropy flux

$$
\begin{aligned}
\rho\eta &= \rho\left[\frac{3}{2}\int\left(\frac{\Phi'}{z} - \frac{5}{3}\frac{\Phi}{z^2}\right)dz + C_3\right] - \frac{\rho_{\langle ij\rangle}\rho_{\langle ij\rangle}}{4T^{7/2}\Phi} \\
&\quad - \frac{\rho_{kki}\rho_{lli}}{10T^{7/2}\Phi\left[7\frac{T}{\Phi}\left(\int \frac{\Phi\Phi'}{z}dz + C_1\right) - 5T\frac{\Phi}{z}\right]} + O(3)\,, \\
h_k &= \frac{1}{T}\frac{1}{2}\rho_{llk} - \frac{2}{5}\frac{\rho_{\langle ki\rangle}\frac{1}{2}\rho_{lli}}{T^{7/2}\Phi} + O(3)\,.
\end{aligned}
\tag{8.58}
$$

There remains the convexity condition with respect to the densities F, F_i, F_{ij}, F_{ill} or $\rho, \rho_{ij}, \rho_{lii}$. In the case of nonequilibrium variables $\rho_{\langle ij\rangle}, \rho_{lli}$, it follows immediately from (8.58)

$$
\Phi(z) > 0, \quad 7\left(\int \frac{\Phi\Phi'}{z}dz + C_1\right) - 5\frac{\Phi^2}{z} > 0. \tag{8.59}
$$

In addition, Hessian of $\rho\eta$ with respect to ρ, ρ_{kk} must be negative definite. After some calculations, it follows

$$
0 < \Phi' < \frac{5}{3}\frac{\Phi}{z}. \tag{8.60}
$$

In the simplest case of a classical ideal gas the thermal state equation has the form

$$p = \frac{k}{m}\rho T \quad \text{i.e.} \quad \Phi(z) = \frac{k}{m} z. \tag{8.61}$$

Nothing is known about the constant C_1 but in the kinetic theory it is zero. Then the function Φ of the relation (8.61) identically satisfies conditions (8.59), (8.60).

Let us write the field equations which we have obtained in the above analysis in the form customary in the classical thermodynamics. In an arbitrary noninertial frame, they are as follows

$$\dot{\rho} + \rho \frac{\partial v_k}{\partial x_k} = 0, \tag{8.62}$$

$$\rho\left[\dot{v}_i - i_i^0 - 2\Omega_{ik} v_k\right] - \frac{\partial}{\partial x_k}\left(t_{\langle ik\rangle} - T^{5/2}\Phi(z)\,\delta_{ik}\right) = 0, \tag{8.63}$$

$$\rho \frac{3}{2}\left(T\frac{\Phi(z)}{z}\right)^{\bullet} + \frac{\partial q_k}{\partial x_k} - \left(t_{\langle ik\rangle} - T^{5/2}\Phi(z)\,\delta_{ik}\right)\frac{\partial v_i}{\partial x_k} = 0, \tag{8.64}$$

$$\rho\left[\left(\frac{t_{\langle ik\rangle}}{\rho}\right)^{\bullet} + 2\frac{t_{j\langle i}}{\rho}\frac{\partial v_{k\rangle}}{\partial x_j} - 4\frac{t_{j\langle i}}{\rho}\Omega_{k\rangle j}\right] - \frac{4}{5}\frac{\partial q_{\langle i}}{\partial x_{k\rangle}} = -\frac{1}{\tau_t} t_{\langle ik\rangle}, \tag{8.65}$$

$$\rho\left[\left(\frac{q_i}{\rho}\right)^{\bullet} + \frac{q_k}{\rho}\frac{\partial v_i}{\partial x_k} - 2\frac{q_k}{\rho}\Omega_{ik}\right]$$

$$-\frac{\partial}{\partial x_j}\left[\frac{7}{2}\left(\frac{T}{\Phi(z)} t_{\langle ij\rangle} - \frac{5}{7}T^{7/2}\delta_{ij}\right)\left(\int \frac{\Phi\Phi'}{z}dz + C_1\right)\right]$$

$$+\frac{2}{5}q_i\frac{\partial v_k}{\partial x_k} + \frac{4}{5}q_k\frac{\partial v_{(i}}{\partial x_{k)}} - \frac{1}{\rho}\left(t_{\langle ij\rangle} - T^{5/2}\Phi(z)\,\delta_{ij}\right)\frac{\partial}{\partial x_k}\left(t_{\langle jk\rangle} - T^{5/2}\Phi(z)\,\delta_{jk}\right)$$

$$+\frac{3}{2}\frac{T^{5/2}\Phi(z)}{\rho}\frac{\partial}{\partial x_k}\left(t_{\langle ik\rangle} - T^{5/2}\Phi(z)\,\delta_{ik}\right) = -\frac{1}{\tau_q}q_i, \quad z = \frac{\rho}{T^{3/2}}. \tag{8.66}$$

In order to see clearly the structure of these equations we rewrite the system in the absolute notation and substitute $\Phi = kz/m$, $p = 2\rho\varepsilon/3$, $C_1 = 0$, $\mathbf{T} = \mathbf{T}^D - p\mathbf{1}$. It follows

$$\dot{\rho} + \rho\,\mathrm{div}\,\mathbf{v} = 0,$$

$$\rho\left[\dot{\mathbf{v}} - \mathbf{i}^0 - 2\mathbf{\Omega}\mathbf{v}\right] - \operatorname{div}\mathbf{T} = 0,$$

$$\rho\dot{\varepsilon} + \operatorname{div}\mathbf{q} - \mathbf{T}\cdot\mathbf{L} = 0, \tag{8.67}$$

$$\rho\left(\frac{\mathbf{T}^D}{\rho}\right)^{\bullet} + \left[2\left(\mathbf{L}-2\mathbf{\Omega}\right)\mathbf{T}\right]^D - \frac{4}{5}\left(\operatorname{grad}\mathbf{q}\right)^D = -\frac{1}{\tau_t}\mathbf{T}^D,$$

$$\rho\left(\frac{\mathbf{q}}{\rho}\right)^{\bullet} + \left(\mathbf{L}-2\mathbf{\Omega}\right)\mathbf{q} - \operatorname{div}\left[\frac{p}{\rho}\left(\frac{7}{2}\mathbf{T} + p\mathbf{1}\right)\right]$$
$$+\frac{2}{5}\mathbf{q}\operatorname{div}\mathbf{v} - \frac{1}{\rho}\left(\mathbf{T} - \frac{3}{2}p\mathbf{1}\right)\operatorname{div}\mathbf{T} = -\frac{1}{\tau_q}\mathbf{q},$$

where $(\operatorname{grad}\mathbf{q})^D = \frac{1}{2}\left(\operatorname{grad}\mathbf{q} + (\operatorname{grad}\mathbf{q})^T\right) - \frac{1}{3}(\operatorname{div}\mathbf{q})\,\mathbf{1}$ and similarly for $\left[2\left(\mathbf{L}-2\mathbf{\Omega}\right)\mathbf{T}\right]^D$.

This completes the thermodynamical analysis of the model. We have obtained results which are explicit up to three constants C_1, C_2, C_3 and the function $\Phi(z)$. The latter is fully determined by the equilibrium thermal state equation. It remains to compare these results with the kinetic theory and with the ordinary thermodynamics. We proceed to do so.

Comparison with the results of the kinetic theory discussed in the previous Chapter (viz. relation (7.63) for the entropy and the entropy flux and (7.67), (7.68) for the remaining quantities) reveals an almost perfect agreement. Indeed if we set the constant C_1 equal to zero we obtain both in the kinetic theory and in the extended thermodynamics of ideal gases with $\Phi(z) = \frac{k}{m}z$ the following relations for the entropy and the entropy flux

$$\eta = \left[\frac{k}{m}\ln\frac{T^{3/2}}{\rho} + C_3\right] - \frac{t_{\langle ij\rangle}t_{\langle ij\rangle}}{4p\rho T} - \frac{q_i q_i}{5p^2 T}, \tag{8.68}$$
$$h_i = \frac{q_i}{T} + \frac{2}{5pT}t_{\langle ij\rangle}q_j.$$

Consequently, it is clear that the continuity of the heat flux q_i and of the entropy flux h_i on the material walls do not yield the continuity of the temperature T as it is the case in the ordinary thermodynamics. Only in the linear approximation we obtain such a result.

There exists a small difference between the two approaches as far as the relaxation times are concerned: in the kinetic theory they are equal to $\tau_t = \tau$ and $\tau_q = \frac{3}{2}\tau$, respectively, while in the extended thermodynamics they are independent. As a matter of fact, they are related to the viscosity and the heat conductivity, respectively. This can be seen if we perform the so-called Maxwell iteration on constitutive relations (8.65), (8.66). Namely, if we evaluate these relations in an infinitesimal neighborhood of equilibrium (i.e. at $t_{\langle ij\rangle} = 0, q_i = 0$) we obtain as the first step of iteration

$$t_{\langle ij\rangle} \approx 2\mu \frac{\partial v_{(i}}{\partial x_{j)}}, \quad \mu = T^{5/2}\Phi\tau_t \quad = p\tau_t \quad \text{for} \quad \Phi = \frac{k}{m}z, \tag{8.69}$$

$$q_i \approx -K\frac{\partial T}{\partial x_i}, \quad K = -\frac{5}{4}T^{5/2}\left[5\frac{\Phi^2}{z} - 7\left(\int \frac{\Phi\Phi'}{z}dz + C_1\right)\right]\tau_q$$

$$= \frac{5}{2}\frac{p^2}{\rho T}\tau_q \quad \text{for} \quad \Phi = \frac{k}{m}z.$$

The second step of the iteration follows when we insert the first step on the right-hand side of relations (8.65), (8.66) and calculate the stress deviator and the heat flux again. We shall not do so in this Chapter but we mention a few properties which follow from this procedure. First of all, the second iteration step contains the dependence on at least squares of relaxation times τ_t, τ_q. This continues in further iteration steps. Hence, the Maxwell procedure is a kind of polynomial expansion with respect to these times. Secondly, already in the second iterate there appear nonobjective contributions depending on Ω_{ij}. Let us demonstrate this nonobjectivity on a simple example. We consider a process of the rigid rotation with the constant temperature of the material element of the fluid, i.e.

$$\frac{\partial v_{(i}}{\partial x_{j)}} = 0, \quad \dot{T} = 0. \tag{8.70}$$

Then the relation for the heat flux (8.66) reduces to the following form

$$q_i = \pi\tau_q\left[\frac{\partial T}{\partial x_i} + 2\tau_q\left(\Omega_{ik} - \frac{\partial v_{[k}}{\partial x_{i]}}\right)\frac{\partial T}{\partial x_k}\right], \tag{8.71}$$

$$\pi = \frac{5}{4}T^{5/2}\left[5\frac{\Phi^2}{z} - 7\int \frac{\Phi\Phi'}{z}dz + C_1\right].$$

The second contribution is the frame-dependent objective vector and it indicates that the heat flux is not parallel to the temperature gradient. Such an effect has been discussed by I. Müller [222] who indicated that the transition from the microscopic description yields these contributions as a consequence of Coriolis forces appearing in microscopic equations of motion.

Finally, the second iterates contain coupling terms between the stress and heat flux relations.

8.3 A few remarks on boundary conditions

Modelling presented above shows clearly advantages of the application of extended thermodynamics. It is well found by the kinetic theory of gases, it defines clearly deviations from the thermodynamical equilibrium and indicates extensions to improve the description of nonequilibrium processes, it yields an advantageous mathematical structure of the hyperbolic hierarchy of differential field equations with locally well-posed Cauchy initial value problems.

However, there are also weak points of this approach which have not been removed as yet. The assumption that field equations follow from balance laws means that for each subsequent level of the hierarchy we need additional boundary conditions. This means that either we have to add some additional control on the boundary (e.g. simultaneously a temperature and a heat flux) or that there exist some microscopical mechanisms which produce constraints on the boundary automatically. These may be, for instance, some conditions imposed on fluctuations as a part of the physical definition of the boundary. As the first possibility seems to be physically meaningless the second one has been tested. These additional boundary data for which we cannot associate any physically meaningful or experimentally measurable quantities are called *noncontrollable data*. Various heuristic approaches have been considered but none of them seems to be indeed a solution of the problem. The one-dimensional stationary heat conduction problem and the one-dimensional Couette flow have been used as bench-mark problems. In order to appreciate difficulties, we present briefly the one-dimensional heat conduction problem used by Struchtrup and Weiss [315] in their motivation of the so-called *minimax principle.*

We consider the stationary heat transfer between rigid walls at rest. We assume that the pressure is fixed and then we can control only four parameters: the temperatures T_L and T_R on the walls and the velocities $v_L = v_R = 0$ on the walls. The heat flux must be adjusted so as to keep the temperatures constant on these walls. Certainly, we can change the controls on the boundaries to heat fluxes but then we cannot control simultaneously temperatures. In order to produce additional boundary conditions Struchtrup and Weiss [315] assume that the local entropy production is minimal

$$\max_{x\in[0,L]} \mathcal{D}(x) \rightarrow \min, \tag{8.72}$$

where L is the distance between the walls. This is the minimax principle. Let us mention that for one-dimensional stationary cases it can be considered to follow from a more general condition formulated in terms of the

L^p-norm of the entropy production

$$\|\mathcal{D}\|_p = \begin{cases} \left(\int_0^L \mathcal{D}^p dx\right)^p & \text{for } 1 \le p < \infty, \\ \max_{1\le x\le L} \mathcal{D} \text{ for } p = \infty, \end{cases} \tag{8.73}$$

which provides a measure of the deviation of the process from thermodynamical equilibrium. The condition (8.72) means that the entropy production is the smallest in this norm for $p = \infty$. This proposition goes back to the principle of minimum entropy production of I. Prigogine [255], who postulated that the global entropy production possesses a minimum in stationary processes. This is equivalent to the minimum principle with $p = 1$. In the work [315] it is shown that Prigogine's postulate does not yield reasonable results for heat conduction.

The set of fields in the general case consists of 14 fields (including the so-called dynamic pressure, Δ) and in the one-dimensional case under considerations it has the form

$$\left\{\rho, v, \sigma = p_{\langle 11\rangle}, T, q = q_1, \Delta = \rho_{iijj}\right\}, \tag{8.74}$$

i.e. six unknown fields: mass density, one component of the velocity, one component of the pressure deviator, temperature, one component of the heat flux. The last quantity, the *dynamic pressure*, Δ, requires in comparison with the set of equations (8.29) an additional equation in the hierarchy. For the stationary case these equations have the form

$$\frac{d}{dx}(\rho v) = 0, \quad 0 \le x \le L,$$

$$\frac{d}{dx}\left(\rho v^2 + \rho RT + \sigma\right) = 0,$$

$$\frac{d}{dx}\left(\frac{1}{2}\rho v^3 + \frac{5}{2}\rho RTv + \sigma v + q\right) = 0,$$

$$\frac{d}{dx}\left[\frac{2}{3}\rho v^3 + \left(\frac{4}{3}\rho RT + \frac{7}{3}\sigma\right)v + \frac{8}{15}q\right] = -\frac{1}{\tau}\sigma, \tag{8.75}$$

$$\frac{d}{dx}\left[\frac{1}{2}\rho v^4 + \left(4\rho RT + \frac{5}{2}\sigma\right)v^2 + \frac{16}{5}qv + \frac{5}{2}\rho R^2T^2 + \frac{7}{2}RT\sigma + \frac{1}{6}\Delta\right]$$
$$= -\frac{1}{\tau}\left(\sigma v + \frac{2}{3}q\right),$$

$$\frac{d}{dx}\left[\rho v^5 + (14\rho RT + 8\sigma)\,v^3 + \frac{84}{5}qv^2 + \left(35\rho R^2T^2 + 28RT\sigma + \frac{7}{3}\Delta\right)v + 28RTq\right] = -\frac{1}{\tau}\left(4\sigma v^2 + \frac{16}{3}qv + \frac{2}{3}\Delta\right).$$

The first three equations are, obviously, the conservation of mass, momentum and energy for ideal gases. The remaining balance equations possess source terms on the right-hand side in which τ is their common relaxation time of the order of magnitude of the mean time of free flight. It is related to the Knutsen number (compare (7.34)) by the relation

$$\tau \approx Kn\frac{L}{\sqrt{RT_L}}. \tag{8.76}$$

R is, obviously, the gas constant.

The boundary conditions for velocities and the first two equations yield

$$v\,(x) = 0. \tag{8.77}$$

Hence, according to the energy balance, the heat flux is constant and the balance of stresses leads to the zero deviatoric stress

$$\frac{dq}{dx} = 0, \quad \sigma = 0. \tag{8.78}$$

Then the balance of momentum yields a constant pressure p

$$p = \rho RT = \text{const.} \tag{8.79}$$

The remaining two equations can be written in the following dimensionless form

$$\frac{5}{2}\frac{d\bar{T}}{d\bar{x}} + \frac{1}{6}\frac{d\bar{\Delta}}{d\bar{x}} = -\frac{2}{3Kn}\frac{\bar{q}}{\bar{T}}, \quad \frac{d\bar{T}}{d\bar{x}} = -\frac{1}{42Kn}\frac{\bar{\Delta}}{\bar{q}\bar{T}}, \tag{8.80}$$

where

$$\bar{x} = \frac{x}{L}, \quad \bar{T} = \frac{T}{T_L}, \quad \bar{\Delta} = \frac{\Delta}{pRT_L}, \quad \bar{\rho} = \frac{\rho RT_L}{p} = \frac{1}{\bar{T}}, \quad \bar{q} = \frac{q}{p\sqrt{RT_L}}. \tag{8.81}$$

We have now the boundary conditions

$$\bar{T}\,(0) = 1, \quad \bar{T}\,(1) = \bar{T}_L. \tag{8.82}$$

However, in order to solve the problem we need one more condition which would determine the constant q.

It can be easily shown that the residual inequality for this problem yields the following dissipation (entropy production)

$$\mathcal{D} = \frac{4}{15Kn}\frac{\bar{q}^2}{\bar{T}^3} + \frac{1}{180Kn}\frac{\bar{\Delta}^2}{\bar{T}^4}. \tag{8.83}$$

It is only for small Knudsen numbers Kn that the nonlinear equations (8.80) reduce to the Fourier law

$$Kn \ll 1 \quad \Rightarrow \quad \bar{q} \approx -\frac{15}{4}Kn\bar{T}\frac{d\bar{T}}{d\bar{x}}. \tag{8.84}$$

The nonlinear problem (8.80), (8.82) has been solved by Struchtrup and Weiss [315] in the following way. For a chosen value $\bar{q}$ they find a numerical solution $\bar{T}(\bar{x}, \bar{q}), \bar{\Delta} = \bar{\Delta}(\bar{x}, \bar{q})$ and then $\mathcal{D}(\bar{x}, \bar{q})$. The minimax principle $\mathcal{D}_{\max}(\bar{q}) = \max_{0 \leq \bar{x} \leq 1} \mathcal{D}(\bar{x}, \bar{q})$ yields the heat flux $\bar{q}_{\min}$ corresponding to the minimum of $\mathcal{D}_{\max}(\bar{q})$.

The solution constructed by this method reveals the existence of boundary layers in the temperature distribution. These are indicated in Fig. 15 which were obtained in a different way by E. Barbera [23], [24]. However, in spite of this qualitative agreement with physical expectations the minimax principle is not acceptable because it is inconsistent with trivial changes of boundary conditions. Namely, if we replace one of the temperatures on the boundary, say T_R, by the heat flux which follows from the above considerations the solutions of the problem will be incompatible. I-Shih Liu, M. A. Rincon and I Müller [197] provide the following example of the exploitation of this principle

$$\begin{aligned} &T_L = 300 \text{ K}, \quad T_R = 387.25 \text{ K}, \quad \Rightarrow \quad q = -4455.00 \text{ W/m}^2, \\ &T_L = 300 \text{ K}, \quad q = -5000 \text{ W/m}^2, \quad \Rightarrow \quad T_R = 387.25 \text{ K}. \end{aligned} \tag{8.85}$$

Obviously, this is physically unacceptable. Similar inconsistencies were pointed out by I-Shih Liu [193] in an alternative approach (variational principle within Grad's kinetic model) of Grmela, Karlin and Zmievski [134].

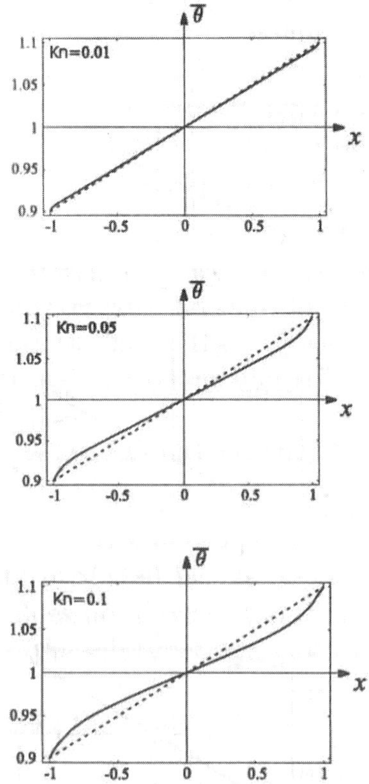

Fig. 15: *Temperature distribution for various values of Knudsen number Numerical results of E. Barbera* [23]; *broken line – classical Fourier's law.*

One of the difficulties with the hierarchic equations seems to be related to orders of magnitude of various contributions describing the deviation from thermodynamical equilibrium. One can expect that the second law of thermodynamics remains valid in truncations of hierarchies if the nonequilibrium variables and their first derivatives remain sufficiently small. However, Brini and Ruggeri [52] have shown that even in stationary processes there appear *critical derivatives* which violate this expectation. Along the same line went the attempt to improve hierarchies of the extended thermodynamics undertaken by E. Barbera, I. Müller, D. Reitebuch, W. Weiss, N.-R. Zhao [224], [24]. One of the results of this analysis is shown in Fig. 15 for the example of heat conduction presented above. They derive boundary conditions as mean values of thermal fluctuations.

It seems that the most successful attempt so far was done by I-Shih

Liu who published a series of papers with A. Rincon on an iterative procedure of derivation solutions without any additional postulates. The method goes back to the work [192] and has been developed in [196] and [193]. It relies on the following observation. The system of governing equations in extended thermodynamics can be divided into two groups: (1) the governing equations of the classical Navier-Stokes-Fourier model and (2) the field equations for the remaining nonequilibrium fields. The solution of this system can be constructed by the iteration: first one finds the solution of (1) under the assumption that all nonequilibrium fields are zero. Then one calculates from the set (2) the first approximation of nonequilibrium fields. These are used to correct the solution of the system (1), etc. In this way, one uses only the condition that nonequilibrium quantities are zero in thermodynamical equilibrium. The use of this property reminds the classical Maxwell iteration.

Similarly to the motivation of the hierarchy of balance equations one can try to motivate the structure of additional macroscopic boundary conditions using the kinetic theory of gases. However, this problem is not so well investigated as the kinetic equation and its moments. Theoretical investigations are difficult due to our lack of knowledge of the structure of boundary layers near surfaces of solids. For instance, in the interaction of a gas with the wall a molecule encountering solid may be impinged upon the surface, it may create chemical bonds, dissociate, form ions, displace surface molecules, etc. This behavior is dependent on the structure of the surface, its cleanliness, temperature, etc. It is the subject of a kinetic theory of surfaces (e.g. see [22]). Some simplified models have been proposed by C. Cercignani and M. Lampis (e.g. [66], [65]) but the field of research is far from being closed.

We return to the discussion of this issue in relation to boundary conditions for porous materials.

These remarks seem to indicate that the general solution of the problem of boundary conditions in extended thermodynamics may not exist at all. Even in the case of ideal gases and some ideal walls when no information on their detailed structure is needed, thermodynamics does not seem to offer any means to formulate such conditions. As one of the founders of extended thermodynamics, Ingo Müller, formulated the problem in his recent book [223]: "In all honesty, however, the problem of assigning data in extended thermodynamics must still be considered open so far. At the present time only such problems have been resolved by extended thermodynamics – with more than 13 moments – which do not need boundary or initial conditions or which possess trivial ones. These include shock waves, which have been treated with moderate success,..., and light scattering, which has been dealt with very satisfactory indeed...".

Chapter 9

Thermodynamical model of viscoelastic materials

9.1 Foundations

In the previous Chapter, we have demonstrated the construction of a macroscopic model of heat conducting viscous ideal gases. The kinetic theory has motivated the macroscopic modelling called extended thermodynamics. This strategy of constructing macroscopic models can be extended to materials for which the kinetic motivation is remote or even none at all. Viscoelastic materials belong to this class. In the first part of the present Chapter, we discuss the thermodynamical construction of the model whose behavior is solid-like and therefore the Lagrangian description is appropriate. We focus the attention on isotropic materials and investigate their general model following from extended thermodynamics. In particular two special cases will be considered: viscous fluids and linear viscoelastic solids. In viscous heat-conducting fluids the resulting equations have the same structure as those presented in the previous Chapter for ideal gases but the theory is less restrictive about some scalar coefficients. As the special case of the model of viscoelastic solids we obtain the standard linear solid of viscoelasticity (e.g. [251]). This construction of the model has been performed by I-Shih Liu in 1989 [188], [189].

We consider the construction of the model for the following 21 fields

- F_{iK} – deformation gradient (9 components),
- v_i – velocity (3 components),
- T – temperature (1 component),
- $t^v_{\langle ij\rangle}$ – viscous (Cauchy) stress deviator (5 components),
- $q_i = J^{-1}Q_K F_{iK}$ – heat flux (3 components).

Both Lagrangian and Eulerian indices refer to Cartesian coordinates.

Field equations follow from the set of balance equations

$$\frac{\partial F_{iK}}{\partial t}-\frac{\partial v_i}{\partial X_K}=0, \tag{9.1}$$

$$\frac{\partial G_i}{\partial t}-\frac{\partial P_{iK}}{\partial X_K}=\rho_0\left(b_i^0+i_i^0\right)+2\Omega_{ij}G_j, \tag{9.2}$$

$$\frac{\partial G_{ij}}{\partial t}+\frac{\partial \Psi_{ijK}}{\partial X_K}-S_{\langle ij\rangle}=2G_{(i}\left(b_{j)}^0+i_{j)}^0\right)-2\Omega_{k(i}\left(G_{j)}v_k+G_{j)k}-P_{j)K}F_{kK}\right), \tag{9.3}$$

$$\frac{\partial G_{ijj}}{\partial t}+\frac{\partial \Psi_{ijjK}}{\partial X_K}-S_{ijj}=3G_{(ij}\left(b_{j)}^0+i_{j)}^0\right) \tag{9.4}$$
$$-3\Omega_{k(i}\left(G_{jj)}v_k+G_{jj)k}+\Psi_{jj)K}F_{kK}\right).$$

The first relation is, obviously, the integrability condition for the deformation gradient. The second one represents the momentum conservation in a noninertial frame with the quantity $G_i=\rho_0 v_i$ representing momentum density. The remaining notation is the same as in the previous Chapters.

The trace of (9.3) represents the energy conservation law. Therefore we identify $\frac{1}{2}G_{ii}=\rho_0\left(\varepsilon+\frac{1}{2}v^2\right)$ as the total energy density and $\frac{1}{2}\Psi_{jjK}=Q_K-P_{jK}v_j$ as its flux.

Bearing Galilean invariance of balance laws in mind, we obtain on the basis of the Ruggeri Theorem (Section 8.1) the following explicit dependence on the velocity

$$\begin{aligned}
&G_{ij}=\rho_0 v_i v_j+\rho_{ij}, \quad G_{ijj}=\rho_{ijj}+3\rho_{(ij}v_{j)}+\rho_0 v_i v^2,\\
&\Psi_{ijK}=R_{ijK}-2v_i P_{j)K}, \quad \Psi_{ijjK}=R_{ijjK}+3v_{(i}R_{jj)K}-3v_{(i}v_j P_{j)K},\\
&S_{\langle ij\rangle}=s_{\langle ij\rangle}, \quad S_{ijj}=s_{ijj}+3s_{(ij}v_{j)},
\end{aligned} \tag{9.5}$$

where $\rho_{ij},\rho_{ijj},R_{ijK},R_{ijjK},P_{jK},s_{\langle ij\rangle},s_{ijj}$ are velocity-independent and Galilean invariant. These relations yield the following identifications

$$\varepsilon=\frac{1}{2\rho_0}\rho_{ii}, \quad Q_K=Jq_i F_{Ki}^{-1}=\frac{1}{2}R_{iiK}, \tag{9.6}$$

where ε is the *specific internal energy* and Q_K denotes the *material heat flux*, i.e. the heat flux vector in the Lagrangian description. In addition, apart from the Piola-Kirchhoff stress tensor P_{iK} we frequently apply the symmetric Cauchy stress tensor $T_{ij}=J^{-1}P_{iK}F_{jK}=T_{ji}$.

In order to construct field equations we have to close the system of balance laws (9.1)-(9.4) by constitutive relations. We select the following set of constitutive variables

$$\mathcal{C}' = \left\{ F_{iK}, T, t^v_{\langle ij \rangle}, q_i \right\}, \tag{9.7}$$

where $t^v_{\langle ij \rangle}$ is supposed to represent the deviatoric viscous stress tensor which we identify later. For technical reasons it is more convenient to choose the variables appearing in (9.5) instead of the above physical variables. We shall do so and after the evaluation of thermodynamical principles we return to this more suggestive choice. Hence, we choose the following set

$$\mathcal{C} = \left\{ F_{iK}, \rho_{ij}, \rho_{ijj} \right\}, \tag{9.8}$$

and assume that the following constitutive quantities

$$\mathcal{F} = \left\{ P_{iK}, R_{ijK}, R_{ijjK}, s_{\langle ij \rangle}, s_{ijj} \right\}, \tag{9.9}$$

are sufficiently smooth functions of constitutive variables

$$\mathcal{F} = \mathcal{F}(\mathcal{C}). \tag{9.10}$$

The velocity v_i or, equivalently, the vector G_i does not appear anymore among constitutive variables which is the consequence of the Galilean invariance. Some other restrictions following from this invariance shall be presented later. Simultaneously, the constitutive dependence contains only fields (comp. (9.7)) or their transformations (9.8). There is no dependence on gradients or time derivatives of fields. This is the characteristic feature of the extended thermodynamics and, for this reason, we call constitutive relations (9.10) local.

9.2 Thermodynamical admissibility

As usual the constitutive relations must satisfy the condition of thermodynamical admissibility which is formulated by means of the entropy inequality

$$\rho_0 \frac{\partial \eta}{\partial t} + \frac{\partial H_K}{\partial X_K} \geq 0, \tag{9.11}$$

where

$$\eta = \eta(\mathcal{C}), \quad H_K = H_K(\mathcal{C}). \tag{9.12}$$

It is obvious that the entropy flux in the Lagrangian description is independent of the velocity. It contains only the nonconvective part which depends on constitutive variables $\mathcal{C}$.

Entropy inequality must hold for all thermodynamical processes, i.e. for all solutions of field equations. As before this constraint is eliminated by Lagrange multipliers which are also constitutive functions. Easy calculations then yield

$$\begin{aligned} \rho_0 d\eta &= \lambda_{iK} dF_{iK} + \lambda_{ij} d\rho_{ij} + \lambda_{ikk} d\rho_{ijj}, \\ dH_K &= -\lambda_i dP_{iK} + \lambda_{ij} dR_{ijK} + \lambda_{ikk} dR_{ijjK}, \end{aligned} \tag{9.13}$$

where the multiplier of the momentum balance equation does not contribute to the above relations and the remaining multipliers $\lambda_{iK}, \lambda_{ij}, \lambda_{ikk}$ are velocity independent parts of multipliers corresponding to constraints imposed by (9.1), (9.3) and (9.4), respectively (compare (7.35)). It follows as well that the multiplier of the momentum balance equation contains the velocity independent part given by the relation

$$\lambda_i = -\frac{3}{\rho_0} \rho_{(ij} \lambda_{j)kk}, \tag{9.14}$$

and the multiplier λ_{iK} has the form

$$\lambda_{iK} = -2\lambda_{ij} P_{jK} + 3\lambda_{kk(j} R_{ij)K}. \tag{9.15}$$

There remains the residual inequality of the following form

$$\lambda_{ij} s_{\langle ij \rangle} + \lambda_{ijj} s_{ijj} + \tag{9.16}$$

$$+ \left[2\lambda_{ij} \left(\rho_{ij} - P_{jK} F_{iK} \right) + \lambda_{ill} \left(\rho_{kjj} + R_{jjK} F_{kK} \right) + 2\lambda_{jll} R_{ijK} F_{kK} \right] \Omega_{ik} \geq 0.$$

Since the inequality must hold for all frames it must be independent of the matrix of angular velocities Ω_{ik}. Consequently, the following identity must be satisfied

$$\left[2\lambda_{ij} \left(\rho_{ij} - P_{jK} F_{iK} \right) + \lambda_{ill} \left(\rho_{kjj} + R_{jjK} F_{kK} \right) + 2\lambda_{jll} R_{ijK} F_{kK} \right] \varepsilon_{ikl} = 0, \tag{9.17}$$

where ε_{ikl} is the permutation symbol. Simultaneously, the residual inequality becomes

$$\mathcal{D} = \lambda_{ij} s_{\langle ij \rangle} + \lambda_{ijj} s_{ijj} \geq 0, \tag{9.18}$$

and this relation defines the dissipation $\mathcal{D}$ in the system.

As we already know it is convenient to change variables $\left\{ F_{iK}, \rho_{ij}, \rho_{ijj} \right\} \to \left\{ F_{iK}, \lambda_{ij}, \lambda_{ikk} \right\}$ which is possible due to the convexity assumption. Then we can define the conjugate four-potential

$$\begin{aligned} h' &= \lambda_{ij}\rho_{ij} + \lambda_{ikk}\rho_{ijj} - \rho_0\eta, \\ H'_K &= -\lambda_i P_{iK} + \lambda_{ij}R_{ijK} + \lambda_{ikk}R_{ijjK} - H_K. \end{aligned} \tag{9.19}$$

It follows

$$\begin{aligned} dh' &= \rho_{ij}d\lambda_{ij} + \rho_{ijj}d\lambda_{ikk} - \lambda_{iK}dF_{ik}, \\ dH'_K &= -P_{iK}d\lambda_i + R_{ijK}d\lambda_{ij} + R_{ijjK}d\lambda_{ikk}. \end{aligned} \tag{9.20}$$

We proceed to analyze the thermodynamical equilibrium defined by the condition $\mathcal{D} = 0$ which corresponds to the requirement of vanishing productions $s_{\langle ij\rangle}\big|_E = 0$, $s_{ijj}\big|_E = 0$. As the inequality (9.18) is the condition for the minimum of the dissipation in the equilibrium, we conclude that the multipliers $\lambda_{\langle ij\rangle}, \lambda_{ijj}$ also vanish in equilibrium. This makes the new choice of variables so attractive from the technical point of view.

Bearing $(9.13)_1$ and (9.15) in mind, we obtain in equilibrium

$$d\eta_E = \frac{2}{3}\lambda_{ii}\left(d\varepsilon_E - \frac{1}{\rho_0}\, P_{iK}\big|_E\, dF_{iK}\right). \tag{9.21}$$

Therefore we can identify

$$\frac{2}{3}\lambda_{ii} = \frac{1}{T}, \tag{9.22}$$

where T is the absolute temperature. The relation (9.21) becomes Gibbs equation. Consequently, we can introduce the equilibrium Helmholtz free energy $\psi_E = \varepsilon_E - T\eta_E$ which yields

$$P_{iK}\big|_E = \rho_0\frac{\partial\psi_E}{\partial F_{iK}}, \quad \varepsilon_E = \psi_E - T\frac{\partial\psi_E}{\partial T}, \quad h'_E = \rho_0\frac{\psi_E}{T}, \tag{9.23}$$

where relations $\frac{1}{2}\rho_{jj} = \rho_0\varepsilon$ and (9.22) have been used. In addition, the evaluation of the identity (9.17) in equilibrium leads to the relation

$$\rho_{\langle ij\rangle}\Big|_E = \frac{\rho_0}{\rho}\, t_{\langle ij\rangle}\big|_E, \quad t_{ij} \equiv J^{-1}P_{iK}F_{jK} \equiv \frac{\rho}{\rho_0}P_{iK}F_{jK}. \tag{9.24}$$

Further analysis will be limited to a few key results of an approximate theory. We construct constitutive relations in the *second order approximation* with respect to the deviation from equilibrium which is measured by multipliers $\lambda_{\langle ij\rangle}$ and λ_{ijj} vanishing in equilibrium. Detailed structure, in particular symmetry properties of material parameters in the nonlinear theory as well as the derivation of restrictions following from the hyperbolicity condition (convexity assumption) are technically very involved and we refer for details to the publications of I-Shih Liu.

In this approximation we have

$$h' = \rho_0 \left(\frac{\psi_E}{T} + \frac{1}{\rho}\, t_{\langle ij \rangle}\big|_E \lambda_{\langle ij \rangle} + \frac{1}{2} h_{ijkl} \lambda_{\langle ij \rangle} \lambda_{\langle kl \rangle} + \frac{1}{2} k_{ij} \lambda_{ikk} \lambda_{jll} \right) + O(3),$$
$$H'_i = J^{-1} H'_K F_{iK} = \left(\chi_{ij} + \chi'_{ijkl} \lambda_{\langle kl \rangle} \right) \lambda_{jmm} + O(3), \tag{9.25}$$

where the coefficients $h_{ijkl}, k_{ij}, \chi_{ij}, \chi'_{ijkl}$ as well as ψ_E and $t_{\langle ij \rangle}\big|_E$ are isotropic functions of the Cauchy-Green deformation tensor $B_{ij} = F_{iK} F_{jK}$ and of the temperature T. The first coefficient, h_{ijkl}, is, in addition, traceless symmetric in both pairs of indices and it is symmetric with respect to these pairs. k_{ij} is symmetric.

The above relations for the four-potential yield constitutive relations for all constitutive quantities of the model except of productions. They enable the exploitation of relations (9.20) which are the generalization of the Gibbs equation to nonequilibrium processes with multipliers as process variables. We proceed to exploit these relations.

In the second order approximation, the first equation (9.20) yields after easy calculations

$$\varepsilon = \varepsilon_E - \frac{T^2}{\rho} \frac{\partial\, t_{\langle ij \rangle}\big|_E}{\partial T} \lambda_{\langle ij \rangle}$$
$$- \frac{T^2}{2} \frac{\partial h_{ijkl}}{\partial T} \lambda_{\langle ij \rangle} \lambda_{\langle kl \rangle} - \frac{T^2}{2} \frac{\partial k_{ij}}{\partial T} \lambda_{ikk} \lambda_{jll} + O(3),$$

$$\rho_{\langle ij \rangle} = \rho_0 \left(\frac{1}{\rho}\, t_{\langle ij \rangle}\big|_E + h_{ijkl} \lambda_{\langle kl \rangle} \right) + O(2),$$
$$\rho_{ijj} = \rho_0 k_{ij} \lambda_{jll} + O(2),$$

$$\lambda_{kK} = - \frac{\rho_0}{T} \frac{\partial \psi_E}{\partial F_{kK}} - \frac{\partial \left(P_{iL}\big|_E F_{jL} \right)}{\partial F_{kK}} \lambda_{\langle ij \rangle}$$
$$- \frac{\rho}{2} \left(\frac{\partial h_{ijlm}}{\partial F_{kK}} \lambda_{\langle ij \rangle} \lambda_{\langle lm \rangle} + \frac{\partial k_{ij}}{\partial F_{kK}} \lambda_{imm} \lambda_{jpp} \right) + O(3).$$

Combination of these results yields the relation for the Helmholtz free energy

$$\psi = \psi_E - \frac{T}{2} h_{ijkl} \lambda_{\langle ij \rangle} \lambda_{\langle kl \rangle} - \frac{T}{2} k_{ij} \lambda_{imm} \lambda_{jpp} + O(3). \tag{9.26}$$

Inspection of (9.15) shows that up to second order terms we have

$$t_{ij}^v = T_{ij} - T_{ij}\big|_E = T\left(-T_{ij}\big|_E + \rho\frac{\partial^2\psi_E}{\partial F_{iK}\partial F_{kL}}F_{jK}F_{lL}\right)\lambda_{\langle kl\rangle} + O(2). \tag{9.27}$$

We have anticipated that R_{ijK} is vanishing in equilibrium. We return later to this point.

Thus the stress is decomposed into an equilibrium part $t_{ij}\big|_E$ and a non-equilibrium part t_{ij}^v. The former is referred to as the *elastic stress* and the latter as the *viscous stress.* One can introduce the following *spatial elasticity tensors (first* and *second)*

$$\begin{aligned} A_{ijkl} &= \rho\frac{\partial^2\psi_E}{\partial F_{iK}\partial F_{kL}}F_{jK}F_{lL}, \\ C_{ijkl} &= A_{ijkl} - t_{jl}\big|_E\,\delta_{ik}. \end{aligned} \tag{9.28}$$

Then

$$t_{ij}^v = TC_{ijkl}\lambda_{\langle kl\rangle} + O(2). \tag{9.29}$$

Now we are in the position to specify the second order approximation of the multiplier λ_i given by (9.14)

$$\lambda_i = \left(r_{ik} + r'_{jlki}\lambda_{\langle jl\rangle}\right)\lambda_{kpp} + O(2), \tag{9.30}$$

where

$$r_{ik} = -\frac{10}{3}\varepsilon_E\delta_{ik} - \frac{2}{\rho}\,t_{\langle ik\rangle}\big|_E, \quad r'_{ijkl} = \frac{10}{3}\frac{T^2}{\rho}\frac{\partial\, t_{\langle ij\rangle}\big|_E}{\partial T}\delta_{kl} - 2h_{ijkl}. \tag{9.31}$$

This completes the constitutive consequences of the first equation (9.20) in terms of multipliers $\lambda_{\langle kl\rangle}, \lambda_{kpp}$ as variables describing the deviation of processes from the equilibrium.

These results make possible the exploitation of the second equation (9.20).

We easily obtain the following relations for quantities transformed to Eulerian description (e.g. $R_{ijk} = J^{-1}R_{ijK}F_{kK}$, $R_{ijjk} = J^{-1}R_{ijjK}F_{kK}$)

$$q_i = \left(\kappa_{ki} + \kappa'_{jlki}\lambda_{\langle jl\rangle}\right)\lambda_{kpp} + O(3), \tag{9.32}$$

$$R_{\langle ijk\rangle} = \alpha_{ijkl}\lambda_{lpp} + O(2), \tag{9.33}$$

$$R_{kppl} = \beta_{kl} + \beta'_{ijkl}\lambda_{\langle ij\rangle} + O(2), \tag{9.34}$$

where

$$\begin{aligned}
\kappa_{ki} &= -T^2 \left(\frac{\partial r_{kj}}{\partial T}\, t_{ji}\big|_E + \frac{\partial \chi_{ki}}{\partial T} \right), \\
\kappa'_{ijkl} &= -T^2 \left(\frac{\partial r'_{ijkp}}{\partial T}\, t_{pl}\big|_E + \frac{\partial \chi'_{ijkl}}{\partial T} + T C_{pl\langle ij\rangle} \frac{\partial r_{kp}}{\partial T} \right), \\
\alpha_{ijkl} &= r'_{ijkp}\, t_{pl}\big|_E + \chi'_{ijkl}, \\
\beta_{ij} &= r_{ip}\, t_{pl}\big|_E + \chi_{ij}, \quad \beta'_{ijkl} = \alpha_{ijkl} + T r_{kp} C_{pl\langle ij\rangle}.
\end{aligned} \tag{9.35}$$

By means of the above relations we can construct the viscous stress in an explicit form. The result is

$$t^v_{ij} = T\left(C_{ijkl}\lambda_{\langle kl\rangle} + C'_{ijkl}\lambda_{kpp}\lambda_{lmm} + C''_{ijklmn}\lambda_{\langle kl\rangle}\lambda_{\langle mn\rangle}\right) + O(3), \tag{9.36}$$

where the following abbreviations for the coefficients of the second order have been introduced

$$\begin{aligned}
C'_{ijkl} &= \frac{\rho}{2}\frac{\partial k_{kl}}{\partial F_{iK}} F_{jK} + 2\alpha_{i\langle kl\rangle j} + \frac{10}{3}\kappa_{i\langle k}\delta_{l\rangle j}, \\
C''_{ijklmn} &= \frac{\rho}{2}\frac{\partial h_{klmn}}{\partial F_{iK}} F_{jK} - 2T\delta_{i\langle k} C_{l\, j\rangle\langle mn\rangle}.
\end{aligned} \tag{9.37}$$

Finally, the comparison of results (9.19), (9.25) and (9.32) yields

$$H_k = \frac{1}{T} q_k + \lambda_{\langle ij\rangle} R_{\langle ij\rangle k} + O(3). \tag{9.38}$$

We conclude that the classical Fourier relation $H_k = q_k/T$ holds only in the first order.

It remains to eliminate the multipliers and return to the physical variables which we have mentioned at the beginning of the Chapter. We obtain from (9.29) and (9.32)

$$\lambda_{\langle ij\rangle} = \frac{1}{T} C^{-1}_{\langle ij\rangle\langle kl\rangle} t^v_{\langle kl\rangle} + O(2), \quad \lambda_{ijj} = \kappa^{-1}_{ij} q_j + O(2), \tag{9.39}$$

where

$$C^{-1}_{\langle ij\rangle\langle mn\rangle} C_{\langle mn\rangle\langle kl\rangle} = \frac{1}{2}\left(\delta_{ik}\delta_{jl} + \delta_{il}\delta_{jk} - \frac{2}{3}\delta_{ij}\delta_{kl}\right). \tag{9.40}$$

The higher order terms can also be obtained by the iteration which we do not present here.

In order to complete field equations of the approximate model we need constitutive relations for productions $s_{\langle ij\rangle}$ and s_{ijj}. We know that they have to satisfy the inequality (9.18) for the dissipation $\mathcal{D}$. Zero value corresponding to equilibrium is the minimum of this function. Consequently, the matrices

$$\sigma_{ijkl} = \frac{1}{\rho_0} \frac{\partial s_{\langle ij \rangle}}{\partial \lambda_{\langle kl \rangle}}\bigg|_E, \quad \tau_{ij} = \frac{1}{\rho_0} \frac{\partial s_{ikk}}{\partial \lambda_{jmm}}\bigg|_E, \tag{9.41}$$

are positive definite, i.e. for arbitrary matrices $A_{\langle ij \rangle}$ and arbitrary vectors B_i

$$\sigma_{ijkl} A_{\langle ij \rangle} A_{\langle kl \rangle} > 0, \quad \tau_{ij} B_i B_j > 0. \tag{9.42}$$

Hence, for linear relations we obtain by means of relations (9.39)

$$s_{\langle ij \rangle} = \frac{\rho_0}{T} \sigma_{ijkl} C^{-1}_{\langle kl \rangle \langle mn \rangle} t^v_{\langle mn \rangle} + O(2), \tag{9.43}$$
$$s_{ijj} = \rho_0 \tau_{ij} \kappa^{-1}_{jm} q_m + O(2).$$

It is helpful to collect in the juxtaposition the material parameters of the second order model for nonequilibrium variables λ_{ijj} and $\lambda_{\langle ij \rangle}$. We do so in the following Table 10.

Table 10: *Nonequilibrium material parameters for constitutive quantities of the second order model in terms of variables* $\lambda_{\langle ij \rangle}$ *and* λ_{ijj}.

entropy	h'	$\frac{1}{2}k_{ij}$	$\frac{1}{2}h_{ijkl}$
entropy flux	H'_i	χ_{ij}	χ'_{ijkl}
specific internal energy	ε	$-T^2/2\ \partial k_{ij}/\partial T$	$-T^2/2\ \partial h_{ijkl}/\partial T$
Helmholtz free energy	ψ	$-T/2\ k_{ij}$	$-T/2\ h_{ijkl}$
viscous stress tensor	t^v_{ij}		TC_{ijkl}
heat flux	q_i	κ_{ij}	κ'_{ijkl}
stress source	$s_{\langle ij \rangle}$		$\rho_0 \sigma_{ijkl}$
heat flux source	s_{ijj}	$\rho_0 \tau_{ij}$	

As shown before, they are not all independent.

9.3 Viscous fluids, linear viscoelastic solids

The above considerations complete the construction of the model. We do not show here the final form of field equations as further results for this set are nonexistent in the general case. Instead, we present two important particular cases: viscous fluids and linear viscoelastic solids.

Let us begin with the *viscous fluid.* For this material, a constitutive dependence on the deformation tensor B_{ij} reduces to its determinant (volume changes) or, equivalently, to the dependence on the current mass density ρ. Simultaneously, all constitutive tensors of the second order model become isotropic functions for which we know the representations. The resulting constitutive relations are as follows

$$\begin{aligned}
t_{ij}\big|_E &= -p\delta_{ij}, \quad t_{ij} = -p\delta_{ij} + t^v_{\langle ij\rangle}, \quad p = p\left(\rho, T\right), \quad \frac{\partial p}{\partial \rho} > 0, \\
\varepsilon &= \varepsilon_E\left(\rho, T\right), \quad \frac{\partial \varepsilon_E}{\partial T} > 0, \\
\rho_{\langle ij\rangle} &= \frac{\rho_0}{\rho}\left(1 + \frac{\gamma}{2\rho}\right) t^v_{\langle ij\rangle}, \quad \gamma = \gamma\left(\rho, T\right), \quad 2p + \gamma < 0, \\
\rho_{ijj} &= \frac{\rho_0}{\rho}\left(\frac{\alpha_2}{\kappa} - \frac{10}{3}\right) q_i, \quad \alpha_2 = \alpha_2\left(\rho, T\right), \quad \alpha_2 - \frac{10}{3}\kappa < 0, \\
R_{\langle ij\rangle k} &= \frac{1}{4}\frac{\alpha_1}{\kappa}\left(q_i\delta_{jk} + q_j\delta_{ik} - \frac{2}{3}q_k\delta_{ij}\right), \quad \alpha_1 = \alpha_1\left(\rho, T\right), \\
R_{ikkj} &= \beta\delta_{ij} + \left(\frac{\alpha_1}{4\rho T} - \frac{10}{3}\varepsilon_E + 2\frac{p}{\rho}\right) t^v_{\langle ij\rangle}, \quad \beta = \beta\left(\rho, T\right), \\
s_{\langle ij\rangle} &= \frac{\rho_0}{2pT}\sigma t^v_{\langle ij\rangle}, \quad \sigma = \sigma\left(\rho, T\right) > 0, \\
s_{ijj} &= \frac{\rho_0}{\kappa}\tau q_i, \quad \tau = \tau\left(\rho, T\right) > 0,
\end{aligned} \tag{9.44}$$

where

$$\kappa = -T^2\left(\frac{\partial \beta}{\partial T} - \frac{10}{3}\varepsilon_E\frac{\partial p}{\partial T}\right). \tag{9.45}$$

Hence, apart from the thermal equation of state $p\left(\rho, T\right)$ and caloric equation of state $\varepsilon_E\left(\rho, T\right)$ the model requires the knowledge of 6 equilibrium material parameters: β, γ, α_1, α_2, σ, τ, the latter two related to the relaxation times τ_t, τ_q, discussed in previous Chapters.

Bearing these constitutive relations in mind, we obtain from the balance equations the following equations

$$\begin{aligned}
t^v_{\langle ij\rangle} &= \frac{4p^2T}{\rho\sigma}\left[\frac{\partial v_{\langle i}}{\partial x_{j\rangle}} + \frac{2p+\gamma}{4p^2}\left(\overset{\bullet}{t^v_{\langle ij\rangle}} - \frac{\partial v_{[i}}{\partial x_{k]}}t^v_{\langle kj\rangle} - \frac{\partial v_{[j}}{\partial x_{k]}}t^v_{\langle ki\rangle}\right)\right. \\
&\left. + \frac{\alpha_1}{4p\kappa}\frac{\partial q_{\langle i}}{\partial x_{j\rangle}}\right] + \frac{\gamma T}{\rho\sigma}\left[\left(\frac{\partial v_{[i}}{\partial x_{k]}} - \Omega_{ik}\right)t^v_{\langle jk\rangle} + \left(\frac{\partial v_{[j}}{\partial x_{k]}} - \Omega_{jk}\right)t^v_{\langle ik\rangle}\right],
\end{aligned} \tag{9.46}$$

$$\begin{aligned} \underline{q_i = -\frac{\kappa^2}{\rho\tau T^2}\left[\frac{\partial T}{\partial x_i}\right.} - \frac{T^2}{\kappa^2}\left(\alpha_2 - \frac{10}{3}\kappa\right)\left(\overset{\bullet}{q}_i - \frac{\partial v_{[i}}{\partial x_{k]}} q_k\right) \\ \left. - \frac{\alpha_1 T}{4p\kappa}\frac{\partial t^v_{\langle ik\rangle}}{\partial x_k}\right] + \frac{1}{\rho\tau}\left(\alpha_2 - \frac{5}{3}\alpha_1\right)\left(\frac{\partial v_{[i}}{\partial x_{k]}} - \Omega_{ik}\right) q_k. \end{aligned} \tag{9.47}$$

Obviously, the underlined terms represent the Navier-Stokes relation and the Fourier relation of the classical thermodynamics. The above generalization contains, in addition, rate terms for $t^v_{\langle ij\rangle}$ and q_i, the coupling of both relations and the contribution of noninertial frame which makes the above relations material frame indifferent.

However, the model does not account properly for the so-called *normal stress effects* (Weissenberg effects) which appear in the classical Maxwell model of viscoelastic fluids (e.g. non-Newtonian fluids) as well as in the so-called Rivlin-Ericksen fluids. We present an appropriate model in the next Section. Unfortunately, there is no uniform approach to this problem of extended thermodynamics which would account for all properties of both classes of models.

We proceed to present the *linear viscoelastic solid.* We assume the initial configuration to be stress-free and, for small deformations,

$$B_{ij} \approx \delta_{ij} + 2\varepsilon_{ij}. \tag{9.48}$$

It follows

$$\begin{aligned} T_{ij} &= \lambda\varepsilon_{kk}\delta_{ij} + 2\mu\varepsilon_{ij} + t^v_{\langle ij\rangle}, \quad \text{i.e.} \\ C_{ijkl} &\approx A_{ijkl} = \lambda\delta_{ij}\delta_{kl} + \mu\left(\delta_{ik}\delta_{jl} + \delta_{jk}\delta_{il}\right), \end{aligned} \tag{9.49}$$

where λ, μ are Lamé moduli, $\varepsilon_0 = \varepsilon_0(T)$ and

$$\begin{aligned} \rho_{\langle ij\rangle} &= 2\mu\varepsilon_{\langle ij\rangle} + \left(1 + \frac{\gamma}{2\mu}\right) t^v_{\langle ij\rangle}, \\ \rho_{ijj} &= -\left(\frac{10}{3} + \frac{\alpha_2}{T}\left(\frac{d\beta}{dT}\right)^{-1}\right) q_i, \\ R_{\langle ij\rangle k} &= \frac{\alpha_1}{4T^2}\left(\frac{d\beta}{dT}\right)^{-1}\left(q_i\delta_{jk} + q_j\delta_{ik} - \frac{2}{3}q_k\delta_{ij}\right), \\ R_{ikkj} &= \beta\delta_{ij} - \frac{10}{3}\varepsilon_0\left(\lambda\varepsilon_{kk}\delta_{ij} + 2\mu\varepsilon_{ij}\right) + \left(\frac{\alpha_1}{4\mu T} - \frac{10}{3}\varepsilon_0\right) t^v_{\langle ij\rangle}, \end{aligned} \tag{9.50}$$

$$s_{\langle ij\rangle} = \frac{\rho}{2\mu T}\sigma t^{v}_{\langle ij\rangle},$$

$$s_{ikk} = -\frac{\rho}{T^2}\left(\frac{d\beta}{dT}\right)^{-1}\tau q_i.$$

All material parameters can be arbitrary functions of temperature and the following restrictions must be satisfied

$$\frac{d\varepsilon_0}{dT} > 0, \quad \lambda + \frac{2}{3}\mu > 0, \quad \mu > 0, \quad 2\mu + \gamma < 0,$$

$$\alpha_2 + \frac{10}{3}T^2\frac{d\beta}{dT} < 0, \quad \sigma > 0, \quad \tau > 0. \tag{9.51}$$

Now the balance equations for stresses and the heat flux yield

$$\tau_t \overset{\bullet}{t}_{\langle ij\rangle} + t_{\langle ij\rangle} - 2\mu\varepsilon_{\langle ij\rangle} - 2\left(\tau_t + \nu\right)\dot{\varepsilon}_{\langle ij\rangle} = 2\nu\frac{\alpha_1}{4\mu T^2}\left(\frac{d\beta}{dT}\right)^{-1}\frac{\partial q_{\langle i}}{\partial x_{j\rangle}}, \tag{9.52}$$

$$\tau_q \overset{\bullet}{q}_i + K\frac{\partial T}{\partial x_i} + q_i = -\frac{\alpha_1 K}{4\mu T}\left(\frac{d\beta}{dT}\right)^{-1}\frac{\partial t^{v}_{\langle ij\rangle}}{\partial x_j}, \quad t^{v}_{\langle ij\rangle} = t_{ij} - 2\mu\varepsilon_{\langle ij\rangle}, \tag{9.53}$$

where

$$\nu = \frac{2\mu^2 T}{\rho\sigma}, \quad K = \frac{T^2}{\rho\tau}\left(\frac{d\beta}{dT}\right)^2 \tag{9.54}$$

$$\tau_t = -\frac{T}{\rho\sigma}\left(2\mu + \gamma\right), \quad \tau_q = -\frac{1}{\rho\tau}\left(\alpha_2 + \frac{10}{3}T^2\frac{d\beta}{dT}\right).$$

The *viscosity coefficient* ν, *the thermal conductivity* K, *the stress relaxation time* τ_t, and *the thermal relaxation time* τ_q are all non-negative.

Equation (9.52) describes the so-called *standard linear solid* in viscoelasticity (e.g. see [73]). For isothermal processes, this can be integrated and we obtain the following stress-strain relation of the *hereditary type (a linear material with fading memory)*

$$t_{\langle ij\rangle} = 2\left(\mu + \frac{\nu}{\tau_t}\right)\varepsilon_{\langle ij\rangle} - 2\frac{\nu}{\tau_t^2}\int_0^t \varepsilon_{\langle ij\rangle}\left(s\right)e^{-\frac{t-s}{\tau_t}}\,ds. \tag{9.55}$$

Simultaneously, equation (9.53) without the right-hand side is the Cattaneo equation.

In any case, the above models are hyperbolic which means that disturbances propagate with finite speeds.

9.4 Maxwell fluid, Rivlin-Ericksen fluids

9.4.1 *Preliminary remarks*

As indicated in the previous Section, the extended thermodynamics model does not account properly for certain effects in non-Newtonian fluids. In particular, this model does not describe the normal stress (Weissenberg) effects.

In order to appreciate this remark let us consider a fluid which a rheologist might describe as a second-order fluid whose Cauchy stress tensor is given by the constitutive relation

$$\mathbf{T} + p\mathbf{1} = \nu \mathbf{A}_1 + \alpha_1 \mathbf{A}_2 + \alpha_2 (\mathbf{A}_1)^2, \tag{9.56}$$

where p is the reaction pressure on incompressibility, $\mathbf{A}_1, \mathbf{A}_2$ are Rivlin-Ericksen tensors defined by relations (3.13), (3.14) and ν is the viscosity coefficient, α_1 and α_2 are the so-called *normal stress coefficients*. Such relations were introduced by Rivlin and Ericksen [267] (for details see the book of Coleman, Markovitz and Noll [78]) through the expansion of memory functionals appearing in Maxwell model of non-Newtonian fluids. Material coefficients have been measured for various materials in Maxwell rheometer in which the nontrivial cylindrical components of the two Rivlin-Ericksen tensors have the form (see Part II of this book)

$$(\mathbf{A}_1)_{r\varphi} = \kappa, \quad (\mathbf{A}_2)_{rr} = 2\kappa^2, \tag{9.57}$$

where κ is the shear rate. Typical values of such measurements are presented in the Table below for 10.5% solution of polyisobutylene in decahydronaphtalene (a typical non-Newtonian fluid)

κ [1/s]	ν [Ns/m^2]	α_1 [Ns^2/m^2]	α_2 [Ns^2/m^2]
0.5	107.8	-27.3	44.6
1.0	98.0	-19.0	28.3
2.0	42.0	-11.9	18.1

They were found by R. F. Ginn and A. B. Metzner [122].

There are difficulties with the second-order fluids because these values are not compatible with the ordinary thermodynamics considerations. Dunn and Fosdick [92] have found the following thermodynamical restrictions following from the condition of stability of the thermodynamical equilibrium

$$\nu > 0, \quad \alpha_1 > 0, \quad \alpha_1 + \alpha_2 = 0. \tag{9.58}$$

Clearly, the second and third conditions contradict experimental results. Fosdick, Rajagopal and many others were aware of the shortcoming of this model (e.g. [113], [114], [257]). They were trying to argue that such models as the second order fluid as well as higher order models[1] should not be considered as approximations of the Maxwell model obtained by the truncation of the series expansion in memory functionals. They should rather be models with their own status. For this reason they have obtained special names – nth *grade fluids*. However, the problem was still not solved as real materials do not seem to satisfy such conditions as (9.58). The situation became even worse when D. D. Joseph found in 1981 that the truncation of the series approximating Maxwell model yields the instability of the rest state [158].

However, no resolution of these flaws were found at the time.

An attempt to construct a thermodynamical model with proper material coefficients has been undertaken by I. Müller and K. Wilmanski in 1986 [227]. They use a method similar to the extended thermodynamics. However, there is a deviation from the strategy presented already in this book. In order to obtain an objective stress relation they have made sources dependent not only on fields but also on a time derivative of stresses. Below we present details and motivation for this deviation.

Let us mention in passing that a simple presentation of the problem can be found in the work [356]. Certain analytical properties of the expansion of the memory functional for Maxwell model can be found in the paper [269]. On a one-dimensional example, it is demonstrated that such expansions yield in some cases also mathematical problems. Namely, for memory integrals on unbounded domains (i.e. memory extends to infinity) the convergence is not uniform and truncations of Maxwell model are unjustified. This example is discussed in details in the book of Wilmanski [364].

[1]e.g. the third order fluid is defined by the constitutive relation

$$\mathbf{T} = -p\mathbf{1} + \nu\mathbf{A}_1 + \alpha_1\mathbf{A}_2 + \alpha_2\mathbf{A}_1^2 + \alpha_3\mathbf{A}_2^2 + \alpha_4\left(\mathbf{A}_1\mathbf{A}_2 + \mathbf{A}_2\mathbf{A}_1\right)$$
$$+\alpha_5\left(\mathbf{A}_1^2\mathbf{A}_2 + \mathbf{A}_2^2\mathbf{A}_1\right) + \alpha_6\left(\mathbf{A}_1\mathbf{A}_2^2 + \mathbf{A}_2^2\mathbf{A}_1\right) + \alpha_7\left(\mathbf{A}_1^2\mathbf{A}_2^2 + \mathbf{A}_2^2\mathbf{A}_1^2\right),$$

where p, $\nu, \alpha_1, ..., \alpha_7$ are functions of mass density and of the invariants

$$\begin{gathered}\operatorname{tr}\mathbf{A}_1, \operatorname{tr}\mathbf{A}_1^2, \operatorname{tr}\mathbf{A}_1^3,\\ \operatorname{tr}\mathbf{A}_2, \operatorname{tr}\mathbf{A}_2^2, \operatorname{tr}\mathbf{A}_2^3,\\ \operatorname{tr}\mathbf{A}_1\mathbf{A}_2, \operatorname{tr}\mathbf{A}_1\mathbf{A}_2^2, \operatorname{tr}\mathbf{A}_1^2\mathbf{A}_2, \operatorname{tr}\mathbf{A}_1^2\mathbf{A}_2^2.\end{gathered}$$

9.4.2 *Extended thermodynamics of non-Newtonian non-conducting fluids*

As for ideal gases (Section 8.2) the objective of thermodynamics of non-Newtonian fluids in Eulerian description is the determination of 13 fields

- mass density ρ,
- momentum density ρv_i,
- momentum flux density F_{ij},
- energy flux density $\frac{1}{2}F_{ijj}$.

We begin with the brief repetition of extended thermodynamics of gases and then indicate necessary extensions for non-Newtonian fluids.

In order to construct field equations we use the balance equations of mass, momentum, momentum flux and energy flux. In a non-inertial frame these balance equations have the form (8.29) which for the purpose of this Chapter we write in the following form

$$\begin{aligned}
\frac{\partial \rho}{\partial t} + \frac{\partial \rho v_k}{\partial x_k} &= 0, \\
\frac{\partial \rho v_i}{\partial t} + \frac{\partial F_{ik}}{\partial x_k} &= \rho\left(b_i^0 + i_i^0\right) + 2\rho v_k \Omega_{ik}, \\
\frac{\partial F_{ij}}{\partial t} + \frac{\partial F_{ijk}}{\partial x_k} &= \hat{f}_{\langle ij\rangle} + 2\rho v_{(i}\left(b_{j)}^0 + i_{j)}^0\right) + 4F_{k(i}\Omega_{j)k}, \\
\frac{\partial F_{ijj}}{\partial t} + \frac{\partial F_{ijjk}}{\partial x_k} &= \hat{f}_{ijj} + 3F_{(ij}\left(b_{j)}^0 + i_{j)}^0\right) + 6F_{kjj}\Omega_{ik}.
\end{aligned} \tag{9.59}$$

This set of equations becomes the set of field equations if we assume the following constitutive relations to hold

$$\begin{aligned}
F_{\langle ijk\rangle} &= F_{\langle ijk\rangle}\left(\rho, v_i, F_{ij}, F_{ijj}\right), \\
F_{ijjk} &= F_{ijjk}\left(\rho, v_i, F_{ij}, F_{ijj}\right), \\
\hat{f}_{\langle ij\rangle} &= \hat{f}_{\langle ij\rangle}\left(\rho, v_i, F_{ij}, F_{ijj}\right), \\
\hat{f}_{ijj} &= \hat{f}_{ijj}\left(\rho, v_i, F_{ij}, F_{ijj}\right),
\end{aligned} \tag{9.60}$$

appropriate for viscous heat conducting gas. We have 13 unknown fields.

For comparison, in ordinary thermodynamics we would seek the fields of mass, momentum and energy, the field equations would follow from balance equations $(9.59)_{1,2}$ (mass and momentum balances) and the trace of equation $(9.59)_3$ (energy balance). Constitutive relations would have the form

$$F_{\langle ij\rangle} = F_{\langle ij\rangle}\left(\rho, v_i, F_{ii}, \frac{\partial \rho}{\partial x_i}, \frac{\partial v_i}{\partial x_j}, \frac{\partial F_{ii}}{\partial x_j}\right), \tag{9.61}$$
$$F_{iij} = F_{iij}\left(\rho, v_i, F_{ii}, \frac{\partial \rho}{\partial x_i}, \frac{\partial v_i}{\partial x_j}, \frac{\partial F_{ii}}{\partial x_j}\right),$$

i.e. they would define the deviatoric part of momentum flux density (consequently, the deviatoric stresses) $F_{\langle ij\rangle}$ and the energy flux (consequently, the heat flux) F_{iij} in terms of mass density ρ, velocity v_i, energy density $\frac{1}{2}F_{ii}$, and gradients of mass density, velocity and energy density (i.e. temperature gradient by an appropriate choice of variables). We would have 5 unknown fields.

According to Ruggeri's Theorem, we can decompose the quantities $\mathbf{F}$ into convective and non-convective parts. They are as follows

$$\begin{aligned} F_{ij} &= \rho_{ij} + \rho v_i v_j, \\ \frac{1}{2}F_{ijj} &= q_i + \frac{3}{2}\rho_{(ij}v_{j)} + \frac{1}{2}\rho v^2 v_i, \\ F_{\langle ijk\rangle} &= \rho_{\langle ijk\rangle} + 3\rho_{\langle ij}v_{k\rangle} + \rho v_{\langle i}v_j v_{k\rangle}, \\ F_{ijjk} &= \rho_{ijjk} + 4\rho_{(ijj}v_{k)} + 6\rho_{(ij}v_j v_{k)} + \rho v_i v^2 v_k. \end{aligned} \tag{9.62}$$

As before, the non-convective quantities were denoted by $\boldsymbol{\rho}$. Thus, for instance, ρ_{ij} is the pressure tensor identical with the negative Cauchy stress tensor $-t_{ij}$, the pressure itself is equal to $p = \frac{1}{3}\rho_{ii}$. The non-convective part of $\frac{1}{2}F_{ijj}$ is the heat flux and, therefore it has been denoted by q_i.

Obviously, there is a one-to-one correspondence between $F_{ij}, F_{ijj}, F_{\langle ijk\rangle}, F_{ijjk}$ and $p, \rho_{\langle ij\rangle}, q_i, \rho_{\langle ijk\rangle}, \rho_{ijjk}$. Hence, the constitutive relations (9.60) can be written in the form

$$\begin{aligned} \rho_{\langle ijk\rangle} &= \rho_{\langle ijk\rangle}\left(\rho, v_i, p, \rho_{\langle ij\rangle}, q_i\right), \\ \rho_{ijjk} &= \rho_{ijjk}\left(\rho, v_i, p, \rho_{\langle ij\rangle}, q_i\right), \\ \hat{f}_{\langle ij\rangle} &= \hat{f}_{\langle ij\rangle}\left(\rho, v_i, p, \rho_{\langle ij\rangle}, q_i\right), \\ \hat{f}_{ijj} - 3v_{(j}\hat{f}_{\langle ij\rangle)} &= \hat{f}_i\left(\rho, v_i, p, \rho_{\langle ij\rangle}, q_i\right). \end{aligned} \tag{9.63}$$

We proceed to generalize the above relations to the case of non-Newtonian fluid. We follow here the work of I. Müller and K. Wilmanski [227]. First of all, in contrast to ideal gases, fields in subsequent equations are in this case not identical with fluxes of the previous equations of the hierarchy. Rather we have the following structure (compare (9.1)-(9.4) in

Lagrangian description)

$$\begin{aligned}
\frac{\partial \rho}{\partial t}+\frac{\partial \rho v_k}{\partial x_k} &= 0,\\
\frac{\partial \rho v_i}{\partial t}+\frac{\partial F_{ik}}{\partial x_k} &= \rho\left(b_i^0+i_i^0\right)+2\rho v_k \Omega_{ik},\\
\frac{\partial G_{ij}}{\partial t}+\frac{\partial F_{ijk}}{\partial x_k} &= \hat{f}_{\langle ij\rangle}+2\rho v_{(i}\left(b_{j)}^0+i_{j)}^0\right)-2\Omega_{k(i}G_{j)k}-2\Omega_{k(i}F_{j)k},\\
\frac{\partial G_{ijj}}{\partial t}+\frac{\partial F_{ijjk}}{\partial x_k} &= \hat{f}_{ijj}+3G_{(ij}\left(b_{j)}^0+i_{j)}^0\right)-3\Omega_{k(j}F_{ij)k}+\Omega_{ik}G_{kjj}.
\end{aligned} \tag{9.64}$$

Now the decomposition into non-convective and convective parts yields

$$\begin{aligned}
F_{ij} &= \rho_{ij}+\rho v_i v_j,\\
G_{ij} &= m_{ij}+\rho v_i v_j,\\
F_{ijk} &= \rho_{ijk}+m_{ij}v_k 2\rho_{k(i}v_{j)}+\rho v_i v_j v_k,\\
G_{ijj} &= m_{ijj}+3m_{(ij}v_{j)}+\rho v^2 v_i,\\
F_{ijjk} &= \rho_{ijjk}+m_{ijj}v_k+3\rho_{k(ij}v_{k)}+3m_{(ij}v_{j)k}+3\rho_{k(j}v_{j)}v_i+\rho v_i v^2 v_k.
\end{aligned} \tag{9.65}$$

Elimination of the **F**'s and **G**'s between (9.64) and (9.65) leads to the following universal balance laws of extended thermodynamics in Eulerian description

$$\begin{gathered}
\overset{\bullet}{\rho}+\rho\frac{\partial v_k}{\partial x_k}=0,\\
\rho\overset{\bullet}{v}_i+\frac{\partial \rho_{ik}}{\partial x_k}=\rho\left(b_i^0+i_i^0\right)+2\rho v_k\Omega_{ik},\\
\rho\overset{\bullet}{\varepsilon}+\frac{\partial q_k}{\partial x_k}+\rho_{ij}\frac{\partial v_i}{\partial x_j}=0,\\
\left[\rho\left(\overset{\bullet}{\frac{m_{\langle ij\rangle}}{\rho}}\right)-m_{\langle ki\rangle}\Omega_{jk}-m_{\langle kj\rangle}\Omega_{ik}\right]\\
+\rho_{\langle ik\rangle}\left[\frac{\partial v_{[j}}{\partial x_{k]}}-\Omega_{jk}\right]+\rho_{\langle jk\rangle}\left[\frac{\partial v_{[i}}{\partial x_{k]}}-\Omega_{ik}\right]\\
+\rho_{\langle ik\rangle}\frac{\partial v_{(j}}{\partial x_{k)}}+\rho_{\langle jk\rangle}\frac{\partial v_{(i}}{\partial x_{k)}}-\frac{2}{3}\rho_{\langle kl\rangle}\frac{\partial v_k}{\partial x_l}\delta_{ij}+2p\frac{\partial v_{\langle i}}{\partial x_{j\rangle}}+\frac{\partial \rho_{\langle ij\rangle k}}{\partial x_k}=\hat{f}_{\langle ij\rangle},
\end{gathered} \tag{9.66}$$

$$\left[\rho\left(\frac{\overset{\bullet}{\frac{1}{2}m_{ijj}}}{\rho}\right)-\frac{1}{2}m_{kjj}\Omega_{ik}\right]+\frac{5}{3}q_k\left[\frac{\partial v_i}{\partial x_k}-\Omega_{ik}\right]+\rho_{\langle ij\rangle k}\left[\frac{\partial v_j}{\partial x_k}-\Omega_{jk}\right]$$
$$+\frac{1}{2}\frac{\partial \rho_{ijjk}}{\partial x_k}-\frac{1}{\rho}\left(m_{ij}+\rho\varepsilon\delta_{ij}\right)\frac{\partial \rho_{jk}}{\partial x_k}=\frac{1}{2}\left(\hat{f}_{ijj}-3v_{(j}\hat{f}_{\langle ij\rangle)}\right),$$

where the notations for the heat flux $q_i = \frac{1}{2}\rho_{ijj}$ and for the specific internal energy $\varepsilon = \frac{1}{2}m_{ii}/\rho$ are used. As before, the dot denotes the material time derivative. The first three of these equations – balance of mass, momentum and internal energy – are common to both ordinary and extended thermodynamics.

The above general equations become field equations if we specify constitutive relations for the following *constitutive quantities*

$$\varepsilon, m_{\langle ij\rangle}, m_{ijj}, \rho_{\langle ij\rangle k}, \rho_{ijjk}, \hat{f}_{\langle ij\rangle}, \hat{f}_{ijj}. \tag{9.67}$$

We do so for a non-conducting fluid. A more general case will be considered in the next Subsection. For the present case we set $q_i = 0$ and ignore the last equation in the set (9.66). The remaining fields which should be determined are

$$\rho, v_i, p, \rho_{\langle ij\rangle} = -t_{\langle ij\rangle} \quad \text{or} \quad T, v_i, p, \rho_{\langle ij\rangle} = -t_{\langle ij\rangle}, \tag{9.68}$$

where the temperature T can always be introduced instead of ρ if the thermal equation of state $p = p(\rho, T)$ is known and invertible. The constitutive quantities of (9.67) reduce now to

$$\begin{aligned}
\rho &= \rho\left(T, v_i, p, \rho_{\langle ij\rangle}\right), \\
\varepsilon &= \varepsilon\left(T, v_i, p, \rho_{\langle ij\rangle}\right), \\
m_{\langle ij\rangle} &= m_{\langle ij\rangle}\left(T, v_i, p, \rho_{\langle ij\rangle}\right), \\
m_{ijj} &= m_{ijj}\left(T, v_i, p, \rho_{\langle ij\rangle}\right), \\
\rho_{\langle ij\rangle k} &= \rho_{\langle ij\rangle k}\left(T, v_i, p, \rho_{\langle ij\rangle}\right), \\
\hat{f}_{\langle ij\rangle} &= \hat{f}_{\langle ij\rangle}\left(T, v_i, p, \rho_{\langle ij\rangle}, \overset{\Delta}{\rho}_{\langle ij\rangle}\right),
\end{aligned} \tag{9.69}$$

where

$$\overset{\Delta}{\rho}_{\langle ij \rangle} = \overset{\bullet}{\rho}_{\langle ij \rangle} + \rho_{\langle ik \rangle} \frac{\partial v_{[k}}{\partial x_{j]}} + \rho_{\langle jk \rangle} \frac{\partial v_{[k}}{\partial x_{i]}}, \tag{9.70}$$

$$\text{i.e.} \quad \overset{\Delta}{\boldsymbol{\rho}} = \overset{\bullet}{\boldsymbol{\rho}} + \boldsymbol{\rho}\mathbf{W} - \mathbf{W}\boldsymbol{\rho}, \quad W_{ij} = \frac{\partial v_{[i}}{\partial x_{j]}} = \frac{1}{2}\left(\frac{\partial v_i}{\partial x_j} - \frac{\partial v_j}{\partial x_i}\right).$$

The choice of the *corotational derivative* as an objective time derivative is immaterial. One can choose as well any other objective derivative with an appropriate correction of objective contributions to constitutive relations.

There is a break of equipresence in relations (9.69). The presence of the time derivative as a constitutive variable is needed because the non-Newtonian fluids possess some memory – the absence of this dependence would yield models of non-conducting viscous fluids similar to these presented in previous Sections. The type of dependence is tolerated here for simplicity. One could construct extended thermodynamics models with higher moments well motivated by molecular considerations but these would be technically much more complicated.

The constitutive functions in (9.69) are restricted, as usual, by the following requirements

– the principle of material frame indifference (material objectivity),

– the entropy principle,

– the thermodynamical stability.

The first condition yields constitutive functions independent of the velocity v_i and isotropic in other variables. We restrict the considerations to processes close to the thermodynamical equilibrium. Then the pressure deviator $\rho_{\langle ij \rangle}$ and its rate of change $\overset{\Delta}{\rho}_{\langle ij \rangle}$ are small. For this reason, we ignore terms with products of these quantities and admit only terms quadratic in $\rho_{\langle ij \rangle}$. Under these assumptions the representation theorems for isotropic functions yield

$$\begin{aligned} \rho &= \rho_0 + \rho_1 P, \\ \varepsilon &= \varepsilon_0 + \varepsilon_1 P, \\ \varepsilon_{\langle ij \rangle} &= e_2 \rho_{\langle ij \rangle} + e_3 \left(\rho_{\langle ik \rangle}\rho_{\langle kj \rangle} - \frac{1}{3} P \delta_{ij}\right), \\ \rho_{\langle ij \rangle k} &= 0, \\ \hat{f}_{\langle ij \rangle} &= \beta_1 \rho_{\langle ij \rangle} + \beta_2 \overset{\Delta}{\rho}_{\langle ij \rangle} + \beta_3 \left(\rho_{\langle ik \rangle}\rho_{\langle kj \rangle} - \frac{1}{3} P \delta_{ij}\right), \end{aligned} \tag{9.71}$$

where

$$P = \rho_{\langle ij \rangle}\rho_{\langle ij \rangle}, \quad \varepsilon_{\langle ij \rangle} = \frac{1}{2}\frac{m_{\langle ij \rangle}}{\rho}, \tag{9.72}$$

and all coefficients ρ_0 through β_3 are functions of T and p. The quantity $\varepsilon_{\langle ij\rangle}$ is called the *internal energy deviator*. Its combination with ε

$$\varepsilon_{ij} = \varepsilon_{\langle ij\rangle} + \frac{1}{3}\varepsilon\delta_{ij}, \tag{9.73}$$

is called the *internal energy tensor.*

We proceed to investigate the restrictions which the entropy principle imposes on constitutive relations. The entropy density η and its flux h_i have to satisfy the constitutive relations

$$\eta = \eta\left(T, v_i, p, \rho_{\langle ij\rangle}\right), \quad h_i = h_i\left(T, v_i, p, \rho_{\langle ij\rangle}\right). \tag{9.74}$$

Under the above introduced simplifying assumptions we obtain the following representations

$$\eta = \eta_0 + \eta_1 P, \quad h_i = 0. \tag{9.75}$$

Then the entropy inequality and the Liu Theorem yield the inequality

$$\begin{aligned} &\rho\overset{\bullet}{\eta} - \Lambda\left(\overset{\bullet}{\rho} + \rho\frac{\partial v_k}{\partial x_k}\right) - \Lambda^i\left[\rho\overset{\bullet}{v}_i + \frac{\partial \rho_{ij}}{\partial x_j} - \rho\left(b_i + i_i^0\right) - 2\rho v_k\Omega_{ik}\right] \\ &-\Lambda^{ij}\left[2\left(\rho\overset{\bullet}{\varepsilon}_{ij} - \varepsilon_{\langle ki\rangle}\Omega_{jk} - \varepsilon_{\langle kj\rangle}\Omega_{ik}\right) + \rho_{\langle ik\rangle}\left(\frac{\partial v_j}{\partial x_k} - \Omega_{jk}\right)\right. \\ &\left.+\rho_{\langle jk\rangle}\left(\frac{\partial v_i}{\partial x_k} - \Omega_{ik}\right) + 2p\frac{\partial v_{(i}}{\partial x_{j)}} - \hat{f}_{\langle ij\rangle}\right] \geq 0, \end{aligned} \tag{9.76}$$

which must hold for all fields T, v_i, p_{ij}. The Lagrange multipliers $\Lambda, \Lambda^i, \Lambda^{ij}$ may be functions of constitutive variables $T, v_i, p, \rho_{\langle ij\rangle}, \overset{\Delta}{\rho}_{\langle ij\rangle}$.

Insertion of (9.71) and (9.75) in (9.76) shows that the left-hand side of this inequality is linear with respect to derivatives

$$\overset{\bullet}{T}, \overset{\bullet}{\rho}_{ij}, \overset{\bullet}{v}_i, \frac{\partial v_i}{\partial x_j}. \tag{9.77}$$

This yields the following identities

$$\begin{aligned} \frac{\partial\eta}{\partial T} - \frac{\Lambda}{\rho}\frac{\partial\rho}{\partial T} - 2\Lambda^{kl}\frac{\partial\varepsilon_{kl}}{\partial T} &= 0, \\ \left(\frac{\partial\eta}{\partial\rho_{ij}} - \frac{\Lambda}{\rho}\frac{\partial\rho}{\partial\rho_{ij}} - 2\Lambda^{kl}\frac{\partial\varepsilon_{kl}}{\partial\rho_{ij}}\right) &= 0, \end{aligned}$$

$$\left(\frac{\partial \eta}{\partial \rho_{ij}} - \frac{\Lambda}{\rho}\frac{\partial \rho}{\partial \rho_{ij}} - 2\Lambda^{kl}\frac{\partial \varepsilon_{kl}}{\partial \rho_{ij}}\right)\left(\delta_{in}\delta_{jm} - \frac{1}{3}\delta_{ij}\delta_{mn}\right) - \frac{\beta_2}{\rho}\Lambda^{\langle nm\rangle} = 0, \tag{9.78}$$

$$\Lambda^i = 0,$$
$$\rho\Lambda\delta_{nm} + 2\Lambda^{in}\rho_{\langle im\rangle} + 2p\Lambda^{nm} - \beta_2\Lambda^{im}\rho_{\langle in\rangle} + \beta_2\Lambda^{in}\rho_{\langle im\rangle} = 0.$$

There remains the residual inequality

$$\Lambda^{ij}\left(4\rho\varepsilon_{\langle ki\rangle}W_{jk} + 2\rho_{\langle ki\rangle}W_{jk} + \beta_1\rho_{\langle ij\rangle} + \beta_3\left(\rho_{\langle ik\rangle}\rho_{\langle kj\rangle} - \frac{1}{3}P\delta_{ij}\right)\right) \geq 0. \tag{9.79}$$

It is clear from (9.78) that Λ and Λ^{ij} are independent of $\overset{\Delta}{\rho}_{\langle ij\rangle}$ and v_i. Hence we can write in the quadratic approximation

$$\frac{\Lambda}{\rho} = l_0 + l_1 P, \tag{9.80}$$
$$\Lambda^{ij} = \frac{1}{3}\left(\lambda_0 + \lambda_1 P\right)\delta_{ij} + \lambda_2\rho_{\langle ij\rangle} + \lambda_3\left(\rho_{\langle ik\rangle}\rho_{\langle kj\rangle} - \frac{1}{3}P\delta_{ij}\right).$$

Now the substitution in identities (9.78) yields the following results within the order of approximation we make in this analysis

$$\frac{\partial \eta_0}{\partial T} + \frac{\partial \eta_1}{\partial T}P - l_0\frac{\partial \rho_0}{\partial T} - l_1 P\frac{\partial \rho_0}{\partial T} - l_0\frac{\partial \rho_1}{\partial T}P - \frac{2}{3}\lambda_0\frac{\partial \varepsilon_0}{\partial T}$$
$$-\frac{2}{3}\lambda_0\frac{\partial \varepsilon_1}{\partial T}P - \frac{2}{3}\lambda_1 P\frac{\partial \varepsilon_0}{\partial T} - 2\lambda_2 P\frac{\partial \varepsilon_2}{\partial T} = 0,$$

$$\frac{\partial \eta_0}{\partial p} + \frac{\partial \eta_1}{\partial p}P - l_0\frac{\partial \rho_0}{\partial p} - l_1 P\frac{\partial \rho_0}{\partial p} - l_0\frac{\partial \rho_1}{\partial p}P - \frac{2}{3}\lambda_0\frac{\partial \varepsilon_0}{\partial p}$$
$$-\frac{2}{3}\lambda_0\frac{\partial \varepsilon_1}{\partial p}P - \frac{2}{3}\lambda_1 P\frac{\partial \varepsilon_0}{\partial p} - 2\lambda_2 P\frac{\partial \varepsilon_2}{\partial p} = 0, \tag{9.81}$$

$$\eta_1\rho_{\langle nm\rangle} - l_0\rho_1\rho_{\langle nm\rangle} - \lambda_2 e_2\rho_{\langle nm\rangle} - \frac{2}{3}\varepsilon_1\lambda_0\rho_{\langle nm\rangle} + \frac{\beta_2}{2\rho_0}\lambda_2\rho_{\langle nm\rangle} = 0,$$

$$\left[\rho^2\left(l_0 + l_1 P\right) + \frac{2}{3}\lambda_2 P + \frac{2}{3}p\left(\lambda_0 + \lambda_1 P\right)\right]\delta_{nm}$$
$$+\left[\frac{2}{3}\lambda_0 + 2p\lambda_2\right]\rho_{\langle nm\rangle} + \left[2\lambda_2 + 2p\lambda_3\right]\left(\rho_{\langle im\rangle}\rho_{\langle in\rangle} - \frac{1}{3}P\delta_{nm}\right) = 0.$$

These are polynomials in the components of $\rho_{\langle nm\rangle}$ and we conclude that their coefficients must vanish. This yields the following set of identities

$$\frac{\partial \eta_0}{\partial T} - l_0 \frac{\partial \rho_0}{\partial T} - \frac{2}{3}\lambda_0 \frac{\partial \varepsilon_0}{\partial T} = 0, \quad \frac{\partial \eta_0}{\partial p} - l_0 \frac{\partial \rho_0}{\partial p} - \frac{2}{3}\lambda_0 \frac{\partial \varepsilon_0}{\partial p} = 0, \tag{9.82}$$

$$\frac{\partial \eta_1}{\partial T} - l_1 \frac{\partial \rho_0}{\partial T} - l_0 \frac{\partial \rho_1}{\partial T} - \frac{2}{3}\lambda_0 \frac{\partial \varepsilon_1}{\partial T} - \frac{2}{3}\lambda_1 \frac{\partial \varepsilon_0}{\partial T} - 2\lambda_2 \frac{\partial e_2}{\partial T} = 0,$$

$$\frac{\partial \eta_1}{\partial p} - l_1 \frac{\partial \rho_0}{\partial p} - l_0 \frac{\partial \rho_1}{\partial p} - \frac{2}{3}\lambda_0 \frac{\partial \varepsilon_1}{\partial p} - \frac{2}{3}\lambda_1 \frac{\partial \varepsilon_0}{\partial p} - 2\lambda_2 \frac{\partial e_2}{\partial p} = 0, \tag{9.83}$$

$$\eta_1 - l_0 \rho_1 - \lambda_2 e_2 - \frac{2}{3}\varepsilon_1 \lambda_0 + \frac{\beta_2}{2\rho_0}\lambda_2 = 0, \tag{9.84}$$

$$\begin{aligned} \rho_0^2 l_0 + \frac{2}{3} p \lambda_0 &= 0, \\ \lambda_2 + p\lambda_1 + \frac{3}{2}\rho_0^2 l_1 + 3\rho_0 \rho_1 l_0 &= 0, \\ \frac{2}{3}\lambda_0 + 2p\lambda_2 &= 0, \\ \lambda_2 + p\lambda_3 &= 0. \end{aligned} \tag{9.85}$$

Simultaneously, the residual inequality up to second order terms in $\rho_{\langle nm\rangle}$ has the form

$$\beta_1 \lambda_2 P \geq 0. \tag{9.86}$$

Conditions (9.82) combine to the relation

$$d\eta_0 = \frac{2}{3}\lambda_0 \left(d\varepsilon_0 - \frac{p}{\rho_0^2} d\rho_0 \right) \quad \Rightarrow \quad \lambda_0 = \frac{3}{2}\frac{1}{T}, \tag{9.87}$$

because in thermodynamical equilibrium this must be, of course, identical with the Gibbs equation. It implies the integrability condition

$$\frac{\partial \rho_0}{\partial T} = \frac{\rho_0^2}{T}\frac{\partial \varepsilon_0}{\partial p} - \frac{p}{T}\frac{\partial \rho_0}{\partial p}. \tag{9.88}$$

Now substitution of $(9.87)_2$ in (9.85) leads to the following relations for multipliers

$$l_0 = -\frac{p}{\rho_0^2 T}, \quad \lambda_2 = -\frac{1}{2pT}, \quad \lambda_3 = \frac{1}{2p^2 T}. \tag{9.89}$$

Hence, for nonequilibrium contributions we obtain

$$\varepsilon_1 - T\eta_1 - \frac{p}{\rho_0^2}\rho_1 = \frac{1}{2p}\left(e_2 - \frac{\beta_2}{2\rho_0}\right). \tag{9.90}$$

Then the remaining identities yield

$$\frac{2}{3}\lambda_1 T c_p = -\frac{\partial \frac{1}{2p}\left(e_2 - \frac{\beta_2}{2\rho_0}\right)}{\partial T} - \eta_1 - \frac{1}{3p\rho_0^2}\frac{\partial \rho_0}{\partial T} + \frac{1}{p}\frac{\partial e_2}{\partial T}, \tag{9.91}$$

$$\frac{2}{3}\lambda_1 T\left(\frac{T}{\rho_0^2}\frac{\partial \rho_0}{\partial T}\right) = -\frac{\partial \frac{1}{2p}\left(e_2 - \frac{\beta_2}{2\rho_0}\right)}{\partial p} - \frac{\rho_1}{\rho_0^2} - \frac{1}{3p\rho_0^2}\frac{\partial \rho_0}{\partial p} + \frac{1}{p}\frac{\partial e_2}{\partial p}, \tag{9.92}$$

$$l_1 T = -\frac{2}{3}\lambda_1 T\frac{p}{\rho_0^2} + 2\frac{p}{\rho_0^2}\rho_1 + \frac{1}{3p\rho_0^2}, \tag{9.93}$$

where

$$c_p = \frac{\partial \varepsilon_0}{\partial T} - \frac{p}{\rho_0^2}\frac{\partial \rho_0}{\partial T}, \tag{9.94}$$

is the *specific heat at constant pressure*. These results yield by means of the residual inequality (9.86) the following inequality

$$\beta_1 \leq 0. \tag{9.95}$$

Summing up, apart from the identification of multipliers we have obtained only two useful results (9.88) and (9.95). However, the relation (9.90) yields additional conclusions if we combine it with the stability analysis of the thermodynamical equilibrium.

As shown, for instance, in the book of I. Müller [222], the Gibbs free energy (free enthalpy)

$$g = \varepsilon - T\eta + \frac{p}{\rho}, \tag{9.96}$$

decreases under conditions of constant temperature and pressure and it reaches its minimum in the thermodynamical equilibrium. The constitutive considerations which we presented above provide the following relation for this function

$$\begin{aligned} g &= \varepsilon_0 + \varepsilon_1 P - T\left(\eta_0 + \eta_1 P\right) + \frac{p}{\rho_0 + \rho_1 P} \\ &\approx g_0 + \left(\varepsilon_1 - T\eta_1 - \frac{p}{\rho_0^2}\rho_1\right)P, \quad g_0 = \varepsilon_0 - T\eta_0. \end{aligned} \tag{9.97}$$

In equilibrium $P = 0$ and we obtain the condition for minimum of g

$$\varepsilon_1 - T\eta_1 - \frac{p}{\rho_0^2}\rho_1 > 0. \tag{9.98}$$

Hence, by means of (9.90)

$$e_2 - \frac{\beta_2}{2\rho_0} > 0. \tag{9.99}$$

In addition, we have the classical conditions for the equilibrium part of the Gibbs function

$$\frac{\partial \eta_0}{\partial T} > 0, \quad \frac{\partial \rho_0}{\partial p} > 0, \quad \left(\frac{\partial \rho_0}{\partial T}\right)^2 < \rho_0^2 \frac{\partial \eta_0}{\partial T}\frac{\partial \rho_0}{\partial p}, \tag{9.100}$$

which follow from the variation of T and p in the vicinity of equilibrium (see Chapter 6). They yield the positivity of the specific heat and of the compressibility as well as an upper bound for the thermal expansion.

In order to appreciate these results let us consider a particular case in which we assume the fluid to be at rest in an inertial frame. We drop the nonlinear terms in (9.76) and use $(9.71)_5$. It follows

$$2\rho_0 e_2 \overset{\bullet}{\rho}_{\langle ij\rangle} = \beta_1 \rho_{\langle ij\rangle} + \beta_2 \overset{\bullet}{\rho}_{\langle ij\rangle}, \tag{9.101}$$

i.e.

$$\left(e_2 - \frac{\beta_2}{2\rho_0}\right)\overset{\bullet}{t}_{\langle ij\rangle} = \frac{\beta_1}{2\rho_0} t_{\langle ij\rangle}. \tag{9.102}$$

Hence, inequalities (9.95) and (9.99) assure the *relaxation of the stress deviator*.

We proceed to compare the results with those of ordinary thermodynamics. Let us first rewrite the field equations (9.66) for the fields $T, v_i, p, \rho_{\langle ij\rangle}$ in the case under considerations. We have

$$\begin{aligned} \overset{\bullet}{\rho} + \rho \frac{\partial v_k}{\partial x_k} &= 0, \\ \rho \overset{\bullet}{v}_i + \frac{\partial \rho_{ik}}{\partial x_k} &= \rho\left(b_i^0 + i_i^0\right) + 2\rho v_k \Omega_{ik}, \\ \rho \overset{\bullet}{\varepsilon} + \frac{\partial q_k}{\partial x_k} + \rho_{ij}\frac{\partial v_i}{\partial x_j} &= 0, \end{aligned} \tag{9.103}$$

$$2\rho\left[\overset{\bullet}{\varepsilon}_{\langle ij\rangle}-\varepsilon_{\langle ki\rangle}\Omega_{jk}-\varepsilon_{\langle kj\rangle}\Omega_{ik}\right]+\rho_{\langle ik\rangle}\left[\frac{\partial v_j}{\partial x_k}-\Omega_{jk}\right]+\rho_{\langle jk\rangle}\left[\frac{\partial v_i}{\partial x_k}-\Omega_{ik}\right]$$
$$-\frac{2}{3}\rho_{\langle kl\rangle}\frac{\partial v_k}{\partial x_l}\delta_{ij}+\underline{2p\frac{\partial v_{\langle i}}{\partial x_{j\rangle}}=\beta_1\rho_{\langle ij\rangle}+\beta_2\overset{\Delta}{\rho}_{\langle ij\rangle}+\beta_3\left(\rho_{\langle ik\rangle}\rho_{\langle kj\rangle}-\frac{1}{3}P\delta_{ij}\right)},$$

where ρ, ε and $\varepsilon_{\langle ij\rangle}$ are given by the constitutive relations $(9.71)_{1,2,3}$. An important characteristic feature of the last equation is its dependence on the frame which, certainly, means that constitutive relations following from the extended thermodynamics do not fulfil the classical principle of material objectivity. Moreover, as we see further, this equation does not permit the assumption on the incompressibility of fluid. Both these features do not appear if we limit the attention to the leading terms in $(9.103)_4$. These were underlined above.

The last equation can be simplified by applying the Maxwell iteration which yields results for the small deviation from the thermodynamical equilibrium. In our case the equilibrium is given by the relation $\rho_{\langle ij\rangle}\Big|_E=0$. Then the first step in iteration follows if we substitute this equilibrium value on the left-hand side of equation $(9.103)_4$. Hence

$$\frac{\beta_2}{\beta_1}\overset{\Delta}{t}_{\langle ij\rangle}+t_{\langle ij\rangle}=-\frac{2p}{\beta_1}\frac{\partial v_{\langle i}}{\partial x_{j\rangle}}+\frac{\beta_3}{\beta_1}\left(t_{\langle ik\rangle}t_{\langle kj\rangle}-\frac{1}{3}P\delta_{ij}\right), \tag{9.104}$$
$$t_{\langle ij\rangle}=-\rho_{\langle ij\rangle},\quad P=t_{\langle ij\rangle}t_{\langle ij\rangle}.$$

This equation is known in the rheological literature as a constitutive equation of the *Maxwell fluid.*

Obviously, if we ignore the rate term and the nonlinear term on the right-hand side, we obtain the *Navier-Stokes equation*

$$t_{\langle ij\rangle}=2\nu\frac{\partial v_{\langle i}}{\partial x_{j\rangle}}, \tag{9.105}$$

provided the viscosity ν is given by

$$\nu=-\frac{p}{\beta_1}. \tag{9.106}$$

In order to obtain the next step in the Maxwell iteration we insert (9.105) in the rate and nonlinear terms of (9.104). It follows

$$t_{\langle ij\rangle}=\nu\frac{\partial v_{\langle i}}{\partial x_{j\rangle}}+2\alpha_1\frac{\overset{\Delta}{\partial v}_{\langle i}}{\partial x_{j\rangle}}+4\alpha_2\left(\frac{\partial v_{\langle i}}{\partial x_{k\rangle}}\frac{\partial v_{\langle k}}{\partial x_{j\rangle}}-\frac{1}{3}\frac{\partial v_{\langle n}}{\partial x_{m\rangle}}\frac{\partial v_{\langle n}}{\partial x_{m\rangle}}\delta_{ij}\right), \tag{9.107}$$

where

$$\alpha_1 = \frac{\beta_2}{p}\nu^2, \quad \alpha_2 = -\frac{\beta_3}{p}\nu^3, \tag{9.108}$$

and the product $\overset{\bullet}{p}$ and $\partial v_{\langle i}/\partial x_{j\rangle}$ has been neglected. According to (9.95) and (9.99), the above parameters satisfy the inequalities

$$\nu \geq 0, \quad \alpha_1 < \frac{\nu^2}{p} 2\rho_0 e_2. \tag{9.109}$$

Consequently, α_1 may well be negative as indicated at the beginning of this Section.

The Maxwell procedure and its result, the constitutive equation (9.107) yield, of course the model which is different from the model which follows from extended thermodynamics. In the latter case we obtain the Maxwell model (9.104) and the set of field equations is hyperbolic while in the first case we obtain a modified fluid of the second grade (9.107) and the set of field equations is parabolic. In order to estimate the order of magnitude of the parameters we consider a simple example of the viscometric experiment in which the flow is described by the velocity field

$$\mathbf{v} = (0, v_2(x_1), 0)^T. \tag{9.110}$$

Then the stress relation (9.107) yields the following normal stresses

$$\begin{aligned} t_{\langle 11\rangle} &= \left(\frac{4}{3}\alpha_1 + \frac{1}{3}\alpha_2\right)\kappa^2, \\ t_{\langle 22\rangle} &= \left(-\frac{2}{3}\alpha_1 + \frac{1}{3}\alpha_2\right)\kappa^2, \\ t_{\langle 33\rangle} &= -\left(\frac{2}{3}\alpha_1 + \frac{2}{3}\alpha_2\right)\kappa^2, \quad \kappa = \frac{\partial v_2}{\partial x_1}. \end{aligned} \tag{9.111}$$

Consequently, the *normal stress differences* which are used in viscometry are as follows

$$t_{\langle 11\rangle} - t_{\langle 33\rangle} = (2\alpha_1 + \alpha_2)\kappa^2, \quad t_{\langle 22\rangle} - t_{\langle 33\rangle} = \alpha_2\kappa^2. \tag{9.112}$$

Hence

$$\alpha_1 = \frac{t_{\langle 11\rangle} - t_{\langle 22\rangle}}{2\kappa^2}, \quad \alpha_2 = \frac{t_{\langle 22\rangle} - t_{\langle 33\rangle}}{2\kappa^2}, \tag{9.113}$$

i.e.

$$\beta_2 = \frac{p}{2\nu^2}\frac{t_{\langle 11\rangle} - t_{\langle 22\rangle}}{\kappa^2}, \quad \beta_3 = -\frac{p}{\nu^3}\frac{t_{\langle 22\rangle} - t_{\langle 33\rangle}}{\kappa^2}. \tag{9.114}$$

For the 6.5 wt.% solution of polyisobutylen in decalin Ginn and Metzner [122] report the following data if $\kappa = 10$ 1/s

$$\nu = 5.6 \text{ Pa s}, \quad t_{\langle 11\rangle} - t_{\langle 33\rangle} = -6 \text{ Pa}, \quad t_{\langle 22\rangle} - t_{\langle 33\rangle} = 47 \text{ Pa}. \tag{9.115}$$

Consequently, for $p = 10^5$ Pa,

$$\frac{\beta_1}{\kappa} = -1786, \quad \beta_2 = -845, \quad \beta_3\nu = -1498. \tag{9.116}$$

These values justify the iteration presented above. It would not be justified if the pressure p would be reduced below some 10^3 Pa or if we made an experiment on a turntable with some 100 revolutions per second.

It is much more difficult to estimate the parameter e_2 appearing in the energy deviator $\varepsilon_{\langle ij\rangle}$. We can get some insight through the relation (9.102) which defines the relaxation time τ_t of the stress deviator $t_{\langle ij\rangle}$. We have for this time

$$\tau_t = \frac{\nu}{p}\left(2\rho_0 e_2 - \beta_2\right). \tag{9.117}$$

For Newtonian fluids in which β_2 is zero this time is estimated on the basis of the time between molecular collisions (time of the free flight) and it is of the order 10^{-12} s. Hence, for water,

$$2\rho_0 e_2 = 10^{-4}. \tag{9.118}$$

For ideal gases we have $2\rho_0 e_2 = 1$. Consequently, for non-Newtonian fluids we may neglect e_2 in combination with β_2.

We complete these considerations with a few remarks concerning the assumption on the incompressibility of the fluid. The common definition of incompressibility in continuum mechanics and thermodynamics reflects two properties of processes:

– only isochoric motions are possible, i.e. $\partial v_i / \partial x_i = 0$,

– superposition of a time-dependent pressure $\pi(t)$ upon an existing pressure field $p(\mathbf{x}, t)$ does not change the fields $T(\mathbf{x}, t), v_i(\mathbf{x}, t), \rho_{\langle ij\rangle}(\mathbf{x}, t)$.

The latter means that neither the pressure nor its time derivative may occur in the field equations. As $\pi(t)$ is $\mathbf{x}$-independent, the spacial derivatives of pressure are, however, admissible.

It is easy to see that this condition is not fulfilled by the field equations (9.103) of extended thermodynamics. Indeed the pressure p appears in (9.103) in some terms in the explicit manner while in some other it does not appear at all. However, as we have argued above, only the underlined terms of $(9.103)_4$ are of importance in the approximation of a small deviation from the thermodynamical equilibrium. If we ignore all other contributions we have a chance to specify the constitutive relations in such a way that the requirements of incompressibility are satisfied. Obviously, $\rho_0, \rho_1, \varepsilon_0, \varepsilon_1$ must then be independent of p and $\beta_1, \beta_2, \beta_3$ must be proportional to p. As a consequence of (9.88) we see that ρ_0 is also independent of the temperature T – the thermal expansion in an incompressible fluid is impossible. We have indicated this property already in Chapter 6 on thermostatics. Simultaneously, the mass balance equation $(9.103)_1$ reduces to

$$\frac{d\rho_1}{dT}\overset{\bullet}{T} + \rho_1 \overset{\bullet}{P} = 0, \tag{9.119}$$

and we conclude that we must have

$$\rho_1 = 0, \quad \frac{d\rho_1}{dT} = 0, \tag{9.120}$$

so as not to place undue restrictions upon the fields $T, v_i, \rho_{\langle ij \rangle}$.

Accounting for these properties, we obtain from (9.92) the following equation

$$\frac{\partial e_2}{\partial p} = -\frac{e_2}{p}, \tag{9.121}$$

with the solution

$$e_2 = \frac{C\,(T)}{p}. \tag{9.122}$$

The function $C\,(T)$ must be found experimentally, for instance by measuring the relaxation time of shear pulses.

9.4.3 *Thermodynamics of a heat conducting Maxwellian fluid*

9.4.3.1 *Evaluation of the entropy inequality*

In the previous Subsection, we have been considering non-conducting fluids. This was done for technical reasons. In order to correct this model, we consider in this Subsection a heat conducting Maxwellian fluid. We follow here the work [357] and limit the attention to small deviations from the thermodynamical equilibrium which is now defined by two conditions $t_{\langle ij \rangle}\big|_E = 0$, $q_i\big|_E = 0$, and we use an inertial frame of reference. The fields

v_i, T, q_i, t_{ij} are described by the following set of field equations

$$\begin{aligned} \overset{\bullet}{\rho} + \rho \frac{\partial v_k}{\partial x_k} &= 0, \\ \rho \overset{\bullet}{v}_i - \frac{\partial t_{ik}}{\partial x_k} &= 0, \\ \rho \overset{\bullet}{\varepsilon} + \frac{\partial q_k}{\partial x_k} - t_{ij} \frac{\partial v_i}{\partial x_j} &= 0, \end{aligned} \tag{9.123}$$

$$\begin{aligned} \tau_q \overset{\triangle}{q}_i + q_i + K \frac{\partial T}{\partial x_i} &= 0, \\ \tau_t \overset{\triangle}{t}_{\langle ij \rangle} + t_{\langle ij \rangle} - 2\nu \frac{\partial v_{\langle i}}{\partial x_{j\rangle}} &= 0, \end{aligned} \tag{9.124}$$

where

$$\begin{aligned} \overset{\triangle}{q}_i &= \overset{\bullet}{q}_i + \xi q_k \frac{\partial v_{[k}}{\partial x_{i]}} + \gamma_q q_k \frac{\partial v_{(k}}{\partial x_{i)}}, \\ \overset{\triangle}{t}_{\langle ij \rangle} &= \overset{\bullet}{t}_{\langle ij \rangle} + t_{\langle ik \rangle} \frac{\partial v_{[k}}{\partial x_{j]}} + t_{\langle jk \rangle} \frac{\partial v_{[k}}{\partial x_{i]}} \\ &+ \gamma_t \left(t_{\langle ik \rangle} \frac{\partial v_{(k}}{\partial x_{j)}} + t_{\langle jk \rangle} \frac{\partial v_{(k}}{\partial x_{j)}} - \frac{2}{3} t_{\langle kl \rangle} \frac{\partial v_{(k}}{\partial x_{l)}} \delta_{ij} \right), \end{aligned} \tag{9.125}$$

and

$$\begin{aligned} \rho &= \rho(T, q_k, t_{kl}), \quad \tau_q = \tau_q(T, p), \quad K = K(T, p), \\ \varepsilon &= \varepsilon(T, q_k, t_{kl}), \quad \tau_t = \tau_t(T, p), \quad p = -\frac{1}{3} t_{kk}. \end{aligned} \tag{9.126}$$

In this model the objective time derivatives are not corotational. They contain contributions of the symmetric part of the velocity gradient which are, obviously, objective, with material coefficients γ_q, γ_t. This is known in rheology that such material parameters may be needed in adjusting the model to experiments. The role of the coefficient ξ will be explained later. Equations (9.125) are here not derived from balance laws of extended thermodynamics even though equation $(9.125)_2$, for instance, can be considered to be the simplification of equation $(9.103)_4$. For $\gamma_t = 0$ the time derivative becomes corotational. Two particular cases considered by Oldroyd [238] and other rheologists $\gamma_t = \pm 1$ have the practical bearing. The upper sign corresponds to the *upper Maxwellian model* and the lower sign – to the *lower Maxwellian model.*

On the other hand, equation $(9.125)_1$ is the generalization of C. Cattaneo equation [61]. In the case $\xi = 1, \gamma_q$ – arbitrary the time derivative is objective. The kinetic theory indicates rather $\xi = -1$. Cattaneo's equation is, on the other hand, of the form

$$\tau_q \frac{\partial q_i}{\partial t} + q_i + K \frac{\partial T}{\partial x_i} = 0, \tag{9.127}$$

which is also non-objective.

We proceed to exploit the entropy inequality for this model. We use Liu's method. The entropy inequality joint with the above field equations has the form

$$\rho \overset{\bullet}{\eta} + \frac{\partial h_k}{\partial x_k} - \Lambda^\rho \left(\overset{\bullet}{\rho} + \rho \frac{\partial v_k}{\partial x_k} \right) - \Lambda^{v_k} \left(\rho \overset{\bullet}{v}_k - \frac{\partial t_{kl}}{\partial x_l} \right) - \Lambda^\varepsilon \left(\rho \overset{\bullet}{\varepsilon} + \frac{\partial q_k}{\partial x_k} - t_{ij} \frac{\partial v_i}{\partial x_j} \right)$$
$$- \Lambda^{q_k} \left(\tau_q \overset{\triangle}{q}_k + q_k + K \frac{\partial T}{\partial x_k} \right) - \Lambda^{\langle kl \rangle} \left(\tau_t \overset{\triangle}{t}_{\langle kl \rangle} + t_{\langle kl \rangle} - 2\nu \frac{\partial v_{\langle k}}{\partial x_{l \rangle}} \right) \geq 0, \tag{9.128}$$

for all fields v_k, T, q_k, t_{kl}. The multipliers $\Lambda^\rho, \Lambda^{v_k}, \Lambda^\varepsilon, \Lambda^{q_k}, \Lambda^{\langle kl \rangle}$ are, of course, functions of these fields as well. Consequently, even though the field equations are not following from balance equations of extended thermodynamics the general strategy remains the same – we deal with the local model in which constitutive relations do not contain dependencies on derivatives of fields.

The linearity of (9.4.3.1) with respect to $\overset{\bullet}{v}_k$ yields

$$\Lambda^{v_k} = 0. \tag{9.129}$$

We split the gradient of velocity in the following manner

$$\frac{\partial v_k}{\partial x_l} = \frac{\partial v_{\langle k}}{\partial x_{l \rangle}} + \frac{1}{3} \frac{\partial v_i}{\partial x_i} \delta_{kl} + \frac{\partial v_{[k}}{\partial x_{l]}}. \tag{9.130}$$

Then the inequality indicates

$$\rho \Lambda^\rho + \rho \Lambda^\varepsilon + \frac{2}{3} \tau_t \gamma_t \Lambda^{\langle kl \rangle} t_{\langle kl \rangle} + \frac{1}{3} \tau_q \gamma_q \Lambda^{q_k} q_k = 0, \tag{9.131}$$

$$\Lambda^\varepsilon t_{\langle kl \rangle} + 2\nu \Lambda^{\langle kl \rangle} - \tau_t \gamma_t \left(\Lambda^{\langle km \rangle} t_{\langle ml \rangle} + \Lambda^{\langle lm \rangle} t_{\langle mk \rangle} - \frac{2}{3} \Lambda^{\langle nm \rangle} t_{\langle mn \rangle} \delta_{kl} \right)$$
$$- \frac{1}{2} \tau_q \gamma_q \left(\Lambda^{q_k} q_l + \Lambda^{q_l} q_k - \frac{2}{3} \Lambda^{q_m} q_m \delta_{kl} \right) = 0, \tag{9.132}$$

$$\tau_t \left(\Lambda^{\langle km \rangle} t_{\langle ml \rangle} - \Lambda^{\langle lm \rangle} t_{\langle mk \rangle} \right) + \frac{1}{2} \tau_q \xi \left(\Lambda^{q_k} q_l - \Lambda^{q_l} q_k \right) = 0. \tag{9.133}$$

Simultaneously, the inequality (9.4.3.1) is linear with respect to the following time derivatives $\dot{T}, \dot{q}_k, \dot{t}_{kl}\delta_{kl}, \dot{t}_{\langle kl\rangle}$ which leads to the following identities

$$\rho\frac{\partial\eta}{\partial T} - \Lambda^{\rho}\frac{\partial\rho}{\partial T} - \Lambda^{\varepsilon}\rho\frac{\partial\varepsilon}{\partial T} = 0,$$

$$\rho\frac{\partial\eta}{\partial q_k} - \Lambda^{\rho}\frac{\partial\rho}{\partial q_k} - \Lambda^{\varepsilon}\rho\frac{\partial\varepsilon}{\partial q_k} - \Lambda^{q_k}\tau_q = 0, \tag{9.134}$$

$$\left(\rho\frac{\partial\eta}{\partial t_{ij}} - \Lambda^{\rho}\frac{\partial\rho}{\partial t_{ij}} - \Lambda^{\varepsilon}\rho\frac{\partial\varepsilon}{\partial t_{ij}}\right)\delta_{ij} = 0,$$

$$\left(\rho\frac{\partial\eta}{\partial t_{ij}} - \Lambda^{\rho}\frac{\partial\rho}{\partial t_{ij}} - \Lambda^{\varepsilon}\rho\frac{\partial\varepsilon}{\partial t_{ij}}\right)\left(\delta_{ik}\delta_{jl} - \frac{1}{3}\delta_{ij}\delta_{kl}\right) - \Lambda^{\langle kl\rangle}\tau_t = 0.$$

In the similar manner, we find the inequality (9.4.3.1) to be linear with respect to the derivatives $\partial T/\partial x_k\, \partial q_l/\partial x_k, \partial t_{ij}/\partial x_k$. This yields

$$\begin{aligned}
\frac{\partial h_k}{\partial T} - K\Lambda^{q_k} &= 0,\\
\frac{\partial h_k}{\partial q_l} - \Lambda^{\varepsilon}\delta_{kl} &= 0,\\
\frac{\partial h_k}{\partial t_{ij}} &= 0.
\end{aligned} \tag{9.135}$$

Finally, there remains the residual inequality

$$\Lambda^{q_k} q_k + \Lambda^{\langle kl\rangle} t_{\langle kl\rangle} \leq 0. \tag{9.136}$$

Integrability condition of equations (9.135) indicates

$$\frac{\partial\Lambda^{\varepsilon}}{\partial q_j}\left(\delta_{ij}\delta_{kl} - \delta_{lj}\delta_{ki}\right) = 0 \quad \Rightarrow \quad \Lambda^{\varepsilon} = \Lambda^{\varepsilon}\left(T, t_{ij}\right). \tag{9.137}$$

Substitution of this relation in $(9.135)_2$ gives the following result for the entropy flux

$$h_k = \Lambda^{\varepsilon} q_k + \hat{h}_k\left(T, t_{ij}\right). \tag{9.138}$$

However, for isotropic fluids the second term must vanish – one cannot construct an isotropic vector function from a scalar and a second rank tensor. As the equilibrium part of the flux must be zero we obtain by substitution in $(9.135)_3$

$$\Lambda^{\varepsilon} = \Lambda^{\varepsilon}\left(T\right). \tag{9.139}$$

On the other hand, it follows from $(9.135)_1$

$$\Lambda^{q_k} = \frac{1}{K}\frac{d\Lambda^\varepsilon}{dT}q_k. \tag{9.140}$$

Relations (9.134) can be written in the following compact form

$$\rho d\eta = \Lambda^\rho d\rho + \Lambda^\varepsilon \rho d\varepsilon + \tau_q \Lambda^{q_k} dq_k + \tau_t \Lambda^{\langle kl\rangle} t_{\langle kl\rangle}. \tag{9.141}$$

In addition, we obtain from (9.131)

$$\Lambda^\rho = -\frac{p}{\rho}\Lambda^\varepsilon - \frac{2}{3}\frac{\tau_t\gamma_t}{\rho}\Lambda^{\langle kl\rangle}t_{\langle kl\rangle} - \frac{1}{3}\frac{\tau_q\gamma_q}{\rho K}\frac{d\Lambda^\varepsilon}{dT}q_k q_k. \tag{9.142}$$

To proceed with the exploitation of the second law we need some properties of the thermodynamical equilibrium. It is the state in which the dissipation defined by the inequality (9.136) vanishes, i.e.

$$q_k|_E = 0, \quad t_{\langle kl\rangle}\big|_E = 0. \tag{9.143}$$

In this state, the relation (9.141) has the form

$$d\eta = \Lambda^\varepsilon|_E \left[d\varepsilon|_E + p\left.\left(\frac{1}{\rho}\right)\right|_E \right], \quad \Lambda^\varepsilon|_E = \Lambda^\varepsilon, \tag{9.144}$$

where (9.139) has been used. Consequently, the classical Gibbs equation indicates

$$\Lambda^\varepsilon = \frac{1}{T}. \tag{9.145}$$

Hence, we obtain

$$h_k = \frac{1}{T}q_k, \quad \Lambda^{q_k} = -\frac{1}{KT^2}q_k, \tag{9.146}$$

i.e. we obtain the classical Fourier relation between entropy and heat flux.

It remains to eliminate the multiplier $\Lambda^{\langle kl\rangle}$. To this aim we use relations (9.132), (9.133). We have

$$t_{\langle kl\rangle} + \frac{1}{2}\frac{\tau_q\gamma_q}{KT}q_{\langle k}q_{l\rangle} = -L_{\langle kl\rangle\langle ij\rangle}\Lambda^{\langle ij\rangle},$$

$$\Lambda^{\langle ik\rangle}t_{\langle il\rangle} - \Lambda^{\langle il\rangle}t_{\langle ik\rangle} = 0, \tag{9.147}$$

where

$$L_{\langle kl\rangle\langle ij\rangle} = \nu T\left(\delta_{ki}\delta_{lj} + \delta_{li}\delta_{kj} - \tfrac{2}{3}\delta_{kl}\delta_{ij}\right) \tag{9.148}$$
$$-\tau_t\gamma_t T\left(\tfrac{1}{2}\left(t_{\langle ik\rangle}\delta_{jl} + t_{\langle il\rangle}\delta_{jk} + t_{\langle jk\rangle}\delta_{il} + t_{\langle jl\rangle}\delta_{ik}\right)\right.$$
$$\left. -\tfrac{2}{3}t_{\langle kl\rangle}\delta_{ij} - \tfrac{2}{3}t_{\langle ij\rangle}\delta_{kl}\right).$$

This linear mapping

$$\Lambda^{\langle ij\rangle} \mapsto t_{\langle kl\rangle} + \frac{1}{2}\frac{\tau_q\gamma_q}{KT}q_{\langle k}q_{l\rangle} \tag{9.149}$$

is non-singular and, consequently, it can be inverted. It is quite simple for the corotational model ($\gamma_t = 0$)

$$\Lambda^{\langle kl\rangle} = -\frac{1}{2\nu T}\left(t_{\langle kl\rangle} + \frac{1}{2}\frac{\tau_q\gamma_q}{KT}q_{\langle k}q_{l\rangle}\right). \tag{9.150}$$

Substitution of (9.146) and (9.150) in (9.133) yields

$$\frac{\tau_t}{2\nu T}\frac{1}{2}\frac{\tau_q\gamma_q}{KT}\left(q_k t_{\langle il\rangle} - q_l t_{\langle ik\rangle}\right) = 0, \tag{9.151}$$

which means that for a non-trivial model, when relaxation times τ_t, τ_q are different from zero, the time derivative of the heat flux (9.125) must also be corotational, $\gamma_q = 0$.

Let us summarize results for multipliers in this particular case

$$\Lambda^{\varepsilon} = \frac{1}{T}, \quad \Lambda^{\rho} = -\frac{p}{\rho T}, \quad \Lambda^{q_k} = -\frac{q_k}{KT^2}, \quad \Lambda^{\langle kl\rangle} = -\frac{t_{\langle kl\rangle}}{2\nu T}. \tag{9.152}$$

Gibbs equation has now the form

$$Td\eta = d\varepsilon + pd\left(\frac{1}{\rho}\right) - \frac{\tau_q}{\rho KT}q_k dq_k - \frac{\tau_t}{2\rho\nu}t_{\langle kl\rangle}dt_{\langle kl\rangle}. \tag{9.153}$$

For the stability analysis it is convenient to introduce the free enthalpy

$$g = \varepsilon - T\eta + \frac{p}{\rho}, \quad g = g\left(T, q_k, t_{kl}\right). \tag{9.154}$$

Then Gibbs equation transforms to the following form

$$dg = -\eta dT + \frac{1}{\rho}dp + \frac{\tau_q}{\rho KT}q_k dq_k + \frac{\tau_t}{2\rho\nu}t_{\langle kl\rangle}dt_{\langle kl\rangle}. \tag{9.155}$$

This equation yields the following relations

$$\frac{\partial g}{\partial T} = -\eta, \quad \frac{\partial g}{\partial p} = \frac{1}{\rho}, \quad \frac{\partial g}{\partial q_k} = \frac{\tau_q}{\rho KT}q_k,$$

$$\frac{\partial g}{\partial t_{ij}}\left(\delta_{ik}\delta_{jl}-\frac{1}{3}\delta_{kl}\delta_{ij}\right)=\frac{\tau_t}{2\rho\nu}t_{\langle kl\rangle}. \tag{9.156}$$

They, in turn, indicate the following set of integrability conditions

$$\frac{\partial}{\partial T}\left(\frac{1}{\rho}\right)=-\frac{\partial\eta}{\partial p},\quad \frac{\partial}{\partial q_k}\left(\frac{1}{\rho}\right)=\frac{\partial}{\partial p}\left(\frac{\tau_q}{\rho KT}\right)q_k,$$

$$\frac{\partial}{\partial T}\left(\frac{\tau_q}{\rho KT}\right)q_k=-\frac{\partial\eta}{\partial q_k},\quad \frac{\partial}{\partial T}\left(\frac{\tau_t}{2\rho\nu}\right)t_{\langle kl\rangle}=-\frac{\partial\eta}{\partial t_{ij}}\left(\delta_{ik}\delta_{jl}-\frac{1}{3}\delta_{kl}\delta_{ij}\right), \tag{9.157}$$

$$\frac{\partial}{\partial p}\left(\frac{\tau_t}{2\rho\nu}\right)t_{\langle kl\rangle}=\frac{\partial}{\partial t_{ij}}\left(\frac{1}{\rho}\right)\left(\delta_{ik}\delta_{jl}-\frac{1}{3}\delta_{kl}\delta_{ij}\right).$$

We can also write the results for the incompressible fluids. In the chosen variables it means that the pressure should enter the constitutive relations only through the explicit presence. Hence, according to $(9.157)_{1,2,5}$,

$$\rho=\text{const}. \tag{9.158}$$

and the free enthalpy is as follows

$$g=g_0(T)+\frac{1}{2}\frac{\tau_q}{\rho KT}q_kq_k+\frac{1}{4}\frac{\tau_t}{\rho\nu}t_{\langle kl\rangle}t_{\langle kl\rangle}. \tag{9.159}$$

Additionally,

$$\begin{aligned}\eta&=-\frac{dg_0}{dT}-\frac{1}{2\rho}\frac{d}{dT}\left(\frac{\tau_q}{KT}\right)q_kq_k-\frac{1}{4\rho}\frac{d}{dT}\left(\frac{\tau_t}{\rho\nu}\right)t_{\langle kl\rangle}t_{\langle kl\rangle},\\ \varepsilon&=\varepsilon_0+\frac{1}{2\rho}\left[\frac{\tau_q}{KT}-T\frac{d}{dT}\left(\frac{\tau_q}{KT}\right)\right]q_kq_k+\frac{1}{4\rho}\left[\frac{\tau_t}{\nu}-T\frac{d}{dT}\left(\frac{\tau_t}{\rho\nu}\right)\right]t_{\langle kl\rangle}t_{\langle kl\rangle},\\ \varepsilon_0&=g_0-T\frac{dg_0}{dT}-\frac{p}{\rho}.\end{aligned} \tag{9.160}$$

Simultaneously, the residual inequality has the form

$$\frac{1}{KT^2}q_kq_k+\frac{1}{2\nu T}t_{\langle kl\rangle}t_{\langle kl\rangle}\geq 0. \tag{9.161}$$

As it should hold for arbitrary fields q_k and $t_{\langle kl\rangle}$ it indicates

$$K>0,\quad \nu>0. \tag{9.162}$$

At the same time, the stability of the thermodynamical equilibrium, i.e. the convexity of g, yields

$$\frac{\tau_q}{\rho K T} > 0 \quad \Rightarrow \quad \tau_q > 0, \quad \frac{\tau_t}{\rho \nu} > 0 \quad \Rightarrow \quad \tau_t > 0. \tag{9.163}$$

These inequalities guarantee the hyperbolicity of the set of field equations because τ_q and τ_t are relaxation times of the heat flux and of the stress deviator, respectively.

It can be shown immediately by use of (9.132) and (9.133) that the condition $\gamma_q = 0$ (with $\tau_q \neq 0$!) implies $\gamma_t = 0$, i.e. again the full corotational form of (9.125).

Let us note that the above thermodynamical considerations do not involve the constant ξ. This is due to the relation $(9.146)_2$ which, substituted in (9.133) eliminates the second term. It means that both objective: $\xi = 1$ and non-objective: $\xi = -1$ forms of the definition (9.125) are thermodynamically admissible.

9.4.3.2 *Transitions to second order fluids*

The complicated form of the general definition (9.148) of the operator $L_{\langle kl\rangle\langle ij\rangle}$ does not allow for similar considerations in the general case. Therefore, we limit the attention to the corotational model.

As we have already indicated, the transition from the hyperbolic model to various nth order fluids may be questionable. For the model which we present in this Section we can easily locate the source of troubles in Maxwell iteration.

Let us recall the constitutive relation (9.56) defining the second order fluid. In Cartesian coordinates we have

$$t_{kl} + p\delta_{kl} = \nu \left(\mathbf{A}_1\right)_{kl} + \alpha_1 \left(A_2\right)_{\langle kl\rangle} + \alpha_2 \left[\left(\mathbf{A}_1\right)_{ki} \left(\mathbf{A}_1\right)_{il} - \frac{1}{3} \left(\mathbf{A}_1\right)_{ij} \left(\mathbf{A}_1\right)_{ij} \delta_{kl}\right], \tag{9.164}$$

where

$$\left(\mathbf{A}_1\right)_{kl} = \frac{\partial v_k}{\partial x_l} + \frac{\partial v_l}{\partial x_k}, \quad \left(A_2\right)_{kl} = \left(\mathbf{A}_1\right)_{kl} + (\dot{\mathbf{A}}_1)_{ki} \frac{\partial v_i}{\partial x_l} + (\dot{\mathbf{A}}_1)_{li} \frac{\partial v_i}{\partial x_k}, \tag{9.165}$$

(compare (3.13), (3.14)).

Now the Maxwell iteration in the evolution equation $(9.124)_2$ yields in the zeroth approximation

$$t^{(0)}_{\langle kl\rangle} = 2\nu \frac{\partial v_{\langle k}}{\partial x_{l\rangle}}, \tag{9.166}$$

which is, of course, identical with the Navier-Stokes fluid. Then the first approximation follows from the equation

$$\tau_t \overset{\triangle}{t}{}^{(0)}_{\langle kl\rangle} + t^{(1)}_{\langle kl\rangle} - 2\nu \frac{\partial v_{\langle k}}{\partial x_{l\rangle}} = 0. \tag{9.167}$$

For incompressible materials: $(\mathbf{A}_1)_{kk} = 0$, we obtain for a constant viscosity ν

$$t^{(1)}_{\langle kl\rangle} = \nu\left(\mathbf{A}_1\right)_{\langle kl\rangle} - \tau_t \nu \Big[\left(\mathbf{A}_2\right)_{\langle kl\rangle} \tag{9.168}$$
$$+\left(\gamma_t - 1\right)\Big(\left(\mathbf{A}_1\right)_{\langle ki\rangle}\left(\mathbf{A}_1\right)_{\langle il\rangle} - \tfrac{1}{3}\left(\mathbf{A}_1\right)_{\langle ij\rangle}\left(\mathbf{A}_1\right)_{\langle ij\rangle}\delta_{kl}\Big)\Big].$$

Comparison of (9.164) and (9.168) yields

$$\alpha_1 = -\nu\tau_t, \quad \alpha_2 = -\alpha_1\left(1 - \gamma_t\right). \tag{9.169}$$

For the corotational derivative $\gamma_t = 0$ the second relation checks with results of Dunn and Fosdick [92] and contradicts experiments (see the beginning of this Section). Simultaneously, $\alpha_1 < 0$ yields the thermodynamical instability of the approximation.

Similar problem arises in the case of Maxwell iteration applied to equation $(9.124)_1$ for the heat flux. We have

$$q_k^{(0)} = -K\frac{\partial T}{\partial x_k},$$

$$q_k^{(1)} = -K\frac{\partial T}{\partial x_k} - \tau_q \overset{\triangle}{\left(K\frac{\partial T}{\partial x_k}\right)}. \tag{9.170}$$

In contrast to the Fourier relation, the first approximation $q_k^{(1)}$ yields the instability of the thermodynamical equilibrium for $K > 0, \tau_q > 0$.

As already mentioned, the field equations in any degree of iteration are parabolic instead of the original set which is hyperbolic.

It is easy to explain the above problems. Expansions of $t_{\langle kl\rangle}$ and q_k into power series in τ_t and τ_q, respectively, and subsequent use of the regular perturbation method reduces necessarily differential equations (9.124) to algebraic relations. This is due to the fact that τ_t and τ_q appear in front of the differential operator. This means, of course, that the proper perturbation should be singular rather than regular. Consequently, the Maxwell iteration is inappropriate for the construction of approximate models which should preserve physical properties of the original hyperbolic model.

Chapter 10

Elasto-viscoplastic materials

10.1 Preliminaries

The subject of elasto-viscoplastic deformations of solids is so extensive that we are not able even to list the main approaches and models in this Chapter. We quote here a few general references, books and monographs, on these topics. However, all these references have one common feature – very limited thermodynamical foundations. The classical linear plasticity is based on the assumption that the evolution equation for deformations should be invariant with respect to rescaling of time (e.g. [162], [145]). As a consequence one obtains the existence of a yield surface which specifies the domain of stresses in the six-dimensional stress space for which deformations are reversible (elastic). This assumption eliminates simultaneously the main effect described by the second law of thermodynamics – the spontaneous evolution of the system to the thermodynamical equilibrium. One can still calculate the full dissipation (plastic work) in a given process of finite duration but the strategy of thermodynamics in the spirit presented in the present book is impossible. In spite of this limitation some thermodynamic-like considerations have been conducted in order to justify the existence of some potentials in plasticity which would extend such notions of thermodynamics as the Helmholtz or Gibbs free energies. One should mention here in particular the works of R. Hill [145] and J. R. Rice [263]. Albeit in the works of M. Kleiber, B. Raniecki [166], Th. Lehmann [184], S. Nemat-Nasser [230], A. R. S. Ponter, J. Bataille, J. Kestin [254] attempts have been made to establish such thermodynamical origins the problem could not be solved up to the standards of modern thermodynamics.

Even though extensions of classical plasticity on viscous effects yield non-trivial thermodynamical processes again, their limitation by the second law is incorporated in a rather arbitrary manner by the application of

Clausius-Duhem inequality and a subsequent use of equilibrium properties such as constitutive laws for stresses in dependence on elastic deformations [201]. As we have demonstrated in the previous Chapter, the applicability of Clausius-Duhem inequality is even for viscoelastic materials limited to a small vicinity of the thermodynamical equilibrium (e.g. compare the relation (9.38) between the entropy and heat fluxes).

In this Chapter, we present main notions of elasto-viscoplasticity. It is claimed (e.g. [185], [186]) that viscoplastic effects (creep) are essential first when the temperature is sufficiently high, approximately one third of the absolute melting point. Below one expects the brittle behavior [229]. However, from thermodynamical point of view plastic deformations in lower temperatures can often be well modelled by viscoplasticity.

The main part of our considerations concerns a particular model, the so-called *crystal (structural) plasticity*. For simplicity, we choose this example for the demonstration of the structure of viscoplasticity rather than any more general approach based on the evolution of internal variables [202]. Crystal plasticity possesses all characteristic features of general models but it allows easier for a physical insight. We return briefly at the end of this Chapter to a general and mathematically well-found model proposed by P. Perzyna [245], [246], [248], [247], [249].

10.2 Local configurations

The classical linear plasticity suggests the additive separation of the elastic and plastic deformations of elastoplastic materials

$$\mathbf{e} = \mathbf{e}^e + \mathbf{e}^p, \tag{10.1}$$

where the former fulfils the constitutive law of linear elasticity: $\mathbf{T} = \mathbf{C}\mathbf{e}^e$, $\mathbf{C} = C_{ijkl}\mathbf{e}_i \otimes \mathbf{e}_j \otimes \mathbf{e}_k \otimes \mathbf{e}_l$ – the fourth rank tensor of elastic constants, and the latter is the residual deformation after the unloading. Such an additive decomposition cannot hold for large deformations. As a nonlinear extension, Ekkehard Kröner [173] has proposed in 1961 the *multiplicative decomposition* of the deformation gradient $\mathbf{F}$

$$\mathbf{F} = \mathbf{F}^e\mathbf{F}^p, \quad \det \mathbf{F}^e \neq 0, \quad \det \mathbf{F}^p \neq 0, \tag{10.2}$$

which, of course, yields (10.1) in the case of small deformations. The relation (10.2) has been incorporated in the frame of continuum mechanics by E. H. Lee [183] and since then extensively studied and, even though sometimes questioned (e.g. M. B. Rubin [274], [275], [277]) commonly used in nonlinear plasticity. Doubts concerning the existence of the decomposition

(10.2) are, for instance, well found in the case of transition from the plasticity of monocrystals (grains) to aggregates (polycrystals). The averaging procedure destroys the multiplicative structure on the level of polycrystals even if we assume such a decomposition for monocrystals.

It can be easily shown that, in contrast to $\mathbf{F} = \partial \mathbf{f}/\partial \mathbf{X} = \mathrm{Grad}\,\mathbf{f}$, the mappings $\mathbf{F}^e$ and $\mathbf{F}^p$ are not integrable. It means that one cannot split the function of motion $\mathbf{f}$ (see (1.1)) in such a way that $\mathbf{f} = \mathbf{f}^e \circ \mathbf{f}^p$ (i.e. $\mathbf{Z} = \mathbf{f}^p(\mathbf{X}, t), \mathbf{x} = \mathbf{f}^e(\mathbf{Z}, t)$ with $\mathbf{x} = \mathbf{f}(\mathbf{X}, t)$) and $\mathbf{F}^p = \partial \mathbf{f}^p/\partial \mathbf{X}, \mathbf{F}^e = \partial \mathbf{f}^e/\partial \mathbf{Z}$. Such mappings $\mathbf{f}^e$ and $\mathbf{f}^p$ do not exist. There also exists no differentiable manifold of points $\mathbf{Z}$. For this reason, the mappings of vectors: $d\mathbf{Z} = \mathbf{F}^p d\mathbf{X}, \quad d\mathbf{x} = \mathbf{F}^e d\mathbf{Z}$ are called *local plastic and elastic configurations*, respectively. The names are misleading because they refer to mappings defined on vector spaces and not to any subsets of global configuration spaces (i.e. differentiable manifolds) but, all the same, they are commonly used in the plasticity.

The decomposition (10.2) yields immediately the following relations for *full, elastic and plastic rates of deformation*

$$\begin{aligned} &\overset{\bullet}{\mathbf{F}} = \overset{\bullet}{\mathbf{F}}{}^e\mathbf{F}^p + \mathbf{F}^e\overset{\bullet}{\mathbf{F}}{}^p \\ \Rightarrow \quad &\mathbf{L} = \mathbf{L}^e + \mathbf{L}^p, \quad \mathbf{L}^e = \overset{\bullet}{\mathbf{F}}{}^e\mathbf{F}^{e-1}, \quad \mathbf{L}^p = \mathbf{F}^e\overset{\bullet}{\mathbf{F}}{}^p\mathbf{F}^{p-1}\mathbf{F}^{e-1}. \end{aligned} \tag{10.3}$$

These definitions lead, obviously to the following rules of transformations for the Euclidean transformation $\mathbf{x}^* = \mathbf{O}(t)\,\mathbf{x} + \mathbf{c}(t)$ (see (3.4))

$$\begin{aligned} \mathbf{L}^* &= \mathbf{O}\mathbf{L}\mathbf{O}^T + \boldsymbol{\Omega}, \quad \boldsymbol{\Omega} = \overset{\bullet}{\mathbf{O}}\mathbf{O}^T, \\ \mathbf{L}^{e*} &= \mathbf{O}\mathbf{L}^e\mathbf{O}^T + \boldsymbol{\Omega}, \quad \mathbf{L}^{p*} = \mathbf{O}\mathbf{L}^p\mathbf{O}^T. \end{aligned} \tag{10.4}$$

Consequently, in contrast to the elastic rate $\mathbf{L}^e$, the plastic rate $\mathbf{L}^p$ is objective. It also means that in the additive decomposition

$$\mathbf{L}^p = \mathbf{D}^p + \mathbf{W}^p, \quad \mathbf{D}^p = \frac{1}{2}\left(\mathbf{L}^p + \mathbf{L}^{pT}\right), \quad \mathbf{W}^p = \frac{1}{2}\left(\mathbf{L}^p - \mathbf{L}^{pT}\right), \tag{10.5}$$

both *plastic stretching* $\mathbf{D}^p$ and *plastic spin* $\mathbf{W}^p$ are objective.

The purpose of modelling of viscoplastic materials is to find field equations for the following fields

$$\{\mathbf{f}, \mathbf{F}^p, T\}\,. \tag{10.6}$$

In contrast to the function of motion $\mathbf{f}$ and the temperature, T, which are macroscopically observable, the plastic configuration, $\mathbf{F}^p$ is the *internal variable*, whose development must be defined by an evolution equation not following from classical macroscopic balance laws. Many different propositions have been made for this equation and, obviously, as they are characteristic for a particular class of materials, they are very different. These equations expose some microscopic features of materials and require a description of some additional variable parameters. Therefore the set of field equations must be extended beyond the fields (10.6). The typical example is the description of hardening of metals which requires additional equations. In the next Section we present a typical example of such a construction. This model called *crystal plasticity* is designed to describe large plastic deformations of metals in the range of relatively low temperatures (app. below 60% of the temperature of point of melting T_m). It illustrates the way in which microstructural parameters are introduced into the macroscopic model and, simultaneously, has the practical bearing in predicting the so-called plastic anisotropy which appears in such processes as forging, deep drawing or rolling of polycrystalline metals. Subsequently, we use this example to demonstrate some thermodynamical features of viscoplasticity.

10.3 Crystal plasticity of monocrystals

In monocrystals, the main mechanism of plastic deformations is a slip along certain crystallographic directions. Such a motion follows from the creation and redistribution of dislocations. These line defects have the property that they may only appear as a closed loop. This means that the motion of the dislocation line can only terminate on the boundary. In the case of monocrystals it yields microscopical changes of surfaces and macroscopically it is demonstrated as a permanent deformation. As the density of dislocation loops on the typical surface of an annealed metal grows from some 10^5 cm^{-2} to some 10^{12} cm^{-2} after large plastic deformations the relation between the plastic deformations and dislocations seems to be well established. For polycrystals, dislocations get stuck on the boundaries (*incoherent boundaries*) of monocrystal grains until the rise of temperature sufficiently increases their mobility and enables jumps of dislocations between neighboring grains.

The motion of dislocations in monocrystals is specified on a mesoscopic level of observation by a *slip system* $\left(\mathbf{B}^{(\alpha)}, \mathbf{N}^{(\alpha)}, \overset{\bullet}{\gamma}^{(\alpha)}\right)$, where the unit vector $\mathbf{B}^{(\alpha)}$ specifies the direction of α-slip, the unit vector $\mathbf{N}^{(\alpha)}$ is perpendicular to the plane of α-slip, and $\overset{\bullet}{\gamma}^{(\alpha)}$ is the so-called *rate of α-slip*. The vectors $\mathbf{B}^{(\alpha)}, \mathbf{N}^{(\alpha)}$, orthogonal to each other, are defined in the reference configuration of the monocrystal by crystallographic considerations ($1 \leq \alpha \leq a$, a runs approximately from one to usually three-four active slip systems) and they are assumed to be material, i.e. their motion is determined by the macroscopic deformation. The physics of this process is specified by the rate $\overset{\bullet}{\gamma}^{(\alpha)}$. We discuss further some details.

The slip system is assumed to define the rate of plastic configuration $\mathbf{F}^p$

$$\overset{\bullet}{\mathbf{F}}^p \mathbf{F}^{p-1} = \sum_{\alpha=1}^{a} \overset{\bullet}{\gamma}^{(\alpha)} \mathbf{B}^{(\alpha)} \otimes \mathbf{N}^{(\alpha)}, \tag{10.7}$$

where a is the number of operating slip systems at a given point $\mathbf{X}$ of the monocrystal. This number depends on process variables and on material properties of the monocrystal. The Eulerian image of (10.7) which is, of course, the plastic rate $\mathbf{L}^p$, has the following form

$$\mathbf{L}^p = \sum_{\alpha=1}^{a} \overset{\bullet}{\gamma}^{(\alpha)} \kappa^{(\alpha)} \mathbf{b}^{(\alpha)} \otimes \mathbf{n}^{(\alpha)}, \tag{10.8}$$

where

$$\mathbf{b}^{(\alpha)} = \frac{\mathbf{F}^e \mathbf{B}^{(\alpha)}}{\left|\mathbf{F}^e \mathbf{B}^{(\alpha)}\right|}, \quad \mathbf{n}^{(\alpha)} = \frac{\mathbf{F}^{e-T} \mathbf{N}^{(\alpha)}}{\left|\mathbf{F}^{e-T} \mathbf{N}^{(\alpha)}\right|}, \quad \kappa^{(\alpha)} = \left|\mathbf{F}^e \mathbf{B}^{(\alpha)}\right| \left|\mathbf{F}^{e-T} \mathbf{N}^{(\alpha)}\right|, \tag{10.9}$$

as the vectors $\mathbf{B}^{(\alpha)}, \mathbf{N}^{(\alpha)}$ are not influenced by the slip.

In many cases of practical interest the factor $\kappa^{(\alpha)}$ can be neglected. It is the case when the elastic part of the deformation is small in comparison with plastic deformations as, for instance, for metal forming. We have

$$\kappa^{(\alpha)2} = \left(\mathbf{B}^{(\alpha)} \mathbf{C}^e \mathbf{B}^{(\alpha)}\right) \left(\mathbf{N}^{(\alpha)} \mathbf{C}^{e-1} \mathbf{N}^{(\alpha)}\right), \quad \mathbf{C}^e = \mathbf{F}^{eT} \mathbf{F}^e. \tag{10.10}$$

Let the eigenvalues of $\mathbf{C}^e$ be $\lambda^{e(r)}, r = 1, 2, 3$, and the corresponding normalized eigenvectors be $\mathbf{K}^{(r)}$. Then (10.10) can be written in the form

$$\kappa^{(\alpha)2} = \left[1 + \sum_{r=1}^{3} \left(\lambda^{e(r)} - 1\right) \left(\mathbf{B}^{(\alpha)} \cdot \mathbf{K}^{(r)}\right)^2\right] \tag{10.11}$$
$$\times \left[1 - \sum_{r=1}^{3} \frac{\lambda^{e(r)} - 1}{\lambda^{e(r)}} \left(\mathbf{N}^{(\alpha)} \cdot \mathbf{K}^{(r)}\right)^2\right],$$

where we have used the relations

$$\sum_{r=1}^{3} \mathbf{B}^{(\alpha)} \cdot \mathbf{K}^{(r)} = 1, \quad \sum_{r=1}^{3} \mathbf{N}^{(\alpha)} \cdot \mathbf{K}^{(r)} = 1. \tag{10.12}$$

For small elastic deformations we have $\left|\lambda^{e(r)} - 1\right| \ll 1$. Consequently,

$$\kappa^{(\alpha)} = 1 + \sum_{r=1}^{3} \left(\sqrt{\lambda^{e(r)}} - 1\right) \left(\left(\mathbf{B}^{(\alpha)} \cdot \mathbf{K}^{(r)}\right)^2 - \left(\mathbf{N}^{(\alpha)} \cdot \mathbf{K}^{(r)}\right)^2\right)$$
$$+ O\left(\max\left|\sqrt{\lambda^{e(r)}} - 1\right|^2\right). \qquad (10.13)$$

In the zeroth approximation $\kappa^{(\alpha)} \approx 1$.

The main part of the relation (10.8) is the rate of plastic slip $\dot{\gamma}^{(\alpha)}$. Motivated by experimental results, it is assumed to satisfy the *evolution equation*

$$\dot{\gamma}^{(\alpha)} = g^{(\alpha)}\left(\tau^{(\alpha)}, \tau_0^{(\alpha)}, T\right), \quad \alpha = 1, ..., a. \qquad (10.14)$$

In this equation, $\tau^{(\alpha)}$ is the so-called *Schmid resolved shear stress* on the α-slip [284], [285] and in the current configuration it is defined by the relation

$$\tau^{(\alpha)} = \mathbf{b}^{(\alpha)} \cdot \mathbf{T}\mathbf{n}^{(\alpha)}, \qquad (10.15)$$

where $\mathbf{T}$ is the macroscopic Cauchy stress. $\tau_0^{(\alpha)}$ describes the state of *hardening* and it is assumed to have the form

$$\dot{\tau}_0^{(\alpha)} = \sum_{\beta=1}^{a} h_{\alpha\beta}\dot{\gamma}^{(\beta)}. \qquad (10.16)$$

The form of the *hardening matrix* $h_{\alpha\beta}$ depends on the history of deformation and temperature.

The above structure of the model contains many simplifying assumptions. One of them concerns the relation (10.15). The resolved shear stress in monocrystals should be specified by real microscopic stresses which, apart from macroscopic stress $\mathbf{T}$ should contain some self-equilibrated stresses which appear in the vicinity of defects. This is to some extend included in the hardening effects which account, for instance, for the so-called dislocation core but still the above model is more phenomenological than really microscopic. Simultaneously, the hardening matrix $h_{\alpha\beta}$ depends on various parameters related to the distribution of dislocations and obstacles in their motion, e.g. the damping by dislocation forests clamped on the slip systems or interaction with dislocations moving on different slip systems. These interactions yield viscoplastic effects.

The above construction yields the set of equations for the fields:

$$\left\{\mathbf{f}, T, \gamma^{(\alpha)}, \mathbf{b}^{(\alpha)}, \mathbf{n}^{(\alpha)}\right\} \tag{10.17}$$

provided the relation $\mathbf{T} = \mathbf{T}(\mathbf{C}^e, T)$ can be inverted with respect to $\mathbf{C}^e = \mathbf{C}^e(\mathbf{T}, T)$. In order to see it, it is sufficient to eliminate $\mathbf{F}^e$ from the relations (10.9). Differentiation with respect to time yields

$$\begin{aligned}
&\overset{\Delta}{\mathbf{b}^{(\alpha)}} + \mathbf{L}^p\mathbf{b}^{(\alpha)} - \mathbf{b}^{(\alpha)}\left(\mathbf{b}^{(\alpha)}\cdot\mathbf{L}^p\mathbf{b}^{(\alpha)}\right) = 0,\\
&\overset{\Delta}{\mathbf{n}^{(\alpha)}} - \mathbf{L}^{pT}\mathbf{n}^{(\alpha)} + \mathbf{n}^{(\alpha)}\left(\mathbf{n}^{(\alpha)}\cdot\mathbf{L}^p\mathbf{n}^{(\alpha)}\right) = 0,\\
&\overset{\Delta}{\kappa^{(\alpha)}} + \left(\mathbf{b}^{(\alpha)}\cdot\mathbf{L}^p\mathbf{b}^{(\alpha)} - \mathbf{n}^{(\alpha)}\cdot\mathbf{L}^p\mathbf{n}^{(\alpha)}\right) = 0,
\end{aligned} \tag{10.18}$$

where

$$\begin{aligned}
\overset{\Delta}{\mathbf{b}^{(\alpha)}} &= \overset{\bullet}{\mathbf{b}^{(\alpha)}} - \mathbf{L}\mathbf{b}^{(\alpha)} + \mathbf{b}^{(\alpha)}\left(\mathbf{b}^{(\alpha)}\cdot\mathbf{L}\mathbf{b}^{(\alpha)}\right),\\
\overset{\Delta}{\mathbf{n}^{(\alpha)}} &= \overset{\bullet}{\mathbf{n}^{(\alpha)}} + \mathbf{L}^T\mathbf{n}^{(\alpha)} - \mathbf{n}^{(\alpha)}\left(\mathbf{n}^{(\alpha)}\cdot\mathbf{L}\mathbf{n}^{(\alpha)}\right),\\
\overset{\Delta}{\kappa^{(\alpha)}} &= \overset{\bullet}{\kappa^{(\alpha)}} - \kappa^{(\alpha)}\left(\mathbf{b}^{(\alpha)}\cdot\mathbf{L}\mathbf{b}^{(\alpha)} - \mathbf{n}^{(\alpha)}\cdot\mathbf{L}\mathbf{n}^{(\alpha)}\right).
\end{aligned} \tag{10.19}$$

To complete the presentation of structural viscoplasticity of monocrystals let us mention a particular model in which the relation (10.14) has the form of a power law, i.e.

$$g^{(\alpha)} = g_0^{(\alpha)}\left(\frac{\tau^{(\alpha)}}{\tau_0^{(\alpha)}}\right)^m, \tag{10.20}$$

with $g_0^{(\alpha)}$– scaling constant and, for metals, $m = 300-500$. Such a model is frequently used in numerical codes for the calculation of textures in metals (e.g. the works of U. F. Kocks *et al.* such as [167], [168]). The relation (10.20) is called the *Norton power law*. The inverse form of this relation

$$\tau^\alpha = \tau_0^{(\alpha)}\left(\frac{g^{(\alpha)}}{g_0^{(\alpha)}}\right)^{1/m}, \tag{10.21}$$

shows that the limit $m \to \infty$ yields the threshold behavior characteristic for the plastic rather than viscoplastic reaction of the material: $\tau^\alpha = 0$ for $g^{(\alpha)} < g_0^{(\alpha)}$ and $\tau^\alpha = \tau_0^{(\alpha)}$ for $g^{(\alpha)} = g_0^{(\alpha)}$.

Let us collect the above equations for the case of *rigid-viscoplastic materials*. Then it is assumed that elastic deformations are so small that they can be neglected entirely and solely the elastic rotations must be incorporated

$$\mathbf{F} = \mathbf{R}^e \mathbf{F}^p, \quad \mathbf{R}^{eT} = \mathbf{R}^{e-1}. \tag{10.22}$$

Hence, we have

$$\mathbf{L} = \mathbf{W}^e + \sum_{\alpha=1}^{a} \overset{\bullet}{\gamma}^{(\alpha)} \mathbf{b}^{(\alpha)} \otimes \mathbf{n}^{\alpha}, \quad \mathbf{W}^e = \overset{\bullet}{\mathbf{R}}^e \mathbf{R}^{eT}, \tag{10.23}$$
$$\mathbf{b}^{(\alpha)} = \mathbf{R}^e \mathbf{B}^{(\alpha)}, \quad \mathbf{n}^{(\alpha)} = \mathbf{R}^e \mathbf{N}^{(\alpha)}.$$

As the plastic deformation defined by slips does not influence crystallographic directions the contribution $\mathbf{W}^e$ is often called the *lattice rotation*.

The set of fields is as follows

- mass density ρ,
- velocity $\mathbf{v}$,
- Cauchy stress $\mathbf{T}$,
- elastic rotation $\mathbf{R}^e$,
- temperature T,

for which we have the following field equations

$$\frac{\partial \rho}{\partial t} + \operatorname{div}(\rho \mathbf{v}) = 0$$

$$\rho \left(\frac{\partial \mathbf{v}}{\partial t} + \mathbf{v} \cdot \operatorname{grad} \mathbf{v} \right) = \operatorname{div} \mathbf{T},$$

$$\mathbf{L} = \mathbf{W}^e + \sum_{\alpha=1}^{a} \overset{\bullet}{\gamma}^{(\alpha)} \mathbf{b}^{(\alpha)} \otimes \mathbf{n}^{\alpha}, \quad \mathbf{L} = \operatorname{grad} \mathbf{v}, \tag{10.24}$$

$$\rho \left(\frac{\partial \varepsilon}{\partial t} + \mathbf{v} \cdot \operatorname{grad} \varepsilon \right) + \operatorname{div} \mathbf{q} = \operatorname{tr} \mathbf{T} \mathbf{L}^T,$$

$$\varepsilon = \varepsilon (T, \mathbf{F}^p), \quad \mathbf{q} = -K (T, \mathbf{F}^p) \operatorname{grad} T,$$

with

$$\overset{\bullet}{\gamma}^{(\alpha)} = g_0^{(\alpha)} \left(\frac{\tau^{(\alpha)}}{\tau_0^{(\alpha)}} \right)^m, \quad \tau^{(\alpha)} = \left(\mathbf{R}^e \mathbf{B}^{(\alpha)} \right) \cdot \mathbf{T} \left(\mathbf{R}^e \mathbf{N}^{(\alpha)} \right),$$

$$\overset{\bullet}{\tau}_0^{(\alpha)} = \sum_{\beta=1}^{a} h_{\alpha\beta} \overset{\bullet}{\gamma}^{(\alpha)}, \tag{10.25}$$

i.e. 14 equations for 14 fields provided constitutive laws for the internal energy and for the heat flux are known.

In many works on crystal plasticity, relation (10.24) is split into two parts. According to (10.15) and (10.23) we can write the following constitutive law

$$\mathbf{D} = g_0 \sum_{\alpha=1}^{a} \mathbf{r}^{(\alpha)} \left(\frac{\mathbf{T} \cdot \mathbf{r}^{(\alpha)}}{\tau_0'^{(\alpha)}} \right)^m, \tag{10.26}$$

$$\mathbf{r}^{(\alpha)} = \operatorname{sym} \left(\mathbf{b}^{(\alpha)} \otimes \mathbf{n}^{(\alpha)} \right), \quad \tau_0'^{(\alpha)} = \left(\frac{g_0}{g_0^{(\alpha)}} \right)^{1/m} \tau_0^{(\alpha)}.$$

Simultaneously

$$\mathbf{W} = \mathbf{W}^e + \frac{1}{2} \sum_{\alpha=1}^{a} \overset{\bullet}{\gamma}^{(\alpha)} \left(\mathbf{b}^{(\alpha)} \otimes \mathbf{n}^{\alpha} - \mathbf{n}^{\alpha} \otimes \mathbf{b}^{(\alpha)} \right). \tag{10.27}$$

In order to appreciate better the structure of these equations we quote here the set for a two-dimensional case (plane stress) under isothermal conditions ($T = const$). Then the objects appearing in the above equations have the following structure

$$\mathbf{T} = \begin{pmatrix} \sigma_{11} & \sigma_{12} & 0 \\ \sigma_{12} & \sigma_{22} & 0 \\ 0 & 0 & 0 \end{pmatrix}, \quad \mathbf{v} = \begin{pmatrix} v_1 \\ v_2 \\ 0 \end{pmatrix},$$

$$\mathbf{B}^{(\alpha)} = \begin{pmatrix} \cos \varphi_0^{(\alpha)} \\ \sin \varphi_0^{(\alpha)} \\ 0 \end{pmatrix}, \quad \mathbf{N}^{(\alpha)} = \begin{pmatrix} -\sin \varphi_0^{(\alpha)} \\ \cos \varphi_0^{(\alpha)} \\ 0 \end{pmatrix}, \tag{10.28}$$

$$\mathbf{R}^e = \begin{pmatrix} \cos \psi^e & \sin \psi^e & 0 \\ -\sin \psi^e & \cos \psi^e & 0 \\ 0 & 0 & 1 \end{pmatrix}, \quad \mathbf{W}^e = \begin{pmatrix} 0 & -\dot{\psi}^e & 0 \\ \dot{\psi}^e & 0 & 0 \\ 0 & 0 & 0 \end{pmatrix},$$

$$\mathbf{R}^e\left(\mathbf{B}^{(\alpha)}\otimes\mathbf{N}^{(\alpha)}\right)\mathbf{R}^{eT}$$

$$=\begin{pmatrix} -\frac{1}{2}\sin 2\left(\varphi_0^{(\alpha)}+\psi^e\right) & \cos^2\left(\varphi_0^{(\alpha)}+\psi^e\right) & 0 \\ -\sin^2\left(\varphi_0^{(\alpha)}+\psi^e\right) & \frac{1}{2}\sin 2\left(\varphi_0^{(\alpha)}+\psi^e\right) & 0 \\ 0 & 0 & 0 \end{pmatrix},$$

where the angles $\varphi_0^{(\alpha)}$ of crystallographic orientations of slip systems are given. For the fields $\left(v^1, v^2, \sigma_{11}, \sigma_{12}, \sigma_{22}, \psi^e\right)$ the set of field equations now has the form

$$\begin{aligned} \frac{\partial v_1}{\partial x^1} &= -\frac{\partial v_2}{\partial x^2} = -\sum_{\alpha=1}^{a}\dot{\gamma}^{(\alpha)}\frac{1}{2}\sin 2\left(\varphi_0^{(\alpha)}+\psi^e\right), \\ \frac{\partial v_1}{\partial x^2} &= \dot{\psi}^e+\sum_{\alpha=1}^{a}\dot{\gamma}^{(\alpha)}\cos^2\left(\varphi_0^{(\alpha)}+\psi^e\right), \\ \frac{\partial v_2}{\partial x^3} &= -\dot{\psi}^e-\sum_{\alpha=1}^{a}\dot{\gamma}^{(\alpha)}\sin^2\left(\varphi_0^{(\alpha)}+\psi^e\right), \\ \frac{\partial \sigma_{11}}{\partial x^1}+\frac{\partial \sigma_{12}}{\partial x^2} &= \rho\left(\frac{\partial v^1}{\partial t}+v^1\frac{\partial v^2}{\partial x^1}+v^2\frac{\partial v^1}{\partial x^2}\right), \\ \frac{\partial \sigma_{12}}{\partial x^1}+\frac{\partial \sigma_{22}}{\partial x^2} &= \rho\left(\frac{\partial v^2}{\partial t}+v^1\frac{\partial v^2}{\partial x^1}+v^2\frac{\partial v^2}{\partial x^2}\right), \end{aligned} \tag{10.29}$$

with the evolution equations

$$\begin{aligned} \dot{\gamma}^{(\alpha)} &= g_0^{(\alpha)}\left(\frac{\tau^{(\alpha)}}{\tau_0^{(\alpha)}}\right)^m, \quad \dot{\tau}_0^{(\alpha)}=\sum_{\beta=1}^{a}h_{\alpha\beta}\dot{\gamma}^{(\alpha)}, \\ \tau^{(\alpha)} &= -\frac{1}{2}\sin 2\left(\varphi_0^{(\alpha)}+\psi^e\right)\left(\sigma_{11}-\sigma_{22}\right)+\cos 2\left(\varphi_0^{(\alpha)}+\psi^e\right)\sigma_{12}. \end{aligned} \tag{10.30}$$

This set has been investigated on stability properties and shear band formation (e.g. A. Molinari [212], [211]) as well as it has been solved numerically for various simple aggregates (e.g. R. J. Asaro, A. Needleman [15], [16], [17], [228]). In the latter case, the pole figures (i.e. diagrams demonstrating changes of crystallographic directions – textures) check reasonably well with experiments in spite of the simplicity of this model.

10.4 Polycrystals and orientation distribution function

Polycrystals are aggregates of grains which, in turn, are monocrystals. Properties of polycrystals are determined by properties of grains and grain boundaries. Grain boundaries usually have a very complex structure entirely different from the crystallographic structure of grains. In some cases, the so-called *low angle boundaries*, mechanical properties can be modelled by ensembles of dislocations. However, this is rather seldom the case. Most boundaries are *incoherent* which means that they are formed from various defects like dislocations, extra matter, vacancies, etc. pinned to the neighboring grains. They can be activated in high temperatures, usually above some 60% of the temperature of melting. Consequently, mechanical properties of polycrystals in the range of relatively low temperatures, in contrast to high temperature range, are given by mechanical properties of grains, their size and the distribution of crystallographic orientation of grains. The latter is specified by the so-called *orientation distribution function*, ODF. Before we proceed with elasto-viscoplastic modelling let us briefly present this notion.

We consider a microscopic domain $\mathcal{M}$ of a polycrystal which we divide into N cubic cells. This domain is prescribed to a chosen material point $\mathbf{X}$ in which we are describing the crystallographic properties. All cells possess the same volume V_c and it is assumed to be small enough for the deformation to be considered as homogeneous within V_c. Grain boundaries are assumed to be sufficiently narrow to be replaced by singular surfaces. This is the *coarse-graining procedure* which we have already mentioned in Chapter 7 on kinetic theories. In each i-cell, $1 \leq i \leq N$, we introduce the triple of unit orthogonal vectors $\mathbf{K}_A^{(i)}, \quad \mathbf{K}_A^{(i)} \cdot \mathbf{K}_B^{(i)} = \delta_{AB}$, connected with the crystallographic lattice. Let us also choose a reference crystallographic basis $\mathbf{K}_A, \quad A = 1, 2, 3$. Then we can define the family of orthogonal tensors $\mathbf{\Pi}^{(i)}, 1 \leq i \leq N$, such that

$$\mathbf{K}_A^{(i)} = \mathbf{\Pi}^{(i)} \mathbf{K}_A, \quad \mathbf{\Pi}^{(i)T} = \mathbf{\Pi}^{(i)-1}. \tag{10.31}$$

The family $\left\{\mathbf{\Pi}^{(i)}\right\}_{i=1}^{N}$ defines the *texture at the point* $\mathbf{X}$ of the viscoplastic material.

In the current configuration, the vectors $\mathbf{K}_A^{(i)}$ transform due to the microscopic elastic deformation ${}^m\mathbf{F}_{(i)}^e$ (plastic part does not influence crystallographic directions!). As customary in metal plasticity [147], we neglect the elastic stretch ${}^m\mathbf{U}_{(i)}^e \approx \mathbf{0}$, i.e.

$$\mathbf{k}_A^{(i)} = {}^m\mathbf{R}_{(i)}^e \mathbf{K}_A^{(i)}, \tag{10.32}$$

where the elastic microrotation ${}^m\mathbf{R}_{(i)}^e$ is constant in each cell i. It is convenient to transform the basis vectors $\mathbf{K}_A$ of the lattice using only the rotation $\mathbf{R}$ as well (recall: $\mathbf{F} = \mathbf{RU}$), i.e. we define the current reference vectors $\mathbf{k}_A$ in the following way

$$\mathbf{k}_A = \mathbf{R}\mathbf{K}_A, \quad \mathbf{k}_A \cdot \mathbf{k}_B = \delta_{AB}. \tag{10.33}$$

Then the family of orthogonal tensors

$$\left\{ \boldsymbol{\pi}^{(i)} \middle| \boldsymbol{\pi}^{(i)} = {}^m\mathbf{R}^e_{(i)} \boldsymbol{\Pi}^{(i)} \mathbf{R}^T \right\}, \quad \text{i.e.} \quad \mathbf{k}_A^{(i)} = \boldsymbol{\pi}^{(i)} \mathbf{k}_A, \tag{10.34}$$

is called the *current texture* at the point $\mathbf{X}$ of the viscoplastic material. Sometimes in metallurgy the following two-point definition of the texture is also used

$$\bar{\boldsymbol{\pi}}^{(i)} = {}^m\mathbf{R}^e_{(i)} \boldsymbol{\Pi}^{(i)}. \tag{10.35}$$

It is usually given in terms of Euler angles (e.g. [147]) but, from the formal point of view, it is rather inconvenient as it transforms quantities from Lagrangian to Eulerian representations.

It is convenient to write changes of texture in the incremental form. Differentiation of (10.34) yields

$$\overset{\bullet}{\boldsymbol{\pi}}{}^{(i)} = {}^m\mathbf{W}^e_{(i)} \boldsymbol{\pi}^{(i)} - \boldsymbol{\pi}^i \mathbf{W}, \quad {}^m\mathbf{W}^e_{(i)} = {}^m\overset{\bullet}{\mathbf{R}}{}^e_{(i)} \, {}^m\mathbf{R}^{eT}_{(i)}. \tag{10.36}$$

Consequently, in order to find the current texture, one has to find spins ${}^m\mathbf{W}^e_{(i)}, \mathbf{W}$. In crystal plasticity it follows from the solution of the boundary value problem. Numerous numerical examples can be found in the literature. For purely macroscopic models of viscoplastic materials the spins ${}^m\mathbf{W}^e_{(i)}$ must be given by some additional macroscopic relations.

The discrete representation of texture described above is rather inconvenient in macroscopic models. For this reason, one has to smear out the model of microstructure. For instance, we can define the following number density

$$n(\boldsymbol{\pi}.t) = \sum_{i=1}^{N} \delta\left(\boldsymbol{\pi} - \boldsymbol{\pi}^{(i)}(t)\right), \quad \int_{O(3)} n(\boldsymbol{\pi}, t)\, dv(\boldsymbol{\pi}) = N, \tag{10.37}$$

where the integration extends over the space $O(3)$ of all three-dimensional orthogonal tensors (e.g. three Euler angles). As the volume of cells V_0 is small we can replace the number density by its limit ($V_0 \to 0, N \to \infty$)

$$f(\boldsymbol{\pi}, t) = \lim_{V_0 \to 0, N \to \infty} \frac{1}{N} \sum_{i=1}^{N} \delta\left(\boldsymbol{\pi} - \boldsymbol{\pi}^{(i)}(t)\right). \tag{10.38}$$

The function f is called the *orientation distribution function* (ODF).

This function as well as the texture itself may be found by the solution of the boundary value problem of structural plasticity. It may also be described in purely macroscopic models of viscoplasticity by an equation of its own. The classical way of doing it is an *ad hoc* assumption that f satisfies a diffusion equation (e.g. [54]) or else it may satisfy an evolution equation following from a microscopic conservation law ([358]). We shall not present this problem in this book.

Let us complete the presentation of crystal plasticity with a brief description of a procedure of solution for polycrystalline problems. We follow here the work of A. Molinari, G. R. Canova and S. Ahzi [213]. The procedure is based on an iteration of spacial heterogeneity of the material. Such procedures are commonly used, for instance, in the description of acoustic waves in heterogeneous elastic materials (e.g. [321]) and stochastic media (e.g. [301]). The constitutive law (10.26) which we can write in the symbolic form

$$\mathbf{D} = \mathbf{\Delta}(\mathbf{T}), \tag{10.39}$$

is inverted by the Taylor expansion of the inverse relation

$$\mathbf{T} = \mathbf{\Upsilon}(\mathbf{D}),$$

around the reference stretching $\mathbf{D}'$

$$\sigma_{ij} = -p\delta_{ij} + A_{ijkl}(\mathbf{D}')\, D_{kl} + \sigma^0_{ij}(\mathbf{D}'), \tag{10.40}$$

where

$$A_{ijkl}(\mathbf{D}') = \frac{\partial \Upsilon_{ij}}{\partial D_{kl}}(\mathbf{D}') = \left[\frac{\partial \mathbf{\Delta}}{\partial \mathbf{T}}\mathbf{\Upsilon}(\mathbf{D})\right]^{-1}_{ijkl}, \tag{10.41}$$
$$\sigma^0_{ij}(\mathbf{D}') = \sigma_{ij}(\mathbf{D}') - A_{ijkl}(\mathbf{D}')\, D'_{kl}.$$

The tensors A_{ijkl} and σ^0_{ij} are now decomposed into uniform parts, A^0_{ijkl}, σ^{00}_{ij}, and space dependent parts $\bar{A}_{ijkl}$ and $\bar{\sigma}_{ij}$. The equilibrium condition (momentum conservation) and the incompressibility relation then yield

$$A^0_{ijkl}\frac{\partial^2 v_k}{\partial x^j \partial x^l} - \frac{\partial p}{\partial x^i} + b_i = 0, \quad \frac{\partial v_k}{\partial x^k} = 0, \tag{10.42}$$

where

$$b_i = \frac{\partial\left(\bar{A}_{ijkl}\frac{\partial v_k}{\partial x^l}\right)}{\partial x^j} + \frac{\partial \bar{\sigma}_{ij}}{\partial x^j}. \tag{10.43}$$

As in the theory of linear elasticity, the problem (10.42) can be solved by means of the Green function. This yields the integral equation for the approximate solution of the original problem. This, in turn, can be solved iteratively. In the self-consistent approach one seeks the solution of the inclusion problem in a homogeneous equivalent medium (Eshelby method).

In the work quoted above (i.e. [213]) the authors have considered the problem of the spherical inclusion for f.c.c. metal and found the pole figures (texture) well fitting experimental results. Many similar results can be found in the literature.

10.5 Thermodynamical admissibility

We return now to some remarks on thermodynamical properties of viscoplasticity. We present here only the simplest possible version as each model of viscoplasticity would require separate thermodynamical research, extensive and not yet available. According to the above considerations on crystal plasticity the macroscopic model in the Lagrangian formulation should describe the fields of motion $\mathbf{f}$, temperature T, and the local plastic configuration $\mathbf{F}^p$ as functions of $\mathbf{X}$ and t. Field equations follow from the balance laws

$$\rho_0 \frac{\partial \mathbf{v}}{\partial t} = \operatorname{Div} \mathbf{P}, \quad \mathbf{P} = J\mathbf{T}\mathbf{F}^{-T},$$

$$\rho_0 \frac{\partial \varepsilon}{\partial t} + \operatorname{Div} \mathbf{Q} = \mathbf{P} \cdot \frac{\partial \mathbf{F}}{\partial t}, \quad \mathbf{Q} = J\mathbf{q}\mathbf{F}^{-T} \tag{10.44}$$

$$\frac{\partial \mathbf{F}}{\partial t} = \operatorname{Grad} \mathbf{v},$$

with the following constitutive relations

$$\begin{aligned} \frac{\partial \mathbf{F}^p}{\partial t} &= \mathcal{F}, \quad \mathbf{F} = \mathbf{F}^e \mathbf{F}^p, \\ \mathbf{P} &= \mathbf{P}\left(\mathbf{C}^e, \mathbf{F}^p, T\right), \quad \mathbf{C}^e = \mathbf{F}^{eT}\mathbf{F}^e, \\ \varepsilon &= \varepsilon\left(\mathbf{C}^e, \mathbf{F}^p, T\right), \\ \mathbf{Q} &= -K\left(\mathbf{C}^e, \mathbf{F}^p, T\right) \operatorname{Grad} T, \\ \mathcal{F} &= \mathcal{F}\left(\mathbf{C}^e, \mathbf{F}^p, T\right). \end{aligned} \tag{10.45}$$

We exploit the second law of thermodynamics by means of Lagrange multipliers

$$\rho_0 \frac{\partial \eta}{\partial t} - \operatorname{Div}\left(H \operatorname{Grad} T\right) - \Lambda^\varepsilon \left(\rho_0 \frac{\partial \varepsilon}{\partial t} - \operatorname{Div}\left(K \operatorname{Grad} T\right) - \mathbf{P} \cdot \frac{\partial \mathbf{F}}{\partial t}\right)$$

$$-\mathbf{\Lambda}^F \cdot \left(\frac{\partial \mathbf{F}}{\partial t} - \operatorname{Grad} \mathbf{v}\right) - \mathbf{\Lambda}^p \cdot \left(\frac{\partial \mathbf{F}^p}{\partial t} - \mathcal{F}\right) \geq 0, \tag{10.46}$$

for all fields. For simplicity, as in the case of the heat flux, we have assumed the linearity of the entropy flux with respect to the temperature gradient $\mathbf{H} = -H \operatorname{Grad} T$. The momentum balance equation has been left out because it does not restrict the entropy inequality due to linearity of the appearance of acceleration. The influence of the temperature gradient has been incorporated in the simplest form (linear isotropic) in order to expose better the influence of viscoplastic deformations.

Coefficients of time derivatives lead to the following identities

$$\frac{\partial \eta}{\partial T} - \Lambda^{\varepsilon} \frac{\partial \varepsilon}{\partial T} = 0,$$
$$2\rho_0 \mathbf{F}^e \left(\frac{\partial \eta}{\partial \mathbf{C}^e} - \Lambda^{\varepsilon} \frac{\partial \varepsilon}{\partial \mathbf{C}^e} \right) + \Lambda^{\varepsilon} \mathbf{P} \mathbf{F}^{pT} = 0, \tag{10.47}$$
$$\rho_0 \left(\frac{\partial \eta}{\partial \mathbf{F}^p} - \Lambda^{\varepsilon} \frac{\partial \varepsilon}{\partial \mathbf{F}^p} \right) + \Lambda^{\varepsilon} \mathbf{F}^{eT} \mathbf{P} - \Lambda^p = 0.$$

We have left out the contribution of $\mathbf{\Lambda}^F$ because this multiplier must be zero due to the linear contribution of $\operatorname{Grad} \mathbf{v}$. The coefficients of spacial derivatives yield

$$\mathbf{\Lambda}^F = 0, \quad H - \Lambda^{\varepsilon} K = 0, \quad \Lambda^{\varepsilon} = \Lambda^{\varepsilon}(T). \tag{10.48}$$

This simple result which, of course, yields the Fourier relation between fluxes, follows from the very restrictive form of constitutive relations chosen for this example. However, for instance, in the case of dependence on higher gradients, theories with couple stresses, etc. which are common in contemporary viscoplasticity, such a result would not hold.

The residual inequality remains

$$\mathcal{D} := \mathbf{\Lambda}^p \cdot \mathcal{F} - \frac{\partial \Lambda^{\varepsilon}}{\partial T} K \left| \operatorname{Grad} T \right|^2 \geq 0, \tag{10.49}$$

where $\mathcal{D}$ is the dissipation. Consequently, in the thermodynamical equilibrium

$$\mathcal{F}|_E = 0 \quad \Rightarrow \left. \frac{\partial \mathbf{F}^p}{\partial t} \right|_E = 0, \quad \operatorname{Grad} T|_E = 0. \tag{10.50}$$

In the thermodynamical equilibrium, the combination of $(10.47)_{1,2}$ yields

$$d\eta|_E = \frac{1}{\Lambda^{\varepsilon}} \left(d\varepsilon|_E - \frac{1}{\rho_0} \left(\mathbf{P} \mathbf{F}^T \right) \cdot \left(\mathbf{F}^{e-T} d\mathbf{F}^e \right) \right), \tag{10.51}$$

and this allows for the identification

$$\Lambda^{\varepsilon} = \frac{1}{T}. \tag{10.52}$$

Summing up, we have obtained the following results

$$\begin{aligned}
&\psi = \varepsilon - T\eta, \quad \eta = -\frac{\partial \psi}{\partial T}, \quad \varepsilon = \psi - T\frac{\partial \psi}{\partial T}, \\
&\mathbf{T} = 2\rho \mathbf{F}^e \frac{\partial \psi}{\partial \mathbf{C}^e} \mathbf{F}^{eT}, \quad \mathbf{h} = \frac{1}{T}\mathbf{q}, \\
&K \geq 0, \quad \mathbf{T} \cdot \mathbf{L}^p \geq \rho \frac{\partial \psi}{\partial \mathbf{F}^p} \cdot \mathcal{F}, \quad \mathbf{L}^p = \mathbf{F}^e \left(\frac{\partial \mathbf{F}^p}{\partial t} \mathbf{F}^{p-1} \right) \mathbf{F}^{e-1},
\end{aligned} \tag{10.53}$$

where $\mathbf{h} = J^{-1}\mathbf{H}\mathbf{F}^T, \mathbf{q} = J^{-1}\mathbf{Q}\mathbf{F}^T$. These are the thermodynamical restrictions which yield the Clausius-Duhem inequality and simultaneously, they are commonly used in the analysis of viscoplastic materials (e.g. [249]). We do not pursue this point any further and present only a few remarks on more general models of viscoplastic materials.

10.6 Two particular classes of models

The first special class of models which we proceed to present is being developed since some 40 years by P. Perzyna and his coworkers. These are purely macroscopic models with some elements of physical motivation for its applicability in description of plastic deformations of metals. Following the early work of P. W. Bridgman [51] who advocated for the description of plasticity by internal variables, Perzyna introduces the so-called *method of preparation.* This notion refers to the initial information on local material properties which is needed in the description of temporal behavior of the system. It is realized by means of *internal variables* described by the set of evolution equations for which the method of preparation specifies initial conditions. In the example of structural plasticity of monocrystals which we have described previously the internal variables are the rates of plastic slip $\overset{\bullet}{\gamma}{}^{(\alpha)}$ and limit stresses $\tau_0^{(\alpha)}$ describing the state of hardening. In the macroscopic description, the plastic local configuration $\mathbf{F}^p$ can be considered to be the internal variable. For all these quantities we need initial conditions in order to describe their evolution. As a consequence, in the model of Perzyna constitutive variables are divided into two groups. The first one, the local deformation-temperature configuration, belongs to the space $\mathcal{P}$ and the second one, the method of preparation, to the space $\mathcal{K}$. For instance, in [248] Perzyna describes a detailed topological structure of these two spaces. For our purposes it is sufficient to notice that the elements of the second space satisfy evolution equations. In the case of a finite dimensional space $\mathcal{K}$ the model reduces to the general model with internal state variables. For example, in the one-dimensional viscoplasticity, presented in

[248] they are chosen in the following manner

$$\{e^p, \gamma, \kappa, \xi\}, \tag{10.54}$$

where e^p is interpreted as an inelastic strain, γ is the viscosity parameter, κ is the work-hardening parameter and ξ denotes the concentration of point defects. They are assumed to satisfy the following equations

$$\begin{aligned} \overset{\bullet}{e^p} &= \gamma \left\langle \Phi \left[\frac{\mathcal{T}}{\varkappa} - 1 \right] \right\rangle, \quad \overset{\bullet}{\gamma} = \Gamma(\sigma^*) e^p, \\ \overset{\bullet}{\kappa} &= K(\sigma^*) \overset{\bullet}{e^p}, \quad \overset{\bullet}{\xi} = \Xi_1(\sigma^*) \overset{\bullet}{e^p} + \Xi_2(\sigma^*), \end{aligned} \tag{10.55}$$

$$\sigma^* := (e, T, \omega),$$

where $\left\langle \Phi \left[\frac{\mathcal{T}}{\varkappa} - 1 \right] \right\rangle$ is the function

$$\left\langle \Phi \left[\frac{\mathcal{T}}{\varkappa} - 1 \right] \right\rangle = \begin{cases} 0 \text{ for } \frac{\mathcal{T}}{\varkappa} \leq 1 \\ \Phi \left[\frac{\mathcal{T}}{\varkappa} - 1 \right] \text{ for } \frac{\mathcal{T}}{\varkappa} > 1 \end{cases}, \tag{10.56}$$

and the form of the function Φ depends on a chosen model [244]. $\Gamma(\sigma^*)$, $K(\sigma^*)$, $\Xi_1(\sigma^*)$, $\Xi_2(\sigma^*)$ are supposed to be given. Obviously, the structure of this model is very similar to this which we have constructed within structural plasticity.

The second model of viscoplastic deformations which we want to mention is the model constructed for soils. Many models of plastic deformations of soils have been developed within the classical nonlinear rate-independent approach. The most frequently used is the Cam-clay model but there also exist many modifications (for the review with the discussion of variations dependent on the kind of soil see the book of D. M. Wood [380]). In 1978 in his PhD-Thesis D. Kolymbas [170] has introduced the model based on the equation for the evolution of stresses. This model called *hypoplasticity* is similar in formal structure to the Maxwell model of viscoelastic media. As in the viscoplastic models of Perzyna, there is no yield surface in this model. Under the assumption of isothermal character of processes the fundamental equation for the evolution of stresses has in the hypoplasticity the following form (e.g. [171], [384], [110])

$$\begin{aligned} \overset{\triangle}{\mathbf{T}} &= C_1 (\operatorname{tr} \mathbf{T}) \mathbf{D} + C_2 \frac{\operatorname{tr}(\mathbf{TD})}{\operatorname{tr} \mathbf{T}} \mathbf{T} + C_3 \frac{\sqrt{\operatorname{tr} \mathbf{D}^2}}{\operatorname{tr} \mathbf{T}} \mathbf{T}^2 + C_4 \frac{\sqrt{\operatorname{tr} \mathbf{D}^2}}{\operatorname{tr} \mathbf{T}} (\operatorname{dev} \mathbf{T})^2 \\ \overset{\triangle}{\mathbf{T}} &:= \overset{\bullet}{\mathbf{T}} - \mathbf{WT} + \mathbf{TW}, \quad \operatorname{dev} \mathbf{T} := \mathbf{T} - \frac{1}{3} (\operatorname{tr} \mathbf{T}) \mathbf{1}. \end{aligned} \tag{10.57}$$

The dimensionless material constants C_1, C_2, C_3 are negative and C_4 is positive. For instance, E. Bauer found in his PhD-Thesis in 1992 [28] the following values for the loose Karlsruhe sand: $C_1 = -33.3$, $C_2 = -308.4$, $C_3 = -306.8$, $C_4 = 321.3$. The equation of the same structure as (10.57) but with different parameters was proposed by von Wolffersdorff [379], the model is still improved and extended to account for, for instance, couple stresses (e.g. [29]) and this is motivated by its faults in cyclic loading.

Nothing has been done on thermodynamical foundations of hypoplasticity. However, in this case such an extension seems to be rather straightforward as we already know thermodynamical properties of Maxwell models.

Chapter 11

Thermodynamics of miscible mixtures

11.1 General structure and field equations

Continuum models of *multicomponent* systems (compare Section 6.2) rely on the assumption that particles of distinguishable species are sufficiently densely distributed to enable the volume averaging. For instance, let us consider a domain $\mathcal{P}(\mathbf{x},t)$ which contains at the instant of time t a chosen point $\mathbf{x}$ of the space. Now let us choose an α-species and denote by $\rho^{\alpha}_{micro}(\mathbf{z},t)$ the real mass density of the α-species at the point $\mathbf{z}$ and the instant of time t. The domain $\mathcal{P}(\mathbf{x},t)$ is assumed to be sufficiently large to contain many subdomains of the substance α. Then the macroscopic average mass density of the α-species over the domain $\mathcal{P}(\mathbf{x},t)$ is defined in the following way

$$\rho^{\alpha}(\mathbf{x},t) = \frac{1}{\operatorname{vol}\mathcal{P}(\mathbf{x},t)} \int_{\mathcal{P}(\mathbf{x},t)} \rho^{\alpha}_{micro}(\mathbf{z},t)\,\chi^{\alpha}(\mathbf{z},t)\,dV_z, \tag{11.1}$$

where $\chi^{\alpha}(\mathbf{z},t)$ is the characteristic function of the α-species, i.e. it equals 1 at the point in which a particle of the α-species appears and it is 0 otherwise. $\operatorname{vol}\mathcal{P}(\mathbf{x},t)$ is the volume of the domain $\mathcal{P}(\mathbf{x},t)$. Obviously, this definition works also in the discrete case when the microscopic density has the form

$$\rho^{\alpha}_{micro}(\mathbf{z},t)\,\chi^{\alpha}(\mathbf{z},t) = \sum_{\mathbf{z}_n \in \mathcal{P}(\mathbf{x},t)} m^{\alpha}\delta\left(\mathbf{z} - \mathbf{z}^{\alpha}(t)\right), \tag{11.2}$$

where $\mathbf{z}^{\alpha}(t)$ is the location of an α-molecule at the instant of time t and m^{α} is the corresponding molecular mass. The quantity $\rho^{\alpha}(\mathbf{x},t)$ can be considered a smooth field of mass density provided the domain $\mathcal{P}(\mathbf{x},t)$ is sufficiently small from the macroscopic point of view and, simultaneously, sufficiently large from the microscopic point of view to include many subdomains of α-species or many α-molecules. We return further to a more detailed discussion of this averaging procedure for porous materials.

In this Chapter we consider *miscible* mixtures which mean that a microstructure described by the characteristic functions $\chi^\alpha(\mathbf{z},t)$ does not influence macroscopic properties of the mixture. It is different from *immiscible* mixtures such as porous and granular materials or suspensions where these characteristic functions extend the set of macroscopic fields on, say, porosity, tortuosity, etc. (see also the discussion of many practical problems of fluid mixtures with discrete particles in the classical book of S. L. Soo [303]). We consider such mixtures later in this book.

It is convenient to describe mixtures of fluids, as we do in the fluid dynamics, in the Eulerian reference. It means that the current configuration is the reference configuration for the motion. All fields of the model are functions of the spacial variable $\mathbf{x} \in \mathcal{B}_t$ and the time variable $t \in \mathcal{T}$. The domain of the mixture $\mathcal{B}_t$ depends on time and in any instant of time each point of this domain is occupied simultaneously by particles of all species of the mixture which means that there are no subdomains of $\mathcal{B}_t$ of the nonzero volume measure in which one of the species does not appear. The species of the mixture on the macroscopic level of description will be called *components*. The above assumption means that each component of the mixture is described by a set of *partial fields* with a common domain of the definition $\mathcal{B}_t$ at any instant of time t. We distinguish the components by the index α running from 1 to A, where A is the number of components.

In the case of thermomechanical systems considered in this Chapter the *thermodynamical process* is assumed to be described by the following fields

$$\{\rho^\alpha, \mathbf{v}^\alpha, T\}, \quad \alpha = 1, ..., A, \tag{11.3}$$

where ρ^α are the current partial mass densities of components referring to the common unit volume in the current configuration $\mathcal{B}_t$, $\mathbf{v}^\alpha$ are the velocity fields of components and T denotes the absolute temperature common for all components. The last quantity requires an assumption that all components possess the same temperature. In many practical applications this assumption is not satisfied. However, the problem of thermodynamical modelling of multicomponent systems with multiple temperatures is still not fully solved (compare [280]).

Field equations for the fields (11.3) follow from the partial balance equations of mass and momentum and from the energy balance equation for the mixture. The partial balance equations for $\alpha = 1, ..., A$ have the form

$$\frac{d}{dt}\int_{\mathcal{P}_t^\alpha} \rho^\alpha dv = \int_{\mathcal{P}_t^\alpha} \hat{\rho}^\alpha dv, \tag{11.4}$$

$$\frac{d}{dt}\int_{\mathcal{P}_t^\alpha} \rho^\alpha \mathbf{v}^\alpha dv = \oint_{\partial\mathcal{P}_t^\alpha} \mathbf{T}^\alpha \mathbf{n} dS_t + \int_{\mathcal{P}_t^\alpha} \left(\rho^\alpha \mathbf{b}^\alpha + \hat{\mathbf{p}}^\alpha\right) dv, \tag{11.5}$$

$$\frac{d}{dt}\int_{\mathcal{P}_t^\alpha}\rho^\alpha\left(\varepsilon^\alpha+\tfrac{1}{2}\mathbf{v}^\alpha\cdot\mathbf{v}^\alpha\right)d\mathcal{S}_t=\oint_{\partial\mathcal{P}_t^\alpha}\left(\mathbf{T}^\alpha\mathbf{v}^\alpha-\mathbf{q}^\alpha\right)\cdot\mathbf{n}d\mathcal{S}_t \qquad (11.6)$$
$$+\int_{\mathcal{P}_t^\alpha}\left(\rho^\alpha\mathbf{b}^\alpha\cdot\mathbf{v}^\alpha+\rho^\alpha r^\alpha+\hat{\varepsilon}^\alpha\right)dv,$$

where $\mathcal{P}_t^\alpha\subset\mathcal{B}_t$ is a measurable subset of the current configuration which moves with the kinematics of the α-component and $\partial\mathcal{P}_t^\alpha$ denotes its surface with the unit outward normal vector $\mathbf{n}$. The quantities with the hat are volume densities of sources (productions) and, according to the fundamental assumption of Truesdell's mixture theory [20], [19] [327] they must satisfy the following conservation laws

$$\sum_{\alpha=1}^{A}\hat{\rho}^\alpha=0,\quad\sum_{\alpha=1}^{A}\hat{\mathbf{p}}^\alpha=0,\quad\sum_{\alpha=1}^{A}\hat{\varepsilon}^\alpha=0. \qquad (11.7)$$

The remaining quantities $\mathbf{T}^\alpha,\mathbf{b}^\alpha,\varepsilon^\alpha,\mathbf{q}^\alpha,r^\alpha$ denote the symmetric partial Cauchy stress tensor in the α-component, the body force per unit mass of the α-component, the partial density of the internal energy of the α-component, the partial heat flux in the α-component, and the density of energy radiation in the α-component, respectively.

The assumption that partial densities satisfy balance laws yields, according to the Cauchy Theorem which we have discussed in Chapter 4, the existence of partial stress tensors $\mathbf{T}^\alpha$ and partial heat flux vectors $\mathbf{q}^\alpha$.

In the case of the single temperature field, the partial energy balance equations are not used and we need only the energy conservation of the mixture. This must have, of course, the classical form

$$\frac{d}{dt}\int_{\mathcal{P}_t}\rho\left(\varepsilon+\tfrac{1}{2}\mathbf{v}\cdot\mathbf{v}\right)dv=\oint_{\partial\mathcal{P}_t}\left(\mathbf{T}\mathbf{v}-\mathbf{q}\right)\cdot\mathbf{n}d\mathcal{S}_t+\int_{\mathcal{P}_t}\rho\left(\mathbf{b}\cdot\mathbf{v}+\mathbf{r}\right)dv. \qquad (11.8)$$

We obtain indeed this relation if we add equations (11.6) and introduce the definitions

$$\rho=\sum_{\alpha=1}^{A}\rho^\alpha,\quad\rho\mathbf{v}=\sum_{\alpha=1}^{A}\rho^\alpha\mathbf{v}^\alpha,\quad\rho\varepsilon=\sum_{\alpha=1}^{A}\rho^\alpha\left(\varepsilon^\alpha+\tfrac{1}{2}\mathbf{u}^\alpha\cdot\mathbf{u}^\alpha\right),$$

$$\mathbf{T}=\sum_{\alpha=1}^{A}\left(\mathbf{T}^\alpha-\rho^\alpha\mathbf{u}^\alpha\otimes\mathbf{u}^\alpha\right),\quad\mathbf{q}=\sum_{\alpha=1}^{A}\left(\mathbf{q}^\alpha+\rho^\alpha\left(\varepsilon^\alpha+\tfrac{1}{2}\mathbf{u}^\alpha\cdot\mathbf{u}^\alpha\right)\mathbf{u}^\alpha-\mathbf{T}^\alpha\mathbf{u}^\alpha\right),$$

$$\rho r=\sum_{\alpha=1}^{A}\rho^\alpha\left(r^\alpha+\mathbf{b}^\alpha\cdot\mathbf{u}^\alpha\right),\quad\mathbf{u}^\alpha=\mathbf{v}^\alpha-\mathbf{v},\quad\sum_{\alpha=1}^{A}\rho^\alpha\mathbf{u}^\alpha=0. \qquad (11.9)$$

The presence of convective terms containing the *diffusion velocities* $\mathbf{u}^\alpha$ follows from the fact that subsets $\mathcal{P}_t$ of the current configuration $\mathcal{B}_t$ possess the kinematics of the *barycentric velocity field* $\mathbf{v}$ and it means that fluxes contain additional contributions – none of these subsets is *material* with respect to any of the components.

It can be easily checked that the above definitions yield as well the classical conservation of mass and momentum for the whole mixture

$$\frac{d}{dt}\int_{\mathcal{P}_t}\rho dv = 0, \quad \frac{d}{dt}\int_{\mathcal{P}_t}\rho\mathbf{v}dv = \oint_{\partial\mathcal{P}_t}\mathbf{T}\mathbf{n}dS_t + \int_{\mathcal{P}_t}\rho\mathbf{b}dv, \quad \rho\mathbf{b} = \sum_{\alpha=1}^{A}\rho^\alpha\mathbf{b}^\alpha. \tag{11.10}$$

Let us remark that the presence of diffusion yields a macroscopic heat flux $\mathbf{q}$ even in the case when the partial heat conduction contributions $\mathbf{q}^\alpha$ are absent. This property is essential for a peculiar form of the second law of thermodynamics of mixtures which we discuss further.

The above global balance equations yield in a standard way local laws. They have different form in *regular* points in which fields are continuous and in *singular* points on surfaces on which fields may suffer jumps. Namely

$$\frac{\partial\rho^\alpha}{\partial t} + \operatorname{div}(\rho^\alpha\mathbf{v}^\alpha) = \hat{\rho}^\alpha,$$

$$\frac{\partial\rho^\alpha\mathbf{v}^\alpha}{\partial t} + \operatorname{div}(\rho^\alpha\mathbf{v}^\alpha\otimes\mathbf{v}^\alpha - \mathbf{T}^\alpha) = \rho^\alpha\mathbf{b}^\alpha + \hat{\mathbf{p}}^\alpha,$$

$$\frac{\partial\rho\varepsilon}{\partial t} + \operatorname{div}(\rho\varepsilon\mathbf{v} + \mathbf{q}) = \mathbf{T}\cdot\operatorname{grad}\mathbf{v} + \rho r, \tag{11.11}$$

in regular points. The last equation, the so-called balance of internal energy, follows by elimination of the contribution of kinetic energy by means of mass and momentum conservation laws.

In singular points the balance equations have the form of jump conditions across the surface. We use them here only for the bulk mass and the bulk energy of the mixture but, of course, they may be easily derived for all other quantities as well. We have for these two quantities (comp. Table 4, Chapter 2)

$$[[\rho(\mathbf{v}\cdot\mathbf{n} - c)]] = 0, \quad [[...]] = (...)^+ - (...)^-,$$

$$[[\mathbf{q}\cdot\mathbf{n}]] = [[\mathbf{T}(\mathbf{v} - c\mathbf{n})]]\cdot\mathbf{n} - \left[\left[\varepsilon + \tfrac{1}{2}(\mathbf{v} - c\mathbf{n})\cdot(\mathbf{v} - c\mathbf{n})\right]\right]\rho(\mathbf{v}\cdot\mathbf{n} - c), \tag{11.12}$$

where c denotes the speed of the surface and $\mathbf{n}$ is the unit normal vector to this surface. As indicated the double parentheses denote the difference of limits on both sides of the singular surface.

As we already know from thermodynamical theories of single component systems the system of balance laws may be transformed into field equations by means of the closure. This means that we have to specify constitutive relations for a set of constitutive quantities.

We begin the discussion of this problem with the specification of mass sources for chemical reactions. Components of the mixture are then not the most fundamental ingredients of the macroscopic picture. These are rather molecules (constituents) which satisfy laws of mass conservation. To see this structure let us consider the simplest example of formation of water from hydrogen H and oxygen O. These two are molecules (constituents) while the mixture consists of the following components: H, O, H_2, O_2, OH, H_2O. They are participating in chemical reactions which can be symbolically written in the form

$$\begin{array}{l} H_2 - 2H = 0, \\ O_2 - 2O = 0, \\ H + O - OH = 0, \\ H + OH - H_2O = 0, \\ 2H_2 + O_2 - 2H_2O = 0, \end{array} \qquad (\gamma_\alpha^r) = \begin{pmatrix} -2 & 0 & 1 & 0 & 0 & 0 \\ 0 & -2 & 0 & 1 & 0 & 0 \\ 1 & 1 & 0 & 0 & -1 & 0 \\ 1 & 0 & 0 & 0 & 1 & -1 \\ 0 & 0 & 2 & 1 & 0 & -2 \end{pmatrix}, \tag{11.13}$$

The matrix of coefficients γ_α^r, $\alpha = 1, ..., 6$, $r = 1, ..., 5$ in this set of relations is called the *stoichiometric matrix*. Its rank determines the number of independent reactions. In our example this rank is equal to 4. The conservation of mass for molecules (constituents) in each reaction can be written in the form

$$\sum_{\alpha=1}^{A} \gamma_\alpha^r M_\alpha \mu_H = 0, \quad r = 1, ..., \mathcal{R}, \tag{11.14}$$

where M_α denotes the molecular weight of the component α, μ_H is the mass of a hydrogen atom and $\mathcal{R}$ is the number of reactions.

Further we consider only the set of independent chemical reactions and we denote their number by ν. The contribution of each reaction to the production of components depends also on the *rate of reaction* which we denote by $\lambda^r, r = 1, ..., \nu$. Then the mass source in partial mass balance equations can be written in the form

$$\hat{\rho}^\alpha = \sum_{r=1}^{\nu} \gamma_\alpha^r M_\alpha \mu_H \lambda^r. \tag{11.15}$$

It is clear that the conservation relations (11.14) imply $(11.7)_1$.

Let us now return to the general closure problem. We transform the balance equations (11.11) into field equations for the fields (11.3). We use the strategy of continuum thermodynamics and assume that the *constitutive quantities*

$$\mathcal{F} = \{\lambda^r, \mathbf{T}^\alpha, \hat{\mathbf{p}}^\alpha, \varepsilon, \mathbf{q}\}, \tag{11.16}$$

are sufficiently smooth functions of *constitutive variables.* The choice of the constitutive variables specifies the class of substances admissible in the model under construction. We consider here only an example of inviscid fluids. However, in contrast to the thermodynamical modelling of the single inviscid (ideal) fluid we include among constitutive variables the gradients of partial mass densities. It has been shown by I. Müller [217], [222] that this yields a desirable structure of interactions of components. We return later to this point. The chosen set of constitutive variables is as follows

$$\mathcal{C} = \{\rho^\alpha, \operatorname{grad}\rho^\alpha, \mathbf{v}^\alpha, T, \operatorname{grad}T\}. \tag{11.17}$$

Hence the *closure* of the set of balance equations has the form

$$\mathcal{F} = \mathcal{F}(\mathcal{C}). \tag{11.18}$$

These relations must satisfy the conservation restrictions $(11.7)_2$.

In order to describe viscous fluids we would have to add a dependence on gradients of velocity to the set of constitutive variables. This can be easily done but the results are not so transparent as in the present case.

The constitutive relations (11.18) in the thermodynamical modelling procedure must satisfy principles which we were already discussing for single component continua. For fluid mixtures two of them will be presented in details

- material objectivity (material frame indifference),
- thermodynamical admissibility.

As we already know the principle of material objectivity (material frame indifference) concerns the behavior of field equations under the so-called Euclidean transformation, i.e the transformation described by the following relation for an arbitrary point of the configuration space

$$\mathbf{x}^* = \mathbf{O}(t)\,\mathbf{x} + \mathbf{c}(t), \quad \mathbf{O}^T = \mathbf{O}^{-1}. \tag{11.19}$$

Let us recall that the scalar, a, the vector, $\mathbf{w}$, and the tensor, $\mathbf{T}$, transform in an objective manner if they satisfy the following transformation rules (comp. (5.79))

$$\varphi^* = \varphi, \quad \mathbf{h}^* = \mathbf{Oh}, \quad \mathbf{T}^* = \mathbf{OTO}^T. \tag{11.20}$$

Obviously, neither the velocity nor the acceleration transform in an objective manner. Differentiation of (11.19) with respect to time yields

$$\mathbf{v}^* = \mathbf{Ov} + \dot{\mathbf{O}}\mathbf{x} + \dot{\mathbf{c}}, \quad \dot{\mathbf{O}} = \frac{d\mathbf{O}}{dt}, \quad \dot{\mathbf{c}} = \frac{d\mathbf{c}}{dt} \tag{11.21}$$

$$\mathbf{a}^* = \frac{\partial \mathbf{v}}{\partial t} + \mathbf{v}^* \cdot \operatorname{grad} \mathbf{v}^* = \mathbf{Oa} + 2\dot{\mathbf{O}}\mathbf{v} + \ddot{\mathbf{O}}\mathbf{x} + \ddot{\mathbf{c}}.$$

Consequently, the momentum balance equations do not transform in an objective manner. The transformation from an inertial to a noninertial frame yields centrifugal, Coriolis, Euler, and translational forces which form together the inertial forces discussed already in previous Chapters. Simultaneously, the mass and energy balance equations transform in an objective manner.

In the case of chemical reactions there appears an additional problem related to the contribution of mass sources to the momentum balance equations. It becomes clear when we use the chain rule of differentiation on the left-hand side of $(11.11)_2$ and substitute mass balance equations $(11.11)_1$. Then the above described structure remains unchanged under the Euclidean transformation if we assume that the following vector

$$\hat{\mathbf{p}}^\alpha - \hat{\rho}^\alpha \mathbf{v}^\alpha, \tag{11.22}$$

is objective rather than the momentum source itself. We do so in further considerations.

The principle of material objectivity (material frame indifference) requires that functions

$$\mathcal{F}_o = \{\lambda^r, \mathbf{T}^\alpha, \hat{\mathbf{p}}^\alpha - \hat{\rho}^\alpha \mathbf{v}^\alpha, \varepsilon, \mathbf{q}\} = \mathcal{F}_o(\mathcal{C}), \tag{11.23}$$

remain unchanged under an arbitrary Euclidean transformation, i.e.

$$\mathcal{F}_o^* = \mathcal{F}_o(\mathcal{C}^*). \tag{11.24}$$

Note that the constitutive *function* $\mathcal{F}_o(.)$ is the same in (11.23) and (11.24).

An immediate consequence of the above principle is the elimination of one of velocities among constitutive variables. Instead of partial velocities this set may contain only relative velocities which are objective. These may be diffusion velocities $\mathbf{u}^\alpha$, or, as they are only $A-1$ independent variables of this sort, relative velocities

$$\mathbf{w}^\alpha = \mathbf{v}^\alpha - \mathbf{v}^A, \tag{11.25}$$

which are more convenient in the calculations.

The full representation of constitutive functions in their objective form is a rather complicated task and we shall not do so in this Chapter. We limit our attention to constitutive functions which are *linear* in $\operatorname{grad}\rho^\alpha, \mathbf{w}^\alpha, \operatorname{grad} T$. In this case the most general form of the constitutive equations compatible with the material objectivity is as follows

$$\begin{aligned}
\lambda^r &= \lambda^r\left(\rho^\beta, T\right), \\
\mathbf{T}^\alpha &= -p^\alpha\left(\rho^\beta, T\right)\mathbf{1}, \\
\hat{\mathbf{p}}^\alpha - \hat{\rho}^\alpha \mathbf{v}^\alpha &= \sum_{\beta=1}^{A} M_\rho^{\alpha\beta} \operatorname{grad}\rho^\beta + \sum_{\beta=1}^{A-1} M_w^{\alpha\beta}\mathbf{w}^\beta + M_T^\alpha \operatorname{grad} T, \\
\mathbf{q}_I &= \sum_{\beta=1}^{A} q_\rho^\beta \operatorname{grad}\rho^\beta + \sum_{\beta=1}^{A-1} q_w^\beta \mathbf{w}^\beta + q_T \operatorname{grad} T, \\
\varepsilon_I &= \varepsilon_I\left(\rho^\beta, T\right),
\end{aligned} \tag{11.26}$$

where the index I denotes the so-called *intrinsic* part of the quantity

$$\begin{aligned}
\mathbf{q}_I &= \mathbf{q} - \sum_{\alpha=1}^{A} \tfrac{1}{2}\rho^\alpha\left(\mathbf{u}^\alpha\cdot\mathbf{u}^\alpha\right)\mathbf{u}^\alpha = \sum_{\alpha=1}^{A}\mathbf{q}^\alpha + \sum_{\alpha=1}^{A}\left(\rho^\alpha\varepsilon^\alpha\mathbf{1} - \mathbf{T}^\alpha\right)\mathbf{u}^\alpha, \\
\varepsilon_I &= \varepsilon - \sum_{\alpha=1}^{A}\frac{1}{2}\frac{\rho^\alpha}{\rho}\mathbf{u}^\alpha\cdot\mathbf{u}^\alpha = \sum_{\alpha=1}^{A}\frac{\rho^\alpha}{\rho}\varepsilon^\alpha.
\end{aligned} \tag{11.27}$$

The coefficients appearing in relations (11.26) may be dependent on ρ^β and T. However, they are not independent due to the previously discussed restrictions on sources. Namely we have

$$\begin{aligned}
&\sum_{\alpha=1}^{A} M_\rho^{\alpha\beta} = 0, \quad \beta = 1, ..., A, \\
&\sum_{\alpha=1}^{A} M_w^{\alpha\beta} + \hat{\rho}^\beta = 0, \quad \beta = 1, ..., A-1, \\
&\sum_{\alpha=1}^{A} M_T^\alpha = 0.
\end{aligned} \tag{11.28}$$

In spite of the above assumed linearity the whole model remains nonlinear due to various nonlinear explicit contributions.

11.2 Thermodynamical admissibility

We proceed to impose the condition of thermodynamical admissibility on the above constitutive relations. As in the case of the single component continua we assume that the *second law of thermodynamics* is based upon the entropy inequality. As discussed earlier, the strategy of continuum thermodynamics in construction of macroscopic models relies on the assumption that solutions of the field equations identically satisfy the second law of thermodynamics also called the entropy principle. This law consists of four parts:

i/ There exist a nontrivial entropy density function η and the entropy flux $\mathbf{h}$ which are both constitutive. In the case of miscible mixtures, considered in this Chapter, they must have the following general form

$$\eta = \eta\left(\rho^{\beta}, T\right), \quad \mathbf{h} = \sum_{\alpha=1}^{A} h_{\rho}^{\alpha} \operatorname{grad} \rho^{\alpha} + \sum_{\alpha=1}^{A-1} h_{w}^{\alpha} \mathbf{w}^{\alpha} + h_{T} \operatorname{grad} T, \tag{11.29}$$

where $h_{\rho}^{\alpha}, h_{w}^{\alpha}$, and h_T may be functions of ρ^{β}, T.

ii/ The entropy density satisfies the balance equation whose form in regular points is as follows

$$\frac{\partial \rho \eta}{\partial t} + \operatorname{div}\left(\rho \eta \mathbf{v} + \mathbf{h}\right) = \hat{\eta}, \tag{11.30}$$

where $\hat{\eta}$ denotes the entropy source.

iii/ The entropy source is nonnegative for all solutions of field equations, i.e. for all thermodynamic processes. Consequently, the following inequality holds

$$\forall \text{ thermodynamic processes } \frac{\partial \rho \eta}{\partial t} + \operatorname{div}\left(\rho \eta \mathbf{v} + \mathbf{h}\right) \geq 0. \tag{11.31}$$

iv/ There exist *ideal walls* on which there is no entropy production and the temperature is continuous across it, i.e.

$$[[\mathbf{h} \cdot \mathbf{n}]] + [[\eta]]\, \rho\left(\mathbf{v} \cdot \mathbf{n} - c\right) = 0, \quad [[T]] = 0. \tag{11.32}$$

The method of Lagrange multipliers applied to the inequality (11.31) yields the following extended form of the inequality

$$\forall \text{ fields } \frac{\partial \rho \eta}{\partial t} + \operatorname{div}\left(\rho \eta \mathbf{v} + \mathbf{h}\right)$$
$$-\sum_{\alpha=1}^{A} \Lambda^{\rho^{\alpha}} \left(\frac{\partial \rho^{\alpha}}{\partial t} + \operatorname{div}\left(\rho^{\alpha} \mathbf{v}^{\alpha}\right) - \sum_{r=1}^{\nu} \gamma_{\alpha}^{r} M_{\alpha} \mu_{H} \lambda^{r} \right)$$

$$-\sum_{\alpha=1}^{A} \boldsymbol{\Lambda}^{v^\alpha} \cdot \left(\frac{\partial \rho^\alpha \mathbf{v}^\alpha}{\partial t} + \operatorname{div}\left(\rho^\alpha \mathbf{v}^\alpha \otimes \mathbf{v}^\alpha - \mathbf{T}^\alpha\right) - \hat{\mathbf{p}}^\alpha \right) \tag{11.33}$$

$$-\Lambda^\varepsilon \left(\frac{\partial \rho\varepsilon}{\partial t} + \operatorname{div}\left(\rho\varepsilon\mathbf{v} + \mathbf{q}\right) - \mathbf{T} \cdot \operatorname{grad}\mathbf{v} \right) \geq 0.$$

The Lagrange multipliers $\Lambda^{\rho^\alpha}, \boldsymbol{\Lambda}^{v^\alpha}, \Lambda^\varepsilon$ are functions of constitutive variables $\rho^\beta, \operatorname{grad}\rho^\beta$, $\mathbf{v}^\beta, T, \operatorname{grad}T$. After the exploitation of the above inequality these multipliers must be eliminated as auxiliary quantities.

Insertion of constitutive relations into the inequality (11.33) leads to an inequality which is explicitly linear in the derivatives

$$\left\{ \frac{\partial T}{\partial t}, \operatorname{grad}\otimes\operatorname{grad}T, \frac{\partial \rho^\beta}{\partial t}, \operatorname{grad}\otimes\operatorname{grad}\rho^\beta, \frac{\partial \mathbf{v}^\beta}{\partial t}, \operatorname{grad}\mathbf{v}^\beta \right\}. \tag{11.34}$$

Since the inequality (11.33) must hold for arbitrary fields, these derivatives can be chosen arbitrarily. Consequently, the inequality can be violated if these contributions do not vanish identically. This yields a series of identities which, on the one hand, determine multipliers and, on the other hand, restrict constitutive relations. In addition, there remains a nonlinear part of the inequality, the so-called *residual* inequality which determines the *dissipation* density of processes.

The identities following from the entropy inequality have the following form

$$\frac{\partial \eta}{\partial T} - \Lambda^\varepsilon \frac{\partial \varepsilon}{\partial T} = 0, \tag{11.35}$$

$$\frac{\partial \mathbf{h}}{\partial \mathbf{g}} - \Lambda^\varepsilon \frac{\partial \mathbf{q}}{\partial \mathbf{g}} = 0, \quad \mathbf{g} = \operatorname{grad}T, \tag{11.36}$$

$$\frac{\partial \eta}{\partial \rho^\alpha} - \Lambda^\varepsilon \frac{\partial \varepsilon}{\partial \rho^\alpha} + \frac{1}{\rho}\left(\eta - \Lambda^\varepsilon \varepsilon\right) - \frac{1}{\rho}\Lambda^{\rho^\alpha} - \frac{1}{\rho}\mathbf{v}^\alpha \cdot \boldsymbol{\Lambda}^{v^\alpha} = 0, \quad \alpha = 1, ..., A-1 \tag{11.37}$$

$$\frac{\partial \mathbf{h}}{\partial \operatorname{grad}\rho^\alpha} - \Lambda^\varepsilon \frac{\partial \mathbf{q}}{\partial \operatorname{grad}\rho^\alpha} = 0, \quad \alpha = 1, ..., A \tag{11.38}$$

$$\frac{\partial \eta}{\partial \mathbf{w}^\alpha} - \Lambda^\varepsilon \frac{\partial \varepsilon}{\partial \mathbf{w}^\alpha} - \frac{\rho^\alpha}{\rho}\boldsymbol{\Lambda}^{v^\alpha} = 0, \quad \alpha = 1, ..., A-1, \tag{11.39}$$

$$-\sum_{\beta=1}^{A}\left(\frac{\partial\eta}{\partial\mathbf{w}^\beta}-\Lambda^\varepsilon\frac{\partial\varepsilon}{\partial\mathbf{w}^\beta}\right)-\frac{\rho^A}{\rho}\mathbf{\Lambda}^{v^A}=0, \tag{11.40}$$

$$\left(\frac{\partial\eta}{\partial\rho^\alpha}-\Lambda^\varepsilon\frac{\partial\varepsilon}{\partial\rho^\alpha}\right)\mathbf{1}-\frac{1}{\rho^2}\Lambda^\varepsilon\mathbf{T}$$
$$-\frac{1}{\rho\rho^\alpha}\left(\frac{\partial\mathbf{h}}{\partial\mathbf{w}^\alpha}-\Lambda^\varepsilon\frac{\partial\mathbf{q}}{\partial\mathbf{w}^\alpha}\right)+\frac{1}{\rho}\mathbf{\Lambda}^{v^\alpha}\otimes\mathbf{u}^\alpha=0,\quad \alpha=1,...,A-1,$$

$$\left(\frac{\partial\eta}{\partial\rho^A}-\Lambda^\varepsilon\frac{\partial\varepsilon}{\partial\rho^A}\right)\mathbf{1}-\sum_{\beta=1}^{A-1}\frac{1}{\rho\rho^A}\left(\frac{\partial\mathbf{h}}{\partial\mathbf{w}^\beta}-\Lambda^\varepsilon\frac{\partial\mathbf{q}}{\partial\mathbf{w}^\beta}\right) \tag{11.41}$$
$$-\frac{1}{\rho}\mathbf{\Lambda}^{v^A}\otimes\mathbf{u}^A-\frac{1}{\rho^2}\Lambda^\varepsilon\mathbf{T}=0.$$

There remains the residual inequality which has the following form

$$\sum_{\beta=1}^{A}\left[\left(\frac{\partial\mathbf{h}}{\partial\rho^\beta}-\Lambda^\varepsilon\frac{\partial\mathbf{q}}{\partial\rho^\beta}\right)-\rho\left(\frac{\partial\eta}{\partial\rho^\beta}-\Lambda^\varepsilon\frac{\partial\varepsilon}{\partial\rho^\beta}\right)\mathbf{u}^\beta\right.$$
$$\left.-\sum_{\alpha=1}^{A}\mathbf{\Lambda}^{v^\alpha}\frac{\partial p^\alpha}{\partial\rho^\beta}+\Lambda^\varepsilon\frac{1}{\rho}\mathbf{T}\mathbf{u}^\beta\right]\cdot\operatorname{grad}\rho^\beta \tag{11.42}$$
$$+\left(\frac{\partial\mathbf{h}}{\partial T}-\Lambda^\varepsilon\frac{\partial\mathbf{q}}{\partial T}-\sum_{\alpha=1}^{A}\mathbf{\Lambda}^{v^\alpha}\frac{\partial p^\alpha}{\partial T}\right)\cdot\mathbf{g}+\sum_{\alpha=1}^{A}\Lambda^{\rho^\alpha}\hat{\rho}^\alpha+\sum_{\alpha=1}^{A}\mathbf{\Lambda}^{v^\alpha}\cdot\hat{\mathbf{p}}^\alpha\geq 0.$$

We use further the following relation indicated by the above identities

$$\Lambda^\varepsilon\sum_{\alpha=1}^{A}p^\alpha=-\sum_{\alpha=1}^{A}\rho^\alpha\Lambda_I^{\rho^\alpha}+\rho\left(\eta-\Lambda^\varepsilon\varepsilon_I\right), \tag{11.43}$$

where

$$\Lambda_I^{\rho^\beta}=\Lambda^{\rho^\beta}-\frac{1}{T}\left(\mathbf{v}^\beta\cdot\mathbf{u}^\beta-\tfrac{1}{2}\mathbf{u}^\beta\cdot\mathbf{u}^\beta\right)=\Lambda_I^{\rho^\beta}\left(\rho^\alpha,T\right). \tag{11.44}$$

Before the discussion of thermodynamical properties of the mixture, let us check the form of the above relations for a single fluid, i.e. we set $A=1$.

We obtain

$$\boldsymbol{\Lambda}^v = 0,$$
$$d\eta = \Lambda^\varepsilon \left(d\varepsilon - \frac{p}{\rho^2} d\rho \right),$$
$$\mathbf{h} = \Lambda^\varepsilon \mathbf{q}, \tag{11.45}$$

$$\frac{\partial \Lambda^\varepsilon}{\partial \rho} \mathbf{q} \cdot \operatorname{grad} \rho + \frac{\partial \Lambda^\varepsilon}{\partial T} \mathbf{q} \cdot \operatorname{grad} T \geq 0.$$

We use now part iv/ of the second law of thermodynamics. Let us consider an impermeable ideal wall which separates two different single fluids I and II. Such a wall moves with the speed $c = \mathbf{v} \cdot \mathbf{n}$ (material surface). The jump condition (11.32) then yields

$$[[\mathbf{h}]] \cdot \mathbf{n} = 0 \quad \text{and} \quad [[T]] = 0. \tag{11.46}$$

Bearing (11.12) and $(11.45)_3$ in mind, we obtain

$$\Lambda_I^\varepsilon (\rho_I, T) = \Lambda_{II}^\varepsilon (\rho_{II}, T) . \tag{11.47}$$

This relation must hold for an arbitrary independent choice of ρ_I and ρ_{II}. Hence it can be satisfied only if it is independent of mass densities at all. We conclude

$$\Lambda_I^\varepsilon (T) = \Lambda_{II}^\varepsilon (T) . \tag{11.48}$$

Consequently, the function $\Lambda^\varepsilon (T)$ is the same for all ideal fluids and, for this reason, we call it *universal.*

In order to determine this function it is sufficient to inspect $(11.45)_2$ for the case of ideal gases. Then it is identical with $1/T$. Hence it follows in general

$$\Lambda^\varepsilon = \frac{1}{T}. \tag{11.49}$$

Relations (11.45) become

$$d\eta = \frac{1}{T} \left(d\varepsilon - \frac{p}{\rho^2} d\rho \right), \quad \mathbf{h} = \frac{1}{T} \mathbf{q}, \quad \mathbf{q} \cdot \operatorname{grad} T \leq 0. \tag{11.50}$$

These are the results identical with those of the classical thermodynamics. The first relation is, of course, the Gibbs equation.

If we consider an impermeable wall separating a mixture I from a single fluid II and account for the jump conditions we obtain immediately

$$\Lambda_I^{\varepsilon}\left(\rho_I^{\alpha}, T\right) = \frac{1}{T}. \tag{11.51}$$

Hence the multiplier Λ^{ε} remains universal also for mixtures.

One cannot argue in the same way in a general model of mixtures in which each component possesses its own temperature. The multipliers of the partial energy balance equations are not in general universal functions of the absolute temperature. This means as well that temperature cannot be measured on the surface of contact between two different systems because such a surface, even if it is an ideal wall, does not yield the continuity of temperatures. Consequently, we cannot construct thermometers.

Let us now consider a semipermeable wall separating two mixtures I and II. The wall is assumed to be permeable for a single chosen component γ. On such a wall

$$\mathbf{v}^{\alpha} \cdot \mathbf{n} = c \quad \text{for} \quad \alpha \neq \gamma, \quad \mathbf{v} = c\mathbf{n} + \frac{\rho^{\gamma}}{\rho}\left(\mathbf{v}^{\gamma} - c\mathbf{n}\right). \tag{11.52}$$

The jump of internal energy $(11.12)_2$ reads

$$[[\mathbf{q}]] \cdot \mathbf{n} - \left[\left[\frac{\mathbf{n} \cdot \mathbf{T}\mathbf{n}}{\rho} - \varepsilon - \tfrac{1}{2}\left(\mathbf{v} - c\mathbf{n}\right) \cdot \left(\mathbf{v} - c\mathbf{n}\right)\right]\right] \rho\left(\mathbf{v} \cdot \mathbf{n} - c\right) = 0,$$

or, bearing the separation (11.27) and the relation

$$\mathbf{T} = -\sum_{\alpha=1}^{A}\left(p^{\alpha}\mathbf{1} + \rho^{\alpha}\mathbf{u}^{\alpha} \otimes \mathbf{u}^{\alpha}\right),$$

in mind, we obtain

$$[[\mathbf{q}_I]] \cdot \mathbf{n} + \left[\left[\frac{1}{\rho}\sum_{\alpha=1}^{A} p^{\alpha} + \varepsilon_I + \tfrac{1}{2}\left(\mathbf{v} - c\mathbf{n}\right) \cdot \left(\mathbf{v} - c\mathbf{n}\right)\right]\right] \rho\left(\mathbf{v} \cdot \mathbf{n} - c\right) = 0. \tag{11.53}$$

Simultaneously, the jump of the entropy is of the following form

$$[[\mathbf{h}]] \cdot \mathbf{n} + [[\eta]]\, \rho\left(\mathbf{v} \cdot \mathbf{n} - c\right) = 0. \tag{11.54}$$

These two relations can now be substituted into one of the identities following from the evaluation of the entropy inequality. The result is

$$\left[\left[\varepsilon_I - T\eta + \tfrac{1}{2}\left(\mathbf{v} - c\mathbf{n}\right) \cdot \left(\mathbf{v} - c\mathbf{n}\right) + \frac{1}{\rho}\sum_{\alpha=1}^{A} p^{\alpha} - T\sum_{\alpha=1}^{A}\left(\delta^{\gamma\beta} - \frac{\rho^{\beta}}{\rho}\right)\Lambda_I^{\rho^{\beta}}\right]\right] = 0. \tag{11.55}$$

Bearing (11.43) in mind, this relation can be transformed as follows

$$[[\mu^{\gamma}]] = 0, \tag{11.56}$$

where

$$\mu^{\gamma} = \mu_I^{\gamma} + \tfrac{1}{2}\left(\mathbf{v}-c\mathbf{n}\right)\cdot\left(\mathbf{v}-c\mathbf{n}\right), \quad \mu_I^{\gamma} = -T\Lambda_I^{\rho^{\gamma}}. \tag{11.57}$$

This quantity continuous across ideal semipermeable walls is called the *chemical potential of the component* γ (comp. (6.56)). This can be determined experimentally. Namely according to the relation (11.43) we have

$$\sum_{\alpha=1}^{A} \rho^{\alpha}\mu_I^{\alpha} = \rho\left(\varepsilon_I - T\eta\right) + \sum_{\alpha=1}^{A} p^{\alpha}, \tag{11.58}$$

and, for the single fluid,

$$\mu_I = \varepsilon - T\eta + \frac{p}{\rho}, \tag{11.59}$$

which is, of course, the free enthalpy, and, hence, a measurable quantity. We can now argue in the same way as we did for the temperature. The contact between the mixture and the fluid identical with the γ-component of the mixture through the semipermeable wall for this component yields the measurement of the chemical potential of this component.

The remaining results of the exploitation of the second law of thermodynamics can now be written in the following compact form

$$\begin{aligned}
\mathbf{\Lambda}^{v^{\alpha}} &= -\frac{1}{T}\mathbf{u}^{\alpha}, \\
d\left(\rho\eta\right) &= \frac{1}{T}\left(d\left(\rho\varepsilon_I\right) - \sum_{\alpha=1}^{A}\mu_I^{\alpha}d\rho^{\alpha}\right), \\
\rho\left(\varepsilon_I - T\eta\right) &= \sum_{\alpha=1}^{A}\rho^{\alpha}\mu_I^{\alpha} - p, \\
\mathbf{h} &= \frac{1}{T}\left(\mathbf{q} - \sum_{\alpha=1}^{A}\rho^{\alpha}\mu_I^{\alpha}\mathbf{u}^{\alpha}\right).
\end{aligned} \tag{11.60}$$

Equation $(11.60)_2$ is the Gibbs equation for the mixtures. We investigate further its consequences. Equation $(11.60)_3$ is called the *Gibbs-Duhem equation*. Finally, the relation $(11.60)_4$ between the entropy flux $\mathbf{h}$ and the heat flux $\mathbf{q}$ demonstrates the fundamental difference between the thermodynamics of single component media for which $\mathbf{h} = \mathbf{q}/T$ and the multicomponent media. The presence of the additional term in the above relation proves that in thermodynamics of mixtures we cannot rely on the classical Clausius-Duhem inequality as the form of the second law of thermodynamics. This is one of the main observations made by Ingo Müller in the construction of thermodynamical mixture theories [217]. In spite of its self-evidence clearly visible in the above derivation it is still sometimes ignored in thermodynamics of mixtures (e.g. [259]).

There is a simple physical explanation of the difference between these fluxes. They describe the transfer through an interface which moves with a velocity either related to the barycentric motion and then the components move with respect to the interface with the diffusion velocity $\mathbf{u}^\alpha$ or related to the motion of one of the components, say β, and then components move with the relative velocity $\mathbf{v}^\alpha - \mathbf{v}^\beta$. In any case the transfer of energy and the transfer of entropy through the interface will be caused not only by conduction in which case the fluxes $\mathbf{h}$ and $\mathbf{q}$ would be proportional but there is a drag transfer through crossing the interface by particles moving with the velocity different from the velocity of the interface. This is exactly what the additional contribution in relation $(11.60)_4$ is describing.

The above presented Gibbs equation yields immediately the following relations

$$\mu_I^\alpha = \frac{\partial \rho\left(\varepsilon_I - T\eta\right)}{\partial \rho^\alpha}, \quad \rho\eta = -\frac{\partial \rho\left(\varepsilon_I - T\eta\right)}{\partial T}, \tag{11.61}$$

as well as the integrability conditions (comp. (6.59))

$$\frac{\partial \rho\varepsilon_I}{\partial \rho^\alpha} = -T^2 \frac{\partial\left(\mu_I^\alpha / T\right)}{\partial T}, \quad \frac{\partial \mu_I^\alpha}{\partial \rho^\beta} = \frac{\partial \mu_I^\beta}{\partial \rho^\alpha}. \tag{11.62}$$

Hence the derivatives of the Helmholtz free energy density $\rho\left(\varepsilon_I - T\eta\right)$ specify constitutive relations for chemical potentials and for the entropy. In addition, integrability relations demonstrate couplings between components: the chemical potential of component α depends on all other mass densities.

There are some additional restrictions due to the fact that the multiplier Λ^ε is dependent only on the temperature. For this reason the residual inequality contains a contribution linear in $\operatorname{grad}\rho^\alpha$ which must vanish according to the same argument we made before. This yields the following identity

$$\sum_{\alpha=1}^{A} \frac{\partial \Lambda_I^{\rho^\alpha}}{\partial \rho^\gamma} \rho^\alpha \mathbf{u}^\alpha + \frac{\partial \Lambda^\varepsilon}{\partial T} q_\rho^\gamma \operatorname{grad} T - \Lambda^\varepsilon \sum_{\alpha=1}^{A} \left(M_\rho^{\alpha\gamma} - \frac{\partial p^\alpha}{\partial \rho^\gamma} \right) \mathbf{u}^\alpha = 0. \tag{11.63}$$

Consequently, we obtain

$$q_\rho^\gamma = 0, \quad M_\rho^{\alpha\gamma} = \frac{\partial p^\alpha}{\partial \rho^\gamma} - \rho^\alpha \frac{\partial \mu_I^\alpha}{\partial \rho^\gamma}. \tag{11.64}$$

This yields the following form of the energy flux and the sources in momentum equations

$$\mathbf{q}_I = q_T \operatorname{grad} T + \sum_{\beta=1}^{A-1} q_w^\beta \mathbf{w}^\beta,$$

$$\hat{\mathbf{p}}^\alpha - \hat{\rho}^\alpha \mathbf{v}^\alpha = M_T^\alpha \operatorname{grad} T + \sum_{\beta=1}^{A-1} M_w^{\alpha\beta} \mathbf{w}^\beta + \sum_{\beta=1}^{A} \left(\frac{\partial p^\alpha}{\partial \rho^\beta} - \rho^\alpha \frac{\partial \mu_I^\alpha}{\partial \rho^\beta} \right) \operatorname{grad} \rho^\beta. \tag{11.65}$$

We see that the only place where the gradient of partial mass densities appears in the model is the source of momentum. However, even this weak form of influence has an important bearing on interactions in the mixture. Namely, if we assume that the gradient of mass densities does not appear at all in the model we obtain from (11.65)

$$\frac{1}{\rho^\alpha} \frac{\partial p^\alpha}{\partial \rho^\beta} = \frac{\partial \mu_I^\alpha}{\partial \rho^\beta}. \tag{11.66}$$

This relation yields the following integrability condition

$$-\frac{1}{\rho^{\alpha 2}} \delta^{\alpha\gamma} \frac{\partial p^\alpha}{\partial \rho^\beta} = -\frac{1}{\rho^{\alpha 2}} \delta^{\alpha\beta} \frac{\partial p^\alpha}{\partial \rho^\gamma}. \tag{11.67}$$

Hence it follows

$$\frac{\partial p^\alpha}{\partial \rho^\beta} = 0 \quad \text{when} \quad \alpha \neq \beta. \tag{11.68}$$

This means that the partial pressure p^α depends in this case on its mass density ρ^α but not on the mass densities of other components. We then say that the mixture is *simple.* Molecules in such mixtures do not feel forces of interactions created in the material due to heterogeneity. This observation made by Ingo Müller in 1968 [217] has solved one of the fundamental problems of the classical theory of mixtures of fluids. We see further that a similar problem arises for immiscible mixtures.

Obviously the relation (11.66) yields for simple mixtures $\mu_I^\alpha = \mu_I^\alpha(\rho^\alpha, T)$. Hence

$$\frac{\partial^2 \rho (\varepsilon_I - T\eta)}{\partial \rho^\alpha \partial \rho^\beta} = 0 \quad \text{for} \quad \alpha \neq \beta. \tag{11.69}$$

Consequently the free energy is the sum of functions which depend on only one mass density and temperature. The entropy density possesses the same property and so does the internal energy. Therefore we can write

$$\rho\varepsilon_I = \sum_{\alpha=1}^{A} \rho^\alpha \varepsilon^\alpha \left(\rho^\alpha, T\right), \quad \rho\eta = \sum_{\alpha=1}^{A} \rho^\alpha \eta^\alpha \left(\rho^\alpha, T\right). \tag{11.70}$$

This decomposition shows that there is no energy of interaction between components in simple mixtures. It yields as well the following relation for chemical potentials

$$\mu_I^\alpha = \varepsilon^\alpha - T\eta^\alpha + \frac{p^\alpha}{\rho^\alpha}, \tag{11.71}$$

which means that in simple mixtures the chemical potential and the partial free enthalpy are identical.

We proceed to investigate the residual inequality which remains after the analysis presented above. It has the form

$$\mathcal{D} := \begin{pmatrix} \operatorname{grad} T \\ \mathbf{w}^\alpha \end{pmatrix} \cdot \begin{pmatrix} -\frac{q_T}{T} & A^\beta \\ A^\alpha & -\frac{1}{T} M_w^{\alpha\beta} \end{pmatrix} \begin{pmatrix} \operatorname{grad} T \\ \mathbf{w}^\beta \end{pmatrix}$$

$$-\frac{1}{T} \sum_{r=1}^{\nu} \left(\sum_{\alpha=1}^{A-1} \left(\mu_I^\alpha - \mu_I^A + \tfrac{1}{2} \mathbf{w}^\alpha \cdot \mathbf{w}^\alpha\right) \gamma_\alpha^r M_\alpha \mu_H \lambda^r \right) \geq 0, \tag{11.72}$$

where

$$A^\alpha = -\frac{1}{2T} \left\{ \frac{q_w^\alpha}{T} + \left[\left(\frac{M_T^\alpha}{\rho^\alpha} - \frac{M_T^\gamma}{\rho^\gamma} \right) - \left(\frac{1}{\rho^\alpha} \frac{\partial p^\alpha}{\partial T} - \frac{1}{\rho^\gamma} \frac{\partial p^\gamma}{\partial T} \right) \right.\right.$$

$$\left.\left. + T \frac{\partial \left(\mu_I^\alpha - \mu_I^\gamma\right)/T}{\partial T} \right] \sum_{\gamma=1}^{A-1} \left(\rho^\alpha \delta^{\alpha\gamma} - \frac{\rho^\alpha \rho^\gamma}{\rho} \right) \right\}. \tag{11.73}$$

The function $\mathcal{D}$ is called the *dissipation density* and it consists of four contributions:

- chemical reactions which are primarily characterized by the difference of chemical potentials $\mu_I^\alpha - \mu_I^A$,
- heat conduction which contributes with the square of the temperature gradient $\operatorname{grad} T$,
- diffusion which contributes with the square of the relative velocity $\mathbf{w}^\alpha$,
- thermal diffusion which contributes with the product of the temperature gradient and relative velocity.

The vanishing dissipation defines the *thermodynamical equilibrium*

$$\operatorname{grad} T|_E = 0, \quad \left(\mu_I^\alpha - \mu_I^A\right)\big|_E = 0, \quad \mathbf{w}^\alpha|_E = 0, \quad \alpha = 1, ..., A-1. \tag{11.74}$$

Such processes correspond to the minimum of dissipation and this yields additional conditions – thermodynamic stability conditions of equilibrium.

Let us add that, in addition to this classical approach to thermodynamics of mixtures there also exists a relativistic extended thermodynamics of mixtures of ideal fluids and this theory gives an important physical insight to the description of chemical reactions. Details can be found in the book of I. Müller and T. Ruggeri [225].

Chapter 12

Thermodynamics of immiscible mixtures: Introduction and models without the field of porosity

The main topic of the following Chapters is the construction of continuous models of multicomponent systems in which one of the components is a solid creating skeleton (a solid confinement) for the motion of fluid components. As we allow for large deformations of the solid phase we shall mostly use the Lagrangian description of motion. Such models are called *porous material models*.

Theories of porous and granular materials can be constructed on different levels of observation. *Microscopic* models rely on Newton's equations of motion of material points or molecules and use methods of molecular dynamics. Such models can be transferred on a *semimacroscopic* level by multiscaling and averaging procedures (e.g. [281], [155], [208], [207], [253], [153]). Some numerical methods such as Monte Carlo are also used in practical applications. On a *macroscopic* level continuum field models are constructed. These may either follow from semimacroscopic models by homogenizing, averaging over Representative Elementary Volumes (REV), construction of moments of kinetic distribution functions or they may be constructed by means of a phenomenological macroscopic approach (e.g. [47], [48]). In these Chapters we present solely the latter type of models with a marginal reference to averaging procedures.

The construction of macroscopic continuous models of systems with a solid component in its most sophisticated form stems from models of multicomponent systems. Differences are primarily connected with an kind of interactions within the solid component. Models must be clearly different in the cases of suspensions (for the theory of mixture gas-fluid, see, for instance [50]), of granular materials or of porous materials. In the first case solid particles or gaseous bubbles interact with each other either through the fluid or through collisions and there is no permanent contact between them. In the second case a granular solid component may behave as a solid which cannot carry a tensile loading (unilateral constraint on constitutive relations) or it may fluidize and then behave as a suspension. Finally, a porous material behaves in average as a usual solid and it forms a deformable carrier for fluid components. Most likely, the first attempt to use systematically a multicomponent model to such porous materials was made by G. Heinrich and K. Desoyer in 1955-1956 [140], [141]. We limit our attention to this case.

The most important feature of *porous materials* is the appearance of different kinematics for the solid component - the *skeleton*, and fluid components in channels of the skeleton. This yields *diffusion processes* characterized by relative velocities of components. In most cases of practical bearing the dependence on the relative velocity is reduced to a linear contribution to momentum balance equations (momentum sources) or even to a simpler form called the *Darcy law*.

The problem of *thermodiffusion* within such models is still very much open. This is related to difficulties with an appropriate definition of the temperature on the macroscopic level of description. The most important property of the classical *thermodynamical temperature*, its continuity on ideal thermal walls and, consequently, its experimental measurability, is not fulfilled in porous materials (e.g. [361]). Simultaneously, such processes as phase transitions or chemical reactions in porous materials are characterized by real thermodynamical temperatures (e.g. melting and freezing points, evaporation, etc.) of components on a semimacroscopic level of description. It means that even if we have introduced a macroscopic notion of temperature we would have to know a rule of transformation of this quantity to the semimacroscopic level. This is mathematically an ill-posed problem. Even though one can formally work with notions such as partial heat fluxes, specific heats etc. their operational meaning is not clarified yet. This seems to be the most important challenge of modelling porous and granular materials.

In addition, we have to deal frequently with the problem of different temperatures for different components. In contrast to gases, a local thermodynamical equilibrium is reached in porous and granular materials after macroscopically long relaxation times. For instance, a hot water flowing through a cold porous material does not reach locally a common temperature with the skeleton within seconds or minutes. Consequently, we should construct thermodynamical models with different temperatures of components. Such a construction, as we have indicated in the previous Chapter, is missing even in the case of fluid mixtures. One of the reasons is again the problem of measurability.

Let us mention in passing that in theories of granular materials stemming from a kinetic equation it is common to work with a *kinetic temperature* rather than a thermodynamical temperature (e.g. [157]). It is defined in a way similar to this of the kinetic theory of gases (7.42) through a mean kinetic energy of granulae. There are numerous difficulties connected with such a notion. For example a natural equilibrium state of a granular material in which particles do not move would have a temperature equal to zero. Consequently, deviations from the equilibrium state which are used in the construction of macroscopic moment equations of the kinetic theory (comp. (7.60)) would have to be constructed by means of a trivial distribution function. Certainly, this cannot give any reasonable physical results. For this reason moment equations are constructed by a reference to a Maxwell-like distribution describing processes of simple shearing flows rather than real equilibrium states. In contrast to, say, Grad's 13 moment method of rarefied gases such procedures are not justified in any way. Moreover the questions of measurability of kinetic temperature, a relation to the thermodynamical temperature etc. are not even asked as yet (e.g. see [33]).

Some progress for granular materials has been achieved by J. D. Goddard [124], [123] who has tried to incorporate arguments of statistical mechanics to quasistatic granular ensembles. However, as he states himself, there are still problems with the identification of the multiplier β of the distribution function as an inverse of temperature: "*As another interpretation, more in keeping with conventional thermodynamics, one might also identify* β^{-1} *with the kinetic energy of some hypothetical state of agitation or "shaking" involved in the preparation of a static granular state. However, the validity of this criterion for static equivalence is less than evident to this investigator*" [124].

Within multicomponent continuous models an exchange of mass is described by mass sources in partial mass balance equations. These contributions, as demonstrated for fluid mixtures, must contain additional *microstructural* variables. This requires an extension of the set of field equations. In many cases additional equations for microstructural variables have the form of evolution equations. Then there is no need to introduce additional boundary conditions. Such microstructural variables cannot be controlled, they develop spontaneously from initial data. On the other hand, the latter can usually be easily formulated because many microstructural variables are defined in such a way that they vanish in thermodynamical equilibria.

Further, in Chapter 13, we present in some details a model of processes of exchange of mass called *adsorption*. These processes appear in cases of components which, in contrast to chemical bindings, form weak van der Waals bindings solely with the skeleton. Such are, for example, processes of transport of many pollutants in soils. According to the simplest model of these processes, developed by Langmuir, they are described by an additional field of the so-called *number of bare sites*. In the case of materials with very small diameters of channels adsorption processes possess a hysteresis loop in the relation between the partial pressure of adsorbate in the fluid phase and an amount of mass adsorbed by the skeleton and this plays a very important role in controlling technological processes in such materials. Such loops are caused by *capillary effects*. For this reason, they do not appear in materials with moderate and large channels which is characteristic for the usual soils but they do appear, for instance, in concrete.

As already mentioned above multicomponent models of porous materials contain more than one velocity field. This yields field equations following from partial momentum balance equations with a corresponding number of partial accelerations. Consequently, one expects that in such models additional modes of weak discontinuity waves (*acoustic waves*) have to appear. This is indeed the case. One of these modes was predicted by M. A. Biot in 1941. Due to the tradition stemming from geophysics this mode is called P2 compressional (Biot's) wave as the usual longitudinal wave registered in seismograms was called P1. Existence of this mode was confirmed in numerous experiments. It has been found out that, most frequently, it is the slowest of three modes P1, S (transversal wave) and P2. It is also very strongly attenuated.

As consequence of existence of additional bulk modes there exist as well additional modes of *surface waves*. Apart from the classical Rayleigh wave there exist the so-called Stoneley waves, various leaky waves and, in general, a number of possible modes of surface waves depends on properties of neighboring systems, i.e. on the structure of *boundary conditions*. Some aspects of this problem we discuss further in these Chapters. Let us mention that surface waves are much weaker attenuated than bulk waves and for this reason they are easier attainable in measurements. In recent years one can observe a vehement progress in these measuring techniques.

Couplings of dynamic properties of porous and granular materials with mass exchange between components play an important role in various combustion and explosion problems. These are connected with the propagation of strong discontinuities such as *shock waves* and *combustion fronts* in combustion of solid fuels or deformations of soils due to impacts of meteorites. Models for such processes are still rather weakly developed. Most important contributions are based on the model proposed by M. A. Goodman and C. Cowin [125] which refers to some additional microstructural properties called the *principle of equilibrated pressures*. The model leads to a quasilinear hyperbolic set of equations which admits the existence of shock waves. Incidentally, a similar model is used in the description of avalanches, landslides and mud flows (e.g. [94], [151]). However, apart from some simple properties of propagation conditions, usually one-dimensional, results are rather scarce.

Some elementary properties of one-dimensional Rankine-Hugoniot conditions have also been investigated within the frame of the model with the porosity balance equation. However a comprehensive theory of shock waves is still missing and one of the reasons is the lack of a selection (entropy) criterion.

Many processes in porous and granular materials are connected with the development of instabilities. They lead to fluidization of saturated sands, to the creation of patterns in porous materials and to some instabilities, such as Saffmann-Taylor, in flows of fluid components. As usual they are connected with nonlinearities appearing in the model.

One of the most spectacular phenomena accompanying earthquakes is the fountain-like explosion of water and mud from the sand. It has been found that prior to this phenomenon the character of permeability of the soil changes in an unstable manner. In the first stage the homogeneity of the system breaks down and a pattern of chimney-like channels with a very high permeability is formed. In the second stage one of these channels becomes dominant and this leads to an explosion-like eruption of water from the ground. This behavior seems to be connected with a nonlinear coupling of the diffusion velocity with the gradient of porosity (e.g. [350]).

Another class of instabilities appears in the model with the balance equation of porosity. These are connected with the coupling between dynamic changes of porosity and partial stresses in components. Let us mention one of those instabilities. In a case of a Riemann problem the system develops soliton-like waves of porosity [378], [377], [256]. These are connected with the loss of symmetry of the front of propagation if the two-dimensional front is concave. Most likely in the vicinity of the symmetry axis the system develops a mushy region. Multicomponent modelling of porous materials is based on the assumption that additionally to the usual fields of theories of fluid mixtures there exists a microstructure which is reflected in the simplest case by a single additional field of porosity (the volume fraction of voids) and by solid-like properties of one of the components. It seems that L. W. Morland [215] was the first researcher who systematically used the volume fraction concept. In some models this microstructural extension is even broader and the corresponding models contain, for instance, the so-called volume fractions of all components, double porosity, tortuosity as a simplest measure of complexity of geometrical structure of channels, couple stresses etc.

We begin the presentation of models of porous materials with an example of a two-component system whose theory is based on the assumption of incompressibility of components. Models of this sort appear quite frequently in applications to soil mechanics or glaciology. For the review of various possible constraint conditions one can consult the work [368].

First of all, let us make a comment on the calculation of averages in modelling porous materials. The microstructure of these materials usually has the characteristic dimensions of an almost macroscopic nature. Typical dimensions of granulae or radii of channels are in the order of micrometers and sometimes even millimeters. This means that we can, in principle, apply a continuum model on this semimacroscopic level and consider either a single component continuum (skeleton) or a mixture of fluids (fluid components in channels of the skeleton). This is not being done because extremely complicated shapes of channels practically rule out the possibility of formulation of boundary value problems for semimacroscopic field equations. Instead we construct volume averages over the so-called *Representative Elementary Volumes* (e.g. [30]). These are three-dimensional sets whose dimensions are sufficiently large to be able to assume the randomness of the microstructural geometry and, simultaneously, sufficiently small when compared with macroscopic dimensions in order to be able to prescribe average properties to values of the corresponding macroscopic fields. In Figure 16, we show schematically such a REV-domain and the distributions of the skeleton and channels on the semimacroscopic level. These distributions are described by characteristic functions $\chi^{\alpha}(\mathbf{Z},t)$ which, for the α-component, have the value 1 if the point $\mathbf{Z}$ is in the instant of time t occupied by the α-component and 0 otherwise. The notation $\mathbf{Z}$ as well as $\mathbf{X}$ appearing further is characteristic for the Lagrangian description of the multicomponent system which we present later in details. In the following derivation, it is immaterial as we consider the time t only as a parameter and the whole considerations concern spatial properties of the system which may be represented in Lagrangian as well as in Eulerian description. This issue will be made precise further in this Chapter.

Let us denote by $\mathcal{M}(\mathbf{X},t)$ the REV-domain which at the instant of time t is located at the point $\mathbf{Z}=\mathbf{X}$. The point $\mathbf{X}$ selecting $\mathcal{M}$ can be chosen arbitrarily but it is usually convenient to make a special choice related to the geometry of REV. Namely, the shape of REV-domains should be chosen in such a way that it does not deviate from natural symmetries of the microstructure. For instance, in the most common case of isotropic microstructure the shape should be spherical. Then $\mathcal{M}(\mathbf{X},t)$ is the ball of the constant radius R whose center lies at $\mathbf{X}$ at the instant of time t

$$\mathcal{M}(\mathbf{X},t)=\{\mathbf{Z}|\,\mathbf{X}\in\mathcal{B}_0\ \&\ |\mathbf{Z}-\mathbf{X}|<R\}. \tag{12.1}$$

By this choice, R must be much larger than, say, typical diameters of channels and much smaller than the macroscopic dimensions of the body.

Obviously, there is a problem of defining REV in the vicinity of the boundary as well as the problem of the definition of the boundary $\partial\mathcal{B}_0$ itself. We return to this question in the discussion of boundary conditions.

However, it should be mentioned that these problems yield the existence of *boundary layers* which are replaced by additional surface quantities.

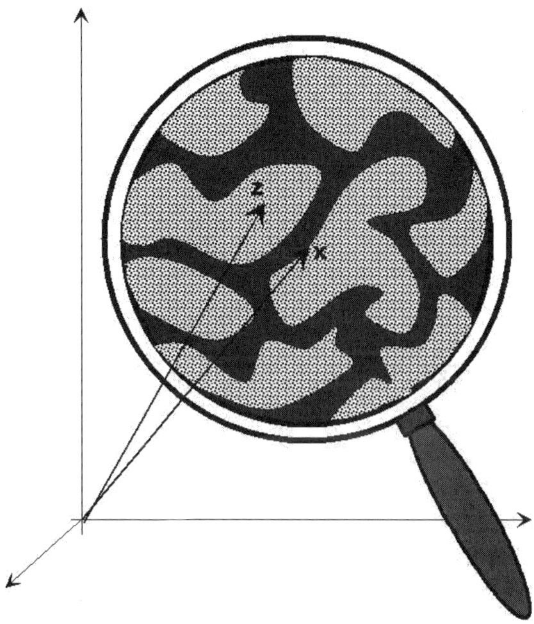

Fig. 16: *Schematic Representative Elementary Volume (REV).*

Now let us consider the construction of volume averages. For an arbitrary quantity $\varphi^\alpha(\mathbf{Z}, t)$ characteristic for the α-component we construct the average

$$\langle \varphi^\alpha \rangle (\mathbf{X}, t) = \frac{1}{V} \int_{\mathcal{M}(\mathbf{X},t)} \varphi^\alpha(\mathbf{Z}, t)\, \chi^\alpha(\mathbf{Z}, t)\, dV_Z, \tag{12.2}$$

where $V = \operatorname{vol} \mathcal{M}(\mathbf{X}, t)$ and, for simplicity it is assumed to be a constant. By means of this definition of the volume average we have to define macroscopic gradients of quantities $\langle \varphi^\alpha \rangle (\mathbf{X}, t)$ and their time derivatives. The latter is straightforward because we construct instantaneous averages over the spatial domain. However the construction of the gradients is more complicated. Let us choose an infinitesimal vector $\delta\mathbf{X}$ which is constant over $\mathcal{M}(\mathbf{X}, t)$. This vector defines the shift of the REV-domain $\mathcal{M}(\mathbf{X}, t)$. We have

$$\langle \varphi^\alpha \rangle (\mathbf{X} + \delta\mathbf{X}, t) = \frac{1}{V} \int_{\mathcal{M}(\mathbf{X}+\delta\mathbf{X},t)} \varphi^\alpha(\mathbf{Z}, t)\, \chi^\alpha(\mathbf{Z}, t)\, dV_Z$$

$$= \frac{1}{V}\left\{\int_{\mathcal{M}(\mathbf{X},t)} \varphi^{\alpha}(\mathbf{Z},t)\,\chi^{\alpha}(\mathbf{Z},t)\,dV_Z\right.$$

$$\left. + \oint_{\partial\mathcal{M}(\mathbf{X},t)} \varphi^{\alpha}(\mathbf{Z},t)\,\chi^{\alpha}(\mathbf{Z},t)\,\mathbf{N}_Z(\mathbf{Z},t)\cdot\delta\mathbf{X}dS_Z\right\} + O\left(|\delta\mathbf{X}|^2\right)$$

$$= \langle\varphi^{\alpha}\rangle(\mathbf{X},t) + \frac{1}{V}\left\{\int_{\mathcal{M}(\mathbf{X},t)} \mathrm{Div}\left[\varphi^{\alpha}(\mathbf{Z},t)\,\chi^{\alpha}(\mathbf{Z},t)\,\delta\mathbf{X}\right]dV_Z\right\} + O\left(|\delta\mathbf{X}|^2\right)$$

$$= \langle\varphi^{\alpha}\rangle(\mathbf{X},t) + \frac{1}{V}\delta\mathbf{X}\cdot\int_{\mathcal{M}(\mathbf{X},t)} \mathrm{Grad}\left[\varphi^{\alpha}(\mathbf{Z},t)\,\chi^{\alpha}(\mathbf{Z},t)\right]dV_Z + O\left(|\delta\mathbf{X}|^2\right)$$

$$= \langle\varphi^{\alpha}\rangle(\mathbf{X},t) + \delta\mathbf{X}\cdot\langle\mathrm{Grad}\,\varphi^{\alpha}\rangle(\mathbf{X},t)$$

$$+\frac{1}{V}\delta\mathbf{X}\cdot\int_{\mathcal{M}(\mathbf{X},t)} \varphi^{\alpha}(\mathbf{Z},t)\,\mathrm{Grad}\left[\chi^{\alpha}(\mathbf{Z},t)\right]dV_Z + O\left(|\delta\mathbf{X}|^2\right), \qquad (12.3)$$

where $\mathbf{N}_Z$ is the unit outward normal vector to the surface $\partial\mathcal{M}(\mathbf{X},t)$. In order to evaluate the last integral let us introduce the set

$$\mathcal{M}^{\alpha}(\mathbf{X},t) = \left\{\mathbf{Z}\in\mathcal{M}(\mathbf{X},t)\,|\,\chi^{\alpha}(\mathbf{Z},t) = 1\right\}. \qquad (12.4)$$

Then the above relation yields

$$\langle\varphi^{\alpha}\rangle(\mathbf{X}+\delta\mathbf{X},t) - \langle\varphi^{\alpha}\rangle(\mathbf{X},t) = \delta\mathbf{X}\cdot\langle\mathrm{Grad}\,\varphi^{\alpha}\rangle(\mathbf{X},t) \qquad (12.5)$$

$$+\frac{1}{V}\delta\mathbf{X}\cdot\int_{\partial\mathcal{M}^{\alpha}(\mathbf{X},t)\cap\mathcal{M}(\mathbf{X},t)} \varphi^{\alpha}(\mathbf{Z},t)\,\mathbf{N}_Z(\mathbf{Z},t)\,dS_Z + O\left(|\delta\mathbf{X}|^2\right).$$

We have used here the formula for the differentiation of the characteristic function $\chi^{\alpha}(\mathbf{Z},t)$. Obviously, the gradient of this function is zero in the interior of $\mathcal{M}^{\alpha}(\mathbf{X},t)$ and $\mathcal{M}(\mathbf{X},t)\setminus\mathcal{M}^{\alpha}(\mathbf{X},t)$, where this function is constant. One has to evaluate only the derivative in the direction orthogonal to the boundary $\partial\mathcal{M}^{\alpha}(\mathbf{X},t)$. We can do this using local coordinates such that at every point of the boundary we choose the origin of the orthogonal frame with the coordinate ξ, whose unit base vector is $\mathbf{N}_Z$. We demonstrate the calculations using the following definition of the characteristic function in the vicinity of the point $\xi = 0$

$$\chi^{\alpha}(\xi,t) = \frac{1}{2}\lim_{\alpha\to\infty}\left[1 - \tanh(\alpha\xi)\right], \qquad (12.6)$$

where the dependence on the other two local coordinates is immaterial. Then

$$\operatorname{Grad}\left[\chi^{\alpha}(\mathbf{Z},t)\right] = \left(\mathbf{N}_Z\cdot\operatorname{Grad}\left[\chi^{\alpha}(\mathbf{Z},t)\right]\right)\mathbf{N}_Z \tag{12.7}$$
$$= \left.\frac{d\chi^{\alpha}(\xi,t)}{d\xi}\right|_{\xi=0}\mathbf{N}_Z = -\frac{1}{2}\lim_{\alpha\to\infty}\left.\frac{\alpha}{\cosh^2(\alpha\xi)}\right|_{\xi=0}\mathbf{N}_Z$$
$$= -\delta_{\partial\mathcal{M}^{\alpha}(\mathbf{X},t)}(\xi)\,\mathbf{N}_Z,$$

where $\delta_{\partial\mathcal{M}^{\alpha}(\mathbf{X},t)}(\xi)$ is the Dirac-delta function for the surface $\partial\mathcal{M}^{\alpha}(\mathbf{X},t)$. This yields the transition from the volume integral in (12.3) to the surface integral in (12.5).

The relation (12.5) for the *directional derivative* of the average $\langle\varphi^{\alpha}\rangle$ gives rise to the following formula

$$\forall\ \delta\mathbf{X}\text{-infinitesimal}\ \delta\,\mathbf{X}\cdot\operatorname{Grad}\langle\varphi^{\alpha}\rangle(\mathbf{X},t)$$

$$= \delta\mathbf{X}\cdot\langle\operatorname{Grad}\varphi^{\alpha}\rangle(\mathbf{X},t) - \frac{1}{V}\delta\mathbf{X}\cdot\int_{\partial\mathcal{M}^{\alpha}(\mathbf{X},t)\cap\mathcal{M}(\mathbf{X},t)}\varphi^{\alpha}(\mathbf{Z},t)\,\mathbf{N}_Z(\mathbf{Z},t)\,dS_Z. \tag{12.8}$$

To demonstrate the meaning of this relation let us consider a special case in which $\varphi^{\alpha}(\mathbf{Z},t)$ is piecewise constant. Then the first contribution vanishes identically and in the second one most contributions vanish due to the outward orientation of the vector $\mathbf{N}_Z(\mathbf{Z},t)$. There remain only those whose counterpart lies outside the domain $\mathcal{M}(\mathbf{X},t)$. We explain this argument in the one-dimensional diagram of Fig. 17. Contributions of $\mathcal{M}^{\alpha}(\mathbf{X},t)$-sets: (2),...,(6) vanish due to the opposite orientation of the normal vector on their ends. There remain the contributions of the set (1) and of the set (7) because their ends lie beyond the set $\mathcal{M}(\mathbf{X},t)$ and this gives the contribution $\varphi_1^{\alpha}-\varphi_7^{\alpha}$ of the integral in (12.8), where $\varphi_1^{\alpha},\varphi_7^{\alpha}$ are values of φ^{α} on the subsets (1) and (7), respectively. As V is in this example identical with the length L of the interval in Fig. 17, we obtain the average value of the gradient identical with $(\varphi_7^{\alpha}-\varphi_1^{\alpha})/L$.

It is obvious that this construction is extremely singular. One cannot ever expect any differentiable macroscopic functions to follow from such a construction. Consequently, it is essential that characteristic dimensions of the microstructure are much smaller than macroscopic increments $|\delta\mathbf{X}|$. In addition, the heterogeneity of the function $\varphi^{\alpha}(\mathbf{Z},t)$ should be sufficiently small for the contribution $\langle\operatorname{Grad}\varphi^{\alpha}\rangle(\mathbf{X},t)$ to be negligible in (12.8). Such a contribution is characterized by the parameter of heterogeneity of the microstructure $\varepsilon_{\text{heter}} = |\langle\operatorname{Grad}\varphi^{\alpha}\rangle(\mathbf{X},t)/\operatorname{Grad}\langle\varphi^{\alpha}\rangle(\mathbf{X},t)|$ which must be much smaller than the parameter $\varepsilon_{\text{micro}} = R/l \ll 1$, where l is the macroscopic characteristic length. If these conditions are satisfied we can speak of macroscopic fields as volume approximations of real semimacroscopic quantities. We shall not discuss any further the mathematical structure of such constructions and assume sufficient smoothness for all operations which we perform on the fields.

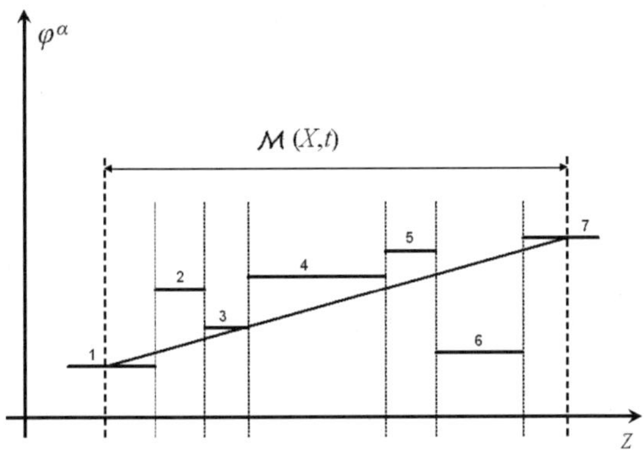

Fig. 17: *Construction of macroscopic gradients in 1D-case.*

We illustrate the above considerations on some simple models of poroelastic materials with an incompressibility assumption. As in all continuum models we define in *Eulerian description* fields on a common domain $\mathcal{B}_t$ which is time dependent and corresponds to a part of the three-dimensional space of motion occupied in a current instant of time by all components. In the case of semipermeable boundary $\partial\mathcal{B}_t$ parts of components which flow out of this domain are considered separately and one has to solve contact problems.

In a purely mechanical model which we want to consider in this Chapter processes are described by two current partial mass densities $\rho_t^S(\mathbf{x},t), \rho_t^F(\mathbf{x},t), \mathbf{x} \in \mathcal{B}_t \subset \Re^3, t \in \mathcal{T} \subset \Re$, for the skeleton and the fluid component, respectively, and by two velocity fields $\mathbf{v}^S(\mathbf{x},t), \mathbf{v}^F(\mathbf{x},t)$ for these two components. All these fields are *macroscopic* which means that they are defined on the common domain $\mathcal{B}_t$ and, for instance the fluid mass and the skeleton mass contained in a subdomain $\mathcal{P}_t \subset \mathcal{B}_t$ are given by the Lebesgue integrals

$$M^F(\mathcal{P}_t) = \int_{\mathcal{P}_t} \rho_t^F dv, \quad M^S(\mathcal{P}_t) = \int_{\mathcal{P}_t} \rho_t^S dv. \tag{12.9}$$

In the definition of incompressible components one uses a "pseudomesoscopic" quantities which are called *realistic mass densities*. We denote them by ρ_t^{SR} and ρ_t^{FR}. They are also defined in each point of the domain $\mathcal{B}_t$ and not in points of the skeleton or of the fluid, respectively. They may be related to mesoscopic (or semimacroscopic) quantities ρ_m^{SR}, ρ_m^{FR} by the following formulae

$$\begin{aligned} \rho_t^{SR}(\mathbf{x},t) &= \frac{1}{V(\mathcal{M}_t^S)} \int_{\mathcal{M}_t} \rho_m^{SR}(\mathbf{z},t)\left(1-\chi^F\right) dv_{\mathbf{z}}, \\ \rho_t^{FR}(\mathbf{x},t) &= \frac{1}{V(\mathcal{M}_t^F)} \int_{\mathcal{M}_t} \rho_m^{FR}(\mathbf{z},t)\,\chi^F dv_{\mathbf{z}}, \end{aligned} \tag{12.10}$$

where

$$\begin{aligned} \mathcal{M}_t &= \mathcal{M}_t^S \cup \mathcal{M}_t^F, \quad \mathcal{M}_t^S \cap \mathcal{M}_t^F = \emptyset, \\ V(\mathcal{M}_t^F) &= \int_{\mathcal{M}_t} \chi^F dv_{\mathbf{z}}, \quad V(\mathcal{M}_t^S) = \int_{\mathcal{M}_t} \left(1-\chi^F\right) dv_{\mathbf{z}}, \end{aligned} \tag{12.11}$$

and $\chi^F(\mathbf{z},t)$ is the characteristic function for $\mathcal{M}_t^F$. $\mathcal{M}_t$ is the image of $\mathcal{M}(\mathbf{X},t)$ if we choose to describe the motion of the skeleton, as we customarily do in the Lagrangian description, by the function of motion $\mathbf{f}^S(.,t) : \mathcal{B}_0 \to \mathcal{B}_t$. Then $\mathcal{M}_t = \mathcal{M}\left(\mathbf{f}^{S-1}(\mathbf{x},t), t\right)$. In contrast to ρ_t^{SR}, ρ_t^{FR} which do not possess any physical interpretation in points of the real fluid for the first quantity and in points of the skeleton for the second one, the mass densities ρ_m^{SR}, ρ_m^{FR} are defined solely in points of the real skeleton, and of the real fluid, respectively. Consequently, they possess the usual physical interpretation. For instance, ρ_m^{FR} is equal to $1000\frac{kg}{m^3}$ for water in normal conditions.

It is easy to check the following relations

$$V(\mathcal{M}_t^S) = (1-n)\,V(\mathcal{M}_t), \quad V(\mathcal{M}_t^F) = nV(\mathcal{M}_t), \tag{12.12}$$

$$n := \frac{1}{V(\mathcal{M}_t)} \int_{\mathcal{M}_t} \chi^F dv, \quad V(\mathcal{M}_t) := \int_{\mathcal{M}_t} dv.$$

The quantity n defined in $(12.12)_3$ is called the *porosity*.
Consequently

$$\begin{aligned} \rho_t^S &:= \frac{1}{V(\mathcal{M}_t)} \int_{\mathcal{M}_t} \rho_m^{SR}\left(1-\chi^F\right) dv = (1-n)\,\rho_t^{SR}, \\ \rho_t^F &:= \frac{1}{V(\mathcal{M}_t)} \int_{\mathcal{M}_t} \rho_m^{FR}\chi^F dv = n\rho_t^{FR}. \end{aligned} \tag{12.13}$$

Clearly, the smeared-out, partial mass densities ρ_t^S, ρ_t^F are related to the common macroscopic volume in the current configuration. Their definitions contain the full volume of REV.

By means of the above relations we are now in the position to introduce the notion of incompressibility appearing in some theories of porous and granular materials. Namely it is assumed for such models that

$$\rho_t^{SR} = \text{const.}, \qquad \rho_t^{FR} = \text{const.} \tag{12.14}$$

Consequently the current mass densities ρ_t^F, ρ_t^S are not independent fields. They can be reduced to the single field of porosity n. In such a case partial mass balance equations (without mass exchange!) reduce to the following form

$$\frac{1}{\rho_t^{SR}}\left\{\frac{\partial \rho_t^S}{\partial t} + \operatorname{div}\left(\rho_t^S \mathbf{v}^S\right)\right\} \equiv -\frac{\partial n}{\partial t} + \operatorname{div}\left((1-n)\,\mathbf{v}^S\right) = 0, \tag{12.15}$$

$$\frac{1}{\rho_t^{FR}}\left\{\frac{\partial \rho_t^F}{\partial t} + \operatorname{div}\left(\rho_t^F \mathbf{v}^F\right)\right\} \equiv \frac{\partial n}{\partial t} + \operatorname{div}\left(n\mathbf{v}^F\right) = 0.$$

We can also combine these two equations to the following one

$$\operatorname{div}\left(n\mathbf{v}^F + (1-n)\,\mathbf{v}^S\right) = 0. \tag{12.16}$$

If we consider equation $(12.15)_1$ as a candidate for the field equation for the porosity n then equation (12.16) is a *constraint condition* of the model. This condition yields certain limitations on constitutive relations appearing in the phenomenological model which are not always physically and mathematically acceptable. For instance, T. J. T. Spanos [304] demonstrates difficulties with the formulation of boundary conditions for constrained porous materials, in particular for dynamic problems. It can be shown as well that the equation $(12.15)_1$ gives solutions for porosity which contradict gedanken experiments of Biot and Willis (compare Section 13.3).

We show here two examples of models which are thermodynamically admissible. The first one is used frequently in soil mechanics and in the description of suspensions. In order to obtain field equations we need momentum balance equations which have the following form in the Eulerian description

$$\frac{\partial\left(\rho_t^S \mathbf{v}^S\right)}{\partial t} + \operatorname{div}\left(\rho_t^S \mathbf{v}^S \otimes \mathbf{v}^S - \mathbf{T}^S\right) = \hat{\mathbf{p}}^S, \tag{12.17}$$

$$\frac{\partial\left(\rho_t^F \mathbf{v}^F\right)}{\partial t} + \operatorname{div}\left(\rho_t^F \mathbf{v}^F \otimes \mathbf{v}^F - \mathbf{T}^F\right) = \hat{\mathbf{p}}^F, \quad \hat{\mathbf{p}}^S + \hat{\mathbf{p}}^F = 0,$$

where $\mathbf{T}^S, \mathbf{T}^F$ denote symmetric partial Cauchy stress tensors, $\hat{\mathbf{p}}^S, \hat{\mathbf{p}}^F$ are momentum sources. We make the assumption that these quantities satisfy the following constitutive relations

$$\begin{aligned} \mathbf{T}^S &= \mathbf{T}^S\left(n, \operatorname{grad} n, \mathbf{e}^S, \mathbf{w}\right), \quad \mathbf{T}^F = \mathbf{T}^F\left(n, \operatorname{grad} n, \mathbf{e}^S, \mathbf{w}\right), \\ \hat{\mathbf{p}}^S &= \hat{\mathbf{p}}^S\left(n, \operatorname{grad} n, \mathbf{e}^S, \mathbf{w}\right), \end{aligned} \tag{12.18}$$

where the symmetric deformation tensor of the skeleton $\mathbf{e}^S$ satisfies for small deformations[1] the equation

$$\frac{\partial \mathbf{e}^S}{\partial t} = \operatorname{sym} \operatorname{grad} \mathbf{v}^S, \tag{12.19}$$

and $\mathbf{w} := \mathbf{v}^F - \mathbf{v}^S$ is the relative velocity of components.

By means of the second law of thermodynamics one can show that the constraint (12.16) is thermodynamically admissible. This would not be the case if constitutive relations (12.18) were independent of $\operatorname{grad} n$. In this sense we deal with a *higher gradient model.* If we assume in addition the isotropy and linearity with respect to both vector variables $\operatorname{grad} n$ and $\mathbf{w}$ then we obtain a fairly explicit form of constitutive relations

$$\begin{aligned} \mathbf{T}^S &= -(1-n)p\mathbf{1} + \mathbf{T}^S_{eff}\left(n, \mathbf{e}^S\right), \quad \mathbf{T}^F = -np\mathbf{1} + \mathbf{T}^F_{eff}\left(n, \mathbf{e}^S\right), \\ \hat{\mathbf{p}}^S &= -\hat{\mathbf{p}}^F = \pi\left(n, \mathbf{e}^S\right)\mathbf{w} - p \operatorname{grad} n, \end{aligned} \tag{12.20}$$

where $\mathbf{T}^S_{eff}, \mathbf{T}^F_{eff}$ are the so called *effective partial stress tensors*. The second one is frequently assumed to be zero and the first one, if it is linear (small deformation of the skeleton $\left\|\mathbf{e}^S\right\| \ll 1$), is given by a Hooke's law with material coefficients depending on the porosity n. The *permeability coefficient* π is also usually assumed to be constant. The *pore pressure* p is the reaction force on the constraint.

In order to account for instabilities of the microstructure one can try to extend the above model by accounting for nonlinear dependence on the relative velocity $\mathbf{w}$. This is justified because such instabilities appear by flows of a high intensity of the fluid component. In a continuum model the latter corresponds to contributions $\rho_t^F \mathbf{w}$. Such an extension yields in the

---

[1] i.e. $\left\|\mathbf{e}^S\right\| \ll 1$, where $\left\|\mathbf{e}^S\right\| = \max\left\{\lambda^{(1)}, \lambda^{(2)}, \lambda^{(3)}\right\}$ and $\lambda^{(1)}, \lambda^{(2)}, \lambda^{(3)}$ are eigenvalues of $\mathbf{e}^S$.

lowest approximation the following constitutive relations [350]

$$\begin{aligned} \mathbf{T}^S &= -(1-n)p\mathbf{1} + \mathbf{T}^S_{eff}\left(n, \mathbf{e}^S\right) + \delta \mathbf{w} \otimes \mathbf{w}, \\ \mathbf{T}^F &= -np\mathbf{1} + \mathbf{T}^F_{eff}\left(n, \mathbf{e}^S\right) - \delta \mathbf{w} \otimes \mathbf{w}, \\ \hat{\mathbf{p}}^S &= -\hat{\mathbf{p}}^F = \pi\left(n, \mathbf{e}^S\right)\mathbf{w} - \left(p + \Gamma \mathbf{w} \cdot \mathbf{w}\right) \operatorname{grad} n, \end{aligned} \tag{12.21}$$

where δ and Γ are additional material parameters.

Such a model seems to be appropriate to describe, for instance, instabilities leading to fluidization and eruption in water saturated sands by earthquakes.

The above described class of models shall not be discussed any further. In spite of their important role in some problems of soil mechanics these models have some faults which do not seem to be acceptable in cases of wave processes. Most important of them is the lack of hyperbolicity (the part of the operator connected with the constraint is elliptic). This leads to a reduced number of real eigenvalues corresponding to speeds of propagation and, consequently, to the lack of certain modes of propagation of weak discontinuity waves. In particular the P2-wave and some important surface waves cannot be described by such models.

Chapter 13

Thermodynamics of poroelastic materials with the balance equation of porosity

13.1 General structure

In this Chapter we present the model of porous materials developed in the recent 15 years for an elastic skeleton and ideal fluid components. Firstly, we present its nonlinear foundations for a system with A fluid components. Only a few basic thermodynamical features will be discussed in order to place the model within the rational extended thermodynamics. Thermodynamical details will be discussed in further Sections for two-component models.

Geometric nonlinearities connected with possibly large deformations of the skeleton indicate that the convenient way to describe processes is to define fields on a *reference configuration* $\mathcal{B}_0$ of the skeleton. For such a configuration, the deformation gradient of the skeleton is defined as the unit matrix: $\mathbf{F}^S = \mathbf{1}$. Hence we formulate the *Lagrangian description* of motion of the porous material. The possibility of this description has been indicated by Coussy [82] and it was systematically introduced to porous models by Wilmanski [361], [364], [363]. Let us mention in passing that there are papers on this subject in which the Lagrangian description is introduced separately for each component. It is not only an unnecessary formal complication but sometimes it yields erroneous results. We do not present this approach in this book.

The aim of the model is to find the following fields defined in points $\mathbf{X}$ of the reference domain $\mathcal{B}_0 \subset \Re^3$ and in instances t of the time interval $\mathcal{T} \subset [0, \infty)$:

(1) mass density of the skeleton referred to a unit reference volume: $\rho^S(\mathbf{X}, t)$,
(2) mass densities of the fluid components referred to a unit reference volume: $\rho^\alpha(\mathbf{X}, t)$, $\alpha = 1, \ldots, A$,
(3) velocity of the skeleton: $\acute{\mathbf{x}}^S(\mathbf{X}, t)$,

(4) deformation gradient of the skeleton: $\mathbf{F}^S(\mathbf{X},t)$, $J^S := \det \mathbf{F}^S > 0$,
(5) velocities of fluid components $\acute{\mathbf{x}}^\alpha(\mathbf{X},t)$, $\alpha = 1,\ldots,A$,
(6) porosity: $n(\mathbf{X},t)$,
(7) temperature common for all components $T(\mathbf{X},t)$.

Consequently a *thermomechanical process* is described by the mapping

$$\mathbf{w} : (\mathbf{X},t) \mapsto \Re^{4A+15}, \quad \mathbf{w} := \left\{\rho^S, \rho^\alpha, \acute{\mathbf{x}}^S, \mathbf{F}^S, \acute{\mathbf{x}}^\alpha, n, T\right\}, \quad \alpha = 1,\ldots,A. \tag{13.1}$$

Field equations for these fields follow from balance equations which we proceed to formulate.

Balance equations are specified in their global form on material domains of components. For porous materials in the Lagrangian description the family of material domains for the skeleton is defined as a class of subsets of $\mathcal{B}_0$ which is time independent and satisfies conditions identical with those of the classical continuum mechanics. Material domains of the skeleton are time independent because the reference configuration $\mathcal{B}_0$ is defined with respect to the deformation gradient of skeleton $\mathbf{F}^S$. For this configuration $\mathbf{F}^S = \mathbf{1}$.

It is not the case any more for fluid components. They have different kinematics than the skeleton. This means that domains, material with respect to a particular fluid component, move in the space of motion with respect to material domains of the skeleton. In the Eulerian description, the velocity field for this motion is given by the difference $\mathbf{v}^\alpha(\mathbf{x},t) - \mathbf{v}^S(\mathbf{x},t)$ for the fluid component α, where $\mathbf{v}^\alpha(\mathbf{x},t)$ is the velocity of the fluid and $\mathbf{v}^S(\mathbf{x},t)$ is the velocity of the skeleton at the same spatial position $\mathbf{x}$ and at the same instant of time t. This relative motion yields the time dependence of material domains of fluid components projected on the reference configuration $\mathcal{B}_0$. The projection is carried by the function of motion of the skeleton

$$\forall \mathbf{X} \in \mathcal{B}_0, t \in \mathcal{T}: \quad \mathbf{x} = \mathbf{f}^S(\mathbf{X},t)$$

$$\Longrightarrow \mathbf{F}^S = \operatorname{Grad} \mathbf{f}^S(\mathbf{X},t), \quad \acute{\mathbf{x}}^S = \frac{\partial \mathbf{f}^S(\mathbf{X},t)}{\partial t}, \tag{13.2}$$

whose existence is assumed in the model. The condition for the existence of the function of motion $\mathbf{f}^S$ shall be formulated later.

In order to describe the kinematics of material fluid domains projected on the reference configuration we consider the mappings shown in Fig. 18. In the current configuration $\mathcal{B}_t = \mathbf{f}^S(\mathcal{B}_0,t)$ we consider an arbitrary subset $\mathcal{P}_t^\alpha \subset \mathcal{B}_t$ which is material with respect to the component

α, i.e. by an infinitesimal increment of time Δt it becomes $\mathcal{P}^{\alpha}_{t+\Delta t} := \{\mathbf{x} \in \mathcal{B}_{t+\Delta t} | \mathbf{x} - \mathbf{v}^{\alpha}(\mathbf{x}, t + \Delta t)\Delta t \in \mathcal{P}^{\alpha}_{t}\}$. These two subsets of the space of current configurations are maps of the subsets $\mathcal{P}^{\alpha}(t), \mathcal{P}^{\alpha}(t + \Delta t)$ of the reference configuration $\mathcal{B}_0$, one at the instant of time t and the other at the instant of time $t + \Delta t$. Obviously, they contain the same particles of the component α. This construction is shown in Fig. 18. Now, we choose an arbitrary particle of the component α which occupies the position $\mathbf{x} \in \mathcal{P}^{\alpha}_{t}$ at the instant of time t, i.e. its position in the reference configuration is $\mathbf{X} = \mathbf{f}^{S-1}(\mathbf{x}, t)$. Its position $\mathbf{x} + \mathbf{\Delta x}$ at the instant $t + \Delta t$ can be written in the form

$$\begin{aligned} \mathbf{x} + \mathbf{\Delta x} &= \mathbf{x} + \acute{\mathbf{x}}^{\alpha}(\mathbf{X}, t)\Delta t = \mathbf{f}^{S}(\mathbf{X} + \Delta\mathbf{X}, t + \Delta t) \\ &= \mathbf{x} + \operatorname{Grad}\mathbf{f}^{S}(\mathbf{X}, t)\Delta\mathbf{X} + \frac{\partial \mathbf{f}^{S}}{\partial t}\Delta t \\ &= \mathbf{x} + \mathbf{F}^{S}\Delta\mathbf{X} + \acute{\mathbf{x}}^{S}\Delta t. \end{aligned} \tag{13.3}$$

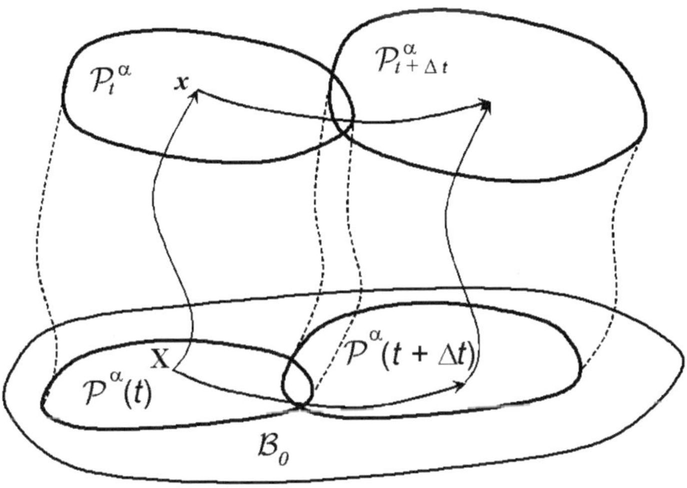

Fig. 18: *Projections of material domains of the α-component (pull-back).*

Consequently, the set $\mathcal{P}^{\alpha}(t) \subset \mathcal{B}_0$ is endowed with the kinematics defined by the following velocity field

$$\forall \mathbf{X} \in \mathcal{B}_0 : \quad \lim_{\Delta t \to 0} \frac{\Delta\mathbf{X}}{\Delta t} =: \acute{\mathbf{X}}^{\alpha}(\mathbf{X}, t) = \mathbf{F}^{S-1}\left(\acute{\mathbf{x}}^{\alpha} - \acute{\mathbf{x}}^{S}\right). \tag{13.4}$$

Obviously, in the above relations we used the following transformations

$$\acute{\mathbf{x}}^{S}(\mathbf{X}, t) = \mathbf{v}^{S}\left(\mathbf{f}^{S}(\mathbf{X}, t), t\right), \quad \acute{\mathbf{x}}^{\alpha}(\mathbf{X}, t) = \mathbf{v}^{\alpha}\left(\mathbf{f}^{S}(\mathbf{X}, t), t\right). \tag{13.5}$$

The fields $\acute{\mathbf{X}}^\alpha$ are called *Lagrangian velocity fields of fluid components.*

In order to appreciate the operational meaning of this transformation, we formulate balance equations of mass for all components. They have the following form

$$\frac{d}{dt}\int_{\mathcal{P}^S} \rho^S dV = \int_{\mathcal{P}^S} \hat{\rho}^S dV, \tag{13.6}$$

for every material domain of the skeleton $\mathcal{P}^S \subset \mathcal{B}_0$, and

$$\frac{d}{dt}\int_{\mathcal{P}^\alpha(t)} \rho^\alpha dV = \int_{\mathcal{P}^\alpha(t)} \hat{\rho}^\alpha dV, \tag{13.7}$$

for every material domain of the α fluid component $\mathcal{P}^\alpha(t) \subset \mathcal{B}_0$, $\alpha = 1, \ldots, A$. In the above relations $\hat{\rho}^S, \hat{\rho}^\alpha$ are the mass sources which satisfy the following bulk conservation law

$$\forall \mathbf{X} \in \mathcal{B}_0, t \in \mathcal{T}: \quad \hat{\rho}^S + \sum_{\alpha=1}^{A} \hat{\rho}^\alpha = 0. \tag{13.8}$$

Time dependence of material domains for fluid components yields the following rules of time differentiation

$$\frac{d}{dt}\int_{\mathcal{P}^S} \rho^S dV = \int_{\mathcal{P}^S} \frac{\partial}{\partial t}\rho^S dV,$$

$$\frac{d}{dt}\int_{\mathcal{P}^\alpha(t)} \rho^\alpha dV = \int_{\mathcal{P}^\alpha(t)} \frac{\partial}{\partial t}\rho^\alpha dV + \oint_{\partial\mathcal{P}^\alpha(t)} \rho^\alpha \mathbf{N}\cdot\acute{\mathbf{X}}^\alpha dS, \tag{13.9}$$

where $\mathbf{N}$ denotes the unit normal vector field of the boundary $\partial\mathcal{P}^\alpha(t)$.

These relations yield the following local form of mass balance equations for $\alpha = 1, \ldots, A$,

$$\frac{\partial \rho^S}{\partial t} = \hat{\rho}^S, \quad \frac{\partial \rho^\alpha}{\partial t} + \operatorname{Div}\left(\rho^\alpha \acute{\mathbf{X}}^\alpha\right) = \hat{\rho}^\alpha, \tag{13.10}$$

in regular points (almost everywhere) of $\mathcal{B}_0$, and

$$U\,[[\rho^S]] = 0, \quad \left[\left[\rho^\alpha\left(\acute{\mathbf{X}}^\alpha\cdot\mathbf{N} - U\right)\right]\right] = 0, \tag{13.11}$$

in points of singular surfaces moving through the reference configuration $\mathcal{B}_0$ with the local speed U. The brackets $[[\cdots]]$ denote the difference of finite limits of quantities in these brackets on the positive and negative sides of the surface (comp. (4.16)).

In a similar manner we obtain the following partial momentum balance equations in their local form. For the regular points of the reference configuration $\mathcal{B}_0$

$$\frac{\partial\left(\rho^S\acute{\mathbf{x}}^S\right)}{\partial t}-\operatorname{Div}\mathbf{P}^S=\hat{\mathbf{p}}^S+\rho^S\mathbf{b}^S,$$

$$\frac{\partial\left(\rho^\alpha\acute{\mathbf{x}}^\alpha\right)}{\partial t}+\operatorname{Div}\left(\rho^\alpha\acute{\mathbf{x}}^\alpha\otimes\acute{\mathbf{X}}^\alpha-\mathbf{P}^\alpha\right)=\hat{\mathbf{p}}^\alpha+\rho^\alpha\mathbf{b}^\alpha,\quad \hat{\mathbf{p}}^S+\sum_{\alpha=1}^{A}\hat{\mathbf{p}}^\alpha=0, \tag{13.12}$$

and for points on singular surfaces

$$\rho^S U\left[\left[\acute{\mathbf{x}}^S\right]\right]+\left[\left[\mathbf{P}^S\right]\right]\mathbf{N}=0,$$

$$\rho^\alpha\left(\acute{\mathbf{X}}^\alpha\cdot\mathbf{N}-U\right)\left[\left[\acute{\mathbf{x}}^\alpha\right]\right]-\left[\left[\mathbf{P}^\alpha\right]\right]\mathbf{N}=0. \tag{13.13}$$

In these relations $\mathbf{P}^S,\mathbf{P}^\alpha$ denote partial Piola-Kirchhoff stress tensors, $\mathbf{b}^S,\mathbf{b}^\alpha$ are partial mass forces, and $\hat{\mathbf{p}}^S,\hat{\mathbf{p}}^\alpha$ denote the momentum sources. Relation $(13.12)_3$ expresses the bulk conservation of momentum.

We do not need to present details of partial energy balance equations. Under the assumption of a single field of temperature, we need solely the bulk energy conservation law. The derivation is based on a principle of the theory of mixtures that bulk quantities must be defined in such a way that balance equations for these quantities have the form of classical conservation laws of the single component continuum thermodynamics.

Bearing this principle in mind we define the following *bulk quantities*

$$\rho:=\rho^S+\sum_{\alpha=1}^{A}\rho^\alpha,\quad \rho\dot{\mathbf{x}}:=\rho^S\acute{\mathbf{x}}^S+\sum_{\alpha=1}^{A}\rho^\alpha\acute{\mathbf{x}}^\alpha,\quad \rho\dot{\mathbf{X}}:=\sum_{\alpha=1}^{A}\rho^\alpha\acute{\mathbf{X}}^\alpha, \tag{13.14}$$

which are the bulk mass density, the bulk momentum, and an objective relative momentum connected with the reference of the motion to the skeleton rather than to local centers of gravity (comp. (11.9) for mixtures of fluids in Eulerian description);

$$\mathbf{P}:=\mathbf{P}_I-\mathbf{F}^S\left\{\rho^S\dot{\mathbf{X}}\otimes\dot{\mathbf{X}}+\sum_{\alpha=1}^{A}\rho^\alpha\left(\acute{\mathbf{X}}^\alpha-\dot{\mathbf{X}}\right)\otimes\left(\acute{\mathbf{X}}^\alpha-\dot{\mathbf{X}}\right)\right\},$$

$$\mathbf{P}_I:=\mathbf{P}^S+\sum_{\alpha=1}^{A}\mathbf{P}^\alpha; \tag{13.15}$$

this is the bulk Piola-Kirchhoff stress tensor with the so-called intrinsic part $\mathbf{P}_I$;

$$\rho\varepsilon := \rho\varepsilon_I + \frac{1}{2}\left\{\rho^S \mathbf{C}^S \cdot \left(\dot{\mathbf{X}} \otimes \dot{\mathbf{X}}\right) + \sum_{\alpha=1}^{A} \rho^\alpha \mathbf{C}^S \cdot \left(\acute{\mathbf{X}}^\alpha - \dot{\mathbf{X}}\right) \otimes \left(\acute{\mathbf{X}}^\alpha - \dot{\mathbf{X}}\right)\right\},$$

$$\rho\varepsilon_I := \rho^S\varepsilon^S + \sum_{\alpha=1}^{A} \rho^\alpha \varepsilon^\alpha, \qquad \mathbf{C}^S := \mathbf{F}^{ST}\mathbf{F}^S, \tag{13.16}$$

this is the bulk specific internal energy with the intrinsic part $\rho\varepsilon_I$. The symmetric tensor $\mathbf{C}^S$ is the *right Cauchy-Green deformation tensor* of the skeleton;

$$\mathbf{Q} := \mathbf{Q}_I + \frac{1}{2}\left\{-\rho^S \dot{\mathbf{X}} \otimes \dot{\mathbf{X}} \otimes \dot{\mathbf{X}} + \sum_{\alpha=1}^{A} \rho^\alpha \left(\acute{\mathbf{X}}^\alpha - \dot{\mathbf{X}}\right) \otimes \left(\acute{\mathbf{X}}^\alpha - \dot{\mathbf{X}}\right) \otimes \left(\acute{\mathbf{X}}^\alpha - \dot{\mathbf{X}}\right)\right\} \mathbf{C}^S,$$

$$\mathbf{Q}_I := \mathbf{Q}^S + \sum_{\alpha=1}^{A} \mathbf{Q}^\alpha - \rho^S\varepsilon^S \dot{\mathbf{X}} + \sum_{\alpha=1}^{A} \rho^\alpha \varepsilon^\alpha \left(\acute{\mathbf{X}}^\alpha - \dot{\mathbf{X}}\right) + \mathbf{P}^{ST}\mathbf{F}^S\dot{\mathbf{X}} - \sum_{\alpha=1}^{A} \mathbf{P}^{\alpha T}\mathbf{F}^S \left(\acute{\mathbf{X}}^\alpha - \dot{\mathbf{X}}\right), \tag{13.17}$$

and this vector describes the bulk heat flux in the Lagrangian description. Again the intrinsic part $\mathbf{Q}_I$ was separated.

The bulk balance equation of energy can now be written in the following form

$$\frac{\partial}{\partial t}\rho\left(\varepsilon + \frac{1}{2}\dot{x}^2\right) + \mathrm{Div}\left\{\rho\left(\varepsilon + \frac{1}{2}\dot{x}^2\right)\dot{\mathbf{X}} + \mathbf{Q} - \mathbf{P}^T\dot{\mathbf{x}}\right\} = \rho\mathbf{b}\cdot\dot{\mathbf{x}} + \rho r, \tag{13.18}$$

where

$$\rho\mathbf{b} := \rho^S\mathbf{b}^S + \sum_{\alpha=1}^{A} \rho^\alpha \mathbf{b}^\alpha,$$

$$\rho r := \rho^S r^S + \sum_{\alpha=1}^{A} \rho^\alpha r^\alpha - \rho^S\mathbf{b}^S \cdot \mathbf{F}^S\dot{\mathbf{X}} + \sum_{\alpha=1}^{A} \rho^\alpha \mathbf{b}^\alpha \cdot \mathbf{F}^S \left(\acute{\mathbf{X}}^\alpha - \dot{\mathbf{X}}\right), \tag{13.19}$$

and r^S, r^α denote the partial radiations.

We skip the presentation of the energy condition on a singular surface because it shall not be used in this book.

In the Lagrangian description and with the choice of fields (13.1) we have at our disposal the following integrability condition

$$\frac{\partial \mathbf{F}^S}{\partial t} = \operatorname{Grad} \acute{\mathbf{x}}^S. \tag{13.20}$$

This condition yields the existence of the function of motion (13.2). By the choice (13.1) of unknown fields this relation plays the role of the field equation for the deformation gradient $\mathbf{F}^S$.

It is useful to write equation (13.20) in the following weaker form

$$\frac{d}{dt} \int_{\mathcal{P}^S} \mathbf{F}^S dV = \oint_{\partial \mathcal{P}^S} \acute{\mathbf{x}}^S \otimes \mathbf{N} dS, \tag{13.21}$$

for every material domain of the skeleton $\mathcal{P}^S \subset \mathcal{B}_0$. This balance equation yields the following condition in points of singular surfaces

$$U \left[\left[\mathbf{F}^S\right]\right] = - \left[\left[\acute{\mathbf{x}}^S\right]\right] \otimes \mathbf{N}. \tag{13.22}$$

This relation is usually derived by means of the Hadamard Theorem for singular surfaces.

Before we present remaining equations of the model let us discuss some properties of the objects which we have introduce above. It is easy to notice a striking similarity of the structure of bulk quantities to that appearing in the classical theory of mixtures. This concerns terms with explicit contributions of velocities. However in contrast to the mixture theory all velocities of the present model are objective because $\acute{\mathbf{X}}^\alpha$ and $\dot{\mathbf{X}}$ are *relative velocities*. Due to constitutive relations these velocities may also be present in the implicit form in intrinsic parts of stress tensors, internal energy and heat flux vector. It is also important to notice that the explicit dependence is at least quadratic. If we consider processes with small deviations from the thermodynamical equilibrium, these contributions can be neglected.

In order to turn over mass balance equations (13.10), momentum balance equations (13.12), energy balance equation (13.18) and compatibility condition (13.20) into field equations for fields (13.1) we need constitutive relations for partial stress tensors, momentum sources, the bulk internal energy and the bulk heat flux. If we had these relations we would have $14\,(A+1)$ equations. Consequently we would be missing one equation. This is connected with the fact that the porosity n is the additional microstructural variable and this requires an additional equation. We proceed to formulate this equation.

We have seen in the previous Chapter that changes of porosity may be described by a balance equation (12.15). This was the consequence of incompressibility of components. If the components are compressible, this equation is not at the disposal for porosity. Compressibility of components is an important feature in the wave analysis and many other problems of practical bearing and linear models (e.g. Biot's model which we present further in this Chapter) yield in such cases a relation for porosity which, in general, does not coincide with this derived for incompressible materials. In addition, the porosity equation following from the mass conservation law does not contain a source. Such a source would describe a spontaneous relaxation of porosity as well as production of porosity by, for instance, damage. We know from experience with other microstructural variables that this is an important property yielding evolution equations for such variables.

All these arguments can be made more precise if we motivate an equation for porosity by a transition from a semimacroscopic model. We do so by means of averaging over the Representative Elementary Volume.

The porosity is the fraction of void spaces in the skeleton. Consequently, if $\chi^S(\mathbf{Z},t)$ denotes the characteristic function of the solid component on the semimacroscopic level, then we have

$$n(\mathbf{X},t) = \frac{1}{V}\int_{\mathcal{M}(\mathbf{X})} \left(1-\chi^S(\mathbf{Z},t)\right) dV_Z, \tag{13.23}$$

where the averaging is performed over the reference configuration of the skeleton, i.e. $\mathbf{X} \in \mathcal{B}_0$ and it means that the points $\mathbf{Z}$ of the real solid move within $\mathcal{M}(\mathbf{X})$ according to semimicroscopic equations of motion. For the purpose of motivation of macroscopic equations it is not necessary to formulate them.

We proceed to investigate time changes of porosity [362], [364]. We have

$$\frac{\partial n}{\partial t} = \frac{1}{V}\int_{\mathcal{M}(\mathbf{X})} \left(-\frac{\partial \chi^S}{\partial t} - \operatorname{Grad}_Z \chi^S \cdot \dot{\mathbf{Z}}\right) dV_Z, \tag{13.24}$$

where Grad_Z denotes the gradient with respect to the variable $\mathbf{Z}$, while $\dot{\mathbf{Z}}$ is the velocity of the real point of the solid. According to (12.7), we can transform this relation in the following way

$$\begin{aligned}
\frac{\partial n}{\partial t} &= \frac{1}{V}\int_{\mathcal{M}}\left(-\frac{\partial \chi^S}{\partial t} + \delta_{\partial\mathcal{M}^\alpha(\mathbf{X},t)}(\xi)\,\mathbf{N}_Z\cdot\dot{\mathbf{Z}}\right)dV_Z \\
&= -\frac{1}{V}\int_{\mathcal{M}}\frac{\partial \chi^S}{\partial t}dV_Z - \frac{1}{V}\int_{\partial\mathcal{M}^S\cap\mathcal{M}}\mathbf{N}_Z\cdot\dot{\mathbf{Z}}dS_Z \\
&= -\frac{1}{V}\int_{\mathcal{M}(\mathbf{X})}\frac{\partial \chi^S}{\partial t}dV_Z - \frac{1}{V}\oint_{\partial\mathcal{M}^S\cap\mathcal{M}\cup(\partial\mathcal{M}\cap\mathcal{M}^S)}\mathbf{N}_Z\cdot\dot{\mathbf{Z}}dS_Z \\
&\quad + \frac{1}{V}\int_{\partial\mathcal{M}\cap\mathcal{M}^S}\mathbf{N}_Z\cdot\dot{\mathbf{Z}}dS_Z \\
&= -\frac{1}{V}\int_{\mathcal{M}(\mathbf{X})}\frac{\partial \chi^S}{\partial t}dV_Z - \frac{1}{V}\int_{\mathcal{M}^S}\operatorname{Div}_Z\dot{\mathbf{Z}}dV_Z + \frac{1}{V}\int_{\partial\mathcal{M}\cap\mathcal{M}^S}\mathbf{N}_Z\cdot\dot{\mathbf{Z}}dS_Z.
\end{aligned} \tag{13.25}$$

These three contributions have the following macroscopic interpretation. The first one describes the *source* of porosity caused, for instance, by microstructural relaxation processes or cracking. We denote this contribution by $\hat{n}$. The second contribution is related to the relative motion of real components on the semimacroscopic level. Bearing the relation (12.8) in mind, we can identify this term with the contribution of the macroscopic *flux of porosity* $\mathbf{J}$, i.e. we can write approximately

$$\frac{1}{V}\int_{\mathcal{M}^S}\operatorname{Div}_Z\dot{\mathbf{Z}}dV_Z = \operatorname{Div}\mathbf{J}. \tag{13.26}$$

On the macroscopic level this contribution results from the relative motion of components. Finally, the last contribution is related to internal deformations of the skeleton. As we consider porous materials whose skeleton is elastic this contribution, in contrast to the other two contributions, must appear in equilibrium as well as nonequilibrium processes. We call it *equilibrium changes of the porosity* and denote by $\partial n_E/\partial t$, i.e.

$$\frac{1}{V}\int_{\partial\mathcal{M}\cap\mathcal{M}^S}\mathbf{N}_Z\cdot\dot{\mathbf{Z}}dS_Z = \frac{\partial n_E}{\partial t}. \tag{13.27}$$

Bearing the above remarks in mind, we introduce in regular points the *balance equation of porosity* in the following general form

$$\frac{\partial \Delta_n}{\partial t} + \operatorname{Div}\mathbf{J} = \hat{n}, \quad \Delta_n = n - n_E, \tag{13.28}$$

where Δ_n is the deviation of porosity from the equilibrium value n_E, the latter together with the flux of porosity $\mathbf{J}$ and the source of porosity $\hat{n}$

must be given by constitutive relations. We expect that n tends to an equilibrium under constant external conditions. The equilibrium value of porosity n_E satisfies the equation (13.28) with the flux and source equal to zero. The latter as we show later follow indeed from the second law of thermodynamics.

Making an assumption that sources of porosity do not carry surface singularities we can write the following compatibility condition for porosity on such surfaces

$$U\left[\left[\Delta_n\right]\right]-\left[\left[\mathbf{J}\right]\right]\cdot\mathbf{N}=0, \tag{13.29}$$

which may suggest the form of natural boundary conditions for porosity.

It is clear that the averaging procedure does not specify the balance equation of porosity but it motivates its structure. This is the typical situation for models which are not based on kinetic microscopical models.

Let us collect balance equations which we have discussed in this Section. They are shown in Tables 11 and 12.

As already indicated we do not quote here the dynamic compatibility relation for the bulk energy.

Table 11: *Balance equations for the* $A+1$-component *porous material in regular points of the reference configuration* $\mathbf{X}\in\mathcal{B}_0$.

mass of S	$\frac{\partial\rho^S}{\partial t}=\hat{\rho}^S$
mass of α	$\frac{\partial\rho^\alpha}{\partial t}+\mathrm{Div}\,\rho^\alpha\acute{\mathbf{X}}^\alpha=\hat{\rho}^\alpha$
momentum of S	$\frac{\partial\left(\rho^S\acute{\mathbf{x}}^S\right)}{\partial t}-\mathrm{Div}\,\mathbf{P}^S=\hat{\mathbf{p}}^S+\rho^S\mathbf{b}^S$
momentum of α	$\frac{\partial(\rho^\alpha\acute{\mathbf{x}}^\alpha)}{\partial t}+\mathrm{Div}\left(\rho^\alpha\acute{\mathbf{x}}^\alpha\otimes\acute{\mathbf{X}}^\alpha-\mathbf{P}^\alpha\right)$ $=\hat{\mathbf{p}}^\alpha+\rho^\alpha\mathbf{b}^\alpha$
bulk energy	$\frac{\partial}{\partial t}\rho\left(\varepsilon+\frac{1}{2}\dot{x}^2\right)+Div\left\{\rho\left(\varepsilon+\frac{1}{2}\dot{x}^2\right)\dot{\mathbf{X}}\right.$ $\left.+\mathbf{Q}-\mathbf{P}^T\dot{\mathbf{x}}\right\}=\rho\mathbf{b}\cdot\dot{\mathbf{x}}+\rho r$
integrability of $\mathbf{F}^S$	$\frac{\partial\mathbf{F}^S}{\partial t}=\mathrm{Grad}\,\acute{\mathbf{x}}^S$
porosity	$\frac{\partial\Delta_n}{\partial t}+\mathrm{Div}\,\mathbf{J}=\hat{n}$

Table 12: *Balance equations (dynamic compatibility conditions) in points of the singular surface.*

mass of S	$U\,[[\rho^S]] = 0$
mass of α	$[[\rho^\alpha \left(\acute{\mathbf{X}}^\alpha \cdot \mathbf{N} - U\right)]] = 0$
momentum of S	$\rho^S U\,[[\acute{\mathbf{x}}^S]] + [[\mathbf{P}^S]]\,\mathbf{N} = 0,$
momentum of α	$\rho^\alpha \left(\acute{\mathbf{X}}^\alpha \cdot \mathbf{N} - U\right)[[\acute{\mathbf{x}}^\alpha]] - [[\mathbf{P}^\alpha]]\,\mathbf{N} = 0$
integrability	$U\,[[\mathbf{F}^S]] = -\,[[\acute{\mathbf{x}}^S]] \otimes \mathbf{N}$
porosity	$U\,[[\Delta_n]] - [[\mathbf{J}]] \cdot \mathbf{N} = 0$

In order to construct field equations for the fields $\mathbf{w}$ listed in relation (13.1) we have to solve the *closure problem*, i.e. we have to add to balance equations of Table 11 constitutive relations. We shall do so for some important particular cases. As the first closure we select the simplest possible one which yields a possibility of exploiting the second law of thermodynamics in a way typical for the rational extended thermodynamics. The other cases shall be investigated by means of the classical approach to the exploitation of the second law of thermodynamics.

In the case of the extended thermodynamics method we proceed as follows (see [366], [370]):

Let us define the following vectors

$$\mathsf{F}_0 := \left\{\rho^S, \rho^\alpha, \rho^S \acute{\mathbf{x}}^S, \rho^\alpha \acute{\mathbf{x}}^\alpha, \rho\left(\varepsilon + \frac{1}{2}\dot{x}^2\right), \mathbf{F}^S, \Delta_n\right\} \in \Re^{4A+15},$$

$$\mathsf{F}_K := \left\{0, \rho^\alpha \acute{\mathbf{X}}^\alpha \cdot \mathbf{G}_K, -\mathbf{P}^S \mathbf{G}_K, \left(\rho^\alpha \acute{\mathbf{x}}^\alpha \otimes \acute{\mathbf{X}}^\alpha - \mathbf{P}^\alpha\right) \mathbf{G}_K,\right.$$

$$\left.\left(\rho\left(\varepsilon + \tfrac{1}{2}\dot{x}^2\right)\dot{\mathbf{X}} + \mathbf{Q} - \mathbf{P}^T \dot{\mathbf{x}}\right) \cdot \mathbf{G}_K, -\acute{\mathbf{x}}^S \otimes \mathbf{G}_K, \mathbf{J} \cdot \mathbf{G}_K\right\} \in \Re^{4A+15}, \quad K = 1, 2, 3,$$

$$\mathsf{f} := \left\{\hat{\rho}^S, \hat{\rho}^\alpha, \hat{\mathbf{p}}^S, \hat{\mathbf{p}}^\alpha, 0, \mathbf{0}, \hat{n}\right\} \in \Re^{4A+15}, \tag{13.30}$$

$$\mathsf{f}_{ext} := \left\{0, 0, \rho^S \mathbf{b}^S, \rho^\alpha \mathbf{b}^\alpha, \rho \mathbf{b} \cdot \dot{\mathbf{x}} + \rho r, \mathbf{0}, 0\right\} \in \Re^{4A+15},$$

where $\mathbf{G}_K$ denote *unit basis vectors* of Lagrangian coordinates. Then the balance equations can be written in the following compact form

$$\frac{\partial \mathsf{F}_0}{\partial t} + \frac{\partial \mathsf{F}_K}{\partial X^K} = \mathsf{f} + \mathsf{f}_{ext}, \tag{13.31}$$

where $\left\{X^K\right\}_{K=1,2,3}$ denote Lagrangian coordinates. For convenience we have chosen again a Cartesian coordinate system.

In procedures of *extended thermodynamics,* as presented in Chapter 8, it is assumed that the vectors defined by (13.30) are sufficiently smooth functions of the vector $\mathbf{w}$ of unknown fields. Then the closure assumption has the form

$$\mathsf{F}_0 = \mathsf{F}_0\left(\mathbf{w}\right), \quad \mathsf{F}_K = \mathsf{F}_K\left(\mathbf{w}\right), \quad \mathsf{f} = \mathsf{f}\left(\mathbf{w}\right). \tag{13.32}$$

In the case under considerations we obtain the model of *poroelastic materials* without heat conduction. In order to incorporate the heat conduction we would have to introduce either an equation for the heat flux or a dependence on Grad T as a constitutive variable. Similarly we would have to extend the model if we wanted to describe viscous or plastic effects - we would have to add equations for partial stresses or a dependence on gradients of partial velocities. Further we present models in which some aspects of such extensions are indeed discussed within a classical approach to the problem of thermodynamical admissibility.

Let us stress that the structure of constitutive relations (13.32) is one of the most characteristic features of extended thermodynamics. Namely the constitutive variables are fields themselves but not their derivatives as it is always the case in ordinary nonequilibrium thermodynamics.

In order to expose the structure characteristic for extended thermodynamics we consider a simpler case of the closure in the form (13.32) under the assumption that processes are *isothermal.* This means that we leave out the temperature in the definition (13.1) of $\mathbf{w}$ and denote it then $\mathbf{w}_{is}$ as well as we ignore the energy balance equation.

The set of equations (13.32) for $\mathbf{w}_{is}$ without energy balance has thermodynamically the same structure as the corresponding set of extended thermodynamics. For this reason we can apply the same principles connected with the *thermodynamical admissibility.* They can be formulated as follows:

- *entropy inequality*: there exist a nontrivial entropy function h_0 and a flux $\mathbf{H} = H_K \mathbf{G}_K$ such that for each thermodynamical process (i.e. for each solution of field equations) the following inequality is satisfied

$$\frac{\partial h_0}{\partial t} + \operatorname{Div}\mathbf{H} \geq 0, \quad h_0 = h_0(\mathbf{w}_{is}) \in \Re, \quad \mathbf{H} = \mathbf{H}(\mathbf{w}_{is}) \in \Re^3, \tag{13.33}$$

- *convexity and causality*: the entropy function $h_0 = h_0(\mathbf{u}_{is})$ is concave, i.e.

$$\forall \mathbf{v} \in \Re^{14(A+1)}, \mathbf{v} \neq \mathbf{0} : \quad \frac{\partial^2 h_0}{\partial \mathbf{w}_{is} \partial \mathbf{w}_{is}} \cdot (\mathbf{v} \otimes \mathbf{v}) < 0, \tag{13.34}$$

- *principle of relativity (Galilean invariance of field equations).*

The last principle yields a decomposition of all quantities of the model into two parts: a convective part which depends explicitly on the absolute velocity fields and a nonconvective part which does not depend on absolute velocities at all. This principle is satisfied identically in the case of Lagrangian description because we deal solely with relative velocities.

Entropy inequality (13.33) is exploited by means of Lagrange multipliers which eliminate the limitation of this inequality to thermodynamical processes. According to this procedure requirement (13.33) is equivalent to the following inequality for *all fields*, and not only for solutions of field equations

$$\forall \mathbf{w}_{is} \in \Re^{14(A+1)} :$$

$$\frac{\partial h_0}{\partial t} + \frac{\partial H_K}{\partial X^K} - \mathbf{\Lambda} \cdot \left(\frac{\partial \tilde{\mathsf{F}}_0}{\partial t} + \frac{\partial \tilde{\mathsf{F}}_K}{\partial X^K} - \tilde{\mathsf{f}} \right) \geq 0, \quad \mathbf{\Lambda} \in \Re^{14(A+1)}, \tag{13.35}$$

$$\mathbf{II} = H_K \mathbf{G}_K,$$

where $\mathbf{\Lambda}$ are the Lagrange multipliers, and functions of $\mathbf{w}_{is}$. As mentioned above $\tilde{\mathsf{F}}_0, \tilde{\mathsf{F}}_K, \tilde{\mathsf{f}}$ are truncations of functions (13.30) to the subspace $\Re^{14(A+1)}$ without the energy balance equation.

The solution of the above problem has the following form

$$dh_0 = \mathbf{\Lambda} \cdot d\tilde{\mathsf{F}}_0, \quad dH_K = d\tilde{\mathsf{F}}_K \cdot \mathbf{\Lambda} \quad \Longrightarrow \quad \frac{\partial \mathbf{\Lambda}}{\partial \mathbf{w}_{is}} = \frac{\partial^2 h_0}{\partial \mathbf{w}_{is} \partial \mathbf{w}_{is}}, \tag{13.36}$$

i.e. according to (13.34), the map $\mathbf{w}_{is} \to \mathbf{\Lambda}$ is globally invertible. Hence after Legendre transformation

$$h'_0(\mathbf{\Lambda}) = \mathbf{\Lambda} \cdot \tilde{\mathsf{F}}_0 - h_0, \quad H'_K(\mathbf{\Lambda}) = \mathbf{\Lambda} \cdot \tilde{\mathsf{F}}_K - H_K$$

$$\Longrightarrow \quad \tilde{\mathsf{F}}_0 = \frac{\partial h'_0}{\partial \mathbf{\Lambda}}, \quad \tilde{\mathsf{F}}_K = \frac{\partial H_K}{\partial \mathbf{\Lambda}}, \quad \mathbf{\Lambda} \cdot \tilde{\mathsf{f}}(\mathbf{\Lambda}) \geq 0. \tag{13.37}$$

Consequently, the functions $\tilde{\mathsf{F}}_0$, and $\tilde{\mathsf{F}}_K$ which determine the left-hand side of the field equations are given if the *four-potential* $(h'_0(\mathbf{\Lambda}), H'_K(\mathbf{\Lambda}))$ is known. It leaves unspecified but restricted by the *dissipation inequality* $(13.37)_5$ only the sources $\tilde{\mathsf{f}}(\mathbf{\Lambda})$ of the field equations. This is one of the most important consequences of the second law within the rational extended thermodynamics. Moreover, relations (13.37) yield the symmetry of field equations for the unknown vector $\mathbf{\Lambda}$:

$$\frac{\partial^2 h_0}{\partial \mathbf{\Lambda} \partial \mathbf{\Lambda}} \frac{\partial \mathbf{\Lambda}}{\partial t} + \frac{\partial^2 H'_K}{\partial \mathbf{\Lambda} \partial \mathbf{\Lambda}} \frac{\partial \mathbf{\Lambda}}{\partial X^K} = \tilde{\mathsf{f}}. \tag{13.38}$$

Consequently under the second part of the condition of thermodynamical admissibility (i.e. nonsingularity of matrix $\frac{\partial^2 h_0}{\partial \mathbf{\Lambda} \partial \mathbf{\Lambda}}$) the system is *symmetric hyperbolic.*

It remains to invert the variables, i.e. to find the map $\mathbf{\Lambda} \rightarrow \mathbf{w}_{is}$. This is usually a very difficult technical problem. For this reason we use further a classical approach which does not require the execution of the last step.

Let us mention that the residual inequality $(13.37)_5$ defines the *dissipation.* This function vanishes in states called *the thermodynamical equilibrium states.* Consequently the necessary and sufficient conditions for the thermodynamical equilibrium within the model discussed in this Section have the form

$$\begin{aligned} \hat{\rho}^\alpha\big|_E &= 0 \quad \text{for } \alpha = 1, \ldots, A, \\ \hat{\mathbf{p}}^\alpha\big|_E &= 0 \quad \text{for } \alpha = 1, \ldots, A, \\ \hat{n}\big|_E &= 0. \end{aligned} \tag{13.39}$$

They follow directly from definition $(13.30)_3$ of the vector f truncated to $\tilde{\mathsf{f}}$.

In the following Sections, we discuss some particular models following from the above thermodynamical scheme.

13.2 Two-component poroelastic materials: dependence on objective accelerations and porosity gradient

In order to see the structure of field equations for nonlinear nonisothermal processes in porous materials, we consider a particular case of the two-component system which describes thermomechanical processes in *saturated poroelastic materials.* This thermodynamical construction is connected with the classical Biot model of poroelastic materials [36], [37] (collection of papers of M. A. Biot on porous materials was published in [325]). Biot's model and some of its linear modifications shall be presented in the next Section. However, we have to mention two characteristic features of Biot's model motivating the construction which we present further. Within a purely isothermal framework, Biot proposed a two-component linear model in which partial momentum balance equations are coupled by three terms. The first one, diffusive forces, following from the relative motion of components, is classical. The second one follows from the assumption that the momentum source contains a contribution of the relative acceleration of components. This yields the consequence that the matrix of partial mass densities is not diagonal. The off-diagonal part is assumed to be symmetric. The third one appears in partial stresses and describes the reaction of one component on volume changes of the other. This coupling is also assumed to be symmetric.

Maurice Biot
1905 – 1985

Jacob Bear
1929–

Oliver Coussy
1953–

It is rather straightforward to show that the second contribution violates the principle of material objectivity [369], [371], [374] and the third contribution contradicts the second law of thermodynamics [375]. However, one can construct a nonlinear model complying with the principle of material objectivity and thermodynamical admissibility whose linear limit is identical with the Biot model. We present this construction in this Section. In order to simplify technicalities, we present this construction in two separate steps. First, we construct the model with a contribution of relative accelerations [374] and then a model with couplings of partial stresses [369].

A nonlinear objective model with a contribution of relative accelerations is thermodynamically admissible if we add some nonlinear contributions to partial stresses and to the free energy. They reflect in the simplest manner the existence of fluctuations of the microstructural kinetic energy caused by the heterogeneity of momentum in the representative elementary volume. The existence of such fluctuations as a result of tortuosity of porous materials has been indicated by O. Coussy in 1989 [83]. However, the constitutive part of a model based on such considerations has not been presented. There exist some attempts to derive Biot's model with the contribution of relative acceleration by means of Hamilton's principle based on the fluctuation kinetic energy. As the true variational principle does not hold for dissipative systems, the dissipation through fluctuation and diffusion is accounted for by a pseudo-potential and a pseudo-variational principle (e.g. [198]). This does not seem to be the right way of handling irreversible processes. For this reason we rely rather on the nonequilibrium thermodynamics in our considerations.

We consider a two-component continuum consisting of a solid skeleton and of a fluid. As in the previous Section, the motion of the skeleton is assumed to be described by the following twice continuously differentiable function

$$\mathbf{x} = \mathbf{f}^S(\mathbf{X}, t), \quad \mathbf{X} \in \mathcal{B}_0, \quad t \in \mathcal{T}, \tag{13.40}$$

where $\mathcal{B}_0$ denotes the reference configuration of the skeleton and $\mathcal{T}$ is the time interval. The velocity, the acceleration and the deformation gradient of the skeleton are defined by the relations

$$\acute{\mathbf{x}}^S := \frac{\partial \mathbf{f}^S}{\partial t}, \quad \acute{\acute{\mathbf{x}}}^S := \frac{\partial \acute{\mathbf{x}}^S}{\partial t}, \quad \mathbf{F}^S := \operatorname{Grad} \mathbf{f}^S. \tag{13.41}$$

Certainly, the value $\mathbf{F}^S = \mathbf{1}$ corresponds to the reference configuration for, say, $t = t_0$ in which $\mathbf{f}^S(\mathbf{X}, t_0) = \mathbf{X}$.

Motion of the fluid is described by the transformation of the Eulerian velocity field $\mathbf{v}^F = \mathbf{v}^F(\mathbf{x}, t)$ defined on the space of current configurations $\mathbf{f}^S(\mathcal{B}_0, t)$ of the skeleton. We have

$$\mathbf{v}^F = \mathbf{v}^F\left(\mathbf{f}^S(\mathbf{X}, t), t\right) =: \acute{\mathbf{x}}^F(\mathbf{X}, t). \tag{13.42}$$

The acceleration of the fluid is then given by

$$\acute{\acute{\mathbf{x}}}^F = \frac{\partial \acute{\mathbf{x}}^F}{\partial t} + \acute{\mathbf{X}}^F \cdot \operatorname{Grad} \acute{\mathbf{x}}^F, \quad \acute{\mathbf{X}}^F := \mathbf{F}^{S-1}\left(\acute{\mathbf{x}}^F - \acute{\mathbf{x}}^S\right), \tag{13.43}$$

where $\acute{\mathbf{X}}^F$ is the Lagrangian velocity of the fluid with respect to the skeleton.

We proceed to determine the transformation rules for the above quantities specified by the Euclidean transformation

$$\mathbf{x}^* = \mathbf{O}(t)\,\mathbf{x} + \mathbf{c}(t). \tag{13.44}$$

The relations (13.41) and the time differentiation of the relation (13.44) yield the following quantities in the new reference system

$$\mathbf{F}^{S*} = \mathbf{O}\mathbf{F}^{S}, \quad \acute{\mathbf{x}}^{S*} = \mathbf{O}\acute{\mathbf{x}}^{S} + \dot{\mathbf{O}}\mathbf{x} + \dot{\mathbf{c}}, \quad \acute{\acute{\mathrm{x}}}^{S*} = \mathbf{O}\acute{\acute{\mathrm{x}}}^{S} + 2\dot{\mathbf{O}}\acute{\mathbf{x}}^{S} + \ddot{\mathbf{O}}\mathbf{x} + \ddot{\mathbf{c}}, \tag{13.45}$$

where the dot denotes the time derivative.

We assume that the transformation rule for the velocity field of the fluid component has the same form as it does for the skeleton

$$\acute{\mathbf{x}}^{F*} = \mathbf{O}\acute{\mathbf{x}}^{F} + \dot{\mathbf{O}}\mathbf{x} + \dot{\mathbf{c}}. \tag{13.46}$$

Consequently

$$\acute{\mathbf{x}}^{F*} - \acute{\mathbf{x}}^{S*} = \mathbf{O}\left(\acute{\mathbf{x}}^{F} - \acute{\mathbf{x}}^{S}\right) \quad \Rightarrow \quad \acute{\mathbf{X}}^{F*} = \acute{\mathbf{X}}^{F}. \tag{13.47}$$

Bearing these relations in mind we can now easily derive the transformation of the acceleration of the fluid. We obtain immediately

$$\begin{aligned} \acute{\acute{\mathrm{x}}}^{F*} &= \frac{\partial}{\partial t}\left(\mathbf{O}\acute{\mathbf{x}}^{F} + \dot{\mathbf{O}}\mathbf{x} + \dot{\mathbf{c}}\right) + \acute{\mathbf{X}}^{F} \cdot \operatorname{Grad}\left(\mathbf{O}\acute{\mathbf{x}}^{F} + \dot{\mathbf{O}}\mathbf{x} + \dot{\mathbf{c}}\right) \\ &= \mathbf{O}\acute{\acute{\mathrm{x}}}^{F} + 2\dot{\mathbf{O}}\acute{\mathbf{x}}^{F} + \ddot{\mathbf{O}}\mathbf{x} + \ddot{\mathbf{c}}, \end{aligned} \tag{13.48}$$

where the definition of the Lagrangian velocity has been used.

Due to the presence of contributions dependent solely on the choice of the frame we say that velocities $\acute{\mathbf{x}}^{S}, \acute{\mathbf{x}}^{F}$ and accelerations $\acute{\acute{\mathrm{x}}}^{S}, \acute{\acute{\mathrm{x}}}^{F}$ are nonobjective. Consequently, their difference is also nonobjective. We have

$$\acute{\acute{\mathrm{x}}}^{F*} - \acute{\acute{\mathrm{x}}}^{S*} = \mathbf{O}\left(\acute{\acute{\mathrm{x}}}^{F} - \acute{\acute{\mathrm{x}}}^{S}\right) + 2\dot{\mathbf{O}}\left(\acute{\mathbf{x}}^{F} - \acute{\mathbf{x}}^{S}\right). \tag{13.49}$$

For this reason the difference of accelerations cannot be used as a constitutive variable in a construction of the macroscopic model of a two-component system.

If we take the gradient of the transformation relations for velocities we obtain

$$\begin{aligned} \operatorname{Grad}\acute{\mathbf{x}}^{S*} &= \mathbf{O}\operatorname{Grad}\acute{\mathbf{x}}^{S} + \dot{\mathbf{O}}\mathbf{F}^{S} \quad \Rightarrow \quad \dot{\mathbf{O}} = \operatorname{Grad}\left(\acute{\mathbf{x}}^{S*} - \mathbf{O}\acute{\mathbf{x}}^{S}\right)\mathbf{F}^{S-1}, \\ \operatorname{Grad}\acute{\mathbf{x}}^{F*} &= \mathbf{O}\operatorname{Grad}\acute{\mathbf{x}}^{F} + \dot{\mathbf{O}}\mathbf{F}^{S} \quad \Rightarrow \quad \dot{\mathbf{O}} = \operatorname{Grad}\left(\acute{\mathbf{x}}^{F*} - \mathbf{O}\acute{\mathbf{x}}^{F}\right)\mathbf{F}^{S-1}. \end{aligned} \tag{13.50}$$

Consequently, we can write

$$\begin{aligned} 2\dot{\mathbf{O}}\left(\acute{\mathbf{x}}^{F} - \acute{\mathbf{x}}^{S}\right) &= (2 - \mathfrak{z})\,\dot{\mathbf{O}}\left(\acute{\mathbf{x}}^{F} - \acute{\mathbf{x}}^{S}\right) + \mathfrak{z}\dot{\mathbf{O}}\left(\acute{\mathbf{x}}^{F} - \acute{\mathbf{x}}^{S}\right) \\ &= (2 - \mathfrak{z})\operatorname{Grad}\left(\acute{\mathbf{x}}^{F*} - \mathbf{O}\acute{\mathbf{x}}^{F}\right)\acute{\mathbf{X}}^{F} + \mathfrak{z}\operatorname{Grad}\left(\acute{\mathbf{x}}^{S*} - \mathbf{O}\acute{\mathbf{x}}^{S}\right)\acute{\mathbf{X}}^{F}, \end{aligned} \tag{13.51}$$

where $\mathfrak{z}$ is arbitrary.

Substitution of this relation in (13.49) yields

$$\overset{''}{\mathbf{x}}^{F*} - \overset{''}{\mathbf{x}}^{S*} - (2-\mathfrak{z})\,\acute{\mathbf{X}}^F \cdot \operatorname{Grad}\acute{\mathbf{x}}^{F*} - \mathfrak{z}\acute{\mathbf{X}}^F \cdot \operatorname{Grad}\acute{\mathbf{x}}^{S*} \tag{13.52}$$

$$= \mathbf{O}\left(\overset{''}{\mathbf{x}}^{F} - \overset{''}{\mathbf{x}}^{S} - (2-\mathfrak{z})\,\acute{\mathbf{X}}^F \cdot \operatorname{Grad}\acute{\mathbf{x}}^{F} - \mathfrak{z}\acute{\mathbf{X}}^F \cdot \operatorname{Grad}\acute{\mathbf{x}}^{S}\right).$$

It means that the quantity

$$\mathbf{a}_r := \frac{\partial}{\partial t}\left(\acute{\mathbf{x}}^F - \acute{\mathbf{x}}^S\right) + \acute{\mathbf{X}}^F \cdot \operatorname{Grad}\acute{\mathbf{x}}^F - (2-\mathfrak{z})\,\acute{\mathbf{X}}^F \cdot \operatorname{Grad}\acute{\mathbf{x}}^F - \mathfrak{z}\acute{\mathbf{X}}^F \cdot \operatorname{Grad}\acute{\mathbf{x}}^S$$

$$\equiv \frac{\partial}{\partial t}\left(\acute{\mathbf{x}}^F - \acute{\mathbf{x}}^S\right) - (1-\mathfrak{z})\,\acute{\mathbf{X}}^F \cdot \operatorname{Grad}\acute{\mathbf{x}}^F - \mathfrak{z}\acute{\mathbf{X}}^F \cdot \operatorname{Grad}\acute{\mathbf{x}}^S, \tag{13.53}$$

is objective, i.e.

$$\mathbf{a}_r^* = \mathbf{O}\mathbf{a}_r. \tag{13.54}$$

We call this quantity an *objective relative acceleration.* As an objective variable it can be incorporated into the set of constitutive variables. Obviously, there exists a class of such accelerations specified by the constitutive coefficient $\mathfrak{z}$.

It is easy to see that a linear momentum source $\hat{\mathbf{p}}$ in an isotropic material may contain a term $\rho_{12}^0\mathbf{a}_r \approx \rho_{12}^0\frac{\partial}{\partial t}\left(\acute{\mathbf{x}}^F - \acute{\mathbf{x}}^S\right)$ as required by relations of Biot's model. The open question is if the second law of thermodynamics admits this type of contribution in a fully nonlinear model.

As we have already mentioned the nonlinear poroelastic two-component model requires the formulation of field equations for the following fields

$$\mathcal{V} := \left\{\rho^F, \acute{\mathbf{x}}^S, \acute{\mathbf{x}}^F, \mathbf{F}^S, T, n\right\} \quad \text{for} \quad (\mathbf{X}, t) \in \mathcal{B}_0 \times \mathcal{T}, \tag{13.55}$$

where ρ^F is the partial mass density of the fluid per unit volume in the reference configuration of the skeleton, i.e. in the current configuration it is given by the relation $\rho_t^F = \rho^F J^{S-1}$, $J^S := \det \mathbf{F}^S$. T is the absolute temperature of the medium common for both components, and n is the current porosity. Other symbols have the same meaning as before.

The partial mass density of the skeleton in the reference configuration, ρ^S, does not appear among the fields because it is constant in a homogeneous material without mass exchange between components.

These fields are assumed to fulfil the following set of balance equations

$$R^F := \frac{\partial \rho^F}{\partial t} + \operatorname{Div}\left(\rho^F \acute{\mathbf{X}}^F\right) = 0, \tag{13.56}$$

$$\mathbf{M}^S := \rho^S \frac{\partial \acute{\mathbf{x}}^S}{\partial t} - \operatorname{Div} \mathbf{P}^S - \hat{\mathbf{p}} = \mathbf{0}, \tag{13.57}$$

$$\mathbf{M}^F := \rho^F \left(\frac{\partial \acute{\mathbf{x}}^F}{\partial t} + \acute{\mathbf{X}}^F \cdot \operatorname{Grad} \acute{\mathbf{x}}^F \right) - \operatorname{Div} \mathbf{P}^F + \hat{\mathbf{p}} = \mathbf{0}, \tag{13.58}$$

$$E := \frac{\partial \rho\varepsilon}{\partial t} + \operatorname{Div} \mathbf{Q} - \mathbf{P}^S \cdot \operatorname{Grad} \acute{\mathbf{x}}^S - \mathbf{P}^F \cdot \operatorname{Grad} \acute{\mathbf{x}}^F - \left(\mathbf{F}^{ST} \hat{\mathbf{p}} \right) \cdot \acute{\mathbf{X}}^F = 0,$$
$$\rho := \rho^S + \rho^F, \tag{13.59}$$

$$F := \frac{\partial \mathbf{F}^S}{\partial t} - \operatorname{Grad} \acute{\mathbf{x}}^S = \mathbf{0}, \tag{13.60}$$

$$N := \frac{\partial \Delta_n}{\partial t} + \operatorname{Div} \mathbf{J} - \hat{n} = 0, \quad \Delta_n := n - n_E, \tag{13.61}$$

where $\mathbf{P}^S, \mathbf{P}^F$ denote the first Piola-Kirchhoff partial stress tensors, $\hat{\mathbf{p}}$ is the momentum source, ε is the specific internal energy per unit mass of the mixture, $\mathbf{Q}$ is the heat flux vector, n_E describes the equilibrium porosity, $\mathbf{J}$ is the porosity flux, and $\hat{n}$ is the porosity source. The porosity balance equation (13.61) yields the model essentially beyond the frame of Biot's model due to the contribution of the source $\hat{n}$.

As we show in the next Section, changes of porosity predicted by the linearized porosity balance equation are identical with those following from Biot's model and Gassmann relations provided the relaxation time of porosity goes to infinity (i.e. $\hat{n} \equiv 0$). However, it should be mentioned that many other approaches to the problem of evolution of volume fractions, porosity, etc. appear in the literature. One of the most popular forms of such an evolution equation follows from the so-called principle of equilibrated pressures introduced by Goodman, Cowin [125], Nunziato, Walsh [233], Passman *et al.* [243] and others. Even though in some applications such an approach may be advantageous to the porosity balance, we do not discuss it any further in this book.

In order to obtain field equations from the above balance equations we have to specify constitutive relations for these quantities, i.e.

$$\mathcal{F} := \left\{ \mathbf{P}^S, \mathbf{P}^F, \hat{\mathbf{p}}, \varepsilon, \mathbf{Q}, n_E, \mathbf{J}, \hat{n} \right\}, \tag{13.62}$$

must be functions of constitutive variables. In the model, we are presently discussing, the set of constitutive variables is chosen as follows

$$\mathcal{C} := \left\{ \rho^F, \mathbf{F}^S, \acute{\mathbf{X}}^F, \Delta_n, T, \mathbf{G}, \mathbf{a}_r \right\}, \quad \mathbf{G} := \operatorname{Grad} T. \tag{13.63}$$

Hence the gradient of porosity does not appear among the variables. As we see further it means that we can expect only the so-called simple mixture model to follow from thermodynamical considerations in which some couplings of components are absent. We do so on purpose to avoid technical difficulties of accounting simultaneously for two new contributions: due to $\mathbf{a}_r$ and due to $\mathbf{N} := \operatorname{Grad} n$. Later we consider the model with the dependence on $\mathbf{N}$ but not on $\mathbf{a}_r$. One can show that the combined model satisfies the second law of thermodynamics.

Once the function

$$\mathcal{F} = \mathcal{F}(\mathcal{C}), \tag{13.64}$$

is given, we obtain a closed system of differential equations for fields $\mathcal{V}$.

Constitutive relations (13.64) satisfy the second law of thermodynamics in the form of the following entropy inequality

$$\frac{\partial \rho\eta}{\partial t} + \operatorname{Div}\mathbf{H} \geq 0, \quad \eta = \eta(\mathcal{C}), \quad \mathbf{H} = \mathbf{H}(\mathcal{C}), \tag{13.65}$$

which must hold for all solutions of field equations. In this inequality η is the specific entropy and $\mathbf{H}$ its flux.

This requirement is equivalent to the following inequality which must hold for all fields

$$\begin{aligned} &\frac{\partial \rho\eta}{\partial t} + \operatorname{Div}\mathbf{H} - \Lambda^{\rho^F} R^F \\ &-\mathbf{\Lambda}^{v^S}\cdot\mathbf{M}^S - \mathbf{\Lambda}^{v^F}\cdot\mathbf{M}^F - \Lambda^{\varepsilon}E - \mathbf{\Lambda}^F\cdot F - \Lambda^n N \geq 0, \end{aligned} \tag{13.66}$$

where

$$\Lambda^{\rho^F}, \mathbf{\Lambda}^{v^S}, \mathbf{\Lambda}^{v^F}, \Lambda^{\varepsilon}, \mathbf{\Lambda}^F, \Lambda^n \tag{13.67}$$

are Lagrange multipliers dependent on constitutive variables $\mathcal{C}$.

The exploitation of the second law of thermodynamics in the general case is technically impossible. Therefore we make simplifying assumptions sufficient for the second law to be satisfied and yielding explicit limitations on constitutive relations. They are as follows [374]:

1° The material is isotropic. Consequently, scalar constitutive functions, for instance, depend on vector and tensor variables solely through invariants. This assumption will be applied in some steps of our proofs. Some relations are simpler in a general form and then we do not introduce this limitation.

2° The dependence on the relative velocity $\acute{\mathbf{X}}^F$ is at most quadratic. This assumption is related to the structure of the nonlinear contribution to the objective relative acceleration. We motivate its form further.

3° The dependence on the temperature gradient $\mathbf{G}$ is linear. We could skip this assumption on the cost of some technicalities but the experience with the thermodynamical construction of poroelastic models shows that this simplifying assumption does not yield any undesired results.

4° The dependence on the deviation of porosity n from its equilibrium value n_E, $\Delta_n \equiv n - n_E$, is quadratic.

5° The dependence on the relative acceleration $\mathbf{a}_r$ is linear.

6° Higher order combinations of variables $\mathbf{G}, \acute{\mathbf{X}}^F, \Delta_n, \mathbf{a}_r$ can be neglected.

As we see further these assumptions limit thermodynamical considerations to a vicinity of the thermodynamical equilibrium similar to this appearing in the classical Onsager thermodynamics.

Bearing these assumptions in mind we can write the following representations of constitutive functions

– partial stresses

$$\mathbf{P}^S = \mathbf{P}_0^S\left(\mathcal{C}_E, \Delta_n\right) + \frac{1}{2}\sigma^S\left(\mathcal{C}_E\right)\mathbf{F}^S\acute{\mathbf{X}}^F \otimes \acute{\mathbf{X}}^F, \quad \mathcal{C}_E := \left\{\rho^F, \mathbf{F}^S, T\right\},$$

$$\mathbf{P}^F = \mathbf{P}_0^F\left(\mathcal{C}_E, \Delta_n\right) + \frac{1}{2}\sigma^F\left(\mathcal{C}_E\right)\mathbf{F}^S\acute{\mathbf{X}}^F \otimes \acute{\mathbf{X}}^F, \quad n_E = n_E\left(\mathcal{C}_E\right), \qquad (13.68)$$

– internal energy and entropy

$$\rho\varepsilon = \rho\varepsilon_0\left(\mathcal{C}_E, \Delta_n\right) + \frac{1}{2}\varepsilon_d\left(\mathcal{C}_E\right)\left(\mathbf{F}^S\acute{\mathbf{X}}^F\right)\cdot\left(\mathbf{F}^S\acute{\mathbf{X}}^F\right),$$

$$\rho\eta = \rho\eta_0\left(\mathcal{C}_E, \Delta_n\right) + \frac{1}{2}\eta_d\left(\mathcal{C}_E\right)\left(\mathbf{F}^S\acute{\mathbf{X}}^F\right)\cdot\left(\mathbf{F}^S\acute{\mathbf{X}}^F\right), \qquad (13.69)$$

– fluxes of energy, entropy, porosity

$$\begin{aligned}\mathbf{Q} &= Q_V\acute{\mathbf{X}}^F - K\mathbf{G} + Q_a\mathbf{F}^{ST}\mathbf{a}_r,\\ \mathbf{H} &= H_V\acute{\mathbf{X}}^F + H_T\mathbf{G} + H_a\mathbf{F}^{ST}\mathbf{a}_r,\\ \mathbf{J} &= \Phi\acute{\mathbf{X}}^F + J_T\mathbf{G} + J_a\mathbf{F}^{ST}\mathbf{a}_r,\end{aligned} \qquad (13.70)$$

where all coefficients are functions of variables $\mathcal{C}_E$,

– momentum source

$$\mathbf{F}^{ST}\hat{\mathbf{p}} = \Pi_V\acute{\mathbf{X}}^F + \Pi_T\mathbf{G} - \rho_{12}^0\mathbf{F}^{ST}\mathbf{a}_r, \qquad (13.71)$$

with coefficients dependent again on variables $\mathcal{C}_E$.

The contributions with the coefficients ε_d, η_d to the energy and entropy are motivated by fluctuations of the microstructural kinetic energy caused by the tortuosity[1]. We do not introduce any additional microstructural variable describing changes of tortuosity. For this reason a macroscopic influence of tortuosity can be solely reflected by the *seepage velocity* which in our model corresponds to the Lagrangian velocity $\acute{\mathbf{X}}^F$. The classical kinetic energy in this model is given by $\frac{1}{2}\left(\rho^S\acute{\mathbf{x}}^S\cdot\acute{\mathbf{x}}^S+\rho^F\acute{\mathbf{x}}^F\cdot\acute{\mathbf{x}}^F\right)$. Consequently, the correction $\frac{1}{2}\varepsilon_d\left(\acute{\mathbf{x}}^F-\acute{\mathbf{x}}^S\right)\cdot\left(\acute{\mathbf{x}}^F-\acute{\mathbf{x}}^S\right)$ may be considered as an *added mass effect* (e.g. [179]) resulting from tortuosity.

As we see further, the dependence of partial stresses on this velocity, introduced in the simplest form by (13.68), is then required by the consistency of the model with the second law of thermodynamics. In other words, we show further that coefficients σ^S, σ^F in the stress relations (13.68) and the coefficient ε_d in the energy relation (13.69) are connected (see formula (13.111)).

The exploitation of the second law of thermodynamics (13.66) is standard. We apply the chain rule of differentiation to constitutive laws. We skip here rather cumbersome technical details.

Linearity of the second law of thermodynamics with respect to time derivatives

$$\left\{\frac{\partial\rho^F}{\partial t},\frac{\partial\mathbf{F}^S}{\partial t},\frac{\partial\Delta_n}{\partial t},\frac{\partial\acute{\mathbf{x}}^S}{\partial t},\frac{\partial\acute{\mathbf{x}}^F}{\partial t},\frac{\partial T}{\partial t},\frac{\partial\mathbf{G}}{\partial t}\right\}$$

yields

$$\Lambda^{\rho^F}=\Lambda_0^{\rho^F}+\frac{1}{2}\left(\frac{\partial\eta_d}{\partial\rho^F}-\Lambda^\varepsilon\frac{\partial\varepsilon_d}{\partial\rho^F}\right)\left(\mathbf{F}^S\acute{\mathbf{X}}^F\right)\cdot\left(\mathbf{F}^S\acute{\mathbf{X}}^F\right),$$

$$\Lambda_0^{\rho^F}:=\frac{\partial\rho\eta_0}{\partial\rho^F}-\Lambda^\varepsilon\frac{\partial\rho\varepsilon_0}{\partial\rho^F}\tag{13.72}$$

$$\mathbf{\Lambda}^F=\frac{\partial\rho\eta_0}{\partial\mathbf{F}^F}-\Lambda^\varepsilon\frac{\partial\rho\varepsilon_0}{\partial\mathbf{F}^F}+\frac{1}{2}\left(\frac{\partial\eta_d}{\partial\mathbf{F}^F}-\Lambda^\varepsilon\frac{\partial\varepsilon_d}{\partial\mathbf{F}^F}\right)\left(\mathbf{F}^S\acute{\mathbf{X}}^F\right)\cdot\left(\mathbf{F}^S\acute{\mathbf{X}}^F\right),\tag{13.73}$$

$$\Lambda^n=\frac{\partial\rho\eta_0}{\partial\Delta_n}-\Lambda^\varepsilon\frac{\partial\rho\varepsilon_0}{\partial\Delta_n},\tag{13.74}$$

[1]Tortuosity is the geometrical property of flows in channels of porous materials and it may be defined as the ratio of the average length of the streamline within REV to the typical linear dimension of REV. Hence, its minimum value is equal to one. Then it corresponds to the straight streamline.

$$\left(\rho^S - \rho_{12}^0\right) \mathbf{\Lambda}^{v^S} + \rho_{12}^0 \mathbf{\Lambda}^{v^F} = -\left(\eta_d - \Lambda^\varepsilon \varepsilon_d\right) \mathbf{F}^S \acute{\mathbf{X}}^F + \rho_{12}^0 \Lambda^\varepsilon \mathbf{F}^S \acute{\mathbf{X}}^F \tag{13.75}$$
$$- \operatorname{Div}\left(H_a \mathbf{F}^S\right) + \Lambda^\varepsilon \operatorname{Div}\left(Q_a \mathbf{F}^S\right) + \Lambda^n \operatorname{Div}\left(J_a \mathbf{F}^S\right) = 0,$$

$$\left(\rho^F - \rho_{12}^0\right) \mathbf{\Lambda}^{v^F} + \rho_{12}^0 \mathbf{\Lambda}^{v^S} = \left(\eta_d - \Lambda^\varepsilon \varepsilon_d\right) \mathbf{F}^S \acute{\mathbf{X}}^F - \rho_{12}^0 \Lambda^\varepsilon \mathbf{F}^S \acute{\mathbf{X}}^F$$
$$+ \operatorname{Div}\left(H_a \mathbf{F}^S\right) - \Lambda^\varepsilon \operatorname{Div}\left(Q_a \mathbf{F}^S\right) - \Lambda^n \operatorname{Div}\left(J_a \mathbf{F}^S\right) = 0, \tag{13.76}$$

$$\frac{\partial \rho \eta_0}{\partial T} - \Lambda^\varepsilon \frac{\partial \rho \varepsilon_0}{\partial T} + \frac{1}{2}\left(\frac{\partial \eta_d}{\partial T} - \Lambda^\varepsilon \frac{\partial \varepsilon_d}{\partial T}\right)\left(\mathbf{F}^S \acute{\mathbf{X}}^F\right) \cdot \left(\mathbf{F}^S \acute{\mathbf{X}}^F\right) = 0. \tag{13.77}$$

These identities still contain linear contributions of $\operatorname{Grad} \mathbf{F}^S$, Δ_n, quadratic contributions of the latter as well as quadratic contributions of Lagrangian velocity. As they should hold for arbitrary fields, coefficients of these contributions must vanish separately. After easy analysis we obtain

$$H_a = 0, \quad Q_a = 0, \quad J_a = 0, \tag{13.78}$$

$$\rho^S \mathbf{\Lambda}^{v^S} = -\rho^F \mathbf{\Lambda}^{v^F} = \mathfrak{y} \mathbf{F}^S \acute{\mathbf{X}}^F, \quad \mathfrak{y} := -\frac{\eta_d - \Lambda^\varepsilon \varepsilon_d - \Lambda^\varepsilon \rho_{12}^0}{\rho^S - \rho_{12}^0 \left(1 + \frac{\rho^S}{\rho^F}\right)}. \tag{13.79}$$

The second law of thermodynamics is also linear with respect to the following spatial derivatives

$$\operatorname{Grad} \acute{\mathbf{x}}^S, \operatorname{Grad} \acute{\mathbf{x}}^F, \operatorname{Grad} \rho^F, \operatorname{Grad} \mathbf{F}^S, \operatorname{Grad} \mathbf{G}, \operatorname{Grad} \Delta_n. \tag{13.80}$$

We have listed them in the order of the further analysis and, simultaneously, skipped the derivative $\operatorname{Grad} \mathbf{a}_r$ because it does not contribute to the second law due to the relations (13.78). The linearity with respect to (13.80) yields a set of identities and leaves a residual inequality which is essentially nonlinear. It defines the *dissipation* in the system and has the following form

$$\begin{aligned} \mathcal{D} := &\left(\frac{\partial H_V}{\partial T} - \Lambda^\varepsilon \frac{\partial Q_V}{\partial T} - \Lambda^n \frac{\partial \Phi}{\partial T} + \Pi_T\right) \acute{\mathbf{X}}^F \cdot \mathbf{G} \\ &+ \left(\frac{\partial H_T}{\partial T} + \Lambda^\varepsilon \frac{\partial K}{\partial T} - \Lambda^n \frac{\partial J_T}{\partial T}\right) \mathbf{G} \cdot \mathbf{G} \\ &+ \Lambda^\varepsilon \Pi_V \acute{\mathbf{X}}^F \cdot \acute{\mathbf{X}}^F + \Lambda^n \hat{n} \geq 0. \end{aligned} \tag{13.81}$$

Hence the state of thermodynamical equilibrium defined by $\mathcal{D} = 0$ appears if

$$\mathbf{G} = 0, \quad \acute{\mathbf{X}}^F = 0, \quad \hat{n} = 0, \tag{13.82}$$

i.e. the temperature gradient, relative motion (diffusion), and the relaxation of porosity cause the deviation from the equilibrium.

Clearly the assumption 4° yields the linearity of $\hat{n}$ and Λ^n with respect to Δ_n. In addition, the above inequality yields homogeneity of these functions, i.e. we can write

$$\hat{n} = -\frac{\Delta_n}{\tau}, \quad \Lambda^n = \lambda^n \Delta_n, \tag{13.83}$$

where τ, λ^n can be solely functions of variables $\mathcal{C}_E$. Consequently, we obtain as well

$$\frac{\partial \Phi}{\partial T} = 0, \quad \frac{\partial J_T}{\partial T} = 0. \tag{13.84}$$

It is worth mentioning that due to (13.78) the relative acceleration does not contribute to the dissipation. This property of the model follows from the fact that the model does not possess any independent field of tortuosity which could relax to the thermodynamical equilibrium.

Now we return to the coefficients of spatial derivatives of fields. The vanishing coefficient of Grad $\acute{\mathbf{x}}^S$ yields the following results

$$\begin{aligned} &\Lambda^\varepsilon \mathbf{P}_0^S \mathbf{F}^{ST} + \left(\frac{\partial \rho \eta_0}{\partial \mathbf{F}^S} - \Lambda^\varepsilon \frac{\partial \rho \varepsilon_0}{\partial \mathbf{F}^S} \right) \mathbf{F}^{ST} \\ &+ \left(-H_V + \Lambda^\varepsilon Q_V + \Lambda^n \Phi + \rho^F \Lambda_0^{\rho^F} \right) \mathbf{1} = 0, \end{aligned} \tag{13.85}$$

$$\frac{\partial \eta_d}{\partial I} - \Lambda^\varepsilon \frac{\partial \varepsilon_d}{\partial I} = 0, \quad \frac{\partial \eta_d}{\partial II} - \Lambda^\varepsilon \frac{\partial \varepsilon_d}{\partial II} = 0, \tag{13.86}$$

$$2 \left(\frac{\partial \eta_d}{\partial III} - \Lambda^\varepsilon \frac{\partial \varepsilon_d}{\partial III} \right) III + \rho^F \left(\frac{\partial \eta_d}{\partial \rho^F} - \Lambda^\varepsilon \frac{\partial \varepsilon_d}{\partial \rho^F} \right) = \mathfrak{y} \left(\sigma^S + \frac{\rho^S}{\rho^F} \sigma^F \right), \tag{13.87}$$

$$\mathfrak{z} = -\frac{1}{2\rho_{12}^0} \frac{\sigma^S \Lambda^\varepsilon - \mathfrak{y} \left(\sigma^S + \frac{\rho^S}{\rho^F} \sigma^F \right)}{\mathfrak{y} \left(1 + \frac{\rho^S}{\rho^F} \right) + \Lambda^\varepsilon}, \tag{13.88}$$

where

$$I := \operatorname{tr} \mathbf{C}^S, \quad II := \frac{1}{2} \left(I^2 - \operatorname{tr} \mathbf{C}^S \right), \quad III := \det \mathbf{C}^S, \quad \mathbf{C}^S := \mathbf{F}^{ST} \mathbf{F}^S, \tag{13.89}$$

are main invariants of the Cauchy-Green deformation tensor $\mathbf{C}^S$.

The coefficient of $\operatorname{Grad}\acute{\mathbf{x}}^F$ yields

$$\Lambda^\varepsilon \left(\mathbf{P}_0^S + \mathbf{P}_0^F\right) = -\left(\frac{\partial \rho\eta_0}{\partial \mathbf{F}^S} - \Lambda^\varepsilon \frac{\partial \rho\varepsilon_0}{\partial \mathbf{F}^S}\right), \tag{13.90}$$

$$\frac{\partial \eta_d}{\partial III} - \Lambda^\varepsilon \frac{\partial \varepsilon_d}{\partial III} = 0, \tag{13.91}$$

$$\frac{1}{2}\Lambda^\varepsilon \left(\sigma^S + \sigma^F\right) = -\rho_{12}^0 \left(1 + \frac{\rho^S}{\rho^F}\right)\mathfrak{y} - \rho^S \mathfrak{y} - \rho_{12}^0 \Lambda^\varepsilon. \tag{13.92}$$

Consequently, bearing (13.86) and (13.91) in mind,

$$\frac{\partial \eta_d}{\partial \mathbf{F}^S} - \Lambda^\varepsilon \frac{\partial \varepsilon_d}{\partial \mathbf{F}^S} = 0. \tag{13.93}$$

Next we consider the coefficient of $\operatorname{Grad}\rho^F$. We have

$$\begin{aligned} &\left(\frac{\partial H_V}{\partial \rho^F} - \Lambda^\varepsilon \frac{\partial Q_V}{\partial \rho^F} - \Lambda^n \frac{\partial \Phi}{\partial \rho^F} - \Lambda^{\rho^F}\right)\acute{\mathbf{X}}^F \\ &+\mathfrak{y}\frac{\partial \mathbf{P}^{ST}}{\partial \rho^F}\mathbf{F}^S\acute{\mathbf{X}}^F - \mathfrak{y}\frac{\rho^F}{\rho^S}\frac{\partial \mathbf{P}^{FT}}{\partial \rho^F}\mathbf{F}^S\acute{\mathbf{X}}^F = 0, \end{aligned} \tag{13.94}$$

$$\frac{\partial H_T}{\partial \rho^F} + \Lambda^\varepsilon \frac{\partial K}{\partial \rho^F} = 0, \quad \frac{\partial J_T}{\partial \rho^F} = 0. \tag{13.95}$$

Similarly, the coefficient of $\operatorname{Grad}\mathbf{F}^S$ yields

$$\begin{aligned} &\operatorname*{sym}_{12}\left\{\acute{\mathbf{X}}^F \otimes \left(\frac{\partial H_V}{\partial \mathbf{F}^S} - \Lambda^\varepsilon \frac{\partial Q_V}{\partial \mathbf{F}^S} - \Lambda^n \frac{\partial \Phi}{\partial \mathbf{F}^S}\right)^T \right. \\ &\qquad \left. + \rho^F \Lambda^{\rho^F} \acute{\mathbf{X}}^F \otimes \mathbf{F}^{S-1} \otimes + \mathfrak{y}\Xi^S + \mathfrak{y}\Xi^F\right\} \end{aligned}$$

$$-\operatorname*{sym}_{12}\left\{\acute{\mathbf{X}}^F \otimes \left(H_V - \Lambda^\varepsilon - \Lambda^n \Phi\right)\mathbf{F}^{-1}\right\} = 0, \tag{13.96}$$

$$\frac{\partial H_T}{\partial \mathbf{F}^S} + \Lambda^\varepsilon \frac{\partial K}{\partial \mathbf{F}^S} = 0, \quad \frac{\partial J_T}{\partial \mathbf{F}^S} = 0, \tag{13.97}$$

where the components of tensors Ξ^S, Ξ^F in Cartesian coordinates are given by the relations

$$\Xi^S_{KLk} = \frac{\partial P^S_{0lL}}{\partial F^S_{kK}} F^S_{lM} \acute{X}^F_M, \quad \Xi^F_{KLk} = -\frac{\rho^S}{\rho^F}\frac{\partial P^F_{0lL}}{\partial F^S_{kK}} F^S_{lM} \acute{X}^F_M. \tag{13.98}$$

Under our assumptions the contribution of Grad Δ_n does not yield any restrictions.

Finally, the last condition follows from the vanishing coefficient of Grad $\mathbf{G}$ and it has the form

$$H_T + \Lambda^\varepsilon K = 0, \quad J_T = 0. \tag{13.99}$$

Inspection of the results (13.95), (13.97) for thermal coefficients yields

$$\Lambda^\varepsilon = \Lambda^\varepsilon(T) \quad \Rightarrow \quad \Lambda^\varepsilon = \frac{1}{T}, \quad \text{i.e.} \quad H_T = -\frac{K}{T}, \tag{13.100}$$

where the ideal wall argument has been used.

It is not quite clear what limitations on partial stress tensors are imposed by conditions (13.94), (13.96). Derivatives of partial stresses with respect to the mass density ρ^F as well as with respect to the deformation gradient $\mathbf{F}^S$ seem to restrict elastic properties of the system in equilibrium. This does not seem very plausible. Hence we assume that the coefficient $\mathfrak{y}$ vanishes, i.e.

$$\mathfrak{y} = 0. \tag{13.101}$$

Then the multipliers of momentum equations vanish as well. As the consequence of (13.77), (13.79), (13.87), (13.93) we obtain

$$-\rho_{12}^0 = \varepsilon_d - T\eta_d = \text{const.} \quad \Rightarrow \quad \varepsilon_d = \text{const.}, \ \eta_d = 0. \tag{13.102}$$

It is convenient to introduce the following notation

$$\begin{aligned} \psi &:= \varepsilon - T\eta, \\ \rho\psi_0 &:= \rho\psi - \tfrac{1}{2}\varepsilon_d \left(\mathbf{F}^S \acute{\mathbf{X}}^F\right) \cdot \left(\mathbf{F}^S \acute{\mathbf{X}}^F\right). \end{aligned} \tag{13.103}$$

Then, for Lagrange multipliers we have

$$\Lambda^{\rho^F} = -\frac{1}{T}\frac{\partial \rho\psi_0}{\partial \rho^F} = \Lambda_0^{\rho^F}, \quad \mathbf{\Lambda}^F = -\frac{1}{T}\frac{\partial \rho\psi_0}{\partial \mathbf{F}^S}, \quad \Lambda^n = -\frac{1}{T}\frac{\partial \rho\psi_0}{\partial \Delta_n} = \lambda^n \Delta_n, \tag{13.104}$$

and the relation (13.77) implies the following classical formula for the internal energy

$$\varepsilon = \psi - T\frac{\partial \psi}{\partial T}. \tag{13.105}$$

Simultaneously the relations (13.94), (13.96) yield

$$\frac{\partial H_V}{\partial \rho^F} - \Lambda^\varepsilon \frac{\partial Q_V}{\partial \rho^F} - \Lambda^{\rho^F} = 0, \quad \frac{\partial \Phi}{\partial \rho^F} = 0,$$

$$2III\left(\frac{\partial H_V}{\partial III}-\Lambda^{\varepsilon}\frac{\partial Q_V}{\partial III}\right)+\rho^F\Lambda^{\rho^F}-\left(H_V-\Lambda^{\varepsilon}Q_V\right)=0, \tag{13.106}$$

$$2III\frac{\partial\Phi}{\partial III}-\Phi=0\quad\Rightarrow\quad\Phi=J^S\Phi_0,\quad\Phi_0=\text{const.} \tag{13.107}$$

These relations yield the following integrability condition for the multiplier Λ^{ρ^F}

$$\rho^F\frac{\partial\Lambda^{\rho^F}}{\partial\rho^F}+J^S\frac{\partial\Lambda^{\rho^F}}{\partial J^S}=0\quad\Rightarrow\quad\Lambda^{\rho^F}=\Lambda^{\rho^F}\left(T,\rho_t^F\right),\quad\rho_t^F:=J^{S-1}\rho^F. \tag{13.108}$$

Consequently, integration of $(13.104)_1$ leads to the following additive splitting of the free energy ψ

$$\rho\psi=\rho^F\psi^F+\rho^S\psi^S-\frac{1}{2}\lambda^nT\Delta_n^2+\frac{1}{2}\varepsilon_d\left(\mathbf{F}^S\acute{\mathbf{X}}^F\right)\cdot\left(\mathbf{F}^S\acute{\mathbf{X}}^F\right),$$
$$\psi^F=\psi^F\left(T,\rho_t^F\right),\quad\psi^S=\psi^S\left(T,\mathbf{F}^S\right). \tag{13.109}$$

The above separation property is characteristic for the so-called *simple mixtures*. It means that the choice of constitutive variables (13.63) rules out the coupling of components through deformations: changes of the current mass density ρ_t^F do not have influence on ψ^S and deformations $\mathbf{F}^S$ of the skeleton do not have influence on ψ^F.

In addition, integration in $(13.106)_1$ yields

$$H_V-\frac{Q_V}{T}=-\frac{\rho^F\psi^F}{T},\quad\text{i.e.}\quad\mathbf{H}=\frac{1}{T}\left(\mathbf{Q}-\rho^F\psi^F\acute{\mathbf{X}}^F\right), \tag{13.110}$$

where we have accounted for the relations (13.99) and (13.100).

Inspection of relations (13.102), (13.88) and (13.92) leads immediately to the following identification of constitutive coefficients coupled to the relative acceleration

$$\varepsilon_d=-\rho_{12}^0,\quad\sigma^S=-2\mathfrak{z}\rho_{12}^0,\quad\sigma^F=-2\left(1-\mathfrak{z}\right)\rho_{12}^0. \tag{13.111}$$

Simultaneously, relation (13.85) with (13.103), (13.104), (13.109) and (13.110) for partial stresses $\mathbf{P}_0^S$ and relation (13.90) for partial stresses $\mathbf{P}_0^F$ yield

$$\mathbf{P}^S=\frac{\partial\rho^S\psi^S}{\partial\mathbf{F}^S}+\beta\Delta_nJ^S\mathbf{F}^{S-T}-\mathfrak{z}\rho_{12}^0\mathbf{F}^S\acute{\mathbf{X}}^S\otimes\acute{\mathbf{X}}^S,$$
$$\beta:=T\lambda^n\Phi_0J^{S-1}, \tag{13.112}$$
$$\mathbf{P}^F=-\rho_t^{F2}\frac{\partial\psi^F}{\partial\rho_t^F}J^S\mathbf{F}^{S-T}-\beta\Delta_nJ^S\mathbf{F}^{S-T}-\left(1-\mathfrak{z}\right)\rho_{12}^0\mathbf{F}^S\acute{\mathbf{X}}^S\otimes\acute{\mathbf{X}}^S.$$

Hence, as mentioned in the introduction, the partial stresses do not possess a coupling term characteristic for Biot's model and this fallacy of the model can be removed by additional constitutive variables.

For practical purposes it is convenient to transform equations of the model to Eulerian coordinates. We write them in an arbitrary noninertial reference system. The set of balance equations (13.56) has then the form

– mass balance for the fluid component

$$\frac{\partial \rho_t^F}{\partial t} + \operatorname{div}\left(\rho_t^F \mathbf{v}^F\right) = 0, \tag{13.113}$$

– momentum balance for the skeleton

$$\begin{gathered} \rho_t^S \left(\frac{\partial \mathbf{v}^S}{\partial t} + \mathbf{v}^S \cdot \operatorname{grad} \mathbf{v}^S \right) \\ = \operatorname{div} \mathbf{T}^S + \rho_t^S \mathbf{b}^S + J^{S-1} \Pi_T \operatorname{grad} \mathbf{T} + \pi \left(\mathbf{v}^F - \mathbf{v}^S \right) \end{gathered} \tag{13.114}$$

$$-\rho_{12} \left[\frac{\partial}{\partial t}(\mathbf{v}^F - \mathbf{v}^S) - (1-\mathfrak{z})(\mathbf{v}^F - \mathbf{v}^S) \cdot \operatorname{grad} \mathbf{v}^F - \mathfrak{z}(\mathbf{v}^F - \mathbf{v}^S) \cdot \operatorname{grad} \mathbf{v}^S \right],$$

– momentum balance for the fluid

$$\begin{gathered} \rho_t^F \left(\frac{\partial \mathbf{v}^F}{\partial t} + \mathbf{v}^F \cdot \operatorname{grad} \mathbf{v}^F \right) \\ = \operatorname{div} \mathbf{T}^F + \rho_t^F \mathbf{b}^F - J^{S-1} \Pi_T \operatorname{grad} \mathbf{T} - \pi \left(\mathbf{v}^F - \mathbf{v}^S \right) \end{gathered} \tag{13.115}$$

$$+\rho_{12} \left[\frac{\partial}{\partial t}(\mathbf{v}^F - \mathbf{v}^S) - (1-\mathfrak{z})(\mathbf{v}^F - \mathbf{v}^S) \cdot \operatorname{grad} \mathbf{v}^F - \mathfrak{z}(\mathbf{v}^F - \mathbf{v}^S) \cdot \operatorname{grad} \mathbf{v}^S \right],$$

– energy balance

$$\begin{gathered} \frac{\partial \rho_t \varepsilon}{\partial t} + \operatorname{div}\left(\rho_t \varepsilon \mathbf{v}^S + \mathbf{q}\right) - \mathbf{T}^S \cdot \operatorname{grad} \mathbf{v}^S - \mathbf{T}^F \cdot \operatorname{grad} \mathbf{v}^F \\ - \left(\mathbf{v}^F - \mathbf{v}^S\right) \cdot \left\{ \left[\pi \left(\mathbf{v}^F - \mathbf{v}^S\right) + J^{S-1} \Pi_T \operatorname{grad} T \right] \right. \\ -\rho_{12} \frac{\partial}{\partial t} \left(\mathbf{v}^F - \mathbf{v}^S\right) \\ \left. - \left[(1-\mathfrak{z}) \left(\mathbf{v}^F - \mathbf{v}^S\right) \cdot \operatorname{grad} \mathbf{v}^F + \mathfrak{z} \left(\mathbf{v}^F - \mathbf{v}^S\right) \cdot \operatorname{grad} \mathbf{v}^S \right] \right\}, \end{gathered} \tag{13.116}$$

– porosity balance

$$\frac{\partial J^{S-1}\Delta_n}{\partial t} + \operatorname{div}\left(J^{S-1}\Delta_n \mathbf{v}^S + \mathbf{j}\right) + \frac{J^{S-1}\Delta_n}{\tau} = 0. \qquad (13.117)$$

The external forces $\rho_t^S \mathbf{b}^S, \rho_t^F \mathbf{b}_t^F$, called *apparent body forces*, contributing to momentum balance equations have the following structure

$$\begin{aligned}
\rho_t^S \mathbf{b}^S &= \rho_t^S \left(\mathbf{b}_b^S + \mathbf{i}^S\right), \quad \rho_t^F \mathbf{b}^F = \rho_t^F \left(\mathbf{b}_b^F + \mathbf{i}^F\right), \\
\mathbf{i}^S &:= \ddot{\mathbf{c}} + 2\boldsymbol{\Omega}\left(\mathbf{v}^S - \dot{\mathbf{c}}\right) + \left(\dot{\boldsymbol{\Omega}} - \boldsymbol{\Omega}^2\right)(\mathbf{x} - \mathbf{c}), \qquad (13.118) \\
\mathbf{i}^F &:= \ddot{\mathbf{c}} + 2\boldsymbol{\Omega}\left(\mathbf{v}^F - \dot{\mathbf{c}}\right) + \left(\dot{\boldsymbol{\Omega}} - \boldsymbol{\Omega}^2\right)(\mathbf{x} - \mathbf{c}), \\
\boldsymbol{\Omega} &:= \dot{\mathbf{O}}\mathbf{O}^T \equiv -\boldsymbol{\Omega}^T,
\end{aligned}$$

where $\rho_t^S \mathbf{b}_b^S, \rho_t^F \mathbf{b}_b^F$ are true (e.g. gravitational) body forces, and $\rho_t^S \mathbf{i}^S, \rho_t^F \mathbf{i}^F$ are *inertial body forces*. In order of appearance in the above relations, they consist of the inertial force of relative translation, Coriolis force, Euler force, centrifugal force. They depend on the matrix of angular velocity $\boldsymbol{\Omega}$ of the noninertial system with respect to an inertial one. Certainly, the inertial body forces vanish in an inertial reference system. It should be mentioned that the partial accelerations appearing in the above partial momentum balance equations combined with apparent body forces are objective, i.e. invariant with respect to the Euclidean transformation.

The remaining notation used above is as follows

$$\rho_t^S = \rho^S J^{S-1}, \quad \rho_t = \rho_t^S + \rho_t^F, \quad \rho_{12} = \rho_{12}^0 J^{S-1}, \quad \pi = \Pi_V J^{S-1}, \qquad (13.119)$$

while the Cauchy stress tensors $\mathbf{T}^S, \mathbf{T}^F$ are given by the following constitutive relations

$$\mathbf{T}^S = J^{S-1}\mathbf{P}^S\mathbf{F}^{ST} = 2\rho_t^S \left[\frac{\partial \psi^S}{\partial I}\mathbf{B}^S + \frac{\partial \psi^S}{\partial II}\left(I\mathbf{1} - \mathbf{B}^S\right)\mathbf{B}^S + \frac{\partial \psi^S}{\partial III} III\mathbf{1}\right] \qquad (13.120)$$

$$+\beta\Delta_n \mathbf{1} - \mathfrak{z}\rho_{12}\left(\mathbf{v}^F - \mathbf{v}^S\right)\otimes\left(\mathbf{v}^F - \mathbf{v}^S\right), \quad \mathbf{B}^S = \mathbf{F}^S\mathbf{F}^{ST},$$

$$\begin{aligned}
\mathbf{T}^F &= J^{S-1}\mathbf{P}^F\mathbf{F}^{ST} = -p^F\mathbf{1} - \beta\Delta_n\mathbf{1} - (1-\mathfrak{z})\rho_{12}\left(\mathbf{v}^F - \mathbf{v}^S\right)\otimes\left(\mathbf{v}^F - \mathbf{v}^S\right), \\
p^F &= \rho_t^{F2}\frac{\partial \psi^F}{\partial \rho_t^F}, \qquad (13.121)
\end{aligned}$$

with the free energy given by

$$\rho_t\psi = \rho_t^S\psi^S(T, I, II, III) + \rho_t^F\psi^F(T, \rho_t^F) - \rho_{12}(\mathbf{v}^F - \mathbf{v}^S)\cdot(\mathbf{v}^F - \mathbf{v}^S). \tag{13.122}$$

The energy ε and the energy flux vector $\mathbf{q}$ are given by

$$\varepsilon = \psi - T\frac{\partial\psi}{\partial T}, \quad \mathbf{q} = J^{S-1}\mathbf{F}^S\mathbf{Q} = J^{S-1}Q_V(\mathbf{v}^F - \mathbf{v}^S) - J^{S-1}K\,\mathrm{grad}\,T, \tag{13.123}$$

and the porosity flux has the form

$$\mathbf{j} = J^{S-1}\mathbf{F}^S\mathbf{J} = \Phi_0(\mathbf{v}^F - \mathbf{v}^S). \tag{13.124}$$

There remains the question of practical estimation of additional parameters ρ_{12}^0 and $\mathfrak{z}$. The added mass coefficient ρ_{12}^0 has been extensively studied in literature concerning Biot's model. The parameter $\mathfrak{z}$ has been introduced in [374]. There seem to exist various possibilities for its estimation connected to the fact that it appears in contributions which may be time independent. As an example let us consider a stationary isothermal process in which, in a chosen inertial frame of reference, the skeleton does not move (i.e. $\mathbf{v}^S = 0$). Such a flow of the fluid through a porous material is described by the mass balance and by the momentum balance for the fluid. For simplicity we neglect changes of porosity. Then we have

$$\mathrm{div}\,(\rho_t^F\mathbf{v}^F) = 0, \tag{13.125}$$

$$\left[\rho_t^F + 2(1-\mathfrak{z})\rho_{12}\right]\mathbf{v}^F\cdot\mathrm{grad}\,\mathbf{v}^F = -\,\mathrm{grad}\,p^F - \left[\pi + (1-\mathfrak{z})\rho_{12}\,\mathrm{div}\,\mathbf{v}^F\right]\mathbf{v}^F,$$
$$p^F = p^F(\rho_t^F).$$

The correction of the permeability coefficient π driven by volume changes of the fluid $\mathrm{div}\,\mathbf{v}^F$ seems to be very small. However the correction of mass density appearing on the left-hand side of this equation may be essential and measurable. For instance, in an irrotational flow ($\mathrm{rot}\,\mathbf{v}^F = 0$) we have approximately

$$\mathrm{grad}\left[n_0(p - p_0) + \frac{1}{2}\left(\rho_t^F + 2(1-\mathfrak{z})\rho_{12}\right)\mathbf{v}^F\cdot\mathbf{v}^F\right] + \pi\mathbf{v}^F = 0, \tag{13.126}$$

where $p = p^F/n$ is the pore pressure and p_0 its constant reference value. If the pressure increment is of the order of, say, 10 kPa the velocity of the fluid must be of the order of 1 m/s to make both contributions of the similar order. Practically measurable would be the influence of $\mathfrak{z}$ for much smaller velocities which seem to be plausible at least for rocks.

We proceed to present basic results for the model which yields couplings not appearing in simple mixtures. As we have seen in Chapter 11, in order to obtain expected couplings between components, the thermodynamics of miscible mixtures requires a dependence on gradients of mass densities. This indicates that such couplings are related to weak nonlocal interactions which are described in the simplest case by the second gradients of deformation[2]. In the case of porous materials it is sufficient to incorporate a dependence on the gradient of porosity [369]. This does not mean that models with other second gradients are not meaningful but, of course, they are much more difficult to handle. For the purpose of construction of a linear model of Biot's type it is sufficient to add the variable $\mathbf{N} = \operatorname{Grad} n$. The set of unknown fields given by (13.55) remains the same, i.e.

$$\mathcal{V} := \left\{\rho^F, \acute{\mathbf{x}}^S, \acute{\mathbf{x}}^F, \mathbf{F}^S, T, n\right\} \quad \text{for} \quad (\mathbf{X}, t) \in \mathcal{B}_0 \times \mathcal{T}. \tag{13.127}$$

Also the set of balance equations (13.56)-(13.60) and the set of constitutive quantities $\mathcal{F}$ (13.62) do not change. There is a change in the set of constitutive variables in which we leave out the relative acceleration $\mathbf{a}_r$ but add $\mathbf{N}$. Hence, we have

$$\mathcal{C} := \left\{\rho^F, \mathbf{F}^S, \acute{\mathbf{X}}^F, \Delta_n, T, \mathbf{G}, \mathbf{N}\right\}, \quad \mathbf{G} := \operatorname{Grad} T, \quad \mathbf{N} := \operatorname{Grad} n,$$
$$\mathcal{F} = \mathcal{F}(\mathcal{C}). \tag{13.128}$$

We have to modify as well the simplifying assumptions. We replace the assumptions 5° and 6° by the following ones

5'° The dependence on the gradient of porosity $\mathbf{N}$ is linear.

6'° Higher order combinations of variables $\mathbf{G}, \acute{\mathbf{X}}^F, \Delta_n, \mathbf{N}$ can be neglected.

The analysis of the second law of thermodynamics reveals that the thermodynamical equilibrium is again defined by the relations

$$\mathbf{G}|_E = 0, \quad \acute{\mathbf{X}}^F\Big|_E = 0, \quad \boldsymbol{\Delta}_n|_E = 0. \tag{13.129}$$

Vector fluxes, instead of (13.70), and momentum sources, instead of (13.71), must have the form

$$\begin{aligned} \mathbf{Q} &= Q_V \acute{\mathbf{X}}^F - K\mathbf{G} + Q_N \mathbf{N}, \\ \mathbf{H} &= H_V \acute{\mathbf{X}}^F + H_T \mathbf{G} + H_N \mathbf{N}, \\ \mathbf{J} &= \Phi \acute{\mathbf{X}}^F + J_T \mathbf{G} + J_N \mathbf{N}, \end{aligned} \tag{13.130}$$

[2] Current partial mass densities ρ_t^α are described by volume changes, say for small deformations $\left(\rho_0^\alpha - \rho_t^\alpha\right)/\rho_0^\alpha$ which, in turn, are given by deformation gradients. Consequently, gradients of current mass densities describe a dependence on *second* gradients of deformation.

$$\mathbf{F}^{ST}\hat{\mathbf{p}} = \Pi_V \acute{\mathbf{X}}^F + \Pi_T \mathbf{G} + \Pi_N \mathbf{N}, \tag{13.131}$$

where the coefficients may be functions of the reduced set of constitutive variables

$$\mathcal{C}' := \left\{ \mathbf{F}^S, T, \Delta_n \right\}. \tag{13.132}$$

The variables $\mathbf{F}^S, T$ characterize equilibrium processes (ρ^F in equilibrium is not independent due to the vanishing relative velocity $\acute{\mathbf{X}}^F\Big|_E = 0$). The dependence on the nonequilibrium variable Δ_n must be treated separately due to its scalar character.

The rest of the analysis is similar to this presented in the previous case. We skip it and present here only some final results for isotropic materials.

The Helmholtz free energy for this model has the following form

$$\begin{aligned} &\rho\psi = \rho^S \psi^S + \rho^F \psi^F, \quad \rho_t^F = \rho^F J^{S-1}, \\ &\psi^S = \psi^S \left(I, II, J^S, \rho_t^F, T, \Delta_n \right), \quad \psi^F = \psi^F \left(J^S, \rho_t^F, T, \Delta_n \right). \end{aligned} \tag{13.133}$$

The partial Piola-Kirchhoff stress tensors are given by the relations

$$\begin{aligned} \mathbf{P}^F &= -\rho_t^{F2} \frac{\partial \psi^F}{\partial \rho_t^F} J^S \mathbf{F}^{S-T} - \Phi_0 \rho \frac{\partial \psi^F}{\partial \Delta_n} J^S \mathbf{F}^{S-T}, \\ \mathbf{P}^S &= \rho^S \frac{\partial \psi^S}{\partial \mathbf{F}^S} + \Phi_0 \rho \frac{\partial \psi^F}{\partial \Delta_n} J^S \mathbf{F}^{S-T}, \end{aligned} \tag{13.134}$$

while fluxes and sources satisfy the following constitutive relations

$$\begin{aligned} &\mathbf{H} = \frac{1}{T} \left(\mathbf{Q} - \rho^F \psi^F \acute{\mathbf{X}}^F \right), \quad \mathbf{J} = \Phi_0 J^S \acute{\mathbf{X}}^F, \quad \Phi_0 = \text{const.}, \\ &\hat{\mathbf{p}} = \pi \mathbf{F}^S \acute{\mathbf{X}}^F - N \mathbf{F}^{S-1} \operatorname{Grad} n, \quad \hat{n} = -\frac{\Delta_n}{\tau}. \end{aligned} \tag{13.135}$$

Consequently, in this model there exist equilibrium couplings of partial stresses through volume changes of components described by J^S and ρ_t^F as well as nonequilibrium couplings described by the dependence on Δ_n. In addition, there are couplings through the momentum source which are due to the diffusion velocity $\acute{\mathbf{X}}^F$ as well as the gradient of porosity $\operatorname{Grad} n$. In the next Section we show that linearization of this model leads to an extension of the classical Biot model of saturated porous materials.

13.3 Linear models of saturated poroelastic materials

Linear modelling of two-component porous materials was primarily initiated by works in the field of soil mechanics. These were investigations of H. Darcy (1803-1858) [84], K. Terzaghi (1883-1963) (e.g. [342], [344], [343]), P. Fillunger (1883-1937) (e.g. [111]), Y. I. Frenkel (1894-1952) (e.g. [115]), M. A. Biot (1905-1985) (e.g. see Tolstoy's collection of Biot's works on porous materials [325]) and many others. The aim was to construct a description of the coupling between mechanical response of skeleton and flows of fluids through the channels of porous and granular materials. Early models were linear and quasistatic but at the end of 1930s the problem of propagation of acoustic waves began to be investigated as well.

These linear models, in particular the so-called Biot model, describe the behavior of the following fields

1. Current *partial* mass densities, ρ_t^S, ρ_t^F or current *real* mass densities, ρ_t^{SR}, ρ_t^{FR}. The subscript t indicating the reference to the current location is usually dropped as in the linear model there is no distinction between Eulerian and Lagrangian description. Descriptions by either set of densities are equivalent if the porosity field n is incorporated: $\rho_t^S = (1-n)\,\rho_t^{SR}$, $\rho_t^F = n\rho_t^{FR}$.

Henry Darcy
1803 – 1858

Irvin Langmuir
1881 – 1957

Karl von Terzaghi
1883 – 1963

2. The displacement of the skeleton, $\mathbf{u}^S$. Its derivative with respect to $\mathbf{x}$ defines the deformation tensor. In the linear theory this is Almansi-Hamel deformation tensor $\mathbf{e}^S$. The time derivative gives the velocity of the skeleton, $\mathbf{v}^S$. Volume changes are specified by

the first invariant of Almansi-Hamel tensor, $e = \operatorname{tr} \mathbf{e}^S$. If there is no mass exchange between components this, in turn, determines uniquely changes of the partial mass density and the field of mass density ρ_t^S becomes superfluous

$$e = \frac{\rho_0^S - \rho_t^S}{\rho_0^S}, \quad |e| \ll 1, \tag{13.136}$$

where ρ_0^S is the initial value of the partial mass density, corresponding to $\mathbf{e}^S = 0$. This relation follows from the following identity (integrability condition)

$$\frac{\partial \mathbf{e}^S}{\partial t} = \operatorname{sym} \operatorname{grad} \mathbf{v}^S \quad \text{i.e.} \quad \frac{\partial e}{\partial t} = \operatorname{div} \mathbf{v}^S. \tag{13.137}$$

Therefore the mass balance equation yields

$$\frac{\partial \rho_t^S}{\partial t} + \operatorname{div}\left(\rho_t^S \mathbf{v}^S\right) \approx \frac{\partial \rho_t^S}{\partial t} + \rho_0^S \operatorname{div} \mathbf{v}^S = \frac{\partial}{\partial t}\left(\rho_t^S + \rho_0^S e\right) = 0. \tag{13.138}$$

Easy integration leads to (13.136).
3. The velocity of the fluid, $\mathbf{v}^F$. It is customary in Biot's model to work with the displacement of the fluid, $\mathbf{U}$. However, as in the fluid mechanics in general, this quantity does not have any physical meaning and it is not sought in solving boundary value problems. It plays only an auxiliary role in the model and defines the velocity of the fluid $\mathbf{v}^F = \partial \mathbf{U}/\partial t$ as well as its volume changes, $\varepsilon = \operatorname{div} \mathbf{U}$. As in the case of the skeleton, through the partial mass balance equation the latter determines uniquely changes of the partial mass density, ρ_t^F, if there is no mass exchange between components

$$\varepsilon = \frac{\rho_0^F - \rho_t^F}{\rho_0^F}, \quad \frac{\partial \varepsilon}{\partial t} = \operatorname{div} \mathbf{v}^F, \quad |\varepsilon| \ll 1, \tag{13.139}$$

where ρ_0^F is the initial value of the fluid partial mass density. Derivation is analogous to that presented above for the skeleton.
4. The porosity n (denoted by ϕ in many works on Biot's model). In Biot's model this quantity does not appear in the explicit form. However, its changes can be calculated in terms of volume changes of both components and material parameters (compressibilities) appearing in the model. Hence, even though not constant, the porosity is not listed among fields of Biot's model.
5. The temperature T does not appear in Biot's model. Processes are assumed to be isothermal. This assumption may be a good approximation in description of soils and rocks saturated by water, oil

or other similar fluids. However, it does not seem to be appropriate for acoustics of porous materials saturated by a gas in which sound propagates rather in adiabatic conditions. This is, for instance, the case in foams damping the noise.

M. A. Biot wrote equations of his model in the form of an extension of displacement equations of the classical theory of isotropic elasticity. In Figure 19, we reproduce the page of the paper of M. A. Biot from 1956 [37] in which the governing equations (4.2) of his model appear.

M. A. BIOT 172

Hence, ρ_{12} is the additional apparent mass with a change in sign. Therefore, the dynamic coefficients may be written

$$\rho_{11}=\rho_1+\rho_a \quad \rho_{22}=\rho_2+\rho_a \quad \rho_{12}=-\rho_a. \qquad (3.18)$$

Further conditions must be satisfied by these dynamic coefficients if the kinetic energy is to be a positive definite quadratic form. The coefficients ρ_{11} and ρ_{22} must be positive

$$\rho_{11}>0 \quad \rho_{22}>0$$

and

$$\rho_{11}\rho_{22}-\rho_{12}^2>0. \qquad (3.19)$$

These inequalities are always satisfied if the coefficients are given by the relations (3.18) where $\rho_1\rho_2$ and ρ_a are positive by their physical nature.

In terms of stresses the force components are expressed as stress gradients, i.e.,

$$q_x=\frac{\partial\sigma_x}{\partial x}+\frac{\partial\tau_z}{\partial y}+\frac{\partial\tau_y}{\partial z}, \quad Q_x=\partial s/\partial x, \text{ etc.} \qquad (3.20)$$

Hence, we have the dynamic equations

$$\frac{\partial\sigma_x}{\partial x}+\frac{\partial\tau_z}{\partial y}+\frac{\partial\tau_y}{\partial z}=\frac{\partial^2}{\partial t^2}(\rho_{11}u_x+\rho_{12}U_x)$$
$$\frac{\partial s}{\partial x}=\frac{\partial^2}{\partial t^2}(\rho_{12}u_x+\rho_{22}U_x), \text{ etc.} \qquad (3.21)$$

Strictly speaking, the generalized forces are defined as the virtual work of the microscopic stresses per unit value of the displacement vector **u** and **U** and not as the average of the microscopic stresses as used in expressions

direction,

$$N\nabla^2u_x+(A+N)\frac{\partial e}{\partial x}+Q\frac{\partial\epsilon}{\partial x}=\frac{\partial^2}{\partial t^2}(\rho_{11}u_x+\rho_{12}U_x)$$
$$Q\frac{\partial e}{\partial x}+R\frac{\partial\epsilon}{\partial x}=\frac{\partial^2}{\partial t^2}(\rho_{12}u_x+\rho_{22}U_x), \qquad (4.1)$$

and two other similar equations, respectively, for the directions y and z. With the vector notation

$$\mathbf{u}=(u_xu_yu_z)$$
$$\mathbf{U}=(U_xU_yU_z).$$

Equations (5.1) are written

$$N\nabla^2\mathbf{u}+\text{grad}[(A+N)e+Q\epsilon]=\frac{\partial^2}{\partial t^2}(\rho_{11}\mathbf{u}+\rho_{12}\mathbf{U})$$
$$\text{grad}[Qe+R\epsilon]=\frac{\partial^2}{\partial t^2}(\rho_{12}\mathbf{u}+\rho_{22}\mathbf{U}). \qquad (4.2)$$

These six equations for the six unknown components of the displacements **u** and **U** completely determine the propagation.

Because of the statistical isotropy of the material, it can be shown that the rotational waves are uncoupled from the dilatational waves and obey independent equations of propagation. This is done in the usual way by introducing the operations div and curl

$$\text{div}\mathbf{u}=e \quad \text{div}\mathbf{U}=\epsilon$$
$$\text{curl}\mathbf{u}=\omega \quad \text{curl}\mathbf{U}=\mathbf{\Omega}. \qquad (4.3)$$

Applying the divergence operation to both equations (4.2), we obtain

$$\nabla^2(Pe+Q\epsilon)=\frac{\partial^2}{\partial t^2}(\rho_{11}e+\rho_{12}\epsilon)$$
$$\nabla^2(Qe+R\epsilon)=\frac{\partial^2}{\partial t^2}(\rho_{12}e+\rho_{22}\epsilon) \qquad (4.4)$$

Fig.19: *Quotation from the paper of* M. A. Biot*: Theory of Propagation of Elastic Waves in a Fluid-Saturated Porous Solid. I. Low-Frequency Range,* JASA, **28**, 168-178, 1956 [37].

Biot himself was frequently changing the notation. In the version presented in Figure 19, the model contains the constant material parameters A, N $(=\mu^S$ – shear modulus) which describe effective (macroscopic, depending on an initial porosity n_0) elastic properties of the skeleton (solid component), and related to, for instance, Lamé coefficients, λ, μ, the parameter R describes the effective (macroscopic) compressibility of the fluid, Q is the effective coupling parameter, characteristic for Biot's model, and ρ_{12} describes the coupling through the relative accelerations and it is usually attributed to the so-called added mass effect. The Biot model in this form is until today the most frequently used model for the investigation of acoustic properties of porous materials.

In some of his works M. Biot uses also the *increment of fluid contents,* ζ, as a field replacing volume changes of the fluid, ε. This field is defined in the following way

$$\zeta = n_0 \left(e - \varepsilon\right) \equiv n_0 \operatorname{div} \left(\mathbf{u} - \mathbf{U}\right). \tag{13.140}$$

The latter relation yields an important conclusion

$$\frac{\partial \zeta}{\partial t} = n_0 \operatorname{div} \left(\mathbf{v}^S - \mathbf{v}^F\right). \tag{13.141}$$

Hence the field ζ cannot change without diffusion: there must be a relative motion of components in order to change the fluid contents. However, it means as well that changes of fluid contents cannot appear in thermodynamical equilibrium processes. This conclusion has paramount consequences for the applicability of variational methods in the theory of poroelastic materials. It indicates that, in spite of such frequent statements in the literature (e.g. [90]), the true variational principle for porous materials cannot be introduced and only the so-called pseudovariational method can be applied. This means that the so-called action functional does not exist and one may only construct variations as nonintegrable differential forms.

We proceed to derive field equations of the linear model on the basis of nonlinear models discussed in the previous Section. As usual we follow the thermodynamical strategy and construct these equations on the basis of balance equations.

In an inertial reference system the partial momentum balance equations have the following linearized form

$$\rho_0^S \frac{\partial \mathbf{v}^S}{\partial t} = \operatorname{div} \mathbf{T}^S + \rho_0^S \mathbf{b}^S + \hat{\mathbf{p}}, \quad \rho_0^F \frac{\partial \mathbf{v}^F}{\partial t} = -\operatorname{grad} p^F + \rho_0^F \mathbf{b}^F - \hat{\mathbf{p}}, \tag{13.142}$$

where the current mass densities are replaced by their reference values, convective parts of momentum fluxes (i.e. $\operatorname{div}\left(\rho^S \mathbf{v}^S \otimes \mathbf{v}^S\right)$ and $\operatorname{div}\left(\rho^F \mathbf{v}^F \otimes \mathbf{v}^F\right)$, respectively) are neglected and it is assumed that the partial stress tensor in the fluid reduces to the partial pressure p^F. This coincides with the nonlinear result (13.134) under constitutive assumptions presented in the previous Section.

The first two simplifications follow from the linearization, while the last one means that we consider the fluid component to be ideal on the macroscopic level of description. This does not have to be the case for the true fluid whose viscosity appears in the microstructural relation for momentum source $\hat{\mathbf{p}}$ (e.g. [127]). We return further to this relation.

The quantities $\rho_0^S \mathbf{b}^S, \rho_0^F \mathbf{b}^F$ are partial external body forces per unit volume.

Frequently the partial description of momentum is replaced by the bulk momentum and only one of the partial momenta. It is usually the partial momentum balance for the fluid because in some cases of practical bearing the inertial force of the fluid can be neglected and this momentum balance yields then a generalization of Darcy's law. The bulk momentum balance which is the sum of partial equations has the form

$$\rho_0 \frac{\partial \mathbf{v}}{\partial t} = \operatorname{div} \mathbf{T} + \rho_0 \mathbf{b}, \tag{13.143}$$

where

$$\begin{aligned} \mathbf{T} \approx \mathbf{T}^S - p^F \mathbf{1}, \quad \rho_0 &= \rho_0^S + \rho_0^F, \\ \rho_0 \mathbf{v} = \rho_0^S \mathbf{v}^S + \rho_0^F \mathbf{v}^F, \quad \rho_0 \mathbf{b} &= \rho_0^S \mathbf{b}^S + \rho_0^F \mathbf{b}^F. \end{aligned} \tag{13.144}$$

Obviously, $\mathbf{T}$ denotes the bulk stress tensor and $\mathbf{v}$ is the so-called barycentric velocity. The latter is not a very useful field because it does not have any global physical interpretation: Diffusion separates particles of components after a finite time so much that one cannot define any reasonable collection of centers of gravity. However, in some quasistatic problems the bulk momentum balance equation may be very useful. It appears, for instance, in the theory of plastic deformations of soils in which the bulk stress tensor plays an important role (e.g. [342]).

The above structure of momentum balance shows that the partial stress tensor in the skeleton usually cannot be identified with the so-called effective stresses introduced to soil mechanics by Terzaghi [342]. The partial pressure in the fluid p^F is in many cases related to the so-called *pore pressure*, p, by the relation $p^F = np$, where n is the current porosity.

The balance equation of porosity in the simplest case of large relaxation times ($\tau \to \infty$, i.e. $\hat{n} = -\frac{\Delta_n}{\tau} \to 0$) for linear isotropic case has the following form

$$\frac{\partial \Delta_n}{\partial t} + \Phi \operatorname{div} \left(\mathbf{v}^F - \mathbf{v}^S \right) = 0, \quad \Delta_n = n - n_E, \quad n_E = n_0 \left(1 + \delta e \right), \tag{13.145}$$

where δ, Φ are material parameters depending on n_0. The lack of relaxation properties of porosity is the common assumption in geotechnical applications. It is not so obvious, if justified at all, in applications to biomechanics, description of tissues, damage of concrete, etc.

The above equation has the following solution

$$n = n_0 \left[1 + \delta e + \frac{\Phi}{n_0} (e - \varepsilon) \right] . \tag{13.146}$$

Such changes of porosity are always present in Biot's model but not stated explicitly. The reason is that porosity in Biot's model is not an *independent* field. According to (13.146), it is given in terms of volumetric strains of both components, e, ε, and the material parameters required by the above equation, δ, Φ, follow from material parameters present in the model (no additional material parameters for porosity!). As the relation (13.146) follows from gedanken experiments introduced by Biot and Willis (see further), the above consideration seems to be a good justification of models based on balance equations of porosity.

In order to construct field equations by means of the above balance equations we need in general constitutive relations for the following *constitutive quantities*

$$\left\{ \mathbf{T}^S, \mathbf{T}^F, \hat{\mathbf{p}}, \mathbf{J}, n_E, \hat{n} \right\} . \tag{13.147}$$

We have already indicated some simplifications in this set. Instead of the tensor $\mathbf{T}^F$ we need a constitutive relation only for the partial pressure p^F. For the flux of porosity $\mathbf{J}$ and the equilibrium porosity n_E, the constitutive assumptions are already incorporated in the porosity balance equation (13.145). Consequently, in order to close the system we need constitutive relations of the form

$$\begin{aligned} \mathbf{T}^S &= \mathbf{T}^S \left(\mathbf{e}^S, \varepsilon, \Delta_n, \mathbf{v}^F - \mathbf{v}^S, \mathbf{a}_r, \operatorname{grad} n \right) , \\ p^F &= p^F \left(\mathbf{e}^S, \varepsilon, \Delta_n, \mathbf{v}^F - \mathbf{v}^S, \mathbf{a}_r, \operatorname{grad} n \right) , \end{aligned} \tag{13.148}$$

and for the momentum source which we present in a linear form with respect to vectorial variables (comp. the combination of (13.71), (13.135) in the nonlinear isothermal case)

$$\hat{\mathbf{p}} = \hat{\mathbf{p}}_{diff} - N \operatorname{grad} n - \rho_{12} \mathbf{a}_r , \tag{13.149}$$

where $\hat{\mathbf{p}}_{diff}$ is the part determined by the relative velocity $\mathbf{v}^F - \mathbf{v}^S$, N is a material parameter. In the simplest case $\hat{\mathbf{p}}_{diff} = \pi \left(\mathbf{v}^F - \mathbf{v}^S \right)$, where π is the constant permeability coefficient.

For the purpose of construction of Biot's model we can leave out the dependence on Δ_n. This assumption is not necessary for the linearity of the model and some wave problems have already been investigated with an influence of Δ_n. It has been found that this influence is indeed very small

[8]. The lack of dependence on grad n would yield the following result (see: (13.120), (13.121) & (13.122))

$$\mathbf{T}^S = \mathbf{T}^S\left(\mathbf{e}^S\right), \quad p^F = p^F\left(\varepsilon\right). \tag{13.150}$$

This is the *simple mixture* model whose linearization yields Biot's constitutive relations for stresses with $Q = 0$. As shown in the previous Section for the nonlinear model this fault of the model can be removed. Essential is the constitutive dependence on grad n which yields the possibility of coupling proposed by Biot.

In the linear model which follows from (13.148) and (13.149) we obtain the following relations for partial stresses

$$\begin{aligned} \mathbf{T}^S &= \mathbf{T}_0^S + \lambda^S e\mathbf{1} + 2\mu^S\mathbf{e}^S + Q\varepsilon\mathbf{1} - N\left(n - n_0\right)\mathbf{1}, \\ p^F &= p_0^F - \rho_0^F\kappa\varepsilon - Qe - N\left(n - n_0\right), \end{aligned} \tag{13.151}$$

where n_0 is a constant reference value of porosity, $\lambda^S, \mu^S, \kappa, Q, N$ are material parameters which depend on the reference value of porosity, n_0. We have incorporated in stresses the contribution of grad n appearing in the momentum source $\hat{\mathbf{p}}$. This is possible for a constant coefficient N which is the case in the linear model.

Bearing the above relations in mind we see that the set of balance equations (13.137), (13.139), (13.142), (13.145), i.e. two scalar, two vector and one tensor equations form the set of field equations for the fields $\left\{\mathbf{e}^S, \varepsilon, n, \mathbf{v}^S, \mathbf{v}^F\right\}$. Clearly, if we introduce the displacement vector $\mathbf{u}^S$ then we can skip the tensor equation for $\mathbf{e}^S$ and the momentum balance equation for the skeleton becomes the second order equation for this displacement.

Consequently, the model is determined by the following set of material parameters

$$\left\{\lambda^S, \mu^S, \rho_0^F\kappa, Q, \delta, \Phi, n_0; N, \pi, \rho_{12}\right\}. \tag{13.152}$$

The second group contains the parameter N which is additional to the set proposed by Biot and two parameters π, ρ_{12} which describe an influence of relative motion of components.

This set may still be reduced to a smaller and more fundamental set and we present further a procedure which yields this reduction. We should also mention that the constant permeability coefficient π has been found not appropriate in the description of acoustic waves. Measurements for monochromatic waves indicate that it should be frequency-dependent. Such a dependence has already been suggested by Biot who proposed a correction following from a microstructural estimates. The problem is that the correction of equations by introducing the frequency dependent complex permeability as suggested by Biot and many others, say in the form $\pi F(\omega)$, is mathematically nonsensical. After such a correction, solutions of the equations of motion would have to be complex. In addition, functions appearing in these equations are defined on different spaces: some of them are time-dependent and some other frequency-dependent and this is, of course, mathematically not correct. Papers concerning this subject avoid the problem by writing Fourier transforms in which the variable of Fourier transformation is identified with the frequency appearing in F.

The above remarks indicate the necessity of a correction in the formulation of the problem and this, as a matter of fact, was already suggested by Biot himself [38]. Namely, the influence of relative velocities in the macroscopic model should have a form reflecting viscous-like effects following from obstacles for the motion of the real fluid in channels. It means that, similarly to viscoelastic materials, one should expect a kind of hereditary integral describing the momentum source.

In spite of their practical importance explicit relations for the bulk permeability π are not needed in the macroscopic construction of such a model. They may be adjusted to experiments or selected by means of some analysis of microstructure in particular problems. It is only essential in the linear model to account for their changes in time as a dynamic process goes on. This can be done in the same way as it is done for viscoelastic properties of solids. We proceed to construct such a model of momentum sources (see [375]).

First of all, as it was pointed out, the second law of thermodynamics indicates the following dissipation inequality

$$\hat{\mathbf{p}}_{diff} \cdot \mathbf{w} \geq 0, \quad \mathbf{w} := \mathbf{v}^F - \mathbf{v}^S. \tag{13.153}$$

Consequently, $\hat{\mathbf{p}}_{diff}$ must be odd in the relative velocity. Within a linear model of isotropic materials it has then the following form for a *constant* relative velocity $\mathbf{w}$

$$\hat{\mathbf{p}}_{diff}(\mathbf{w}, t) = \pi(t)\,\mathbf{w}, \tag{13.154}$$

which indicates a smallness of the relative velocity. Knowledge of this single-step response function $\pi(t)$ is sufficient to predict an output $\hat{\mathbf{p}}_{diff}$ for an arbitrary constant input $\mathbf{w}$. Let us note that the time dependence of the permeability coefficient π should be invariant with respect to the translation of the origin of time axis. This means that t is the time lag since application of the constant relative velocity $\mathbf{w}$. An input $\mathbf{w}(t) = \mathbf{w}_0 H(t - t_0)$, where $H(t - t_0)$ is the Heaviside function would lead to the output $\hat{\mathbf{p}}_{diff} = \pi(t - t_0)\,\mathbf{w}_0$. Consequently, for the two-step history

$$\mathbf{w}(t) = H(t - t_1)\,\Delta\mathbf{w}_1 + H(t - t_2)\,\Delta\mathbf{w}_2, \tag{13.155}$$

the output must have the following form

$$\hat{\mathbf{p}}_{diff}(t) = \pi_1(t, t_1, t_2)\,\Delta\mathbf{w}_1 + \pi_2(t, t_1, t_2)\,\Delta\mathbf{w}_2. \tag{13.156}$$

This expression is valid for all constant increments of the relative velocity. Hence, it holds in particular for $\Delta\mathbf{w}_2 = 0$. Then the momentum source must be $\pi(t - t_1)\,\Delta\mathbf{w}_1$. In the same way we identify the second contribution and it follows

$$\mathbf{w}(t) = \sum_{\alpha=1}^{2} H(t - t_\alpha)\,\Delta\mathbf{w}_\alpha \quad \Longrightarrow \quad \hat{\mathbf{p}}_{diff}(t) = \sum_{\alpha=1}^{2} \pi_\alpha(t - t_\alpha)\,\Delta\mathbf{w}_\alpha. \tag{13.157}$$

Certainly, the same argument applies for an arbitrary number of steps. In rheology this is called *Boltzmann's superposition principle.* By means of some continuity assumptions which are similar to these of the theory of viscoelasticity we can approximate by step history the response to an arbitrary integrable history of the relative velocity. We have then

$$\mathbf{w}(t) = \int_{-\infty}^{t} H(t - s)\,d\mathbf{w}(s) \quad \Longrightarrow \quad \hat{\mathbf{p}}_{diff}(t) = \int_{-\infty}^{t} \pi(t - s)\,d\mathbf{w}(s). \tag{13.158}$$

As we assume the invariance with respect to the translation the initial point of the relative motion can be chosen arbitrarily. Therefore it is convenient to set the lower limit of integration at $-\infty$.

Integration by parts and the change of variables yields

$$\hat{\mathbf{p}}_{diff}(t) = \pi(0)\,\mathbf{w}(t) + \int_0^\infty \dot{\pi}(s)\,\mathbf{w}(t - s)\,ds, \quad \dot{\pi}(s) := \frac{\partial \pi}{\partial s}. \tag{13.159}$$

This constitutive relation contains the memory functional typical for viscous effect. The instantaneous reaction is characterized by the initial permeability $\pi(0)$. We demonstrate here only some simple properties of the above relation but it is obvious that the relation for the momentum source can be easily extended on nonlinear models. Such extensions are known for memory materials and we shall not present any details in this Section.

Let us write the relation (13.159) for the harmonic disturbance of the frequency ω. We obtain immediately

$$\mathbf{w} = \mathbf{w}^*(\mathbf{x})\,e^{-i\omega t}, \quad \hat{\mathbf{p}}_{diff} = \hat{\mathbf{p}}^*_{diff}(\mathbf{x})\,e^{-i\omega t} \tag{13.160}$$
$$\Longrightarrow \quad \hat{\mathbf{p}}^*_{diff} = \left(-i\omega \frac{1}{\pi(0)} \int_0^\infty \pi(s)\,e^{i\omega s} ds\,\pi(0)\right) \mathbf{w}^*.$$

The transform of permeability coefficient in the parentheses can be now identified with the relation $F(\omega)\,\pi(0)$ proposed for the permeability as a function of frequency. This contribution appears in Fourier transformed momentum balance equations rather than in the original momentum balance equations which hold for an arbitrary time dependent input.

The above hereditary form of the momentum source is, as already indicated, attributed to viscosity of the true fluid and to the tortuosity. However, the influence of tortuosity is assumed to lead also to the added mass effect and, consequently, to the nondiagonal form of the matrix of mass densities. This yields contributions of the relative accelerations which were discussed in the previous Section. We return to this interpretation of the added mass in the next Section on monochromatic waves.

This completes the general remarks on the construction of a two-component model for poroelastic materials. However, we demonstrate here additionally a procedure of identification of some effective materials parameters for linear poroelastic materials which seems to be rather unique in the transition from semimacroscopic to macroscopic description. This procedure has been initiated by works of Gassmann in 1951 [117] and developed by M. A. Biot and D. G. Willis in 1957 [39].

Before we proceed to discuss this procedure we quote here another form of Biot's constitutive relations which possesses a particular practical bearing. First of all, we can use the variable ζ, the increment of fluid contents, instead of the volume change of fluid, ε. Then the relations for the bulk stress $\mathbf{T}$ and the pore pressure p have the form

$$\mathbf{T} \approx \mathbf{T}^S + \mathbf{T}^F = \mathbf{T}_0 + \left(\lambda^S + \rho_0^F \kappa + 2Q\right) e\mathbf{1} + 2\mu^S \mathbf{e}^S - \frac{Q + \rho_0^F \kappa}{n_0} \zeta \mathbf{1},$$

$$\mathbf{T}_0 = \mathbf{T}_0^S - p_0^F \mathbf{1}, \tag{13.161}$$

$$p = \frac{1}{n} p^F \approx \frac{1}{n_0} \left(1 - \frac{n - n_0}{n_0}\right) \left[n_0 p_0 - \left(Q + \rho_0^F \kappa\right) e + \frac{\rho_0^F \kappa}{n_0} \zeta\right], \quad p_0 = \frac{p_0^F}{n_0},$$

and changes of porosity are given by the following relation

$$\frac{n - n_0}{n_0} = \delta e + \frac{\Phi}{n_0^2} \zeta. \tag{13.162}$$

If the initial pore pressure p_0 is much smaller than material parameters $\frac{Q+\rho_0^F \kappa}{n_0}$, $\frac{\rho_0^F \kappa}{n_0^2}$ then we can neglect the contribution following from changes of porosity. We obtain finally

$$\mathbf{T} = \mathbf{T}_0 + K e \mathbf{1} + 2G \operatorname{dev} \mathbf{e}^S - C\zeta \mathbf{1}, \quad \operatorname{dev} \mathbf{e}^S = \mathbf{e}^S - \tfrac{1}{3} e \mathbf{1}, \tag{13.163}$$

$$K = \lambda^S + \tfrac{2}{3} \mu^S + \rho_0^F \kappa + 2Q, \quad C = \frac{Q + \rho_0^F \kappa}{n_0}, \quad G = \mu^S,$$

$$p = p_0 - Ce + M\zeta, \quad M = \frac{\rho_0^F \kappa}{n_0^2}. \tag{13.164}$$

This form shows clearly that stresses consist of equilibrium parts given by the deformation $\mathbf{e}^S$ and nonequilibrium parts containing ζ. The set of parameters

$$\{K, G, M, C, \delta, \Phi, n_0; N, \pi, \rho_{12}\} \tag{13.165}$$

replaces the set (13.152).

We apply these relations to some simple static processes in granular materials. These processes yield purely volumetric deformations. For such processes we introduce a procedure stemming from the averaging method over a representative elementary volume (REV) which yields relations between $\lambda^S + \frac{2}{3}\mu^S$ or K, $\rho_0^F\kappa$, or M, Q or C, and N, δ, Φ on the one hand and true compressibilities of substances, K_s, K_f of the solid frame and fluid components, respectively, the initial porosity, n_0 and the undrained compressibility modulus K_d on the other hand. Certainly, any other combination of macroscopic parameters can then be also easily calculated. The compressibilities K_s, K_f are measured in experiments on the substance of the skeleton and on the fluid outside porous materials and they are, of course, independent of any particular morphology of the porous material. Simultaneously, the modulus K_d is measured in very simple undrained compression experiments and is used in soil mechanics on equal footing with the true compressibilities.

We follow here the strategy proposed in the works [372], [376].

Description of granular materials by means of a macroscopic model is particularly easy when a deformation is homogeneous, spherically symmetric and the mechanical reactions of the system reduce to pressures. Such a system will be considered now.

For such a deformation, we have

$$\mathbf{e}^S = \tfrac{1}{3}e\mathbf{1}, \quad p^S = -\tfrac{1}{3}\operatorname{tr}\mathbf{T}^S, \quad p^F = -\tfrac{1}{3}\operatorname{tr}\mathbf{T}^F, \tag{13.166}$$

and the *macroscopic constitutive relations* for partial pressures are as follows (comp. (13.151))

$$p^S - p_0^S = -\left(\lambda^S + \tfrac{2}{3}\mu^S\right)e - Q\varepsilon + N(n - n_0), \tag{13.167}$$

$$p^F - p_0^F = -\rho_0^F\kappa\varepsilon - Qe - N(n - n_0).$$

For porosity changes we have (see (13.146) or (13.162))

$$\frac{n-n_0}{n_0} = \delta e + \frac{\Phi}{n_0}(e-\varepsilon). \tag{13.168}$$

This relation for the particular case of equilibrium (i.e. $e = \varepsilon$) has been derived by Gassmann [117]. Momentum balance equations are in the homogeneous equilibrium identically satisfied. External loading is applied through the boundary. Hence, in the static case we have the *equilibrium condition of* the full pressure change with a given excess (external) pressure Δp, i.e.

$$\Delta p = \left(p^S - p_0^S\right) + \left(p^F - p_0^F\right). \tag{13.169}$$

In order to solve the problem, i.e. to find five quantities $e, \varepsilon, p^S - p_0^S, p^F - p_0^F, n$ for a given excess pressure Δp we need, in addition to the above four equations (13.167), (13.168), (13.169), another boundary condition. This will be specified later for three different problems which we call *gedanken experiments.*

However, this simple linear problem can also be formulated in terms of semimacroscopic quantities e^R, ε^R, $p^{SR} - p_0^{SR}$, $p^{FR} - p_0^{FR}$ and n. The first two quantities describe real volume changes of the solid material and of the fluid, respectively, and they are related to the real mass densities ρ^{SR}, ρ^{FR} by relations analogous to (13.136) and (13.139), i.e.

$$e^R = \frac{\rho_0^{SR} - \rho^{SR}}{\rho_0^{SR}}, \quad \varepsilon^R = \frac{\rho_0^{FR} - \rho^{FR}}{\rho_0^{FR}}. \tag{13.170}$$

The constitutive relations on this level are very simple

$$p^{SR} - p_0^{SR} = -K_s e^R, \quad p^{FR} - p_0^{FR} = -K_f \varepsilon^R. \tag{13.171}$$

The construction of the boundary condition requires an assumption which we justify further. Namely, it must be assumed that surface contributions of components to the common traction are given by the porosity. We have then

$$\begin{aligned} p^S &= (1-n)\, p^{SR}, \quad p_0^S = (1-n_0)\, p_0^{SR}, \\ p^F &= n p^{FR}, \quad p_0^F = n_0 p_0^{FR}, \end{aligned} \tag{13.172}$$

i.e.

$$\begin{aligned} p^S - p_0^S &= (1-n_0)\left(p^{SR} - p_0^{SR}\right) - \frac{n-n_0}{n_0} n_0 p_0^{SR} \approx (1-n_0)\left(p^{SR} - p_0^{SR}\right), \\ p^F - p_0^F &= n_0\left(p^{FR} - p_0^{FR}\right) + \frac{n-n_0}{n_0} n_0 p_0^{FR} \approx n_0\left(p^{FR} - p_0^{FR}\right). \end{aligned} \tag{13.173}$$

Then the condition (13.169) becomes

$$\Delta p = \left((1-n)\, p^{SR} - (1-n_0)\, p_0^{SR} \right) + \left(n p^{FR} - n_0 p_0^{FR} \right) \qquad (13.174)$$
$$= (1-n)\left(p^{SR} - p_0^{SR} \right) + n \left(p^{FR} - p_0^{FR} \right) - (n - n_0)\left(p_0^{SR} - p_0^{FR} \right).$$

The last contribution vanishes if we assume that the initial state is self-equilibrated on the semimacroscopic level, i.e. $p_0^{SR} = p_0^{FR}$. The unknown five quantities of the semimacroscopic description $e^R, \varepsilon^R, p^{SR} - p_0^{SR}, p^{FR} - p_0^{FR}, n$ can be determined from (13.168), (13.171), (13.174), provided, as in the macroscopic description, we formulate an additional boundary condition.

However, we still have at the disposal relations between macroscopic and semimacroscopic description. These are two dynamic compatibility relations (13.173) and the following geometrical compatibility relations

$$e = \frac{\rho_0^S - \rho^S}{\rho_0^S} = \frac{(1-n_0)\, \rho_0^{SR} - (1-n)\, \rho^{SR}}{(1-n_0)\, \rho_0^{SR}} \approx e^R + \frac{n - n_0}{1 - n_0},$$
$$\varepsilon = \frac{\rho_0^F - \rho^F}{\rho_0^F} = \frac{n_0 \rho_0^{FR} - n \rho^{FR}}{n_0 \rho_0^{FR}} \approx \varepsilon^R - \frac{n - n_0}{n_0}. \qquad (13.175)$$

If we collect all the above equations, we obtain 1 equilibrium condition (13.169), 1 equation for porosity (13.168), two macroscopic constitutive relations (13.167), two semimacroscopic constitutive relations (13.171), two dynamic compatibility conditions (13.173) and two geometrical compatibility conditions (13.175). Additionally, we expect one additional boundary condition for each gedanken experiment we construct. We shall see that there are three such experiments in the model we are considering. Consequently, we have 13 equations. We assume as well that material parameters K_s, K_f, K_d are given.

Simultaneously, we have five geometrical unknowns, $e, \varepsilon, e^R, \varepsilon^R, n$, four dynamic unknowns, $p^S - p_0^S$, $p^F - p_0^F$, $p^{SR} - p_0^{SR}$, $p^{FR} - p_0^{FR}$ and the following six macroscopic material parameters: K, M, C, N, δ, Φ. Altogether these are 15 quantities. This means that we can find solutions of all three problems in terms of the excess pressure Δp and, additionally, if we eliminate Δp we obtain relations between the set of six material parameters and three known parameters K_s, K_f, K_d and the initial porosity n_0.

Three gedanken experiments were described by Biot and Willis [39]. They are called: 1° *jacketed undrained,* 2° *jacketed drained* and 3° *unjacketed* and they are schematically shown in Figure 20.

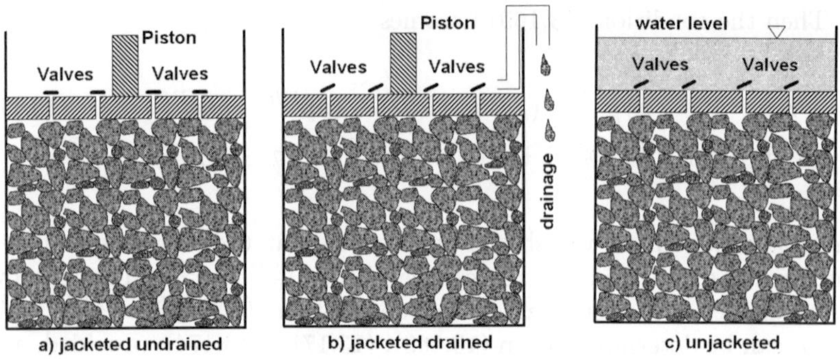

Fig. 20: *Schemes of gedanken experiments:* a) *jacketed undrained,* b) *jacketed drained,* c) *unjacketed.*

In experiment 1), the granular material is contained in the impermeable spherical jacket which is assumed to be ideal, i.e. it does not contribute to the balance of forces in the system. In experiment 2), the jacket is permeable, i.e. the pipes may supply or drain the fluid from the sample. In experiment 3) a heap of the granular material is in contact with the fluid in the container.

These experiments are called *gedanken experiments* (*ger.* = thought experiments) because they are in principle possible but in their evaluation we have to neglect many effects such as saturation and surface tension, nonisothermal character, boundary layers, flows and viscosity, etc.

The conditions which describe them are as follows:

$$1^\circ : \zeta = 0, \quad 2^\circ : p^{FR} = p_0^{FR}, \quad 3^\circ : p^{FR} - p_0^{FR} = \Delta p. \tag{13.176}$$

We shall not present here the full analysis of the problem (see [372]). It can be shown that it has two distinct real solutions. One of them possesses $N \neq 0$, in the other one $N = 0$. In the latter case we can construct the solution in the closed form. The formulae for the material parameters are then called the *Gassmann relations* and they have the following form

$$K = \frac{(K_s - K_d)^2}{\frac{K_s^2}{K_W} - K_d} + K_d, \quad C = \frac{K_s (K_s - K_d)}{\frac{K_s^2}{K_W} - K_d}, \quad M = \frac{K_s^2}{\frac{K_s^2}{K_W} - K_d}, \tag{13.177}$$

where

$$\frac{1}{K_W} := \frac{1 - n_0}{K_s} + \frac{n_0}{K_f}. \tag{13.178}$$

Simultaneously, for parameters describing changes of porosity it follows

$$\delta = \frac{K_V - K}{n_0 (K_s - K_f)}, \quad \Phi = \frac{n_0 C - n_0 K_f}{K_s - K_f}, \tag{13.179}$$

where

$$K_V := (1 - n_0) K_s + n_0 K_f, \tag{13.180}$$

and K and C are given by (13.177).

An extensive discussion of material relations for the linear Biot model as well as numerous practical applications can be found in the book of H. F. Wang [345].

Below we present the numerical results for the relations between material parameters. We show in Figures 21 and 22 the macroscopic parameters as functions of the porosity n_0. The following data were used in these calculations

$$K_s = 48 \times 10^9 \text{ Pa}, \quad K_f = 2.25 \times 10^9 \text{ Pa}, \quad K_d = \frac{K_s}{1 + g n_0}. \tag{13.181}$$

The values of compressibilities K_s and K_f are typical – the first one for the sandstone, the second one for water. The last relation is called the *Geertsma equation* [348] and it specifies K_d in function of the empirical parameter g. The typical value of g for rocks is 50.

It is seen that the compressibility $K = \lambda^S + \frac{2}{3}\mu^S + \rho_0^F \kappa + 2Q$ is a decaying function of n_0 and the results for the Biot model ($N = 0$) and for the full model coincide quantitatively. The compressibility coefficient of the fluid $M = \rho_0^F \kappa / n_0^2$ behaves in the similar way but there exists a quantitative difference between the models in the range of small porosities. The coupling coefficient $C = \left(Q + \rho_0^F \kappa\right) / n_0$ is stronger in the Biot model than in the full model and this is the result of an additional coupling through changes of porosity in the full model. Finally, it is seen that values of the parameter N in the full model are much smaller than values of other parameters. Obviously, in the Biot model this parameter is equal to zero.

In Figure 22, we demonstrate the behavior of the coupling parameter Q for both models. Clearly the values for Biot's model are almost twice as big as these for the full model. This is the indication that interactions between components are in the same range of values because the values of N in the full model are almost identical as these for Q in this model which means that the full interaction term is also identical in both models. This Figure shows as well an instability of the full model for large values of porosity. This may be the indication of a structural instability of granular materials for porosities bigger than app. 0.58 and indicated in some structural works on this subject. This problem has not been investigated any further.

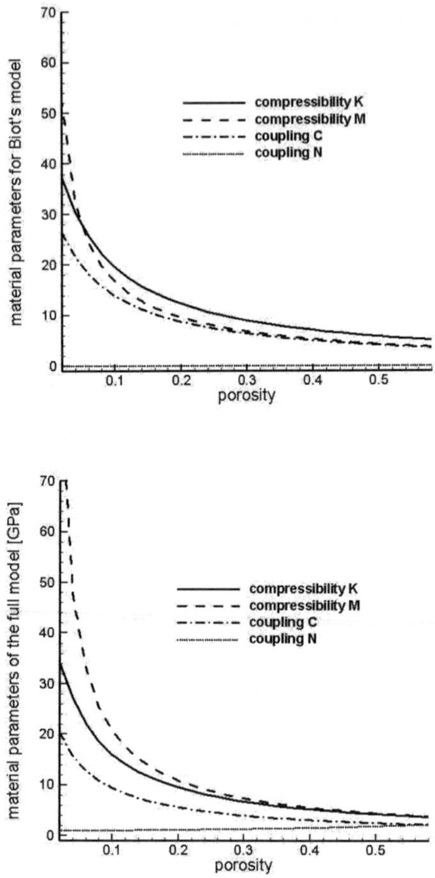

Fig. 21: *Comparison of material parameters for Biot's model (the upper panel) and the full model with the porosity gradient (the lower panel) for Geertsma parameter* $g = 50$.

The above results have a very important bearing in the nondestructive testing of soils which we indicate in the next Section.

Unfortunately the remaining material parameters $G = \mu^S$, π and ρ_{12} do not follow from such simple considerations and require real experiments on porous materials.

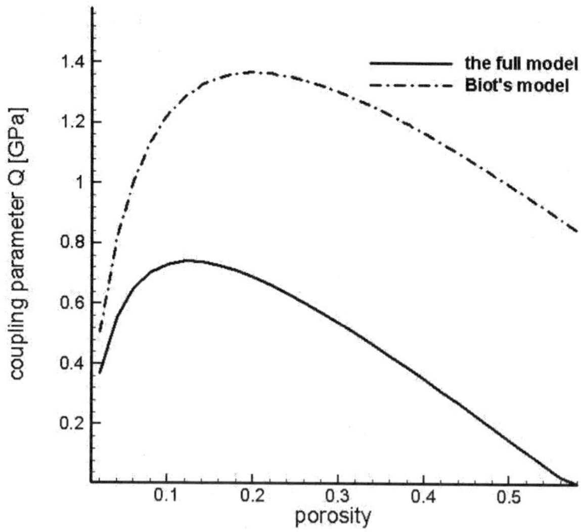

Fig. 22: *Comparison of the coupling parameter Q (right) for the full model and Biot's model with Geertsma parameter $g = 50$.*

13.4 Waves in poroelastic materials

Due to its practical bearing, continuous modelling of waves is one of the most important subjects of continuum thermodynamics. Such wave phenomena as propagation of acoustic waves (sound) [1], [49], surface waves (Rayleigh, Stoneley, Love, water waves) [337], [348], [177], shock waves (collision with meteorites, underground nuclear explosions, avalanches, landslides) [178], [349], solitons (nonlinear water waves, porosity waves in granular materials) [313], [365], are commonly appearing and require modelling for practical purposes. They are also used in laboratory and *in situ* investigations of material properties of systems. Speeds of propagation and attenuations of acoustic waves, decay and growth of amplitudes, transmissions and reflections on boundaries, mode couplings, transitions from weak to strong discontinuity waves, propagation of shock and solitons, all these features can be measured and deliver information on material and transient properties of continua. In most cases of practical bearing mathematical models of waves are formed by hyperbolic systems of partial differential equations. Description by such systems motivated the creation of extended thermodynamics which we presented earlier in this book.

The purpose of this Section is threefold. First, we show on the example of the two-component linear poroelastic materials how the propagation of acoustic fronts can be investigated. Then we demonstrate very briefly on the same example of the model the method of spectral analysis for acoustic waves. Finally, we illustrate typical boundary conditions for porous materials by investigating some surface waves. We limit our attention to a purely mechanical modelling and leave out heat conduction and other thermodynamical effects. Still processes described in this Section are irreversible. This is due to diffusion processes described by the permeability and this yields to attenuation of waves.

We begin with the analysis of propagation of weak discontinuity waves and investigate the propagation of the front. The front of such a wave is defined as a smooth moving surface on which the fields are continuous but their derivatives may have finite discontinuities. This means in the case of the two-component poroelastic materials that the following jump conditions are satisfied

$$\begin{gathered} [[\mathbf{v}^S]] = 0, \quad [[\mathbf{v}^F]] = 0, \quad [[\mathbf{e}^S]] = 0, \quad [[\varepsilon]] = 0, \\ \mathbf{a}^S := \left[\left[\frac{\partial \mathbf{v}^S}{\partial t}\right]\right] \neq 0, \quad \mathbf{a}^F := \left[\left[\frac{\partial \mathbf{v}^F}{\partial t}\right]\right] \neq 0, \end{gathered} \tag{13.182}$$

where the double bracket denotes the jump through the surface $[[\cdots]] := (\cdots)^+ - (\cdots)^-$, the quantities on the right-hand side are limits of the quantities on the positive and negative sides of the surface, respectively.

The fields satisfying the above conditions must fulfil the Maxwell kinematic compatibility conditions on the front which follow from the general Hadamard conditions (see Section 3.3.4). They have the form

$$\begin{aligned} [[\operatorname{grad} \mathbf{v}^S]] &= -\frac{1}{c}\mathbf{a}^S \otimes \mathbf{n}, \quad [[\operatorname{grad} \mathbf{v}^F]] = -\frac{1}{c}\mathbf{a}^F \otimes \mathbf{n}, \\ [[\operatorname{grad} \mathbf{e}^S]] &= -\frac{1}{c}\left[\left[\frac{\partial \mathbf{e}^S}{\partial t}\right]\right] \otimes \mathbf{n}, \quad [[\operatorname{grad} \varepsilon]] = -\frac{1}{c}\left[\left[\frac{\partial \varepsilon}{\partial t}\right]\right]\mathbf{n}, \end{aligned} \tag{13.183}$$

where c is the *speed of propagation* of the front and $\mathbf{n}$ is the unit vector perpendicular to the front.

Bearing the integrability conditions in mind

$$\frac{\partial \mathbf{e}^S}{\partial t} = \frac{1}{2}\left[\operatorname{grad} \mathbf{v}^S + \left(\operatorname{grad} \mathbf{v}^S\right)^T\right], \quad \frac{\partial \varepsilon}{\partial t} = \operatorname{div} \mathbf{v}^F, \tag{13.184}$$

we obtain immediately

$$\left[\left[\frac{\partial \mathbf{e}^S}{\partial t}\right]\right] = -\frac{1}{2c}\left(\mathbf{a}^S \otimes \mathbf{n} + \mathbf{n} \otimes \mathbf{a}^S\right),$$
$$\left[\left[\operatorname{grad} \mathbf{e}^S\right]\right] = \frac{1}{2c^2}\left(\mathbf{a}^S \otimes \mathbf{n} + \mathbf{n} \otimes \mathbf{a}^S\right) \otimes \mathbf{n}, \tag{13.185}$$
$$\left[\left[\frac{\partial \varepsilon}{\partial t}\right]\right] = -\frac{1}{c}\mathbf{a}^F \cdot \mathbf{n}, \quad [[\operatorname{grad} \varepsilon]] = \frac{1}{c^2}\mathbf{a}^F \cdot \mathbf{n}\mathbf{n}.$$

Now we form the jump of field equations on the front. These follow from the balance equations

$$\rho_0^S \frac{\partial \mathbf{v}^S}{\partial t} = \operatorname{div} \mathbf{T}^S + \hat{\mathbf{p}}, \tag{13.186}$$
$$\rho_0^F \frac{\partial \mathbf{v}^F}{\partial t} = \operatorname{div} \mathbf{T}^F - \hat{\mathbf{p}},$$

and constitutive relations

$$\mathbf{T}^S = \mathbf{T}_0^S + \lambda^S e\mathbf{1} + 2\mu^S \mathbf{e}^S + Q\varepsilon\mathbf{1} - N(n - n_0)\mathbf{1},$$

$$\mathbf{T}^F = \mathbf{T}_0^F + \rho_0^F \kappa\varepsilon\mathbf{1} + Qe\mathbf{1} + N(n - n_0)\mathbf{1}, \tag{13.187}$$

$$\hat{\mathbf{p}} = \pi\left(\mathbf{v}^F - \mathbf{v}^S\right), \quad n - n_0 = n_0\delta e + \Phi\left(e - \varepsilon\right).$$

After easy calculations it follows

$$\left\{\rho_0^S c^2 \mathbf{1} - \lambda^S \mathbf{n} \otimes \mathbf{n} - \mu^S\left(\mathbf{1} + \mathbf{n} \otimes \mathbf{n}\right) + N\left(n_0\delta + \Phi\right)\mathbf{n} \otimes \mathbf{n}\right\}\mathbf{a}^S$$

$$-\left\{Q\mathbf{n} \otimes \mathbf{n} + N\Phi\mathbf{n} \otimes \mathbf{n}\right\}\mathbf{a}^F = 0, \tag{13.188}$$

$$-\left\{Q\mathbf{n} \otimes \mathbf{n} + N\left(n_0\delta + \Phi\right)\mathbf{n} \otimes \mathbf{n}\right\}\mathbf{a}^S \tag{13.189}$$
$$+\left\{\rho_0^F c^2 \mathbf{1} - \rho_0^F \kappa\mathbf{n} \otimes \mathbf{n} + N\Phi\mathbf{n} \otimes \mathbf{n}\right\}\mathbf{a}^F = 0.$$

This is an eigenvalue problem. We split it into the normal part parallel to $\mathbf{n}$ and the transversal part perpendicular to $\mathbf{n}$.

In the first case we obtain

$$\left\{\rho_0^S c^2 - \left(\lambda^S + 2\mu^S\right) + N\left(n_0\delta + \Phi\right)\right\}a_n^S - \left\{Q + N\Phi\right\}a_n^F = 0,$$
$$-\left\{Q + N\left(n_0\delta + \Phi\right)\right\}a_n^S + \left\{\rho_0^F c^2 - \rho_0^F \kappa + N\Phi\right\}a_n^F = 0, \tag{13.190}$$

$$a_n^S := \mathbf{a}^S \cdot \mathbf{n}, \quad a_n^F := \mathbf{a}^F \cdot \mathbf{n}.$$

This homogeneous set of equations possesses a nontrivial solution if the determinant vanishes. We obtain the following *propagation condition*

$$c^4 - c^2 \left\{ \frac{\lambda^S + 2\mu^S - N(n_0\delta + \Phi)}{\rho_0^S} + \frac{\rho_0^F \kappa - N\Phi}{\rho_0^F} \right\}$$

$$+ \left\{ \frac{\left(\lambda^S + 2\mu^S\right) - N(n_0\delta + \Phi)}{\rho_0^S} \right\} \left\{ \frac{\rho_0^F \kappa - N\Phi}{\rho_0^F} \right\}$$

$$- \frac{1}{\rho_0^S \rho_0^F} \{Q + N\Phi\} \{Q + N(n_0\delta + \Phi)\} = 0. \qquad (13.191)$$

This biquadratic equation for c has two positive and two negative solutions which define two *longitudinal waves*: P1-wave and P2-wave (Biot's wave). Namely

$$c^2 = \frac{1}{2} \left\{ \frac{\lambda^S + 2\mu^S - N(n_0\delta + \Phi)}{\rho_0^S} + \frac{\rho_0^F \kappa - N\Phi}{\rho_0^F} \pm \sqrt{\Delta} \right\},$$

$$\Delta := \left\{ \frac{\lambda^S + 2\mu^S - N(n_0\delta + \Phi)}{\rho_0^S} - \frac{\rho_0^F \kappa - N\Phi}{\rho_0^F} \right\}^2 + 4 \frac{\{Q + N\Phi\} \{Q + N(n_0\delta + \Phi)\}}{\rho_0^S \rho_0^F}. \qquad (13.192)$$

In the second case we multiply equations (13.188), (13.189) by a unit vector $\mathbf{n}_\perp$ perpendicular to $\mathbf{n}$. We obtain

$$\left\{ \rho_0^S c^2 - \mu^S \right\} a_\perp^S = 0, \quad a_\perp^F = 0, \qquad (13.193)$$

$$a_\perp^S := \mathbf{a}^S \cdot \mathbf{n}_\perp, \quad a_\perp^F := \mathbf{a}^F \cdot \mathbf{n}_\perp.$$

Hence

$$c = \sqrt{\frac{\mu^S}{\rho_0^S}}. \qquad (13.194)$$

This is the velocity of propagation of the *transversal wave*. It looks like the corresponding formula in the classical elasticity.

Let us mention that all these modes are attenuated due to the diffusion. This property cannot be demonstrated by means of such simple arguments for the jumps on the front.

Instead of a simple but cumbersome analytical investigation of the above solutions we present a numerical example which has been analyzed by Albers and Wilmanski [7]. For this analysis we choose the following numerical data typical for sandstone and water

$$\rho_0^{SR} = 2500\,\frac{\text{kg}}{\text{m}^3}, \quad \rho_0^{FR} = 1000\,\frac{\text{kg}}{\text{m}^3}, \qquad (13.195)$$
$$K_s = 48 \text{ GPa}, \quad K_f = 2.25 \text{ GPa}, \quad \pi = n_0 \times 10^7\,\frac{\text{kg}}{\text{m}^3\text{s}},$$

which we use in the evaluation of macroscopic material parameters. The value of the permeability coefficient π corresponds to the hydraulic conductivity $\kappa = 10^{-3}\,\text{m/s} = 0.1\,\text{darcy}$[3]. We present the results in dependence on varying porosity n_0 and varying Poisson's ratio ν.

In Figure 23 we show the behavior of velocities of propagation of fronts of P1-, S-, and P2-waves. These three-dimensional plots illustrate the general qualitative character of the dependence on the initial porosity n_0 and on the Poisson's ratio ν.

In order to show the whole surfaces in these 3D-pictures we collected the points in bars in the direction of the ν-axis. Separation of these bars reflects the steps of numerical evaluation – for n_0 this is 0.02, and for ν – 0.01.

First of all let us point out certain general features of these plots. They were calculated by means of the relation (13.192), with Gassmann-like relations presented in the previous Section. In both cases we have used a classical relation between the shear modulus μ^S and the Poisson's ratio ν.

Values of porosity vary between 0.1 and 0.58. This range has been chosen due to the applicability of the full model. As already mentioned in the previous Section micro-macro relations for this model yield instabilities below $n_0 = 0.1$ and above the value $n_0 \approx 0.58$. Instabilities in the first region, for $n_0 < 0.1$, follow from negative values of the shear modulus for $N \neq 0$.

Clearly, all the above restrictions of the full model are related to the extension of Gassmann relations and, consequently, to the applicability of

[3]These two parameters are related in the following way

$$\pi = n_0 \frac{\rho_0^{FR} g_{earth}}{\kappa},$$

where g_{earth} is the earth acceleration.

the full model to a certain limited class of granular and porous materials for which we can claim Gassmann relations to hold. They may not fully apply to modelling of rocks, concrete and similar materials with a compact skeleton. Plots in the left panel follow for Biot's model and for its simplification which we call "simple mixture" ($Q = 0, N = 0$). For the latter model all material parameters except Q are identical with those of Biot's model. As we can see the velocity of propagation of the front of P1-waves has a similar shape for both models but it is lower for the simple mixture model than for Biot's model. The difference is app. 10%. It is the opposite for P2-waves where fronts propagate faster for the simple mixture model than for Biot's model and the difference is app. 20%. Certainly, velocities of P2-waves decay to zero in both models as n_0 approaches zero.

The plot for velocities of S-waves is identical for both models.

These plots indicate already the important conclusion that the qualitative behavior of Biot's model and the simple mixture model is the same and quantitative differences are small enough to be accepted in practical applications.

Plots for the full model and its simplification with the same values of material parameters except for $Q = 0$ and $N = 0$ are shown in the right panel. For medium values of porosities and Poisson's ratio, the behavior of these models is similar to Biot's model and the simple mixture model. Large deviations appear for P1-waves for small values of porosity where a dependence on the Poisson's ratio is different from Biot's model. Also for P2-waves substantial qualitative differences appear for small values of porosity and small values of the Poisson's ratio. In particular, below $n_0 \approx 0.14$ the velocity of the P2-wave within the full model decays more rapidly than within other models.

Again the velocity of the S-wave is almost identical for both models but it behaves differently from the Biot's model and simple mixture model for small porosities where it is decaying rather than growing and it possesses inflection points at app. $n_0 \approx 0.25$.

All these qualitative differences of both full models may be attributed to the singular behavior of the shear modulus μ^S. They would not appear if values of the parameter N were smaller than these predicted by Gassmann relations.

The attenuation of waves cannot be discussed by means of the simple analysis of propagation of fronts. For completeness we show in Figure 24 the numerical results following from the evaluation of *dispersion relations* for monochromatic waves in the limit of infinite frequencies, $\omega \rightarrow \infty$.

In spite of qualitative similarity in the whole range of parameters, quantitative differences between attenuations calculated for different models are much larger than the differences between velocities. For instance, the attenuation of the P1-wave as given by Biot's model is a few times smaller than in the simple mixture model (the left panel, the upper picture). It is the other way around for P2-waves which are much stronger attenuated according to Biot's model than to the simple mixture model. However, these differences become much smaller for medium values of parameters.

Again the attenuation of transversal waves is the same for both models – Biot's model and the simple mixture model.

For the full model the attenuation of P1-waves becomes negative in the case of large porosities and small Poisson's ratio. The range of instability is approximately limited by the line: $n_0 = \nu$.

The attenuation of P2-waves becomes also much bigger than in Biot's model for small values of porosity.

In contrast to a singular behavior near the limits of n_0 and ν results seem to agree well in practically relevant medium values of porosity and of Poisson's ratio.

We do not discuss in this Section any details of the analysis of monochromatic waves. They are described by the following relations for fields

$$\mathbf{v}^S = \mathbf{V}^S \mathcal{E}, \quad \mathbf{v}^F = \mathbf{V}^F \mathcal{E}, \quad , \mathbf{e}^S = \mathbf{E}^S \mathcal{E}, \quad \varepsilon = E^F \mathcal{E}, \qquad (13.196)$$
$$\mathcal{E} := \exp\left[i\left(k\mathbf{n}\cdot\mathbf{x} - \omega t\right)\right],$$

where ω is a given frequency, k is the so-called *wave number* and it may be complex, $\mathbf{n}$ is the unit vector in the direction of propagation. $\mathbf{V}^S, \mathbf{V}^F, \mathbf{E}^S, E^F$ are constant *amplitudes.* Field equations determine the so-called *dispersion relation* $k = k(\omega)$ which, in turn, determines *phase and group speeds* of propagation, $c_{ph} = \omega / \operatorname{Re} k$, $c_g = \left(d \operatorname{Re} k / d\omega\right)^{-1}$, respectively, and *attenuation* of waves, $\operatorname{Im} k$. However, it is worth mentioning that a simple analysis of the dispersion relation yields some important conclusions concerning parameters of the model. Two of them, Q and ρ_{12} have a particular bearing. As we know from the previous Section the coupling parameter Q can be calculated by means of Gassmann relations. Parameter ρ_{12} interpreted as the tortuosity coefficient is used in soil mechanics in a transformed form proposed by Berryman

$$\rho_{12} = \rho_0^F (1 - a), \quad a = \frac{1}{2}\left(\frac{1}{n_0} + 1\right) \quad \text{i.e.} \quad a \geq 1, \quad \rho_{12} \leq 0, \qquad (13.197)$$

where a is called the *tortuosity.*

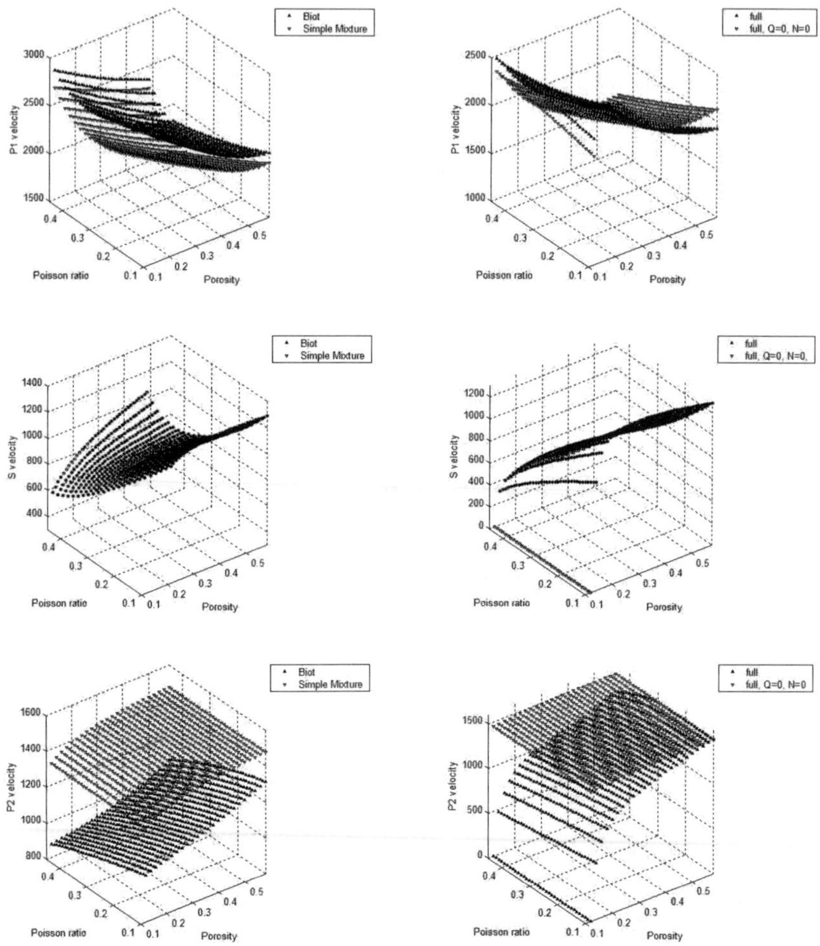

Fig. 23: *Front velocities of P1-, S-, and P2-waves (from top to bottom)*
Left panel: Biot's model and "simple mixture" model $(Q = 0, N = 0)$
Right panel: The full model and the simplified full model.

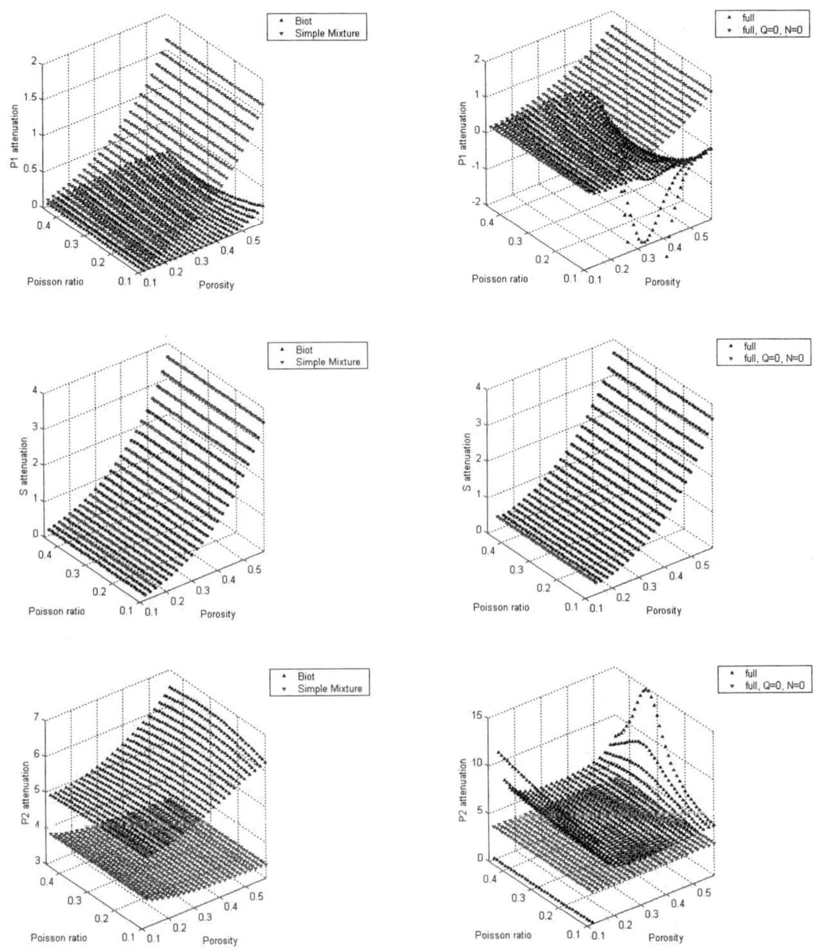

Fig. 24: *Attenuation of fronts of P1-, S-, and P2-waves (from top to bottom)*
Left panel: Biot's model and "simple mixture" model $(Q = 0, N = 0)$
Right panel: The full model and the simplified full model.

Let us apply the monochromatic wave analysis to the Biot model and some of its simplifications. The constitutive relations (13.187) are now replaced by the following ones

$$\mathbf{T}^S = \mathbf{T}_0^S + \lambda^S e\mathbf{1} + 2\mu^S \mathbf{e}^S + Q\varepsilon\mathbf{1},$$

$$\mathbf{T}^F = \mathbf{T}_0^F + \rho_0^F \kappa\varepsilon\mathbf{1} + Qe\mathbf{1}, \tag{13.198}$$

$$\hat{\mathbf{p}} = \pi\left(\mathbf{v}^F - \mathbf{v}^S\right) - \rho_{12}\left(\frac{\partial \mathbf{v}^F}{\partial t} - \frac{\partial \mathbf{v}^S}{\partial t}\right), \quad n - n_0 = n_0\delta e + \Phi\left(e - \varepsilon\right),$$

i.e. $N = 0$ and we account for the influence of the relative acceleration.

We present numerical results for the fastest bulk wave, P1-wave, in dependence on these parameters. We choose the data which practically coincide with those used for the previous diagrams. In order to expose better the damping the permeability π has been chosen to be bigger than before.

$$\begin{aligned} c_{P1} &= \sqrt{\frac{\lambda^S + 2\mu^S}{\rho_0^S}} = 2500\ \tfrac{\mathrm{m}}{\mathrm{s}}, \quad c_{P2} = \sqrt{\kappa} = 1000\ \tfrac{\mathrm{m}}{\mathrm{s}}, \\ c_S &= \sqrt{\frac{\mu^S}{\rho_0^S}} = 1500\ \tfrac{\mathrm{m}}{\mathrm{s}}, \\ \rho_0^S &= 2500\ \tfrac{\mathrm{kg}}{\mathrm{m}^3}, \quad r = \frac{\rho_0^F}{\rho_0^S} = 0.1, \quad \pi = 10^8\ \tfrac{\mathrm{kg}}{\mathrm{m}^3\mathrm{s}}, \\ Q &= 0.8\ G\ \mathrm{Pa}, \quad n_0 = 0.4, \quad a = 1.75. \end{aligned} \tag{13.199}$$

Four particular cases are considered. In the full Biot model both Q and ρ_{12} are different from zero, i.e. a is different from one. In the so-called simple mixture model $Q = 0$ and $a = 1$. In the remaining two cases either $Q = 0$ or $a = 1$. These two cases were investigated in order to clear what is the influence of each of these parameters.

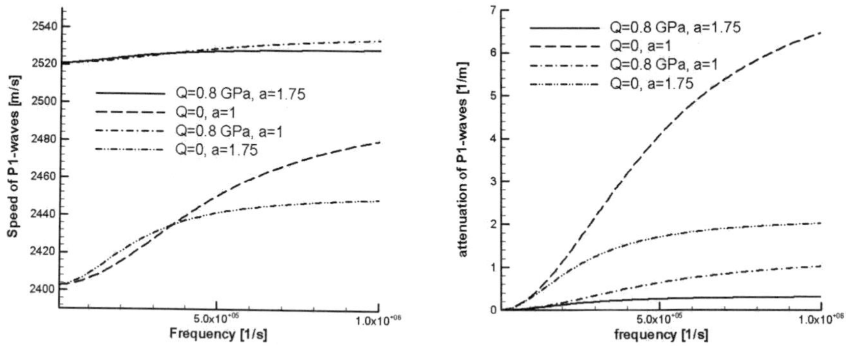

Fig. 25: *Speeds and attenuations of monochromatic P1-waves for four different models.*

In the left panel of Figure 25 we present speeds of propagation of monochromatic P1-waves as functions of the frequency. The simple mixture model ($Q = 0, a = 1$) leads to the biggest difference in the speed between two limits of the frequency $\omega \to 0$ and $\omega \to \infty$. This difference is approximately 100 m/s and it was indeed observed in experiments on soils. Both models with $Q = 0.8\ GPa$ yield almost no difference between these limits. This may indicate that values of the coupling coefficient Q given by Gassmann's relation are too large. Even worse is the influence of tortuosity a. It is seen in diagrams of the right panel. For a moderate value of $a = 1.75$ the attenuation of waves decays more than 10 times. We expect the influence of tortuosity, i.e. complicated shapes of channels of the porous materials to have the exactly opposite effect. Such curvy channels must scatter waves much stronger than straight channels and, consequently, increase rather than decrease the attenuation. Consequently, in spite of numerous papers with such a claim, the added mass coefficient ρ_{12} does not seem to have much to do with the tortuosity. The latter has, of course, a strong influence on values of the permeability and, therefore, it seems to be physically justified to prescribe the influence of tortuosity only to the hereditary character of the diffusion contribution. It should be stressed that this issue is still very controversial.

The propagation conditions change dramatically if a boundary appears. It is known from the simplest elastic model that due to the boundary condition on an elastic half-space in addition to the normal longitudinal and transversal bulk waves the so-called Rayleigh surface wave appears which propagates along the boundary with the speed smaller than both speeds of bulk waves (see the review work on surface waves [373]).

Similar effects are observed for all boundaries. These waves have a very big practical bearing. Their spatial dispersion is much smaller than this of bulk waves because the latter have spherical and the former cylindrical character. This is the reason, for instance, that damages in earthquakes are caused by surface and not by bulk waves.

Before we discuss some properties of surface waves in saturated porous materials we have to formulate boundary conditions for such media.

The construction of boundary value problems for porous materials requires a definition of the boundary. In the macroscopic description which we presented in previous Sections it is a smooth orientable surface $\partial\mathcal{B}_0$ of the reference domain $\mathcal{B}_0$ of the skeleton. It means that its current image $\partial\mathcal{B}_t$ moves with the velocity $\mathbf{v}^S$ of particles of the skeleton. This definition does not relate the boundary to fluxes of fluid components and does not specify any properties related to its morphology like, for instance, a fraction of the area of channels approaching the boundary *(surface porosity)*. These properties are essential for the formulation of boundary conditions. The problem with the transition from the real morphology of porous materials to the macroscopic description is demonstrated in Figure 26. It is clear that there are infinitely many possibilities to envelope a porous or granular material in a smooth surface which models the boundary. Variations of the geometry, if they are sufficiently small in comparison with dimensions of REV, do not have any influence on the macroscopic geometry. However, it is easy to see that these small changes may lead to large variations of the surface porosity, i.e. the fraction of the area of openings of channels on a chosen surface to the total area within REV.

From the physical point of view the only reasonable way to define a smooth macroscopic surface modelling the boundary is to locate it somewhere in the layer encompassing boundary solid elements and whose thickness is not larger than the characteristic linear dimension of REV. This choice should be random. Then it was proved by A. Delesse in 1848 [86] that the surface and volume porosities are equal. Certainly, it means that the boundary surface cannot be tangent to elements of the skeleton on the boundary but it rather cuts the material and leaves some parts of the real skeleton beyond the domain of continuous model of the porous material. This indicates the existence of boundary layers which have to be incorporated in boundary conditions. Similar remarks concern interfaces between two different porous materials, a porous material and a fluid, etc.

One of the consequences of the equal porosities is the condition (13.172) $\rho^S = (1-n)\,\rho^{SR}$, $\rho^F = n\rho^{FR}$, etc. which describes equilibrium properties on the boundary without flow effects, i.e. without an influence of boundary layer effects.

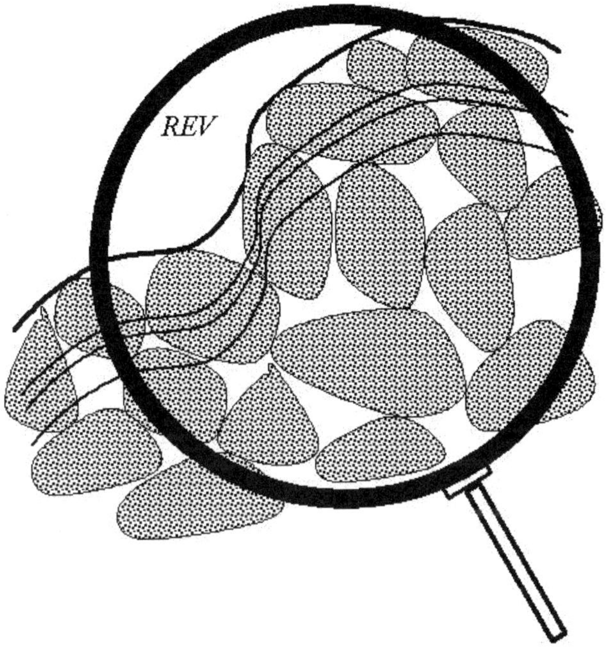

Fig. 26: *On the definition of the boundary of porous materials.*

Now we are in the position to introduce conditions on $\mathcal{B}_t$ which yield boundary conditions. As mentioned above, the boundary $\partial\mathcal{B}_t$ is assumed to be material with respect to the skeleton. Then partial mass balance equation for the skeleton does not give rise to any conditions and the flow of the fluid must satisfy the following jump condition

$$m^F := \rho_t^{F-}\left(\mathbf{v}^{F-}-\mathbf{v}^S\right)\cdot\mathbf{n} = \rho_t^{F+}\left(\mathbf{v}^{F+}-\mathbf{v}^S\right)\cdot\mathbf{n}. \qquad (13.200)$$

Obviously, ρ_t^{F-} is the current partial mass density of the fluid in the porous material (the negative side of the boundary by the outward orientation of the normal vector $\mathbf{n}$) and ρ_t^{F+} is the partial mass density of the fluid outside the porous material. If this is vacuum, this quantity is zero and the condition implies that normal components of velocities of both components are equal. If the material outside is the fluid flowing from the porous material, then $\rho_t^{F+} = \rho_t^{FR}$. Other possibilities are obvious.

One cannot require a continuity of partial tractions in porous materials. Terzaghi has constructed an instructive *gedanken experiment* in which he has shown that the external load is divided between components in a time dependent manner following from the flow of the fluid through the surface.

We can require only the continuity of the momentum flux for the mixture, i.e.

$$\left(\mathbf{T}^{S-} - p^{F-}\mathbf{1}\right)\mathbf{n} = \mathbf{t}_{ext}, \tag{13.201}$$

where $\mathbf{t}_{ext}$ is the full traction applied from exterior to the unit surface.

It means that we have to replace the second vectorial condition by some other conditions. One of them reflects the existence of the boundary layer indicated in the above discussion. Namely we assume that the amount of mass of the fluid which is transported through a unit area of permeable boundary in a unit time is driven by the difference of real pressures in the fluid on both sides of the boundary. Hence, in general,

$$m^F = \alpha' \left(\frac{p^{F-}}{n^-} - \frac{p^{F+}}{n^+}\right), \tag{13.202}$$

where $p^{F\pm}$ are partial pressures in the fluid on both sides of the boundary and $n^{\pm}$ denotes the porosity on both sides of the boundary. In the case of the porous materials on only the negative side of the boundary, i.e. for $n^+ = 1$ we have

$$m^F = \alpha\left(p^{F-} - n^- p^{FR+}\right), \quad \alpha := \frac{\alpha'}{n^-}. \tag{13.203}$$

Obviously, coefficients α', α describe material properties of the boundary. The coefficient α, in some works denoted as $1/T$, is called the *surface permeability*. It has been introduced to the theory of surface waves in porous materials by H. Deresiewicz in 1961 [87] [88]. It reflects jointly all properties of the boundary layer. In the particular case of the impermeable boundary $\alpha = 0$ and conditions (13.200) and (13.203) coincide. However, the boundary value problem does not degenerate because we do not have to solve the exterior problem. Another particular case of an ideal permeable boundary $\alpha \to \infty$ yields the condition of continuity of pressure: $p^{F-} = n^- p^{FR+}$ commonly used by mathematicians in the theory of diffusion.

In addition to this scalar condition, we have to introduce two other conditions. As (13.203) reflects only the behavior of the normal velocity $\left(\mathbf{v}^{F-} - \mathbf{v}^S\right) \cdot \mathbf{n}$ they are supposed to describe the behavior of the tangent component of the relative velocity. It has been shown by G. S. Beavers and D. D. Joseph [32] that, for macroscopically viscous fluids, these components of the velocity have to satisfy a condition similar to Deresiewicz condition. In the case of ideal fluids which we discuss in this Chapter, the tangent part must be continuous. In the case of fluid outside porous material this condition has the form

$$\left(\mathbf{v}^{F-} - \mathbf{v}^S\right) - \left[\left(\mathbf{v}^{F-} - \mathbf{v}^S\right) \cdot \mathbf{n}\right]\mathbf{n} = 0. \tag{13.204}$$

We are ready to formulate the problem of surface waves. In the classical two-dimensional case, we have to solve the field equations for the semiinfinite medium (13.186), (13.187) under the condition that the solution decays with the depth and it moves as a wave in the direction of the boundary.

It is convenient to introduce the displacement vector $\mathbf{u}^S$ for the skeleton, and, formally, the displacement vector $\mathbf{u}^F$ for the fluid. The latter is introduced only for the technical symmetry of considerations and it does not have any physical bearing. Then

$$\mathbf{u}^S = \operatorname{grad}\varphi^S + \operatorname{rot}\boldsymbol{\psi}^S, \quad \mathbf{v}^S = \frac{\partial \mathbf{u}^S}{\partial t}, \quad \mathbf{e}^S = \operatorname{sym}\operatorname{grad}\mathbf{u}^S, \tag{13.205}$$
$$\mathbf{u}^F = \operatorname{grad}\varphi^F + \operatorname{rot}\boldsymbol{\psi}^F, \quad \mathbf{v}^F = \frac{\partial \mathbf{u}^F}{\partial t},$$

where $\varphi^S, \boldsymbol{\psi}^S, \varphi^F, \boldsymbol{\psi}^F$ are two pairs of potentials analogous to those appearing in the classical elasticity.

We choose the axes with the downward orientation of the z-axis and the x-axis in the direction of propagation of the wave. As the problem is assumed to be two-dimensional, we make the following ansatz for solutions

$$\begin{aligned}
\varphi^S &= A^S(z)\exp[i(kx-\omega t)], \quad \varphi^F = A^F(z)\exp[i(kx-\omega t)], \\
\psi_y^S &= B^S(z)\exp[i(kx-\omega t)], \quad \psi_y^F = B^F(z)\exp[i(kx-\omega t)], \\
\psi_x^S &= \psi_z^S = \psi_x^F = \psi_z^F = 0,
\end{aligned} \tag{13.206}$$

and

$$\begin{aligned}
\rho^S - \rho_0^S &= A_\rho^S(z)\exp[i(kx-\omega t)], \quad \rho^F - \rho_0^F = A_\rho^F(z)\exp[i(kx-\omega t)], \\
n - n_0 &= A^\Delta \exp[i(kx-\omega t)].
\end{aligned} \tag{13.207}$$

Substitution in the field equations yields relations for the amplitudes which, in the case under considerations, have the exponential form $\exp(-\gamma z)$, where γ is different for each amplitude. However, these exponents should have the common feature to be positive. We know that this is indeed the case in the classical elasticity and, as a result, we obtain Rayleigh waves. For porous materials, the problem is very complicated and only some solutions indeed possess this property of surface waves. In some cases, such solutions do not exist, in some other one obtains the so-called pseudosurface or *leaky* waves.

In addition, boundary conditions yield the dispersion relation $k = k(\omega)$ which specifies the speeds and attenuations of surface waves. Due to the diffusion, surface waves in porous materials are always attenuated. Only in the limit $\omega \to 0$ attenuation becomes very small and some surface waves in porous materials remind, for instance, classical Rayleigh or Stoneley waves.

In the case of the model with $N = 0, Q = 0, \rho_{12} = 0$ (the simple mixture model) the following dispersion relation follows in this limit of frequencies

$$\left(\frac{\omega}{k}\right)\left\{\left(2-\frac{r+1}{c_s^2}\left(\frac{\omega}{k}\right)^2\right)^2 \right. \tag{13.208}$$
$$\left. -4\sqrt{1-\frac{r+1}{c_s^2}\left(\frac{\omega}{k}\right)^2}\sqrt{1-\frac{r+1}{rc_f^2+1}\left(\frac{\omega}{k}\right)^2}\right\}+O\left(\sqrt{\omega}\right)=0.$$

It shows the existence of the single surface mode – Rayleigh-like wave. All other modes characteristic for porous materials do not propagate. Some further details can be found, for instance, in the work [6].

We have shown the above example to illustrate technical problems which must be overcome in the wave analysis in porous materials. Simultaneously, it is of paramount importance for many practical applications and one can expect an extensive research in this field in the near future.

13.5 On adsorption in porous materials

13.5.1 *Adsorption for large channel diameter*

Mass exchange between components of porous, and granular materials belongs to the most important problems of practical bearing within theories of such materials. This is connected primarily with a very large internal surface per unit volume on which the exchange takes place. For instance in sandstones it reaches the value of 1.5×10^5 m^2/m^3 in comparison with 6 m^2/m^3 for the external surface. This property is used in many technological processes. For instance, in the growth of SiC crystals by sublimation, the vapor of silicium flows through a porous graphite wall in which it forms various carbide connections. A charcoal granular material is also used in gas masks. Lungs, many filters and chemical reactors are made of porous materials for the same reason. One of the most important applications is related to changes of rates of chemical catalytic reactions by porous catalyzers. Transport of pollutants by ground water in soils is an important example of such processes appearing spontaneously in nature.

Classically processes of mass exchange in porous materials were divided into two classes of *chemical and physical adsorption*[133]. This classification seems to be not very sharp but it is still useful by construction of various macroscopic models. In the case of chemical adsorption, we deal with exchange of mass between skeleton and one or more fluid components in which particles build chemical bindings. In contrast to such processes in a physical adsorption particles of skeleton and of a fluid component form weak van der Waals bindings. The first type of exchange is connected with essential thermal effects connected with a release or absorption of energy, i.e. with the latent heat of reaction. On the other hand, the heat of reaction by physical adsorption is very small and we use the approximation of isothermal processes. We shall not present here chemical adsorption processes and limit the attention in this Section to isothermal physical adsorption.

The physical adsorption model of a mass exchange between a fluid component, and a solid in porous, and granular materials is based on the classical work of Langmuir (see [4], [3] for references). In the original works of Langmuir [180], [181], [182], the theory of adsorption was limited to flat solid surfaces interacting with a gas. However, for porous materials whose pores are large - their diameter is greater than app. 500 Å = 50 nm - one can still rely on the assumption that the influence of the curvature of the surface is small. We discuss the influence of the curvature in the next Subsection. This is why we say that we describe materials with a *large diameter of channels.*

We assume as well that the fluid component consists of two phases. The bulk phase does not exchange mass, and it is solely a carrier of the adsorbate. The second phase - the adsorbate - moves with the velocity of the bulk fluid and its concentration is small

$$c := \frac{\rho_t^A}{\rho_t^F + \rho_t^A} \ll 1, \tag{13.209}$$

where ρ_t^A, ρ_t^F denote current partial mass densities of the adsorbate, and of the fluid carrier, respectively.

According to Langmuir a transfer of particles of adsorbate from the fluid phase to the surface of the solid depends on the number of available *bare sites* on this surface. The notion of bare sites is introduced in connection with the landscape of interaction energy between adsorbate and the solid skeleton. This landscape depends on a crystallographic or structural symmetry of the skeleton and it has the form of a regular pattern with distributed spots of a maximum interaction energy. On these spots one can expect particles of adsorbate to settle down most frequently. This process is controlled in the first place by the partial pressure of adsorbate in the fluid phase. Certainly there are also particles which desorbate again if they overcome an energy barrier through thermal fluctuations. In the equilibrium one expects these two processes to have the same rate.

Such an adsorption process is called the *monolayer adsorption.* If in a certain domain all available bare sites on the surface of the skeleton are already occupied, the adsorption may take place due to interaction of adsorbate particles in their fluid phase and those which are already bound to the skeleton. We deal then with the *multilayer adsorption.* In the case of a small concentration of the fluid adsorbate, such exchange of mass between the two phases may be assumed to be less important. We limit attention in these notes solely to the monolayer adsorption model.

In order to construct a continuous model in the spirit of this book we have to construct the mass sources in partial mass balance equations [367]. On the semimacroscopic level (i.e. in the *representative elementary volume* (REV) of a porous or granular material), the normalized fraction of bare sites per unit volume is denoted by $1 - x$, i.e. x is the fraction of *occupied sites*. If the area of the internal surface contained in the representative elementary volume is denoted by f_{int}, and the mass of adsorbate per unit area of the internal surface by m^A, then the amount of mass which is already adsorbed in the representative elementary volume is on average equal to the product $m^A x f_{int}$.

Let us denote by V the volume of the representative elementary volume. Then the amount of mass of adsorbate transferred in unit time from the fluid phase to the solid skeleton is given by the macroscopic balance relation

$$\hat{\rho}_t^A = -m^A \frac{d(xy)}{dt}, \qquad y := \frac{f_{int}}{V}, \tag{13.210}$$

where $\hat{\rho}_t^A$ denotes the current intensity of mass source per unit time, and unit macroscopic volume.

In order to construct the model, we have to specify the rates in this relation. The quantities x, y play in the model the role of additional *microstructural variables.* Field equations for these quantities are assumed to have a form of evolution equations which are specified by relations for the above-mentioned rates.

For the rate dx/dt we assume that changes of the fraction x are described by the Langmuir relation

$$\frac{dx}{dt} = a(1-x)p^A - bxe^{-\frac{E_b}{kT}}, \tag{13.211}$$

where p^A denotes the partial pressure of the adsorbate in the fluid phase, E_b is the energy barrier for particles adsorbed on the solid surface due to the van der Waals interaction forces, and it is assumed to be constant, a and b are material parameters which within the present model may depend solely on the temperature, k is the Boltzmann constant, and T is the absolute temperature. In the case of full phase equilibrium in which the adsorption rate (the first contribution with coefficient a), and the desorption rate (the second contribution with coefficient b) are equal but of opposite sign we obtain from equation (13.211) the following relation for the equilibrium fraction of occupied sites

$$x = x_L := \frac{\frac{p^A}{p_0}}{1 + \frac{p^A}{p_0}}, \qquad p_0 := \frac{b}{a} e^{-\frac{E_b}{kT}}. \tag{13.212}$$

This relation describes the so-called the *Langmuir isotherm*. It begins in the origin $\frac{p^A}{p_0} = 0$ with the zero value of occupied sites and saturates at the value 1 for $\frac{p^A}{p_0} \to \infty$. At any given partial pressure p^A the fraction x is uniquely determined and it may change its value if we vary the pressure. This corresponds to a slow transition from one thermodynamical equilibrium to another one. In reality such processes are conducted through nonequilibrium states which are described by the rate equation (13.211) and are connected with the dissipation.

In the mass source (13.210) we also have another contribution connected with the change of the internal surface. Consequently we must formulate a relation for the rate $\frac{dy}{dt}$. We make the assumption that changes of the internal surface are coupled with dissipative changes of the porosity n which in turn describe relaxation processes of semimacroscopic changes of volume of the skeleton.

First of all let us notice that for sufficiently smooth internal surfaces of porous, and granular materials with a random geometry of pore spaces a change of an average characteristic linear dimension of the internal surface, and this of pores in the elementary representative volume can be assumed to be proportional: $\delta f_{int}^{\frac{1}{2}} \sim \delta (nV)^{\frac{1}{3}}$. Simultaneously dissipative changes of the porosity are given by a source $\hat{n}$ which describes the intensity of these changes per unit time and volume of the porous material. Bearing the above assumption in mind we obtain immediately

$$\frac{1}{y}\frac{dy}{dt} = \varsigma \frac{\hat{n}}{n}, \tag{13.213}$$

where the proportionality factor ς is assumed to be constant for the purpose of this presentation.

Obviously in a thermodynamical phase equilibrium $\hat{n} \equiv 0$, and the equilibrium fraction x is connected with the partial pressure p^A through the relation (13.212). Then the mass source (13.210) vanishes identically.

The behavior of the continuous model based on the above assumptions has been checked on a simple bench-mark homogeneous problem [367]. It was found that results are indeed qualitatively in agreement with observations.

We present here the set of field equations which covers a much more extensive class of problems [4], [5], [3]. In particular we can describe couplings of adsorption and diffusion as well as we can incorporate boundary conditions on permeable boundaries which are characteristic for the majority of practical problems. According to the above remarks we have to determine the following fields

$$\left\{\rho_t^S, \rho_t^L, c, \mathbf{v}^S, \mathbf{e}^S, \mathbf{v}^F, n, x, y\right\}, \quad \rho_t^L := \rho_t^F + \rho_t^A, \tag{13.214}$$

where the concentration c is defined by (13.209). The velocity of the third component does not appear because the adsorbate in the fluid phase moves with the same velocity as the other fluid component. Therefore we use only two momentum balance equations, for the skeleton and for both fluid components together.

Field equations follow from three mass balance equations, two momentum balance equations, the balance equation of porosity, integrability condition for the deformation of the skeleton, and two evolution equations for two additional microstructural variables. In the case of small deformations of the skeleton, small changes of the fluid mass density, small concentrations of adsorbate, and small changes of porosity, i.e.

$$\left\|\mathbf{e}^S\right\| \ll 1, \quad \left|\frac{\rho_t^L - \rho_0^L}{\rho_0^L}\right| \ll 1, \quad 0 < c \ll 1, \quad |\Delta_n| \ll 1, \tag{13.215}$$

they have the form

– mass balance

$$\frac{\partial \rho_t^S}{\partial t} + \operatorname{div}\left(\rho_t^S \mathbf{v}^S\right) = -\rho_t^L \hat{c}, \quad \frac{\partial \rho_t^L}{\partial t} + \operatorname{div}\left(\rho_t^L \mathbf{v}^F\right) = \rho_t^L \hat{c}, \tag{13.216}$$

$$\frac{\partial c}{\partial t} + \mathbf{v}^F \cdot \operatorname{grad} c = (1-c)\,\hat{c}, \quad \hat{c} := \frac{\hat{\rho}_t^A}{\rho_t^L} = -\frac{m^A}{\rho_0^L}\frac{d\,(xy)}{dt},$$

– momentum balance

$$\frac{\partial \rho_t^L \mathbf{v}^F}{\partial t} + \operatorname{div}\left(\rho_t^L \mathbf{v}^F \otimes \mathbf{v}^F + p^L \mathbf{1}\right) + \pi\left(\mathbf{v}^F - \mathbf{v}^S\right) = 0$$

$$\rho_t^S \frac{\partial \mathbf{v}^S}{\partial t} = \operatorname{div} \mathbf{T}^S + \pi\left(\mathbf{v}^F - \mathbf{v}^S\right), \tag{13.217}$$

– porosity balance

$$\frac{\partial \Delta_n}{\partial t} + \Phi_0 \operatorname{div}\left(\mathbf{v}^F - \mathbf{v}^S\right) = -\frac{\Delta_n}{\tau},$$

where

$$\mathbf{T}^S = \mathbf{T}_0^S + \lambda^S \operatorname{tr} \mathbf{e}^S \mathbf{1} + 2\mu^S \mathbf{e}^S + \beta \Delta_n \mathbf{1}, \tag{13.218}$$
$$p^L = p_0^L + \varkappa \left(\rho_t^L - \rho_0^L\right) + \beta \Delta_n, \quad p^F = (1-c)\, p^L, \quad p^A = c p^L,$$

with material parameters $\lambda^S, \mu^S, \varkappa, \beta, \pi$ being constant. They depend parametrically on the constant initial porosity n_0. In addition we have

– integrability condition

$$\frac{\partial \mathbf{e}^S}{\partial t} = \operatorname{sym} \operatorname{grad} \mathbf{v}^S, \tag{13.219}$$

– evolution equations for microstructural variables

$$\frac{d \ln \frac{y}{y_0}}{dt} = -\varsigma \frac{\Delta_n}{n_E}, \qquad y(t=0) = y_0 \equiv \frac{f_{int}(t=0)}{V},$$
$$\frac{dx}{dt} = \frac{1}{\tau_{ad}} \left[(1-x) \frac{c p^L}{p_0} - x \right], \quad x(t=0) = \frac{\frac{c_0 p_0^L}{p_0}}{1 + \frac{c_0 p_0^L}{p_0}}, \tag{13.220}$$
$$\tau_{ad} := \frac{1}{b} e^{\frac{E_b}{kT}}, \qquad c_0 := c(t=0).$$

Again the material parameters $\varsigma, p_0, \tau_{ad}$ are assumed to be constant.

General results for this system of equations have not been obtained as yet. However some important particular problems have been solved under the assumptions of negligible accelerations, and a negligible explicit time dependence of porosity. Their discussion can be found in the above quoted works of B. Albers: Ph-D Thesis [4] and [5], [3]. We quote here solely the most important conclusions of these works. The latter assumption yields the following relation for changes of porosity, and the constitutive relations for partial stresses

$$\Delta_n = -\tau \Phi_0 \operatorname{div} \left(\mathbf{v}^F - \mathbf{v}^S\right),$$
$$\mathbf{T}^S = \mathbf{T}_0^S + \lambda^S \operatorname{tr} \mathbf{e}^S \mathbf{1} + 2\mu^S \mathbf{e}^S - \beta \tau \Phi_0 \operatorname{div} \left(\mathbf{v}^F - \mathbf{v}^S\right) \mathbf{1}, \tag{13.221}$$
$$p^L = p_0^L + \varkappa \left(\rho_t^L - \rho_0^L\right) - \beta \tau \Phi_0 \operatorname{div} \left(\mathbf{v}^F - \mathbf{v}^S\right).$$

Hence changes of porosity appear in equations of motion in a similar way as contributions of bulk viscosities in mechanics of fluids. Relation $(13.221)_1$ allows to eliminate dynamical changes of porosity from field equations.

Let us note that the initial value problem has already been formulated in the above relations. It remains to add boundary conditions. In the two-component case it has the following form

$$\left(\mathbf{T}^S - p^L \mathbf{1}\right) \mathbf{n}\big|_{\partial \mathcal{B}_t} = \mathbf{t}_{ext},$$

$$\rho^L \left(\mathbf{v}^F - \mathbf{v}^S\right) \cdot \mathbf{n}\big|_{\partial \mathcal{B}_t} = \alpha \left(p^L - n p_{ext}\right)\big|_{\partial \mathcal{B}_t}, \qquad (13.222)$$

$$\left(\mathbf{v}^F - \mathbf{v}^S\right) - \left(\mathbf{v}^F - \mathbf{v}^S\right) \cdot \mathbf{nn}\big|_{\partial \mathcal{B}_t} = 0.$$

It is easy to check that these conditions are of the third kind.

Most important results for applications of the model are connected with couplings of adsorption and diffusion. For instance, it has been shown [3] that the mass source as a function of relative velocity of components possesses a maximum. It is zero for the relative velocity equals to zero and it tends again to zero when this velocity goes to infinity. In the first case the system is in equilibrium and the number of bare sites remains constant and determined by the partial pressure p^A. The internal surface is also constant because the porosity does not change. In the second extreme case relaxation times for internal variables: $\tau_{ad}, 1/\varsigma$ are too long for adsorbate to settle down on the skeleton. This indicates the way for an optimal design of filters. Particularly useful is a dependence of the relative velocity from the surface permeability coefficient α because this parameter can be controlled.

13.5.2 *Adsorption for small channel diameter; capillarity*

As we have already mentioned processes in materials with the so-called mesopores, i.e. in materials in which the diameter of channels lies in the range $20 - 500\,\text{Å}$ are connected with the appearance of hysteresis loops in adsorption isotherms. Measurements of their size are used in practical application to find microgeometrical properties of such materials. The reason for the appearance of loops is a capillary condensation. In contrast to the previous case a macroscopic multicomponent model for adsorption with hysteresis has not been constructed yet. However, there are attempts to construct models of unsaturated materials as a three-component porous medium in which these effects, in particular the capillary pressure, play an important role. We shall not present these questions in this book.

We proceed to explain the notion of capillary condensation in terms of semimacroscopic description by means of the Kelvin equation.

We begin with mechanical and thermodynamical equilibrium conditions on an interface between two different phases of a liquid. These conditions determine properties of an interface between a fluid and its vapor in the analysis of condensation.

Let us consider a surface in a three-dimensional Euclidean space described by the position vector $\mathbf{r}$

$$\mathbf{r} = \mathbf{r}\left(\xi^1, \xi^2\right), \qquad (13.223)$$

where (ξ^1, ξ^2) are surface coordinates (compare Sections 3.3.2 and 3.3.3). Base vectors, a normal vector, and the first and second metric tensors of this surface are defined by the relations

$$\mathbf{g}_\alpha := \frac{\partial \mathbf{r}}{\partial \xi^\alpha}, \quad \mathbf{n} := \frac{\mathbf{g}_1 \times \mathbf{g}_2}{|\mathbf{g}_1 \times \mathbf{g}_2|}, \tag{13.224}$$

$$a_{\alpha\beta} := \mathbf{g}_\alpha \cdot \mathbf{g}_\beta, \quad b_{\alpha\beta} := \frac{\partial^2 \mathbf{r}}{\partial \boldsymbol{\xi}^\alpha \partial \xi^\beta} \cdot \mathbf{n}.$$

It can be shown (see Section 4.3) that the normal component of the mechanical equilibrium condition on such a surface has the following general form

$$[[\mathbf{Tn}]] \cdot \mathbf{n} = -S^{\alpha\beta} b_{\alpha\beta}, \tag{13.225}$$

where $\mathbf{T}$ denotes the Cauchy stress tensor in bulk substances on both sides of the surface, and $S^{\alpha\beta}$ are components of the surface stress tensor, both on the level of semimacroscopic description.

In the case of ideal fluids on both sides of the interface which is itself a membrane we have $\mathbf{T} = -p\mathbf{1}$, and $S^{\alpha\beta} = \sigma a^{\alpha\beta}$, where σ is a *surface tension*. Then relation (13.225) reduces to the following one

$$p^G - p^F = \sigma \sum_{\alpha=1}^{2} \kappa^{(\alpha)} \boldsymbol{\beta}^{(\alpha)} \cdot \boldsymbol{\beta}^{(\alpha)}, \tag{13.226}$$

where $\kappa^{(\alpha)}$ are eigenvalues, and $\boldsymbol{\beta}^{(\alpha)}$ eigenvectors of the second metric tensor

$$\left(b_{\alpha\beta} - \kappa a_{\alpha\beta}\right) \beta^\beta = 0. \tag{13.227}$$

On the left-hand side of (13.226) we have the difference of pressures between the gaseous phase, and the fluid phase, respectively.

In addition to the mechanical equilibrium condition (13.226), an equilibrium on the interface is determined by the thermodynamical condition of equal chemical potentials (Gibbs free energies) of both phases

$$\mu^G\left(T^G, p^G\right) = \mu^F\left(T^F, p^F\right), \tag{13.228}$$

where p^G, p^F are pressures in the gaseous phase, and in the fluid phase, respectively. We assume the temperatures to be equal and constant: $T^G = T^F = T = \text{const}$.

We apply the above conditions to a small change of the equilibrium. We have then

$$d\mu^G \equiv V^G dp^G = d\mu^F \equiv V^F dp^F, \tag{13.229}$$

and, hence, by substitution in (13.226)

$$dp^G - dp^F = \left(1 - \frac{V^G}{V^F}\right) dp^G \approx -\frac{V^G}{V^F} dp^G = \sigma d \left(\frac{1}{r_1} + \frac{1}{r_2}\right), \qquad (13.230)$$

where $1/r_1, 1/r_2$ denote the main curvatures. It was assumed that the volume of fluid phase is much smaller than that of the gaseous phase: $V^F \ll V^G$ and that the surface tension σ is constant. For ideal gases we have

$$\frac{RT}{V^F} d \ln p^G = -\sigma d \left(\frac{1}{r_1} + \frac{1}{r_2}\right), \qquad (13.231)$$

where R is the gas constant.

Integration of this relation yields the *Kelvin relation*

$$\ln \frac{p}{p_0} = -\frac{\sigma V^F}{RT} \left(\frac{1}{r_1} + \frac{1}{r_2}\right), \qquad (13.232)$$

in which p_0 is the integration constant which we identify with the saturation pressure.

We demonstrate the properties described by relation (13.232) on a simple example of a cylindrical channel.

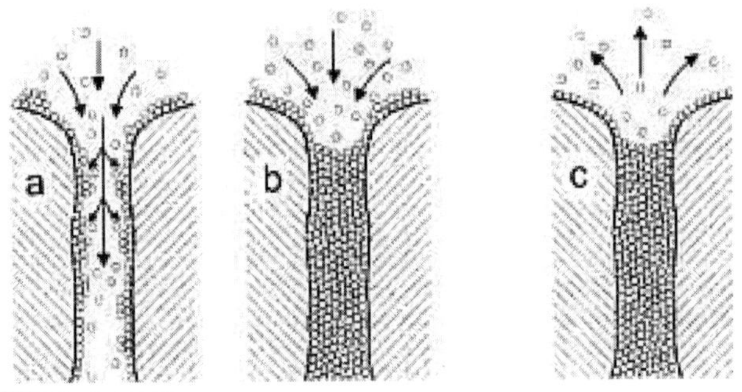

Figure 27: *Scheme of capillary condensation and evaporation in a cylindrical channel* [146].

Imagine a carrier of adsorbate filling the cylindrical channel in the initial state with the partial pressure p of the adsorbate given by (13.232) for $r_1 = r, r_2 = \infty$, where r denotes the radius of the channel (Figure 27). The process of condensation may occur below the saturation pressure p_0 if there is a thin film of adsorbate on the internal wall of the channel. This yields the existence of a surface tension and, according to Kelvin formula, the partial pressure p at which condensation occurs is lower than p_0. Once the nucleation on this thin meniscus begins, the partial pressure of adsorbate decreases and this yields a decrement of the radius of the cylindrical space still filled with the gas (panel (a) in Fig. 27). This process is, of course, unstable and it means that it terminates when the channel contains solely a condensed fluid form of the adsorbate (panel (b) of Fig. 27).

In Figure 28 it corresponds to the lower curve of the hysteresis loop reflecting the process of condensation. On the other hand, the evaporation does not require a nucleation. It proceeds from the heads of the channel by moving hemispherical meniscus from both ends of the channel to its interior (panel (c) of Fig. 27). In this process $r_1 = r_2 = r$, which means that the partial pressure is lower than in the case of condensation. Consequently, we observe a hysteresis shown on the example of Figure 28.

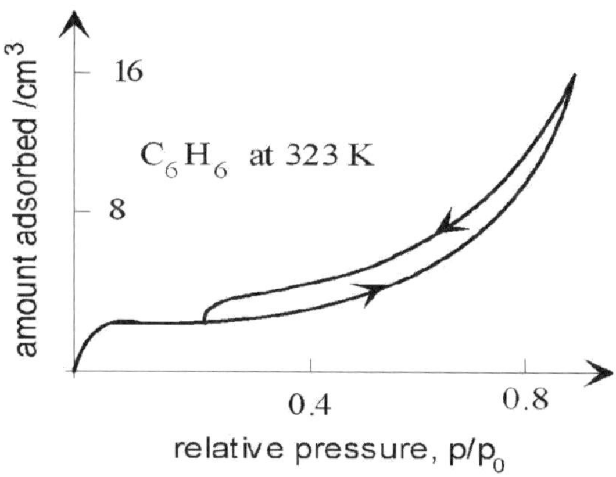

Figure 28: *Condensation and evaporation on a sodium-rich montmorillonite* [133].

Processes in channels of real materials with mesopores are similar to this described in the above example. However the appearance of hysteresis loops is connected with various geometrical properties and their macroscopic description usually requires essential modifications of the model for particular materials.

Chapter 14

Final remarks

Topics of continuum thermodynamics presented in this book were chosen according to my own taste and related to my own research without any pretense to the indication of importance or objective judgement of correctness of different approaches. The main aim was to demonstrate that thermodynamics is a powerful tool to design continuum models for very complex macroscopic processes. However, certain important fields of research in continuum thermodynamics were not mentioned and in this brief Chapter they will be named with some references to the most important contributions.

1. **Materials with memory**

Most likely the first paper on nonlinear memory materials has been published by A. E. Green and R. S. Rivlin in 1956 [129], [130]. In the model presented in this work the current reaction of the material, say the Cauchy stress, depends in a nonlinear way on the *history* of the deformation. This functional dependence has the form of multiple time integrals and it has been almost immediately successfully applied in the description of certain clays. At the beginning of '60s this approach has been generalized and applied to thermodynamical systems. In 1960, B. D. Coleman and W. Noll [79] considered properties of constitutive functionals with *mollifiers* yielding *fading memory*. For example, in the simplest case of a material with memory – the Maxwell fluid, the mollifier has the form of an exponential function: $\exp(-s/\tau)$, where s measures the distance from the present instant of time, t, and τ is the relaxation time. The fading memory means that, due to the mollifier, an influence of events far in the past is smaller than these in the near past. Then in 1963, they published a paper [80] in which the first thermodynamical influences were accounted for. Finally, two big memoirs of B. D. Coleman which have appeared in 1964 [77], [76] have established the main points of this thermodynamical theory. Its most important point is the assumption that constitutive relations contain the whole history of thermodynamical processes. For instance, the Helmholtz free energy is assumed to have the following form

$$\begin{aligned} \psi(\mathbf{X},t) &= \ominus_{s=0}^{\infty}\left(\mathbf{C}^t(\mathbf{X},s), T^t(\mathbf{X},s)\right), \\ \mathbf{C}^t(\mathbf{X},s) &= \mathbf{C}(\mathbf{X},t-s), \quad T^t(\mathbf{X},s) = T(\mathbf{X},t-s), \end{aligned} \tag{14.1}$$

where $\ominus_{s=0}^{\infty}$ is the *constitutive functional* with certain smoothness properties which we shall not discuss here (see [332]). These ideas were put in their final form by M. E. Gurtin [135].

Various assumptions are made about thermodynamical properties of materials with memory but two of them seem to be particularly restrictive. The first one is the second law of thermodynamics which is assumed to have the form of the Clausius-Duhem inequality. We know already that this is not always true and, in particular, for processes far from thermodynamical equilibrium this inequality is violated even by ideal gases. The second assumption concerns the evaluation of this inequality. This is done under the assumption that for any process its linear extension is also an admissible process. This assumption yields the linearity of the entropy inequality with respect to certain Fréchet derivatives of constitutive functionals and, consequently, the existence of certain thermodynamical potentials.

It should be stressed that many thermodynamical models presented in this book, in particular Maxwell fluids, viscoelastic and viscoplastic materials, poroelastic materials with the balance equation of porosity possess memory features. However, they usually possess as well only one or two relaxation times. It is claimed – for instance by D.D. Joseph (private communication) – that many fluids must have an infinite spectrum of relaxation times. This does not follow from a local formulation with memory functionals which result from evolution equations. In this respect, an approach similar to this of B. D. Coleman would be desirable. However, not much progress has been done in that direction since '70s.

2. Phase transitions, hysteresis, non-convex problems

At the end of the last Chapter it was indicated that phase transitions are characterized by the hysteresis. At various places of the presentation in this book, we have also indicated that among desirable thermodynamical properties one of the most essential is the property of thermodynamical stability. However, we know from simplest examples of such substances as vapor described by van der Waals equation that this property is violated when the system undergoes the phase transition (see also remarks on thermodynamical stability of various systems in the book of J. L. Ericksen [103]). This may be the process of condensation-evaporation, freezing-melting or a change of symmetry of crystalline materials, e.g. martensitic phase transformation. Consequently, we should construct thermodynamical models in which such phase changes are admissible. This is a very difficult challenge. As pointed out in the last Chapter, phase changes yield hysteresis phenomena (see A. Visintin [340], and M. Brokate, J. Sprekels [53] for the review of modern mathematical one-dimensional models of this phenomenon). Simultaneously, again an experience with the van der Waals model shows that potentials of such systems are nonconvex functions. This makes the analysis very difficult indeed (e.g. see the application and numerical evaluation of such a nonconvex analysis in various problems of continuum mechanics in the books of P. D. Panagiotopoulos [242] and M. Kuczma [174]). Some progress in development of such thermodynamics has been achieved in numerous works of M. Šilhavý (see many references in his book [297]) who is using again Clausius-Duhem inequality. Most of the models possess an additional scalar variable – the phase fraction, and they are often called phase field models [55]. There are attempts to extend these models to cases with many phase parameters, for instance P. Krejči describes properties of some vector models (e.g. [172]) but thermodynamics of such processes still does not exist. It should be mentioned that there may exist a way motivated by physics on which one could avoid the hysteretic phenomena and nonconvex potentials. This can be achieved by the extension of the dimension of the internal variables. A one-dimensional example, concerning the martensitic phase transformation [360] has been considered but it does not refer to thermodynamical restrictions.

3. **Coupled fields**

One of the most important couplings of thermomechanical processes appears in cases in which materials react on the action of *electromagnetic fields*. Piezoelectricity, ferroelectricity, magneto- and electrostriction, thermodielectric phenomena are typical coupled effects in solids. On the other hand, the description of plasma and electrolytes by magnetohydrodynamics (MHD) is an example of the coupling between electromagnetic fields and flows of fluids. For this reason, the construction of thermodynamical models for such processes is of primary importance. The presentation of the research in this field can be found in the two-volume monograph of C. Eringen and G. Maugin [105], [106] from 1989. A thermodynamical analysis in the spirit of the strategy presented in the previous Chapters appears already in the early book of I. Müller [221] and in the papers of K. Hutter [148], [149]. A modernized version appeared in the book [222] and it has been brought up to date in 2006 by K. Hutter, A. A. F. van de Ven, A. Ursescu [152].

There are two problems which make the thermodynamical formulation of coupled fields theory difficult. The first one is again the form of the second law of thermodynamics. Such effects as Joule heating make the application of Calusius-Duhem inequality not appropriate. The second one is the problem of invariance. The theory of coupled fields requires two additional vector fields to appear in the model: an *electric field* **E** and a *magnetic field* **H**. These two fields have to satisfy *Maxwell equations*

$$\frac{\partial \mathbf{B}}{\partial t} + \operatorname{rot} \mathbf{E} = 0, \quad \operatorname{div} \mathbf{B} = 0, \quad -\frac{\partial \mathbf{D}}{\partial t} + \operatorname{rot} \mathbf{H} = \mathbf{J}, \quad \operatorname{div} \mathbf{D} = \hat{q},$$
$$\mathbf{D} = \varepsilon_0 \mathbf{E}, \quad \mathbf{H} = \frac{1}{\mu_0} \mathbf{B}. \tag{14.2}$$

These are *Faraday equation*, *Gauss equation*, *Ampère-Oersted equation*, and again *Gauss equation* while the second group is the so-called *Maxwell-Lorentz relations* and ε_0, μ_0 are a *dielectric constant* and a *constant of permeability*, respectively. They satisfy the identity $\varepsilon_0 \mu_0 = c^2$, where c is the speed of light. **D** denotes the *dielectric displacement* and **B** is the *magnetic flux*. $\hat{q}$ denotes the *electric charge density* and **J** is the *current*.

The above equations can be written in the invariant tensorial form. If we introduce the following four-matrices

$$\left(F^{AB}\right) = \begin{pmatrix} 0 & B_3 & -B_2 & -E_1 \\ -B_3 & 0 & B_1 & -E_2 \\ B_2 & -B_2 & 0 & -E_3 \\ E_1 & E_2 & E_3 & 0 \end{pmatrix}, \tag{14.3}$$

$$\left(H^{AB}\right) = \begin{pmatrix} 0 & H_3 & -H_2 & -D_1 \\ -H_3 & 0 & H_1 & -D_2 \\ H_2 & -H_1 & 0 & -D_3 \\ D_1 & D_2 & D_3 & 0 \end{pmatrix}, \tag{14.4}$$

$$\left(J^A\right)^T = \left(J_1/c \; J_2/c \; J_3/c \; \hat{q}\right), \tag{14.5}$$

then Maxwell equations have the following form

$$\frac{\partial F_{AB}}{\partial x^C} + \frac{\partial F_{BC}}{\partial x^A} + \frac{\partial F_{CA}}{\partial x^B} = 0, \quad \frac{\partial H^{AB}}{\partial x^B} = J^A, \quad \left(x^A\right)^T = \left(x^0 = ct, x^i\right)^T. \tag{14.6}$$

As we have demonstrated, in the classical mechanics we require an invariance of equations of motion with respect to Galilean transformation and the objectivity principle as an invariance of constitutive relations with respect to Euclidean group of transformations (see Table 8 in Chapter 8). Maxwell equations (14.6), which are for coupled field theories a part of field equations, are invariant with respect to Lorentz group [209]

$$x^{A^*} = \Lambda^{A^*}_{.A} x^A, \quad \left(\Lambda^{A^*}_{.A}\right) = \begin{pmatrix} \Lambda^{0^*}_{.0} & \Lambda^{0^*}_{.i} \\ \Lambda^{i^*}_{.0} & \Lambda^{i^*}_{.i} \end{pmatrix}, \tag{14.7}$$

$$\Lambda^{0^*}_{.0} = \gamma \equiv \frac{1}{\sqrt{1-\beta^2}}, \quad \Lambda^{0^*}_{.i} = \Lambda^{i^*}_{.0} = -\beta\gamma n_i, \quad \Lambda^{i^*}_{.i} = (\gamma - 1) n_{i^*} n_i + \delta_{i^* i},$$

where $\beta c = v$ is the speed of the star reference system with respect to the reference system without star, and $\mathbf{n} = (n_1, n_2, n_3)^T$ is the unit vector in the reference system such that $v\mathbf{n}$ is the velocity of the reference system with the star in the system without star.

This contradiction in invariance requirements can be eliminated by a construction of a model within the special relativity theory rather than within Newtonian mechanics. Both groups of equations would then be invariant with respect to Lorentz transformations. However, at least in some special cases one can preserve the classical formulation of thermomechanics. We shall not present any details of such models and refer to, for instance, books of I. Müller [222] or K. Hutter, A. A. F. van de Ven, A. Ursescu [152].

4. **Relativistic thermodynamics**

The origin of the subject of relativistic thermodynamics lies in cosmology [324]. It has been constructed to describe a gas of classical particles as well as degenerate gases of bosons and fermions. Such models are used, for instance, in the description of white dwarfs which are made of fully ionized particles. Active particles are electrons which form a degenerate gas of fermions [68].

The construction of extended thermodynamics of relativistic gases, as in the nonrelativistic case, can be motivated again by moments of a kinetic equation. In the relativistic case it is the *Boltzmann-Chernikov equation* for the distribution function f which, roughly speaking, specifies the number of atoms with momentum between p^a and $p^a + dp^a$ as $f\left(x^A, p^a\right)\sqrt{-g}/p_0 dp^1 dp^2 dp^3$. This equation has been proposed in 1964 in a series of papers of N. A. Chernikov [72], [70], [71]. Let p^A, $A = 0, 1, 2, 3$, denotes the four-momentum of the particle. Then the kinetic equation has the form

$$p^A \frac{\partial f}{\partial x^A} - \Gamma^d_{AB} p^A p^B \frac{\partial f}{\partial p^d} = \int \left(f\left(p'_C\right) f\left(q'_C\right) - f\left(p_C\right) f\left(q_C\right)\right) h \left\langle pq\right\rangle de d\mathbf{Q}. \tag{14.8}$$

In this equation $\langle pq\rangle$ is the scalar relative speed of particles and h is the scattering factor for scattering into the solid angle de.

Relativistic extended thermodynamics was proposed in 1986 by I-Shih Liu, I. Müller and T. Ruggeri [195]. In the mean time numerous works have been published on a general thermodynamical formulation and on various limits of such a relativistic theory. These were compared with moments following from the kinetic equation (14.8). There are also some results on waves in relativistic gases (e.g. [287]). All these results, presented in the book of I. Müller and T. Ruggeri [225] seem to exhaust the main topics of this subject.

5. **Non-local problems**

Long range interactions in continua were introduced within the variational formulation of continuum mechanics. This procedure followed an analogy to the construction of equations of motion for discrete systems by the use of the action functional and the Lagrangian. One of the first attempts of such a construction is extensively shown in the book of Cz. Wozniak [381] who is also discussing invariance properties, conservation laws, relations to Noether theorems, etc. A similar nonlocal problem in the form of the action functional constructed by means of a multiple space integral appears in the work [352]. The main aim of those constructions was to describe a microstructure created, for instance, in continuum modelling of discrete systems such as dense grids or some composite materials. Approximations of these nonlocal models were models with couple stresses, Cosserat media, media with higher gradients, etc. These constructions were later widely developed by D. Edelen [104], [96], [97], [95]. There was no thermodynamics involved in those models but the main flaw of them was already visible on the level of pure mechanics. Namely, strict nonlocal models possess a feature that interactions are not additive. For instance, for two separate bodies $\mathcal{P}_1$ and $\mathcal{P}_2, \mathcal{P}_1 \cap \mathcal{P}_2 = \varnothing$, the energy of the combination of those two bodies has the structure

$$E(\mathcal{P}_1 \cup \mathcal{P}_2) = E(\mathcal{P}_1) + E(\mathcal{P}_2) + E_b(\mathcal{P}_1, \mathcal{P}_2), \tag{14.9}$$

where $E(\mathcal{P}_1), E(\mathcal{P}_2)$ are energies of $\mathcal{P}_1$ and $\mathcal{P}_2$ and $E_b(\mathcal{P}_1, \mathcal{P}_2)$ is the so-called *binding energy*. Such contributions mean that, for instance, the Cauchy Theorem about the existence of the stress tensor does not hold. One is unable to separate uniquely the surface interaction of two bodies from the volume long range interactions. The whole strategy of a continuum which we presented in this book collapses. In spite of this property, there were attempts to construct thermodynamics of such systems (e.g. [137]). Not only the densities of nonadditive functions were introduced, such as

$$E_b(\mathcal{P}_1, \mathcal{P}_2) = \int_{\mathcal{P}_1} \int_{\mathcal{P}_2} e_b(\mathbf{X}, \mathbf{Y}) \, dV_X dV_Y, \tag{14.10}$$

but even the extension of the second law of thermodynamics on the *binding entropy* was proposed. There are some claims that such two-point functions may be even justified by a kinetic theory with a two-point distribution function. Such models based on a nonequilibrium extension of the work of F. A. Blood [40] and called correlational thermodynamics were constructed by J. Schröter [286] and then applied in modelling of electrolytes.

These constructions seem to be rather artificial within continuum models. However, certain extensions of classical local constitutive laws yield models in which nonlocal interactions are replaced by higher gradients of deformation and even by additional local degrees of freedom. In this way, for instance, the Cosserat models are constructed. They fit the frame of continuum and are quite useful in modelling complex materials. They also smooth out some singularities of numerical procedures (see works of M. Rubin [276], [278]). Continuum thermodynamics of such models is still missing.

6. **Surface thermodynamics**

We have made a few remarks on modelling of surface phenomena in Section 4.3. In spite of the practical importance, particularly in description of biological phenomena and nanomaterials, very little has been done on the subject of nonequilibrium surface thermodynamics.

7. **Axiomatic foundations**

This branch of thermodynamics has been well developed for classical thermostatics. Some references to this subject may be found in [359]. The modern presentation originating from the book of R. Giles published in 1964 [121] is the subject of the book of G. Emch, Chuang Liu [101]. The situation is much worse for the nonequilibrium thermodynamics. There were some attempts to define thermodynamical processes as subsets of some infinite dimensional space of states (e.g. [353], [355], [291], [81]) but the development of this area ceased.

Appendix A

Vectors and tensors on Euclidean spaces

A.1 Reminder of tensor calculus

Continuous modelling is based on the vector and tensor analysis on Euclidean spaces. In this Appendix, we present some basic mathematical properties of this tool.

There are many references to the subject of tensor calculus. Even though not complete and up-to-date, it is worth to recommend the classical works [320], [205] and in reference to applications in continuum mechanics [102]. Modern brief presentation can be found in the Appendix: *Elementary Tensor Analysis* to the book [191].

Continuum mechanics presented in these Notes is based on the geometry of the three-dimensional Euclidean point space $\mathcal{E}^3$. Points of this space, say, $\mathbf{x}, \mathbf{y}$ define vectors $\mathbf{v} = \mathbf{x} - \mathbf{y}$ which belong to a three-dimensional vector space $\mathcal{V}^3$ called the *translation space*. For this reason, the main notions of the vector calculus will be presented for such spaces.

Definition: A *vector space* $\mathcal{V}$ is a set with two operations:

1) $\forall \mathbf{u}, \mathbf{v} \in \mathcal{V}: \quad \mathbf{u} + \mathbf{v} \in \mathcal{V}$ and this operation is called *addition*,

2) $\forall \alpha \in \Re, \mathbf{v} \in \mathcal{V}: \quad \alpha \mathbf{v} \in \mathcal{V}$ and this operation is called *multiplication* of the vector $\mathbf{v}$ by the real number α.

These operations satisfy the following rules for any vectors $\mathbf{u}, \mathbf{v}, \mathbf{w} \in \mathcal{V}$ and any real numbers $\alpha, \beta \in \Re$:

1) $\mathbf{u} + \mathbf{v} = \mathbf{v} + \mathbf{u}$,

2) $\mathbf{u} + (\mathbf{v} + \mathbf{w}) = (\mathbf{u} + \mathbf{v}) + \mathbf{w}$,

3) there exists a null vector $\mathbf{0} \in \mathcal{V}$ such that $\forall \mathbf{v} \in \mathcal{V}: \quad \mathbf{v} + \mathbf{0} = \mathbf{v}$,

4) for any $\mathbf{v} \in \mathcal{V}$ there exists $-\mathbf{v} \in \mathcal{V}$ such that $\mathbf{v} + (-\mathbf{v}) = \mathbf{0}$,

5) $\forall \alpha, \beta \in \Re, \mathbf{v} \in \mathcal{V}: \quad \alpha (\beta \mathbf{v}) = (\alpha\beta) \mathbf{v}$,

6) $\forall \alpha, \beta \in \Re, \mathbf{v} \in \mathcal{V}: \quad (\alpha + \beta) \mathbf{v} = \alpha \mathbf{v} + \beta \mathbf{v}$,

7) $\forall \alpha \in \Re, \mathbf{u}, \mathbf{v} \in \in \mathcal{V}: \quad \alpha (\mathbf{u} + \mathbf{v}) = \alpha \mathbf{u} + \alpha \mathbf{v}$,

8) $1\mathbf{v} = \mathbf{v}$.

As already mentioned, in theories of continua we deal with three-dimensional vector spaces. In general, the dimension n of the space $\mathcal{V}$ is introduced by means of the notion of the set of linearly independent vectors which forms the basis.

Definition: A set of vectors $\{\mathbf{v}_1, ..., \mathbf{v}_n\}$ is said to be a *basis* of $\mathcal{V}$, if

1) the vectors of this set are *linearly independent*, i.e. for any $\alpha_1, ..., \alpha_n \in \Re$, if $\alpha_1 \mathbf{v}_1 + ... + \alpha_n \mathbf{v}_n = \mathbf{0}$ then $\alpha_1 = ... = \alpha_n = 0$,

2) it *spans* the space $\mathcal{V}$, i.e. for any vector $\mathbf{u}$, this vector can be written as a linear combination of $\{\mathbf{v}_1, ..., \mathbf{v}_n\}$.

The latter means that this set cannot possess more than n elements. The number n is called the dimension of the space and, to indicate this property we often write $\mathcal{V}^n$ instead of $\mathcal{V}$.

Obviously, there are many choices of the basis. We usually use for the basis vectors the notation $\{\mathbf{g}_1, ..., \mathbf{g}_n\}$ and write for an arbitrary vector $\mathbf{u}$ which in the n-dimensional space must be a linear combination of basis vectors,

$$\mathbf{u} = \sum_{i=1}^{n} u^i \mathbf{g}_i, \quad \mathbf{u}, \mathbf{g}_1, ..., \mathbf{g}_n \in \mathcal{V}^n, \quad \dim \mathcal{V}^n = n. \tag{A.1}$$

In addition to the above two operations the vector spaces are endowed with a *bilinear* operation which we call the inner (scalar) product. This operation allows to introduce the notions of the length of the vector and of the angle between two vectors.

Definition: An *inner (scalar) product* is the map

$$g : \mathcal{V} \times \mathcal{V} \rightarrow \Re, \tag{A.2}$$

with the following properties

1)$\forall \mathbf{u}, \mathbf{v}, \mathbf{w} \in \mathcal{V}, \alpha \in \Re : \quad g(\mathbf{u}+\alpha\mathbf{v}, \mathbf{w}) = g(\mathbf{u}, \mathbf{w}) + \alpha g(\mathbf{v}, \mathbf{w}),$

2) $\forall \mathbf{u}, \mathbf{v} \in \mathcal{V} : \quad g(\mathbf{u}, \mathbf{v}) = g(\mathbf{v}, \mathbf{u}),$

3) $\forall \mathbf{u} \in \mathcal{V} : \quad g(\mathbf{u}, \mathbf{u}) > 0, \quad \text{if } \mathbf{u} \neq \mathbf{0}.$

We use the following standard notation for this operation

$$g(\mathbf{u}, \mathbf{v}) = \mathbf{u} \cdot \mathbf{v}. \tag{A.3}$$

Definition: The *length* (*norm*) of the vector $\mathbf{v} \in \mathcal{V}$ is defined as

$$|\mathbf{v}| \equiv \|\mathbf{v}\| = \sqrt{\mathbf{v} \cdot \mathbf{v}}. \tag{A.4}$$

Spaces with such a norm are called *Euclidean vector spaces*.

The notion of the *angle* between two vectors is introduced on the basis of the following Schwarz (triangle) inequality

$$|\mathbf{u} \cdot \mathbf{v}| \leq |\mathbf{u}| \, |\mathbf{v}| \, . \tag{A.5}$$

Due to this inequality we can define the cosine of an angle θ between vectors $\mathbf{u}, \mathbf{v}$ in the following manner:

Definition: For any non-zero vectors $\mathbf{u}, \mathbf{v}$, the angle between $\mathbf{u}$ and $\mathbf{v}$, $\theta(\mathbf{u}, \mathbf{v}) \in [0, \pi]$, is defined by

$$\cos\theta(\mathbf{u}, \mathbf{v}) = \frac{\mathbf{u} \cdot \mathbf{v}}{|\mathbf{u}|\,|\mathbf{v}|}. \tag{A.6}$$

Two vectors are *orthogonal* if $\theta = \pi/2$. Obviously, two vectors $\mathbf{u}, \mathbf{v}$ are orthogonal if and only if $\mathbf{u} \cdot \mathbf{v} = 0$.

A vector $\mathbf{v}$ is called the *unit vector* if $|\mathbf{v}| = 1$. The *projection of a vector* $\mathbf{u}$ on the vector $\mathbf{v}$ is then defined as $|\mathbf{u}| \cos\theta(\mathbf{u}, \mathbf{v})$, or as $\mathbf{u} \cdot \mathbf{e}$, where $\mathbf{e}$ is the unit vector in the direction of the vector $\mathbf{v}$, defined by the relation $\mathbf{e} = \mathbf{v}/|\mathbf{v}|$.

Let $\{\mathbf{g}_i, i = 1, ..., n\}$ be a basis of $\mathcal{V}$. Then the scalar product

$$g_{ij} = \mathbf{g}_i \cdot \mathbf{g}_j \tag{A.7}$$

is called the *covariant metric tensor*. This name is related to metric properties of the point space $\mathcal{E}$ for which $\mathcal{V}$ is the translation vector space. We come back to this point later.

Let us consider the scalar product of two vectors $\mathbf{u}, \mathbf{v}$ in the above basis. We have

$$\mathbf{u} \cdot \mathbf{v} = \left(u^i \mathbf{g}_i\right) \cdot \left(\mathbf{v}^j \mathbf{g}_j\right) = u^i v^j g_{ij}. \tag{A.8}$$

We have used the following summation convention.

Summation convention. If an index in the multiplicative term of the expression is repeated once (and only once!), a summation over the range of this index is assumed.

For instance

$$u^i \mathbf{g}_i = \sum_{i=1}^{n} u^i \mathbf{g}_i, \quad u^i v^j g_{ij} = \sum_{i=1}^{n} \sum_{j=1}^{n} u^i v^j g_{ij}. \tag{A.9}$$

The basis vectors, $\{\mathbf{g}_i, i = 1, ..., n\}$, are usually related to the system of coordinates in the point space $\mathcal{E}$. They are chosen to be tangent to parametric lines along which all but one coordinate remain constant. We return later to this relation. However, such a relation indicates as well that one can also introduce another set of vectors, $\{\mathbf{g}^i, i = 1, ..., n\}$ which are perpendicular to parametric surfaces on which all but one coordinate change. Such vectors are chosen to be perpendicular to these surfaces. Consequently, they must satisfy the following relation

$$\mathbf{g}^i \cdot \mathbf{g}_j = \delta^i_{.j}, \tag{A.10}$$

where $\delta^i_{\cdot j}$ is called the *Kronecker delta* defined by

$$\delta^i_{\cdot j} = \begin{cases} 0, & \text{if} \quad i \neq j, \\ 1, & \text{if} \quad i = j. \end{cases} \tag{A.11}$$

From this construction we obtain

$$\mathbf{v} \cdot \mathbf{g}^i = \left(v^j \mathbf{g}_j\right) \cdot \mathbf{g}^i = v^i. \tag{A.12}$$

It can be easily shown that the set $\left\{\mathbf{g}^i, i = 1, ..., n\right\}$ forms the basis of $\mathcal{V}$. It is called *dual to* $\{\mathbf{g}_i, i = 1, ..., n\}$. The first one is the *contravariant basis* and, for this reason, the corresponding components v^i of the vector $\mathbf{v}$ are called *contravariant* while the second one is called *covariant basis* and the corresponding components $v_i = \mathbf{v} \cdot \mathbf{g}_i$ are called *covariant.* The *contravariant metric tensor*

$$g^{ij} = \mathbf{g}^i \cdot \mathbf{g}^j \quad \Longrightarrow \quad g^{ij} g_{jk} = \delta^i_{\cdot k}, \tag{A.13}$$

together with the covariant metric tensor g_{ij} allow to raise and to lower the indices

$$u^i = g^{ij} u_j, \quad u_i = g_{ij} u^j. \tag{A.14}$$

In a particular case of the *orthogonal basis of unit vectors* $\{\mathbf{e}_i, i = 1, ..., n\}$ – the basis is then called *orthonormal,*

$$\mathbf{e}_i \cdot \mathbf{e}_j = \delta_{ij}, \tag{A.15}$$

we do not have to distinguish between contravariant and covariant components: $u_i \stackrel{*}{=} u^i$ ($\stackrel{*}{=}$ means that the relation holds only for a particular choice of the basis or of the frame). This is the characteristic feature of the basis of Cartesian coordinate systems.

Now we proceed to construct objects which are of primary importance for continuous models. They describe, for example, deformations and tensions in materials. Let us consider two vector spaces $\mathcal{U}$ and $\mathcal{V}$ possessing properties described above. A function $\mathbf{T} : \mathcal{U} \to \mathcal{V}$ is called the *linear transformation from* $\mathcal{U}$ to $\mathcal{V}$ if for any $\mathbf{u}, \mathbf{v} \in \mathcal{U}$ and $\alpha \in \Re$

$$\mathbf{T}\left(\mathbf{u}+\alpha\mathbf{v}\right) = \mathbf{T}\left(\mathbf{u}\right) + \alpha\mathbf{T}\left(\mathbf{v}\right). \tag{A.16}$$

If $\mathbf{T}, \mathbf{S}$ are two such linear transformations we then define their addition $\mathbf{S} + \mathbf{T}$ and scalar multiplication $\alpha\mathbf{T}$ which satisfy the following conditions

$$\begin{aligned} \left(\mathbf{T} + \mathbf{S}\right)\left(\mathbf{v}\right) &= \mathbf{T}\left(\mathbf{v}\right) + \mathbf{S}\left(\mathbf{v}\right), \\ \left(\alpha\mathbf{T}\right)\left(\mathbf{v}\right) &= \alpha\mathbf{T}\left(\mathbf{v}\right). \end{aligned} \tag{A.17}$$

With these operations the space of all such linear mappings also becomes a vector space.

The simplest linear transformation satisfying the above conditions is the *tensor product* of two vectors. It is defined by the relation

$$\forall \mathbf{w} \in \mathcal{U}: \quad (\mathbf{v} \otimes \mathbf{u})(\mathbf{w}) = (\mathbf{u} \cdot \mathbf{w})\, \mathbf{v}, \quad \mathbf{v} \in \mathcal{V}, \quad \mathbf{u} \in \mathcal{U}. \tag{A.18}$$

Such products are also called *simple tensors* (sometimes called the *dyadic products*). One can show that any linear transformation can be represented by a linear combination of simple tensors:

Proposition: Let $\{\mathbf{g}_i, i = 1, ..., n\}$ and $\{\mathbf{G}_\alpha, \alpha = 1, ...m\}$ be bases of $\mathcal{V}$ and $\mathcal{U}$, respectively. Then any linear mapping $\mathbf{T}$: $\mathcal{V} \to \mathcal{U}$ can be written in the form

$$\mathbf{T} = T^{i\alpha} \mathbf{g}_i \otimes \mathbf{G}_\alpha. \tag{A.19}$$

The coefficients $T^{i\alpha}$ are called the *contravariant components of the mapping* $\mathbf{T}$. Certainly, we can easily introduce covariant and mixed components using the relations

$$T^{i}_{.\alpha} = T^{i\beta} G_{\alpha\beta}, \quad T_{i}^{.\alpha} = T^{j\alpha} g_{ij}, \quad T_{i\alpha} = T^{j\beta} g_{ij} G_{\alpha\beta}, \quad G_{\alpha\beta} = \mathbf{G}_\alpha \cdot \mathbf{G}_\beta. \tag{A.20}$$

Linear transformations of the vector space $\mathcal{V}$ into the same vector space $\mathcal{V}$ are called the *second-order tensors*. We skip further the parentheses in writing second-order tensors. For instance, relations for components of the mapping $\mathbf{T}$ will be written in the form

$$T^{i\alpha} = \mathbf{g}^i \cdot \mathbf{T}(\mathbf{G}^\alpha) \equiv \mathbf{g}^i \cdot \mathbf{T}\mathbf{G}^\alpha. \tag{A.21}$$

Now, the bases $\{\mathbf{g}_i, i = 1, ..., n\}$ and $\{\mathbf{G}_\alpha, \alpha = 1, ..., m\}$ are the bases of the same space $\mathcal{V}$. The components of the tensor $\mathbf{T} = T^{ij} \mathbf{g}_i \otimes \mathbf{g}_i$ form the *matrix representation* of the tensor

$$\left[T^{ij}\right] = \begin{bmatrix} T^{11} & T^{12} & T^{13} \\ T^{21} & T^{22} & T^{23} \\ T^{31} & T^{32} & T^{33} \end{bmatrix}. \tag{A.22}$$

We can define the *composition* of two second-order tensors $\mathbf{T}, \mathbf{S}$ by the relation

$$\forall \mathbf{v} \in \mathcal{V}: \quad \mathbf{S} \circ \mathbf{T}(\mathbf{v}) = \mathbf{S}(\mathbf{T}(\mathbf{v})) \equiv \mathbf{S}\mathbf{T}\mathbf{v} = S^{ij} T_{j}^{.k} v_k \mathbf{e}_i, \tag{A.23}$$

where the orthonormal basis $\{\mathbf{e}_i, i = 1, ..., n\}$ is used.

We shall not present any further details of the tensor calculus referring the reader to, for instance, the book of I-Shih Liu. We use the standard notation for the *transpose*, the *trace* and the *determinant* of the second-order tensor

$$\mathbf{T}^T = T^{ij}\mathbf{e}_j \otimes \mathbf{e}_i, \quad \operatorname{tr}\mathbf{T} = T^i_{.i}, \quad \det\mathbf{T} = \det\left[T_{ij}\right]. \tag{A.24}$$

In this book we use only vectors and tensors for three-dimensional Euclidean point spaces. Then we can use the Cartesian frame of coordinates with the orthonormal basis $\{\mathbf{e}_i, i = 1, 2, 3\}$. For such tensors we have

$$T^{ij} \stackrel{*}{=} T^i_{.j} \stackrel{*}{=} T_i^{.j} \stackrel{*}{=} T_{ij}. \tag{A.25}$$

We consider the eigenvalue problems of the form

$$(\mathbf{T} - \lambda_T \mathbf{1})\,\mathbf{k}_T = \mathbf{0}, \quad \mathbf{1} = \delta^{ij}\mathbf{e}_i \otimes \mathbf{e}_j, \tag{A.26}$$

where λ_T are eigenvalues following from the equation

$$\begin{aligned} \det(\mathbf{T} - \lambda_T \mathbf{1}) &= -\lambda_T^3 + I_T\lambda_T^2 - II_T\lambda_T + III_T = 0, \\ I_T = \operatorname{tr}\mathbf{T}, \quad II_T &= \frac{1}{2}\left(I_T^2 - \operatorname{tr}\mathbf{T}^2\right), \\ III_T = \det\mathbf{T}, \quad \mathbf{T}^2 &= \mathbf{T}\mathbf{T}, \end{aligned} \tag{A.27}$$

where I_T, II_T, III_T are called the *principal invariants* of the tensor $\mathbf{T}$, and $\mathbf{k}_T$ are the *right eigenvectors* of the tensor $\mathbf{T}$.

Obviously, if $\lambda_T^{(\alpha)}, \alpha = 1, 2, 3$, denote solutions of the *characteristic equation* (A.27) then

$$\begin{aligned} I_T &= \sum_{\alpha=1}^{3} \lambda_T^{(\alpha)}, \quad II_T = \lambda_T^{(1)}\lambda_T^{(2)} + \lambda_T^{(2)}\lambda_T^{(3)} + \lambda_T^{(1)}\lambda_T^{(3)}, \\ III_T &= \lambda_T^{(1)}\lambda_T^{(2)}\lambda_T^{(3)}. \end{aligned} \tag{A.28}$$

In the three-dimensional spaces the second-order tensors satisfy in the Cartesian frame the following

Cayley-Hamilton Theorem:

$$\mathbf{T}^3 - I_T\mathbf{T}^2 + II_T\mathbf{T} - III_T\mathbf{1} = \mathbf{0}. \tag{A.29}$$

This Theorem shows that all powers of the second-order tensor higher than two can be expressed by the tensors $\mathbf{1}, \mathbf{T}, \mathbf{T}^2$ and the principal invariants. The theorem also holds true for the higher dimension n with the corresponding change of powers $\mathbf{1}, ..., \mathbf{T}^{n-1}$ and invariants.

We use often the notion of the vector product. This notion can be easily introduced in the three-dimensional vector spaces but some problems appear, for instance, in the case of two-dimensional spaces. For this reason, we introduce first the notion of exterior product.

Definition (W. Gibbs): For any $\mathbf{v}, \mathbf{u} \in \mathcal{V}$, the *exterior product* of $\mathbf{u}$ and $\mathbf{v}$, denoted $\mathbf{u} \wedge \mathbf{v}$, is defined by

$$\mathbf{u} \wedge \mathbf{v} = \mathbf{u} \otimes \mathbf{v} - \mathbf{v} \otimes \mathbf{u}. \tag{A.30}$$

Obviously, this operation is *skew-symmetric*

$$\mathbf{v} \wedge \mathbf{u} = -\mathbf{u} \wedge \mathbf{v}. \tag{A.31}$$

Matrix representation of this tensor in the Cartesian frame

$$\mathbf{u} \wedge \mathbf{v} = \left(u^i v^j - v^i u^j\right) \mathbf{e}_i \otimes \mathbf{e}_j, \tag{A.32}$$

shows that the dimension of the space of these mappings is $n\,(n-1)\,/2$, i.e. for $n = 2$ it is 1 and for $n = 3$ it is 3. Consequently, in the second case, we can introduce a vector representing the exterior product. We can define

$$\begin{aligned} \mathbf{w} &= \epsilon_{ijk} u^j v^k \mathbf{e}^i \\ &= \frac{1}{2}\epsilon_{ijk}\left(u^j v^k - u^k v^j\right)\mathbf{e}^i, \end{aligned} \tag{A.33}$$

where

$$\epsilon_{ijk} = \begin{cases} 1 & \text{for even permutations } 1 \to 2 \to 3 \\ -1 & \text{for odd permutations } 1 \to 3 \to 2 \\ & \text{otherwise,} \end{cases} \tag{A.34}$$

is the permutation symbol. The following properties of the permutation symbol are useful in calculations

$$\begin{aligned} \epsilon_{ijk}\epsilon_{imn} &= \delta_{jm}\delta_{kn} - \delta_{jn}\delta_{km}, \\ \epsilon_{ijk}\epsilon_{ijl} &= 2\delta_{kl}, \\ \epsilon_{ijk}\epsilon_{ijk} &= 6. \end{aligned} \tag{A.35}$$

Obviously

$$\mathbf{u} \wedge \mathbf{v} = \left(u^j v^k - u^k v^j\right) \mathbf{e}_j \otimes \mathbf{e}_k, \tag{A.36}$$

which means that (A.33) has the form

$$\mathbf{w} = \frac{1}{2}\mathbf{e}\,(\mathbf{u} \wedge \mathbf{v}), \quad \mathbf{e} = \epsilon_{ijk}\mathbf{e}^i \otimes \mathbf{e}^j \otimes \mathbf{e}^k. \tag{A.37}$$

From the definition the vector **w** is perpendicular to both **u** and **v**. Bearing the identity $(A.35)_1$ in mind, we have from (A.33)

$$\begin{aligned} \mathbf{w}\cdot\mathbf{w} = |\mathbf{w}|^2 &= (\mathbf{u}\cdot\mathbf{u})(\mathbf{v}\cdot\mathbf{v}) - (\mathbf{u}\cdot\mathbf{v})(\mathbf{u}\cdot\mathbf{v})\cos^2\theta(\mathbf{u},\mathbf{v}) \\ &= |\mathbf{u}|^2|\mathbf{v}|^2\sin^2\theta(\mathbf{u},\mathbf{v}) \quad \Rightarrow \quad |\mathbf{w}| = |\mathbf{u}||\mathbf{v}|\sin\theta(\mathbf{u},\mathbf{v}). \end{aligned} \tag{A.38}$$

Hence, the length of the vector **w** is given by the relation known from the elementary vector calculus. Certainly, we use the following notation for the vector (cross) product

$$\mathbf{w} = \mathbf{u}\times\mathbf{v}. \tag{A.39}$$

The relation (A.37) allows to define the vector product of vectors also in the two-dimensional vector space $\mathcal{V}^2$. Obviously, this operation produces a vector which does not belong to $\mathcal{V}^2$. A simple example is the vector product of two vectors tangent to a surface at a given point. Such a product is a vector perpendicular to the surface.

A.2 Coordinates

We present now a few remarks on frames of reference in $\mathcal{E}^3$. For mathematical details, one can consult one of many books on differential geometry, e.g. [312]. Any given point of this space, **x**, can be identified with a *position vector* **x** and it can be written as a function of three coordinates (Fig. 29)

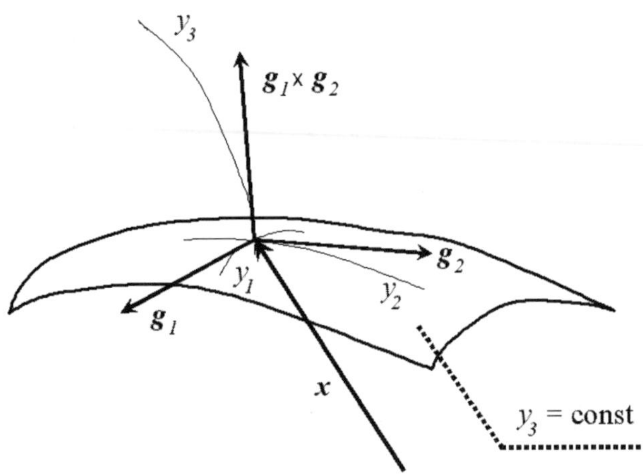

Fig. 29: *Curvilinear coordinates in the Euclidean space.*

$$\mathbf{x} = \mathbf{x}\left(y^1, y^2, y^3\right). \tag{A.40}$$

Parametric lines C^i are curves which contain points for which two of the three coordinates are constant

$$\mathcal{C}^1 = \left\{\mathbf{x} | \, y^2 = const, y^3 = const\right\}, \quad etc. \tag{A.41}$$

Local covariant basis vectors $\mathbf{g}_i$ are defined as derivatives of equations of parametric curves

$$\mathbf{g}_i = \frac{\partial \mathbf{x}}{\partial y^i}. \tag{A.42}$$

Consequently, they are tangent to parametric curves. The infinitesimal *line element* ds at a point $\mathbf{x} \in \mathcal{E}^3$ in an arbitrary direction is defined by the differential

$$ds^2 = d\mathbf{x} \cdot d\mathbf{x} = g_{ij} dy^i dy^j, \quad g_{ij} = \mathbf{g}_i \cdot \mathbf{g}_j. \tag{A.43}$$

This justifies the name: metric of the tensor g_{ij}. It specifies the length of an arc between two values of the parameter s, say s_1 and s_2, in the space $\mathcal{E}^3 : \int_{s_1}^{s_2} ds$.

Parametric surfaces are defined by relations

$$\mathcal{S}^i = \left\{\mathbf{x} | \, y^i = const\right\}, \quad i = 1, 2, 3. \tag{A.44}$$

Obviously, parametric lines on which $y^i \doteq const$ lie on this surface. Consequently, local covariant basis vectors $\mathbf{g}_j, j \neq i$ are tangent to the surface $\mathcal{S}^i$. For instance, the vectors $\mathbf{g}_1, \mathbf{g}_2$ at a given point $\mathbf{x}$ are tangent to the parametric surface $y^3 = const$ going through this point. For this reason, all vectors parallel to $\mathbf{g}_1 \times \mathbf{g}_2$ are perpendicular to the parametric surface. One of them, say $\mathbf{g}^3$, can be chosen as the element of the *local contravariant basis vectors* $\left\{\mathbf{g}^1, \mathbf{g}^2, \mathbf{g}^3\right\}$. These vectors are perpendicular to corresponding parametric surfaces. Hence

$$\mathbf{g}^i \cdot \mathbf{g}_j = \delta^i_j. \tag{A.45}$$

The above definitions of the basis vectors yield natural transformation rules for changes of basis at any point of the space $\mathcal{E}^3$. Namely, if we choose two coordinate systems $\left\{y^k\right\}$ and $\left\{y^{k^*}\right\}$ then the corresponding basis vectors at the point $\mathbf{x}$ are related in the following way

$$\begin{aligned} \mathbf{g}_k &= \frac{\partial \mathbf{x}}{\partial y^k}(\mathbf{x}) = \frac{\partial \mathbf{x}}{\partial y^{k^*}}\frac{\partial y^{k^*}}{\partial y^k}(\mathbf{x}) = \frac{\partial y^{k^*}}{\partial y^k}\mathbf{g}_{k^*}, \\ \mathbf{g}_{k^*} &= \frac{\partial \mathbf{x}}{\partial y^{k^*}}(\mathbf{x}) = \frac{\partial \mathbf{x}}{\partial y^k}\frac{\partial y^k}{\partial y^{k^*}}(\mathbf{x}) = \frac{\partial y^k}{\partial y^{k^*}}\mathbf{g}_k, \end{aligned} \tag{A.46}$$

where

$$y^{k^*} = y^{k^*}\left(y^k\right), \quad y^k = y^k\left(y^{k^*}\right), \tag{A.47}$$

are the transformation functions and their inverse. For contravariant basis vectors, we have

$$\begin{aligned} \mathbf{g}^k \cdot \mathbf{g}_l = \mathbf{g}^k \cdot \left(\frac{\partial y^{k^*}}{\partial y^k}\mathbf{g}_{k^*}\right) = \delta^k_l \quad &\Rightarrow \quad \mathbf{g}^{k^*} = \frac{\partial y^{k^*}}{\partial y^k}\mathbf{g}^k, \\ \mathbf{g}^{k^*} \cdot \mathbf{g}_{l^*} = \mathbf{g}^{k^*} \cdot \left(\frac{\partial y^k}{\partial y^{k^*}}\mathbf{g}_k\right) = \delta^{k^*}_{l^*} \quad &\Rightarrow \quad \mathbf{g}^k = \frac{\partial y^k}{\partial y^{k^*}}\mathbf{g}^{k^*}, \end{aligned} \tag{A.48}$$

where we have used the obvious identities

$$\frac{\partial y^{k^*}}{\partial y^k}\frac{\partial y^k}{\partial y^{l^*}} = \delta^{k^*}_{l^*}, \quad \frac{\partial y^k}{\partial y^{k^*}}\frac{\partial y^{k^*}}{\partial y^l} = \delta^k_l. \tag{A.49}$$

Transformation matrices $\partial y^{k^*}/\partial y^k$ and $\partial y^k/\partial y^{k^*}$ specify transformation rules for vectors and tensors. We have, for instance, for an arbitrary vector $\mathbf{v}(\mathbf{x})$ at the point $\mathbf{x}$

$$\begin{aligned} \mathbf{v} = v^k\mathbf{g}_k = v^k\frac{\partial y^{k^*}}{\partial y^k}\mathbf{g}_{k^*} \quad &\Rightarrow \quad v^{k^*} = \frac{\partial y^{k^*}}{\partial y^k}v^k, \\ \mathbf{v} = v_k\mathbf{g}^k = v_k\frac{\partial y^k}{\partial y^{k^*}}\mathbf{g}^{k^*} \quad &\Rightarrow \quad v_{k^*} = \frac{\partial y^k}{\partial y^{k^*}}v_k. \end{aligned} \tag{A.50}$$

Obviously, these rules are *local*, i.e. they change from one point of the space $\mathcal{E}^3$ to another according to the definition of transformation of curvilinear coordinates (A.47). They are the same only in the case of transformations of Cartesian coordinates for which transformation matrices $\partial y^{k^*}/\partial y^k$ and $\partial y^k/\partial y^{k^*}$ are constant.

Similar derivation can be made for arbitrary tensors.

In the case of Cartesian frames of reference we have

$$\mathbf{x} = x^i\mathbf{e}_i, \quad \mathbf{e}_i \cdot \mathbf{e}_j = \delta_{ij}, \tag{A.51}$$

and the basis vectors $\mathbf{e}_i$ are independent of the point $\mathbf{x} \in \mathcal{E}^3$. This is the characteristic property of Euclidean spaces in which the so-called *parallel transport of vectors* (parallel displacement, shifting; see the next Section) allows to identify all vector spaces $\mathcal{V}^3$ in contrast to general manifolds where they are defined individually for all points $\mathbf{x}$ of the manifold (compare vector spaces of a two-dimensional surface which form tangent planes at each point of the surface).

A.3 Euclidean shifters, differentiation

Vector and tensors in point spaces are local notions related to particular tangent spaces. In Cartesian coordinates, the sum of components of, say, two tensors of the same order but evaluated at different points of space, or, more generally, the integral of a tensor field over a curve, surface or volume should yield tensor components of the tensor of the same order. Under general transformations of coordinates such objects do not transform as tensors. Therefore, it is necessary to define invariant operations which yield the above properties. Such a definition is possible in Euclidean point spaces and it may not be possible in more general spaces (e.g. two-dimensional Riemann spaces in which, for instance, a sum of two vectors tangent to the space at two different points is not a vector tangent to the space). For such purposes, the operation of finite (global) parallel displacement, or *shifting*, is important.

Let us consider the vector $\mathbf{V}(\mathbf{X})$ at a chosen point $\mathbf{X}$ of the space $\mathcal{E}^3$ and another vector $\mathbf{v}(\mathbf{x})$ at a chosen point $\mathbf{x}$ of this space. In a chosen frame of reference these vectors have components v^k and V^K, respectively. We choose them in this way that these components have the same numerical values, i.e.

$$v^k = \delta^k_K V^K, \quad V^K = \delta^K_k v^k. \tag{A.52}$$

In a Cartesian frame, this would mean that the vector $\mathbf{v}$ is the parallel shift of the vector $\mathbf{V}$. However, this property is, of course, not invariant with respect to an arbitrary change of frames. Bearing the property (A.52) in mind, we can write in arbitrary frames $\{y^l\}$ and $\{Y^L\}$

$$v^l(\mathbf{x}) \frac{\partial x^k}{\partial y^l}(\mathbf{x}) = \delta^k_K V^L(\mathbf{X}) \frac{\partial X^K}{\partial Y^L}(\mathbf{X}), \quad \mathbf{v} = v^l \mathbf{g}_l, \quad \mathbf{V} = V^L \mathbf{G}_L, \tag{A.53}$$

where $\{\mathbf{g}_l\}$ are basis vectors of the frame $\{y^l\}$ at the point $\mathbf{x}$ and $\{\mathbf{G}_L\}$ basis vectors of the frame $\{Y^L\}$ at the point $\mathbf{X}$. Consequently

$$v^l(\mathbf{x}) = g^l_{.L}(\mathbf{x}, \mathbf{X}) V^L(\mathbf{X}), \quad V^L(\mathbf{X}) = g^L_{.l}(\mathbf{x}, \mathbf{X}) v^l(\mathbf{X}), \tag{A.54}$$

where

$$g^{l}_{.L}(\mathbf{x},\mathbf{X})=\delta^{k}_{K}\frac{\partial y^{l}}{\partial x^{k}}\frac{\partial X^{K}}{\partial Y^{L}},\quad g^{L}_{.l}(\mathbf{x},\mathbf{X})=\delta^{K}_{k}\frac{\partial Y^{L}}{\partial X^{K}}\frac{\partial x^{k}}{\partial y^{l}}. \tag{A.55}$$

It follows easily from these definitions

$$g^{k}_{.M}g^{M}_{.l}=\delta^{k}_{l},\quad g^{k}_{.K}g^{L}_{.k}=\delta^{L}_{K}. \tag{A.56}$$

The relations (A.54) indicate that the above objects shift vectors from the point $\mathbf{X}$ to the point $\mathbf{x}$ and *vice versa.* For this reason, we call them *shifters.* This operation of transport in the space $\mathcal{E}^3$ is invariant in this sense that $g^{k}_{.K}$ transform as three (contravariant) vectors for $K=1,2,3$ when we perform the transformation $y^{k^*}=y^{k^*}(y^k)$, and they transform as three (covariant) vectors when we perform the transformation $Y^{K^*}=Y^{K^*}(Y^K)$.

For shifters, we can easily prove also the following relations

$$\begin{aligned} g_{k}^{.K}&=g_{kl}g^{l}_{.L}g^{LK},\quad g^{.k}_{K}=g_{KL}g^{L}_{.l}g^{lk}, \\ g_{k}^{.K}&=g^{K}_{.k},\quad g^{K}_{.k}=g_{k}^{.K}. \end{aligned} \tag{A.57}$$

Hence, it is not necessary to mark the place of the index.

A more extensive discussion of properties of shifters can be found in the article of J. Ericksen [102].

Having the technique of parallel transport of vectors and tensors in the Euclidean space at hand, we can introduce invariant differentiation of vector and tensor fields. We proceed to do so.

Let us consider an arbitrary vector field $\mathbf{v}(\mathbf{x})$. In a chosen frame of reference $\{y^k\}$ its local changes are defined by the relation

$$\begin{aligned} d\mathbf{v}&=\frac{\partial\mathbf{v}}{\partial y^{k}}dy^{k}=\frac{\partial}{\partial y^{k}}\left(v^{l}\mathbf{g}_{l}\right)dy^{k} \\ &=\left(\frac{\partial v^{l}}{\partial y^{k}}\mathbf{g}_{l}+v^{l}\frac{\partial\mathbf{g}_{l}}{\partial y^{k}}\right)dy^{k}. \end{aligned} \tag{A.58}$$

Obviously, each vector $\partial\mathbf{g}_l/\partial y^k, k,l=1,2,3$ can be written in components with respect to the covariant basis vectors $\mathbf{g}_m$, i.e.

$$\frac{\partial\mathbf{g}_{l}}{\partial y^{k}}=\Gamma^{m}_{kl}\mathbf{g}_{m},\quad\Gamma^{m}_{kl}=\Gamma^{m}_{lk}, \tag{A.59}$$

where the symmetry follows from the definition of the basis vector (A.42). The quantities Γ^{m}_{kl} are called *Christoffel symbols of the second kind.* They can be easily calculated from the relations

$$\Gamma^{m}_{kl}\equiv\left\{ {m \atop kl} \right\}=\frac{\partial^{2}\mathbf{x}}{\partial y^{k}\partial y^{l}}\cdot\mathbf{g}^{m}. \tag{A.60}$$

Table 13: *Description of cylindrical and spherical coordinates*

Cylindrical coordinates	**Spherical coordinates**
Position vector	
$\mathbf{x}=y^1\cos y^2\mathbf{e}_1+y^1\sin y^2\mathbf{e}_2+y^3\mathbf{e}_3$	$\mathbf{x}=y^1\cos y^2\sin y^3\mathbf{e}_1 + y^1\sin y^2\sin y^3\mathbf{e}_2+y^1\cos y^3\mathbf{e}_3$
Covariant basis vectors	
$\mathbf{g}_1=\cos y^2\mathbf{e}_1+\sin y^2\mathbf{e}_2$ $\mathbf{g}_2=y^1\left(-\sin y^2\mathbf{e}_1+\cos y^2\mathbf{e}_2\right)$ $\mathbf{g}_3=\mathbf{e}_3$	$\mathbf{g}_1=\cos y^2\sin y^3\mathbf{e}_1 +\sin y^2\sin y^3\mathbf{e}_2+\cos y^3\mathbf{e}_3$ $\mathbf{g}_2=-y^1\sin y^2\sin y^3\mathbf{e}_1 +y^1\cos y^2\sin y^3\mathbf{e}_2$ $\mathbf{g}_3=y^1\cos y^2\cos y^3\mathbf{e}_1 +y^1\sin y^2\cos y^3\mathbf{e}_2-y^1\sin y^2\mathbf{e}_3$
Contravariant basis vectors	
$\mathbf{g}^1=\cos y^2\mathbf{e}_1+\sin y^2\mathbf{e}_2$ $\mathbf{g}^2=\frac{1}{y^1}\left(-\sin y^2\mathbf{e}_1+\cos y^2\mathbf{e}_2\right)$ $\mathbf{g}^3=\mathbf{e}_3$	$\mathbf{g}^1=\cos y^2\sin y^3\mathbf{e}_1 +\sin y^2\sin y^3\mathbf{e}_2+\cos y^3\mathbf{e}_3$ $\mathbf{g}^2=\frac{1}{y^1\sin^2 y^3}\left(-\sin y^2\sin y^3\mathbf{e}_1 +\cos y^2\sin y^3\mathbf{e}_2\right)$ $\mathbf{g}^3=\frac{1}{y^1}\left(\cos y^2\cos y^3\mathbf{e}_1 +\sin y^2\cos y^3\mathbf{e}_2-\sin y^3\mathbf{e}_3\right)$
Metric tensors	
$\mathbf{g}=\mathbf{g}_1\otimes\mathbf{g}_1+ +\left(y^1\right)^2\mathbf{g}_2\otimes\mathbf{g}_2+\mathbf{g}_3\otimes\mathbf{g}_3$ $\mathbf{g}=\mathbf{g}^1\otimes\mathbf{g}^1+ +\left(\frac{1}{y^1}\right)^2\mathbf{g}^2\otimes\mathbf{g}^2+\mathbf{g}^3\otimes\mathbf{g}^3$	$\mathbf{g}=\mathbf{g}_1\otimes\mathbf{g}_1 +\left(y^1\sin y^2\right)^2\mathbf{g}_2\otimes\mathbf{g}_2 +\left(y^1\right)^2\mathbf{g}_3\otimes\mathbf{g}_3$ $\mathbf{g}=\mathbf{g}^1\otimes\mathbf{g}^1 +\left(\frac{1}{y^1\sin y^3}\right)^2\mathbf{g}^2\otimes\mathbf{g}^2 +\left(\frac{1}{y^1}\right)^2\mathbf{g}^3\otimes\mathbf{g}^3$
Christoffel symbols	
$\Gamma^2_{12}=\frac{1}{y^1},\Gamma^1_{22}=-y^1$	$\Gamma^1_{22}=-y^1\sin^2 y^3,$ $\Gamma^1_{33}=-y^1,\Gamma^2_{12}=\Gamma^3_{13}=\frac{1}{y^1},$ $\Gamma^3_{22}=-\cos y^3\sin y^3,$ $\Gamma^2_{32}=\cot y^3$

Note: The basis vectors in the above Table are written by means of components in the Cartesian basis $\{\mathbf{e}_1,\mathbf{e}_2,\mathbf{e}_3\}$.

Usually, one calculates the *Christoffel symbols of the first kind* which are defined by the following formula

$$[kl,m] = \frac{1}{2}\left(\frac{\partial g_{km}}{\partial y^l} + \frac{\partial g_{lm}}{\partial y^k} - \frac{\partial g_{kl}}{\partial y^m}\right) \equiv [kl,\overset{\mathbf{g}}{}m], \tag{A.61}$$

and then the symbols of the second kind follow from the relation

$$\Gamma^m_{kl} = g^{mn}[kl,n] \equiv \overset{\mathbf{g}}{\Gamma}{}^m_{kl}, \tag{A.62}$$

(e.g. see [320]). The index above the symbol indicates the metric with respect to which the Christoffel symbols are calculated.

Bearing the definition (A.59) in mind, we obtain

$$\frac{\partial \mathbf{v}}{\partial y^l} = \mathbf{v}^k_{;l}\mathbf{g}_k, \quad v^k_{;l} \equiv \nabla_l v^k = \frac{\partial v^k}{\partial y^l} + \Gamma^k_{lm}v^m. \tag{A.63}$$

The object $v^k_{;l}$ is called the *covariant derivative* of the vector field $\mathbf{v}$ in the frame of reference $\{y^k\}$. If we consider the covariant representation of the vector field: $\mathbf{v} = v_k\mathbf{g}^k$ then we have

$$\frac{\partial \mathbf{v}}{\partial y^l} = \frac{\partial v_k}{\partial y^l}\mathbf{g}^k + v_k\frac{\partial \mathbf{g}^k}{\partial y^l} = v_{k;l}\mathbf{g}^k, \quad v_{k;l} = \frac{\partial v_k}{\partial y^l} - \Gamma^m_{kl}v_m, \tag{A.64}$$

which follows from the identity

$$\frac{\partial}{\partial y^l}(\mathbf{g}_k \cdot \mathbf{g}^m) = 0 = \frac{\partial \mathbf{g}_k}{\partial y^l}\cdot\mathbf{g}^m + \mathbf{g}_k\cdot\frac{\partial \mathbf{g}^m}{\partial y^l} = \Gamma^m_{kl} + \mathbf{g}_k\cdot\frac{\partial \mathbf{g}^m}{\partial y^l}$$
$$\Rightarrow \quad \frac{\partial \mathbf{g}^m}{\partial y^l} = -\Gamma^m_{kl}\mathbf{g}^k. \tag{A.65}$$

For the metric tensor g_{ij}, we have

$$\frac{\partial g_{ij}}{\partial y^k} = \frac{\partial}{\partial y^k}(\mathbf{g}_i\cdot\mathbf{g}_j) = \Gamma^m_{ik}\mathbf{g}_m\cdot\mathbf{g}_j + \Gamma^m_{jk}\mathbf{g}_m\cdot\mathbf{g}_i,$$

i.e.

$$g_{ij;k} = \frac{\partial g_{ij}}{\partial y^k} - \Gamma^m_{ik}g_{mj} - \Gamma^m_{jk}g_{mi} = 0. \tag{A.66}$$

Similarly

$$g^{ij}_{..;k} = 0. \tag{A.67}$$

Hence, the metric tensors behave like constants in covariant differentiation even though they are, in general, functions of coordinates y^k.

Below we present in Table 13 properties of two most common curvilinear frames of reference and the Christoffel symbols related to them.

The relation (A.63) defines, of course, the *local shifting* of the vector. In cases more general than Euclidean spaces, i.e. for general differentiable manifolds Christoffel symbols are called the *metric connection*. It is motivated by the relation (A.61) between Christoffel symbols of the first kind and derivatives of the metric tensor. This object defines local parallel transport of vectors. In some applications to mechanics (for instance, theory of dislocations) one uses also connections which are not metric. Then $\Gamma^m_{kl} \neq \Gamma^m_{lk}$ and the antisymmetric part $\Omega^m_{kl} = \Gamma^m_{[kl]} = \frac{1}{2}\left(\Gamma^m_{kl} - \Gamma^m_{lk}\right)$, the so-called torsion, defines the density of dislocations (e.g. [231], [386]).

The above relations can be immediately extended on tensors. For example, for tensors of the second order we have

$$\frac{\partial \mathbf{T}}{\partial y^k} = T^{mn}{}_{;k}\mathbf{g}_m \otimes \mathbf{g}_n, \quad T^{mn}{}_{;k} = \frac{\partial T^{mn}}{\partial y^k} + \Gamma^m_{kl}T^{ln} + \Gamma^n_{kl}T^{ml}. \qquad \text{(A.68)}$$

The quantities $v^k_{;l}$ and $T^{mn}{}_{;k}$ behave like tensors, respectively, of the second and of the third order.

A.4 Physical components

Scalars, vectors and tensors written in a chosen Cartesian frame of reference possess physical units following from their physical interpretation. For instance, the linear element ds of the Euclidean space has the unit of length, say, meter [m], the mass of the material system m has the unit of mass, say, kilogram [kg], the time has the unit second [s], the vector of velocity $\mathbf{v}$ has components whose unit is meter/second [m/s], etc. This is not an invariant property with respect to the transformation of coordinates – scalars do not change their units but, for instance, the vector of velocity in cylindrical coordinates has components whose units are [1/s].

This is the major problem as dimensional analysis is one of the main tools of the analysis of physical systems. The choice of the set of independent variables, the comparison of the orders of magnitude of different quantities, asymptotic expansions – all these problems are related to physical dimensions of quantities appearing in the model. There are many books dealing systematically with this problem. It has been set by G. Ricci and T. Levi-Civita in 1901 [262]. We mention here only the work of G. Barenblatt [25] who presents most of the modern topics of this subject.

For the purpose of this book, we limit the attention to a simple problem of the representation of vectors and tensors in frames which yield the same dimension for all components. First of all, let us note that Cartesian coordinates possess dimensionless basis vectors. We assume that each coordinate x^k possesses the dimension of length and it is also the assumed dimension of the position vector $\mathbf{x}$, i.e.

$$\mathbf{e}_k = \frac{\partial \mathbf{x}}{\partial x^k} \quad \Rightarrow \quad [\mathbf{e}_k] = \frac{[\mathbf{x}]}{[x^k]} = [-], \tag{A.69}$$

where [...] is the dimension of the quantity. Operations on dimensions are governed by the same rules as products of numbers. For instance $[ab] = [a]\,[b]$. Hence $\left[a^2\right] = [a]^2$, etc. This yields the famous Π-Theorem of the dimensional analysis (Π like the first letter of power!).

Simultaneously, we have

$$[|\mathbf{x}|] = \left[\sqrt{\mathbf{x}\cdot\mathbf{x}}\right] = \left[\sqrt{\delta_{ik}x^i x^k}\right] = \left[\sqrt{[x^k]^2}\right] = \left[x^k\right] = [\mathbf{x}]\,. \tag{A.70}$$

For arbitrary coordinates we have

$$\mathbf{g}_k = \frac{\partial \mathbf{x}}{\partial y^k} \quad \Rightarrow \quad [\mathbf{g}_k] = \frac{[\mathbf{x}]}{[y^k]} \quad [g_{kl}] = \frac{[\mathbf{x}]^2}{[y^k]\,[y^l]}. \tag{A.71}$$

Consequently, in order to obtain a similarity to Cartesian basis we have to normalize arbitrary basis vectors

$$\mathbf{e}_{(k)} = \frac{\mathbf{g}_k}{|\mathbf{g}_k|} = \frac{\mathbf{g}_k}{\sqrt{g_{kk}}} \quad \Rightarrow \quad \left[\mathbf{e}_{(k)}\right] = [-], \quad \text{no sum!} \tag{A.72}$$

where we make the exception from the summation convention. Similarly

$$\mathbf{e}^{(k)} = \frac{\mathbf{g}^k}{|\mathbf{g}^k|} = \frac{\mathbf{g}^k}{\sqrt{g^{kk}}} \quad \Rightarrow \quad \left[\mathbf{e}^{(k)}\right] = [-], \quad \text{no sum!} \tag{A.73}$$

Simultaneously, the product of covariant and covariant basis vectors is dimensionless

$$\mathbf{g}_k \cdot \mathbf{g}^l = \delta_k^l \quad \Rightarrow \quad [\mathbf{g}_k]\left[\mathbf{g}^l\right] = [-] \quad \Rightarrow \quad [\mathbf{g}_k] = \frac{1}{[\mathbf{g}^l]}$$
$$\Rightarrow \quad [g_{kk}] = \frac{1}{[g^{kk}]}, \quad \text{no sum!} \tag{A.74}$$

The normalization of basis vectors is not sufficient for a consistent definition of the physical dimension of vectors and tensors. A transformation of covariant into contravariant indices involves the off-diagonal terms of the metric tensors. However, if the basis $\{\mathbf{g}_i\}$ is orthogonal, i.e.

$$g_{ij} = 0 \quad \text{for} \quad i \neq j, \tag{A.75}$$

then we can introduce a definition of the physical dimension of an arbitrary tensor. In this case, the metric tensors can be written in the form

$$g_{ij} = \delta_{ij}\sqrt{g_{ii}}\sqrt{g_{jj}}, \quad \text{no sum!} \tag{A.76}$$
$$g^{ij} = \delta^{ij}\sqrt{g^{ii}}\sqrt{g^{jj}}, \quad \text{no sum!}$$

Then

$$\mathbf{e}_{(i)} \stackrel{*}{=} \mathbf{e}^{(i)}, \tag{A.77}$$

i.e. the basis $\{\mathbf{e}_{(i)}\}$ is Cartesian. This diagonal form yields, for instance, for a tensor $\mathbf{T}$ of the second order

$$\mathbf{T} = t^{ij}\mathbf{g}_i \otimes \mathbf{g}_j = t^{(i)(j)}\mathbf{e}_{(i)} \otimes \mathbf{e}_{(j)}, \tag{A.78}$$

where

$$t^{(i)(j)} \stackrel{*}{=} t^{(i)}{}_{(j)} \stackrel{*}{=} t_{(i)}{}^{(j)} \stackrel{*}{=} t_{(i)(j)} = t^{ij}\sqrt{g_{ii}}\sqrt{g_{jj}}, \quad \text{no sum!} \tag{A.79}$$

and definitions through other components, $t^{i}_{\cdot j}, t_{i}^{\cdot j}, t_{ij}$, yield, due to orthogonality, the same result. These quantities are called *physical components of the tensor* $\mathbf{T}$. Obviously, as the basis vectors $\{\mathbf{e}_{(i)}\} \stackrel{*}{=} \{\mathbf{e}^{(i)}\}$ are dimensionless all these components have the same physical dimension and the dimension of the tensor $\mathbf{T}$ is defined by the dimension of any of its components, e.g.

$$[\mathbf{T}] = \left[t^{(i)(j)}\right]. \tag{A.80}$$

In the case of nonorthogonal basis one has to perform the transformation to the orthogonal system in order to define the physical dimension.

Extensive comments and discussion of physical components and their relation to the general dimensional analysis can be found in the article of Ericksen [102]. Some basic physical units of quantities appearing in thermomechanics are listed in Appendix B.

A.5 Transformations of Green and Kelvin

In the book we often employ the following transformations

$$\begin{aligned}\int_{\mathcal{P}} \operatorname{grad} \varphi dV &= \oint_{\partial\mathcal{P}} \mathbf{n} \otimes \varphi dS,\\ \int_{\mathcal{P}} \operatorname{div} \varphi dV &= \oint_{\partial\mathcal{P}} \mathbf{n} \cdot \varphi dS,\\ \int_{\mathcal{P}} \operatorname{rot} \varphi dV &= \oint_{\partial\mathcal{P}} \mathbf{n} \times \varphi dS,\end{aligned} \tag{A.81}$$

where $\mathcal{P}$ is an arbitrary measurable domain, $\partial\mathcal{P}$ its boundary which is assumed to be orientable, and $\mathbf{n}$ denotes the outward normal vector to $\partial\mathcal{P}$. These relations were proved by Green in 1828 [132], some special cases were considered by Lagrange in 1762 and Gauss in 1813. An equivalent result was proved by Ostrogradsky in 1831. For this reason, the above relations are also called "Gauss theorem", "Ostrogradsky theorem" or simply "divergence theorem".

Another result frequently used in continuum mechanics is Kelvin's transformation

$$\int_{\mathcal{S}} \epsilon_{mij} n_j \frac{\partial \varphi_{i\ldots n}}{\partial x^k} dS = \oint_{\mathcal{C}} \varphi_{m\ldots n} h_k dl, \tag{A.82}$$

where $\mathcal{C}$ is the curve which encloses the surface $\mathcal{S}$ and $\mathbf{h}$ the vector tangent to the curve $\mathcal{C}$ with the usual right-hand screw convention. This is also called Stokes Theorem.

Appendix B

Basic physical units

A. *Units of pressure and stresses*
(p and partial pressures in mixtures, Piola-Kirchhoff $\mathbf{P}$ and Cauchy $\mathbf{T}$)

pressure in SI: 1Pascal=1kg/s^{2}m

Pa=$\frac{N}{m^2}$	at=$\frac{kp}{cm^2}$	atm	bar	torr	mmWs=$\frac{kp}{m^2}$
1	$1.02\cdot10^{-5}$	$9.87\cdot10^{-6}$	10^{-5}	$75\cdot10^{-4}$	0.102
$9.81\cdot10^{4}$	1	0.968	0.981	736	10^{4}
$1.013\cdot10^{5}$	1.033	1	1.013	760	$1.033\cdot10^{4}$
10^{5}	1.02	0.987	1	750	$1.02\cdot10^{4}$
133	$1.36\cdot10^{-3}$	$1.32\cdot10^{-3}$	$1.33\cdot10^{-3}$	1	13.6
9.81	10^{-4}	$9.68\cdot10^{-5}$	$9.81\cdot10^{-5}$	$7.36\cdot10^{-2}$	1

B. *Units of the force*
(body forces $\rho\mathbf{b}$ have the SI unit: $\left[\text{Newton/m}^3\right]$)

force in SI: 1Newton=1kg·m/s^2

N	kp	Mp	p	dyna
1	0.102	$1.02\cdot10^{-4}$	102	10^{5}
9.81	1	10^{-3}	10^{3}	$9.81\cdot10^{5}$
$9.81\cdot10^{3}$	10^{3}	1	10^{6}	$9.81\cdot10^{8}$
$9.81\cdot10^{-3}$	10^{-3}	10^{-6}	1	981
10^{-5}	$1.02\cdot10^{-6}$	$1.02\cdot10^{-9}$	$1.02\cdot10^{-3}$	1

C. *Units of energy and work*

(energy density $\rho\varepsilon$ has the SI unit: $\left[\text{Joule/m}^3\right]$)

energy in SI: 1 Joule=1kg·m^2/s^2

J	kpm	kWh	kcal	erg	eV
1	0.102	$2.78\cdot10^{-7}$	$2.39\cdot10^{-4}$	10^7	$6.24\cdot10^{18}$
9.81	1	$2.72\cdot10^{-6}$	$2.34\cdot10^{-3}$	$9.81\cdot10^7$	$6.12\cdot10^{19}$
$3.6\cdot10^6$	$3.67\cdot10^5$	1	860	$3.6\cdot10^{13}$	$2.25\cdot10^{25}$
$4.19\cdot10^3$	427	$1.16\cdot10^{-3}$	1	$4.19\cdot10^{10}$	$2.61\cdot10^{22}$
10^{-7}	$1.02\cdot10^{-8}$	$2.78\cdot10^{-14}$	$2.39\cdot10^{-11}$	1	$6.24\cdot10^{11}$
$1.6\cdot10^{-19}$	$1.63\cdot10^{-20}$	$4.45\cdot10^{-26}$	$3.83\cdot10^{-23}$	$1.6\cdot10^{-12}$	1

D. *Units of power*

(working of body forces $\rho\mathbf{b}\cdot\mathbf{v}$, energy radiation ρr, working of stresses $\mathbf{P}\cdot(\partial\mathbf{F}/\partial t)$
and $\mathbf{T}\cdot\mathbf{L}$ have SI unit: $\left[\text{Watt/m}^3\right]$)

power in SI: 1Watt=1kg·m^2/s^3

W	kW	$\frac{\text{kpm}}{\text{s}}$	PS	$\frac{\text{cal}}{\text{s}}$	$\frac{\text{kcal}}{\text{h}}$
1	10^{-3}	0.102	$1.36\cdot10^{-3}$	0.239	0.86
10^3	1	102	1.36	239	860
9.81	$9.81\cdot10^{-3}$	1	$1.33\cdot10^{-2}$	2.34	8.43
736	0.736	75	1	176	632
4.19	$4.19\cdot10^{-3}$	0.427	$5.69\cdot10^{-3}$	1	3.6
1.16	$1.16\cdot10^{-3}$	0.119	$1.58\cdot10^{-3}$	0.278	1

Remark: Heat fluxes $\mathbf{Q}$ and $\mathbf{q}$ have SI unit: $\left[\text{Watt/m}^2\right]$;
entropy has SI unit: $\left[\text{Joule/m}^3\cdot\text{K}\right]$; entropy fluxes $\mathbf{H}$ and $\mathbf{h}$ have SI unit: $\left[\text{Watt/m}^2\cdot\text{K}\right]$;
entropy radiation ρs has the SI unit: $\left[\text{Watt/m}^3\cdot\text{K}\right]$.

Bibliography

[1] ACHENBACH, J. D. *Wave Propagation in Elastic Solids.* North-Holland, Amsterdam, 1973.

[2] ADKINS, J., AND RIVLIN, R. Large elastic deformations of isotropic materials, x. reinforcement by inextensible cords. *Phil. Trans. Roy. Soc. Lond., A 248* (1955), 201–223.

[3] ALBERS, B. Coupling of adsorption and diffusion in porous and granular materials. a 1-d example of the boundary value problem. *Arch. Appl. Mech. 70* (2000), 519–531.

[4] ALBERS, B. *Makroskopische Beschreibung von Adsorptions-Diffusions- Vorgängen in porösen Körpern.* PhD thesis, TU Berlin, Logos-Verlag, 2000.

[5] ALBERS, B. On adsorption and diffusion in porous media. *ZAMM 81*, 10 (2001), 683–690.

[6] ALBERS, B. Modelling of surface waves in poroelastic saturated materials by means of a two-component continuum. In *Surface Waves in Geomechanics: Direct and Inverse Modelling for Soils and Rocks*, C. Lai and K. Wilmanski, Eds., CISM Courses and Lectures. Springer, Wien, 2005, pp. 277–323.

[7] ALBERS, B., AND WILMANSKI, K. On modeling acoustic waves in saturated poroelastic media. *J. Engrg. Mech. 131*, 9 (2005), 974–985.

[8] ALBERS, B., AND WILMANSKI, K. Influence of coupling through porosity changes on the propagation of acoustic waves in linear poroelastic materials. *Arch. Mech. 58*, 4-5 (2006), 313–325.

[9] ALT, W. H. An evolution principle and the existence of entropy. first order systems. *Cont. Mech. Thermodynamics 10* (1988), 81–106.

[10] ALT, W. H., AND PAWLOW, I. On the entropy principle of phase transition models with a conserved order parameter. *Adv. Math. Sci. Appl. 6* (1966), 291–376.

[11] ALTS, T. *Thermodynamik elastischer Körper mit thermo-kine-*

matischen Zwangsbedingungen – fadenverstärkte Materialien. TU-Berlin, 1979.

[12] ANTMAN, S. S. Material constraints in continuum mechanics. *Atti della Accademia Nazionale dei Lincei, Classe diScienze Fisiche, Matematiche e Naturali 70* (1982), 256–264.

[13] ANTMAN, S. S. *Nonlinear Problems of Elasticity.* Springer, N.Y., 2005.

[14] ANTMAN, S. S., AND MARLOW, R. S. Material constraints, lagrange multipliers, and compatibility. applications to rod and shell theories. *Arch. Rat. Mech. Anal. 116*, 3 (1991), 257–299.

[15] ASARO, R. J. Crystal plasticity. *J. Appl. Mech. 50* (1983), 921–934.

[16] ASARO, R. J. Micromechanics of crystals and polycrystals. In *Advances in Applied Mechanics*, vol. 23. Academic Press, 1983, pp. 2–115.

[17] ASARO, R. J., AND NEEDLEMAN, A. Texture development and strain hardening in rate dependent polycrystals. *Acta Metall 33* (1985), 923–953.

[18] ASIMOV, I. *Asimov's Biographical Encyclopedia of Science and Technology.* Doubleday & Co, N.Y., 1964.

[19] ATKIN, R., AND CRAINE, R. Continuum theories of mixtures. applications. *J. Inst. Maths. Applics. 17* (1976).

[20] ATKIN, R. J., AND CRAINE, R. E. Continuum theories of mixtures. basic theory and historical development. *Quaterly J. Mech. Appl. Math. 29* (1976) 209–244.

[21] BAESU, E., AND CASEY, J. A treatment of internally constrained elastic-plastic materials. *Int. J. Engng. Sci. 38* (2000), 1677–1698.

[22] BARANZEV, R. *Interaction of rarefied gases with surfaces (in Russian).* Nauka, Moskwa, 1975.

[23] BARBERA, E. Boundary conditions in extended thermodynamics and applications. In *Trends in Applications of Mathematics to Mechanics* (Aachen, 2005), K. H. Yongqi Wang, Ed., Shaker Verlag, pp. 31–40.

[24] BARBERA, E., MÜLLER, I., REITEBUCH, D., AND ZHAO, N.-R. Determination of boundary conditions in extended thermodynamics via fluctuation theory. *Cont. Mech. Thermodyn. 16*, 5 (2004), 411–425.

[25] BARENBLATT, G. I. *Scaling, Selfsimilarity, and Intermediate Asymptotics.* Cambridge University Press, Cambridge, 1996.

[26] BARENBLATT, G. I., AND JOSEPH, D. D., Eds. *Collected Papers of R. S. Rivlin, vols. I & II.* Springer, N.Y., 1997.

[27] BATRA, R. C. *Elements of Continuum Mechanics.* Amer. Inst. of Aeronautics and Anstronautics, Reston, 2006.

[28] BAUER, E. *Zum mechanischen Verhalten granularer Stoffe unter vor-*

wiegend ödometrischer Beanspruchung. PhD thesis, Fridericiana in Karlsruhe, 1992.

[29] Bauer, E., and Huang, W. Evolution of polar quantities in a granular cosserat material under shearing. In *Proc. of the 5th Int. Workshop on Bifurcation and Localisation in Geomechanics, Perth, Australia (1999)* (2001), H.-B. Mühlhaus and E. Pasternak, Eds., Balkema, pp. 227–238.

[30] Bear, J. *Dynamics of Fluids in Porous Media.* Dover Publications, New York, 1988.

[31] Beatty, M. Topics in finite elasticity. Hyperelasticity of rubber, elastomers and biological tissues - with examples. *Appl. Mech. Rev. 40* (1987), 1699.

[32] Beavers, G. S., and Joseph, D. D. Boundary conditions at a naturally permeable wall. *J. Fluid Mech. 30*, 1 (1967), 197–207.

[33] Bellomo, N., and Pulvirenti, M. *Generalized Boltzmann Models in Applied Sciences: A Kinetic Theory Approach (Modeling and Simulation in Science, Engineering and Technology).* Birkhäuser, Boston, 2000.

[34] Bhatnagar, P. L., Gross, E. P., and Krook, M. A model for collision processes in gases. I. Small amplitude processes in charged and neutral one-component systems. *Phys. Rev. 94* (1954), 511–525.

[35] Biot, M. A. Non-linear theory of elasticity and the linearized case for a body under initial stress. *Phil. Mag. 27*, 7 (1939), 468–489.

[36] Biot, M. A. General theory of three dimensional consolidation. *J. App. Phys. 12*, 2 (1941), 155–164.

[37] Biot, M. A. Theory of propagation of elastic waves in a fluid saturated porous solid, I. Low frequency range. *J. Acoust. Soc. Am. 28*, 2 (1956), 168–178.

[38] Biot, M. A. Theory of propagation of elastic waves in a fluid saturated porous solid, II. Higher frequency range. *J. Acoust. Soc. Am. 28*, 2 (1956), 179–191.

[39] Biot, M. A., and Willis, D. G. The elastic coefficients of the theory of consolidation. *J. Appl. Mech. 24* (1957), 594–601.

[40] Blood, F. A. The r-particle distribution function in classical physics. *J. Math. Phys. 7*, 9 (1966), 1613–1619.

[41] Bogoliubov, N. N. *The Dynamical Theory in Statistical Physics.* Hindustan Publishing Corporation (India), Delhi, 1965.

[42] Boltzmann, L. Studien über das Gleichgewicht der lebend. Kraft zwischen bewegten materiellen Punkten. *Wien. Ber. 58* (1868), 517.

[43] Boltzmann, L. Weitere Studien über Wärmegleichgewicht unter Gasmolekülen (h-theorem). *Wien. Ber. 66* (1872), 275.

[44] Born, M., and Green, H. S. A general theory of liquids. i: The

molecular distribution function. *Proc. Roy. Soc. A 188* (1946), 10–18.

[45] BORN, M., AND GREEN, H. S. A general kinetic theory of liquids. ii: Equilibrium properties (without M. Born). *Proc. Roy. Soc. A189* (1946), 103–117.

[46] BORN, M., AND GREEN, H. S. A general theory of liquids. iii. dynamical properties. *Proc. Roy. Soc. A190* (1946), 455–474.

[47] BOWEN, R. M. Theory of mixtures. In *Continuum Physics*, A. Eringen, Ed., vol. 3. Academic Press, New York, 1976, pp. 2–127.

[48] BOWEN, R. M. Compressible porous media models by use of the theory of mixtures. *Int. J. Engng. Sci. 20*, 6 (1982), 697–763.

[49] BREKHOVSKIKH, L. M., AND GONCHAROV, V. *Mechanics of Continua and Wave Dynamics (Second Edition)*. Springer, Berlin, 1994.

[50] BRENNEN, C. E. *Cavitation and Bubble Dynamics*. Oxford University Press, 1995.

[51] BRIDGMAN, P. The thermodynamics of plastic deformation and generalized entropy. *Rev. Mod. Phys. 22* (1950), 56–63.

[52] BRINI, F., AND RUGGERI, T. Entropy principle for the moment systems of degree α associated to the boltzmann equation. critical derivatives and non-controlable boundary data. *Cont. Mech. Thermodyn. 14*, 2 (2002), 165–189.

[53] BROKATE, M., AND SPREKELS, J. *Hysteresis and Phase Transitions*. Springer, N.Y., 1996.

[54] BUNGE, H. *Texture Analysis in Materials Science*. Butterworth, London, 1982.

[55] CAGINALP, G. An analysis of phase field model of a free boundary. *Arch. Rat. Mech. Anal. 92* (1986), 205–245.

[56] CALLEN, H. *Thermodynamics and an Introduction to Thermostatics (2nd ed.)*. John Wiley and Sons, 1985.

[57] CARATHÉODORY, C. Untersuchungen über die Grundlagen der Thermodynamik. *Math. Ann. 67*, 3 (1909), 355–386.

[58] CARNOT, S. *Reflexions sur la puissance motrice du feu at sur les machines propres a develpper cette puissance*. Chez Bachelier, Paris, 1824.

[59] CASEY, G. A treatment of internally constrained materials. *J. Appl. Mech. 62* (1995), 542–544.

[60] CASEY, G., AND KRISHANASWAMY, S. On constrained thermoelastic materials. In *Contemporary Research in the Mechanics and Mathematics of Materials* (Barcelona, 1996), R. C. Batra and M. F. Beatty, Eds., CIMNE, pp. 359–371.

[61] CATTANEO, C. Sulla conduzione del calore. *Atti Sem. Mat. Fis. Modena 3* (1948), 83–101.

[62] CAUCHY, A. Recherches sur l'équikibreet le mouvement intérieur des

corps solides ou fluides, élastiques ou non élastiques. *Bull. Soc. Philomath.* (1823), 9–13.

[63] CERCIGNANI, C. *The Boltzmann Equation and Its Applications.* Springer, N.Y., 1988.

[64] CERCIGNANI, C. *Ludwig Boltzmann. The Man Who Trusted Atoms.* Oxford University Press, Oxford, 1998.

[65] CERCIGNANI, C. *Rarified Gas dynamics; From Basic Consepts to Actual Calculations.* Cambridge University Press, Cambridge, 2000.

[66] CERCIGNANI, C., LAMPIS, M., AND LENTATI, A. A new scattering kernel in kinetic theory of gases. *Transp. Theory Stat. Phys. 24* (1995), 1319–1336.

[67] CESARO, E. Sulle formole del volterra fondamentali nella teoria delle distorsioni elastiche. *Rend. dell'Acad. delle Scienza Fisiche e Matematiche* (1906), 311.

[68] CHANDRASEKHAR, S. *An Introduction to the Study of Stellar Structure.* Dover, N. Y., 1957.

[69] CHAPMAN, S., AND COWLING, T. G. *The Mathematical Theory of Non-Uniform Gases (Third Edition).* Cambridge University Press, Cambridge, 1970.

[70] CHERNIKOV, N. A. Equilibrium distribution of the relativistic gas. *Acta Phys. Polonica 26* (1964).

[71] CHERNIKOV, N. A. Microscopic foundation for relativistic hydrodynamics. *Acta Phys. Polonica 27* (1964).

[72] CHERNIKOV, N. A. The relativistic gas in the gravitational field. *Acta Phys. Polonica 23* (1964).

[73] CHRISTENSEN, R. M. *Theory of Viscoelasticity.* Academic Press, N. Y., 1971.

[74] CLAUSIUS, R. Über die bewegende Kraft der Wärme und die Gesetze, welche sich daraus für die Wärmelehre selbst ableiten lassen. *Ann. Phys. 19* (1850), 368–398.

[75] CLAUSIUS, R. Über die veränderte Form des zweiten Hauptsatzes der mechanischen Wärmetheorie. *Ann. Phys. 93* (1854), 481–506.

[76] COLEMAN, B. D. On thermodynamics, strain impulses and viscoelasticity. *Arch. Rat. Mech. Anal. 17* (1964), 230–254.

[77] COLEMAN, B. D. Thermodynamics of materials with memory. *Arch. Rat. Mech. Anal. 17* (1964), 1–46.

[78] COLEMAN, B. D., MARKOVITZ, H., AND NOLL, W. *Viscometric Flows on Non-Newtonian Fluids.* Springer, Berlin, 1966.

[79] COLEMAN, B. D., AND NOLL, W. An approximation theorem for functionals, with applications in continuum mechanics. *Arch. Rat. Mech. Anal. 6* (1960), 355–370.

[80] COLEMAN, B. D., AND NOLL, W. The thermodynamics of elastic

materials with heat conduction and viscosity. *Arch. Rat. Mech. Anal. 13* (1963), 167–178.

[81] COLEMAN, B. D., AND OWEN, D. R. A mathematical foundation for thermodynamics. *Arch. Rat. Mech. Anal. 54* (1974), 1–104.

[82] COUSSY, O. *Mechanics of Porous Continua.* John Wiley, Chichester, 1995.

[83] COUSSY, O. *Poromechanics.* Wiley and Sons, Chichester, 2003.

[84] DARCY, H. *Les Fontaines Publiques de la Ville de Dijon.* Dalmont, Paris, 1856.

[85] DE GROOT, S. R., AND MAZUR, P. *Non-Equilibrium Thermodynamics.* North-Holland, Amsterdam, 1962.

[86] DELESSE, A. Pour determiner la composition des roches. *Ann. Mines 4*, 13 (1848), 379–388.

[87] DERESIEWICZ, H. The effect of boundaries on wave propagation in a liquid filled porous solid: II. Love waves in a porous layer. *Bull. Seismol. Soc. Am. 51*, 1 (1961), 51–59.

[88] DERESIEWICZ, H. The effect of boundaries on wave propagation in a liquid filled porous solid. IV. Surface waves in a half-space. *Bull. Seismol. Soc. Am. 52*, 3 (1962), 627–638.

[89] DOYLE, T. C., AND ERICKSEN, J. Nonlinear theory of elasticity. In *Advances in Applied Mechanics*, H. L. Dryden and T. von Kármán, Eds., vol. IV. Academic Press, New York, 1956.

[90] DRUMHELLER, D. S., AND BEDFORD, A. A thermomechanical theory for reacting immiscible mixtures. *Arch. Rat. Mech. Anal. 73* (1980), 257–284.

[91] DUHEM, P. Recherches sur l'hydrodynamique. *Ann. Toulouse 3*, 2 (1901), 315–377.

[92] DUNN, E., AND FOSDICK, R. L. Thermodynamics, stability and boundedness of fluids of complexity 2 and fluids of second grade. *Arch. Rat. Mech. Anal. 56* (1974), 191–252.

[93] ECKART, C. The thermodynamics of irreversible processes. I. The simple fluid. *Phys. Rev. 58* (1940), 267–269.

[94] ECKART, W., FARIA, S., HUTTER, K., KIRCHNER, N., PUDASAINI, S., AND WANG, Y. *Continuum Description of Granular Materials.* Spring School at the Department of Structural and Geotechnical Engineering, Politechnical Institute, Turin, Italy, 2002.

[95] EDELEN, D. G. B. Nonlocal field theories. In *Continuum Physics*, A. Eringen, Ed., vol. IV. Academic Press, 1976.

[96] EDELEN, D. G. B., GREEN, A. E., AND LAWS, N. Nonlocal continuum mechanics. *Arch. Rat. Mech. Anal. 43*, 1 (1971), 36–44.

[97] EDELEN, D. G. B., AND LAWS, N. On the thermodynamics of systems with nonlocality. *Arch. Rat. Mech. Anal. 43*, 1 (1971), 24–35.

[98] EHRENFEST, P., AND T. Über zwei bekannte einwände gegen das boltzmannsche h-theorem. *Physik Z. 67*, 3 (1907), 355–386.

[99] EHRENFEST, P., AND T. *The Conceptual Foundations of the Statistical Approach in Mechanics.* Dover, N.Y., 1990.

[100] ELWELL, D., AND POINTON, A. J., Eds. *Classical Thermodynamics.* Penguin Books, Harmondsworth, Middlesex, 1972.

[101] EMCH, G., AND LIU, C., Eds. *The Logic of Thermostatistical Physics.* Springer, Berlin, 2002.

[102] ERICKSEN, J. L. *Tensor Fields*, vol. III. Springer, Berlin, 1960, ch. 1, pp. 794–858. Flügge, S. (ed.).

[103] ERICKSEN, J. L. *Introduction to the Thermodynamics of Solids.* Springer, N.Y., 1998.

[104] ERINGEN, A., Ed. *Continuum Physics. Vol. I: Mathematics*, vol. 1. Academic Press, N. Y., 1971.

[105] ERINGEN, A., AND MAUGIN, G. *Electrodynamics of Continua I. Foundations and Solid Media.* Springer, N.Y., 1989.

[106] ERINGEN, A., AND MAUGIN, G. *Electrodynamics of Continua II. Fluids and Complex Media.* Springer, N.Y., 1989.

[107] EULER, L. Continuation des recherches sur la théorie du mouvement des fluides. *Hist. Acad. Berlin 11* (1755), 316–361.

[108] EULER, L. Institutionum calculi integralis volumen tertium. *Petropolis Opera 14* (1769), 270–386.

[109] EULER, L. Sectio secunda de principiis motus fluidorum. *Novi Comm. Petrop. 14* (1770), 270–386.

[110] FELLIN, W., AND OSTERMANN, A. Objective integration and geometric properties of hypoplasticity. In *Advanced Mathematical and Computational Geomechanics* (Berlin, 2003), D. Kolymbas, Ed., Springer, pp. 195–210.

[111] FILLUNGER, P. D. *Erdbaumechanik.* Selbstverlag des Verfassers, Wien, 1936.

[112] FISHER, G. M. C., AND LEITMAN, M. J. On continuum thermodynamics with surfaces. *Arch. Rat. Mech. Anal. 30* (1968), 225–262.

[113] FOSDICK, R. L., AND RAJAGOPAL, K. R. Anomalous features in the model of second-order fluid. *Arch. Rat. Mech. Anal. 70* (1979), 145 – 152.

[114] FOSDICK, R. L., AND RAJAGOPAL, K. R. Thermodynamics and stability of fluids of third grade. *Proc. R. Soc. London 339* (1980), 351–377.

[115] FRENKEL, J. On the theory of seismic and seismoelectric phenomena in a moist soil. *Journal of Physics 3*, 4 (1944, Nachdruck), 230–241.

[116] FUNG, Y. *Foundations of Solid Mechanics.* Englewood Cliffs, N.Y., 1965.

[117] GASSMANN, F. Über die Elastizität poröser Medien. *Vierteljahresschrift der Naturforschenden Gesellschaft in Zürich 96*, 1 (1951), 1–23.

[118] GENT, A. N. *Engineering with Rubber (Second Edition).* Hanser Fachbuch, 2001.

[119] GIBBS, J. W. On the equilibrium of heterogeneous substance. *Transactions of the Conneticut Academy 3* (1876-1878), 108–248, 343–524.

[120] GIBBS, J. W. *The Scientific Papers. Vol.I: Thermodynamics*, vol. 1. Dover, 1960.

[121] GILES, R. *Mathematical Foundations of Thermodynamics.* Pergamon, Oxford, 1964.

[122] GINN, R., AND METZNER, A. Measurement of stresses developed in steady laminar shearing flows of viscoelastic media. *Trans. Soc. Rheol. 13*, 4 (1969).

[123] GODDARD, J. On entropy estimates of contact forces in static granular assemblies. *Int. J. Solids and Structures 42*, 21 (2004), 5851–5861.

[124] GODDARD, J. On maximum-entropy estimates for granular statics. In *Proc. 2nd Intl. Symp. on Continuous and Discontinuous Modelling of Cohesive Frictional Materials* (2004), J. Vermeer, Ed., CDM2004, Balkema.

[125] GOODMAN, M., AND COWIN, S. A continuum theory of granular materials. *Arch. Rat. Mech. Anal. 44* (1972), 249–266.

[126] GRAIFF, F. Sulle condizioni di congruenza per deformazioni anche finite. *Rend. Lincei 24*, 8 (1958), 415–422.

[127] GRAY, W. G. Elements of a systematic procedure for the derivation of macroscale conservation equations for multiphase flow in porous media. In *Kinetic and Continuum Theories of Granular and Porous Media*, K. Hutter and K. Wilmanski, Eds., no. 400 in CISM Courses and Lectures. Springer, Wien - New York, 1999, pp. 67–130.

[128] GREEN, A. E., AND ADKINS, J. E. *Large Elastic Deformations and Non-Linear Continuum Mechanics.* Oxford University Press, Fair Lawn, N.J., 1960.

[129] GREEN, A. E., AND RIVLIN, R. S. The mechanics of non-linear materials with memory, part I. *Arch. Rat. Mech. Anal. 1* (1957), 1–21.

[130] GREEN, A. E., AND RIVLIN, R. S. The mechanics of non-linear materials with memory, part II. *Arch. Rat. Mech. Anal. 4* (1957), 387–404.

[131] GREEN, A. E., AND ZERNA, W. *Theoretical Elasticity (Second Edition).* Dover, N.Y., 1968.

[132] GREEN, G. An essay on the application of mathematical analysis to the theories of electricity and magnetism. *J. Reine Angew. Math. 39* (1850), 73–89.

[133] GREGG, S. J., AND SING, K. S. W. *Adsorption, surface area and porosity.* Academic Press, London, 1982.
[134] GRMELA, M., KARLIN, I. V., AND ZMIEVSKI, V. B. Boundary layer variational principles: A case study. *Phys. Rev. E 66*, 011201 (2002), 1–12.
[135] GURTIN, M. E. On the thermodynamics of materials with memory. *Arch. Rat. Mech. Anal. 28* (1968), 40–59.
[136] GURTIN, M. E. *The Linear Theory of Elasticity*, vol. VIa. Springer, S. Flügge, C. A. Truesdell (eds.), Berlin, 1972, ch. 2.
[137] GURTIN, M. E., AND WILLIAMS, W. O. On continuum thermodynamics with mutual body forces and internal radiation. *Appl. Math. Phys. (ZAMP) 22* (1971), 293–298.
[138] HADAMARD, J. *Leçons sur la Propagation des Ondes et les Équations de l'Hydrodynamique.* Hermann, Paris, 1903.
[139] HAUSER, R. A., AND KIRCHNER, N. A historical note on the entropy principle of müller and liu. *Cont. Mech. Thermodyn. 14*, 2 (2002), 223–226.
[140] HEINRICH, G., AND DESOYER, K. Hydromechanische Grundlagen für die Behandlung von stationären und instationären Grundwasserströmungen. *Ingeneur-Archiv 23*, 2 (1955), 73–84.
[141] HEINRICH, G., AND DESOYER, K. Hydromechanische Grundlagen für die Behandlung von stationären und instationären Grundwasserströmungen. *Ingeneur-Archiv 24* (1956), 81–84.
[142] HELMHOLTZ, H. Über Integrale der hydrodynamischen Gleichungen, welche den Wirbelbewegungen entsprechen. *J. Reine Angew. Math. 55* (1859), 285–313.
[143] HENCKY, H. Über die Form des Elastizitätsgestzes bei ideal elastischen Stoffen. *Z. Tech. Phys. 9* (1928), 214–223.
[144] HERSHEY, A. V. A review of the definitions of finite strain. In *Proc. 1st U.S. Nat. Congr. Appl. Mech. (1951)* (1952), pp. 473–478.
[145] HILL, R., Ed. *The Mathematical Theory of Plasticity.* Clarendon Press, Oxford, 1964.
[146] HOHMANN, R., SETZER, M., AND WEHLING, M., Eds. *Bauphysikalische Formeln und Tabellen.* Werner Verlag, München, 2004.
[147] HONEYCOMBE, R. W. K. *The Plastic Deformation of Metals.* E. Arnold Ltd., 1968.
[148] HUTTER, K. On thermodynamics and thermostatics of viscous thermoelastic solids in the electromagnetic fields. a lagrangian formulation. *Arch. Rat. Mech. Anal. 58* (1975), 339–368.
[149] HUTTER, K. A thermodynamic theory of fluids and solids in the electromagnetic fields. *Arch. Rat. Mech. Anal. 64* (1977), 269–298.
[150] HUTTER, K. Avalanche dynamics. In *Hydrology of Disasters,*

V. Singh, Ed. Kluwer Academic Publ., Dodrecht, Boston, London, 1996, pp. 317–394.

[151] HUTTER, K. Order and disorder in granular materials: Experiment and theory. In *Kinetic and Continuum Theories of Granular and Porous Media*, K. Hutter and K. Wilmanski, Eds., no. 400 in CISM Courses and Lectures. Springer, Wien - New York, 1999, pp. 1–66.

[152] HUTTER, K., VAN DE VEN, A., AND URSESCU, A. *Electromagnetic Field Matter Interactions in Thermoelastic Solids and Viscous Fluids.* Springer, Berlin, 2006.

[153] HUTTER, K., AND WILMANSKI, K., Eds. *Kinetic and Continuum Theories of Granular and Porous Media.* Springer, Wien, 1999.

[154] JACOBI, C. Theoria novi multiplicatoris systemati aequationum differentialium vulgarium applicandi. *J. Reine Angew. Math. 27* (1844), 199–268.

[155] JÄGER, W., AND MIKELIĆ, A. On the boundary conditions at the contact interface between two porous media. In *Partial Differential Equations, Theory and Numerical Solution*, W. Jäger, Ed., no. 406 in CRC Research Notes in Mathematics. Chapman and Hall, 1999, pp. 175–186.

[156] JARIĆ, J. *Mehanika Kontinuuma.* IRO "Gradevinska knjiga", Beograd, 1988.

[157] JENKINS, J., AND RICHMAN, M. Grad's 13-moment system for a dense gas of inelastic spheres. *Arch. Rat. Mech. Anal. 87* (1985), 355–377.

[158] JOSEPH, D. D. Instability of the rest state of fluids of arbitary grade greater than one. *Arch. Rat. Mech. Anal. 75* (1981), 251–256.

[159] JOU, D. Lagrange multipliers in nonequilibrium thermodynamics. Tech. Rep. 18, WIAS, 2000.

[160] JOU, D., CASAS-VÁZQUEZ, J., AND LEBON, G. *Extended Irreversible Thermodynamics (Second Edition).* Springer, Berlin, 1996.

[161] KAC, M. *Some Stochastic Problems in Physics and Mathematics.* Magnolia Petroleum Company, Dallas, 1957.

[162] KACZANOV, L. *Osnovy Teorii Plasticznosti (in Russian).* Moskwa, 1956.

[163] KIRCHHOFF, G. Uber die Gleichungen des Gleichgewichts eines elastischen Körpers bei nicht unendlich kleinen Verschiebungen seiner Theile. Sitzsbericht, Akademie der Wissenschaften in Wien, 1852.

[164] KIRKWOOD, J. The statistical mechanical theory of transport processes, i: General theory. *J. Chem. Phys. 14*, 3 (1946), 180–201.

[165] KIRKWOOD, J. The statistical mechanical theory of transport processes, ii: Transport in gasaes. *J. Chem. Phys. 15*, 1 (1947), 72–76.

[166] KLEIBER, M., AND RANIECKI, B. *Elastic-Plastic Materials at Finite*

Strains. Elsevier, 1985, pp. 3–46. Sawczuk, A. and Bianchi, G. (eds.).

[167] KOCKS, U. F., ARGON, A., AND ASHBY, M. *Thermodynamics and Kinetics of Slip*, vol. 19. Pergamon Press, Oxford, 1975. Chalmers, B. and Christian, J.W. and Massalski, T.W. (eds.).

[168] KOCKS, U. F., DAWSON, P. R., AND FOLLANSBEE, P. S. Physical and phenomenological plasticity. report, Los Alamos, 1986.

[169] KOGA, T. *Introduction to Kinetic Theory; Stochastic Processes in Gaseous Systems*. Pergamon Press, Oxford, 1970.

[170] KOLYMBAS, D. *Ein nichtlineares viskoplastisches Stoffgesetz für Böden*. PhD thesis, Fridericiana in Karlsruhe, 1978.

[171] KOLYMBAS, D. *Introduction to Hypoplasticity*. Balkema, 2000.

[172] KREJCI, P. Vector hysteresis models. *European J. Appl. Math. 9* (1991), 155–161.

[173] KROENER, E. Zur plastischen verformung des vielkristalls. *Acta Metall. 9* (1961), 155–161.

[174] KUCZMA, M. *Application of Variational Inequalities in the Mechanics of Plastic Flow and Martensitic Phase Transformations*. Politechnika Poznanska, Poznan, 1999.

[175] LAGRANGE, J. L. Nouvelles recherches sur la nature et la propagation du son. *Misc. Taur. 2* (1760-1761), 11–172.

[176] LAGRANGE, J. L. *Mechanique Analitique*. Paris, 1788.

[177] LAI, C., AND WILMANSKI, K., Eds. *Surface Waves in Geomechanics, Direct and Inverse Modelling for Soils and Rocks*. Springer, Wien, New York, 2005. CISM Courses and Lectures No. 481.

[178] LAMB, G. L. *Elements of soliton theory*. J. Wiley, New York, 1980.

[179] LANDAU, L. D., AND LIFSCHITZ, E. M. *Lehrbuch der Theoretischen Physik. VI. Hydrodynamik*, vol. 6. Akademie-Verlag, Berlin, 1991.

[180] LANGMUIR, I. The constitution and fundamental properties of solids and liquids, part 1. *J. Am. Chem. Soc. 38* (1916), 2221–2295.

[181] LANGMUIR, I. The constitution and fundamental properties of solids and liquids, part 2. *J. Am. Chem. Soc. 39* (1917), 1848–1906.

[182] LANGMUIR, I. The adsorption of gases on plane surfaces of glass, mica and platinum. *J. Amer. Chem. Soc. 40* (1918), 1361–1403.

[183] LEE, E. H. Elasti-plastic deformation at finite strains. *J. Appl. Mech. 36* (1969), 1–6.

[184] LEHMANN, T. *On a Generalized Constitutive Law in Thermoplasticity*. Elsevier, 1985, pp. 115–134. Sawczuk, A. and G. Bianchi, G. (eds.).

[185] LEMAITRE, J., AND CHABOCHE, J. *Mechanics of Solid Materials*. Cambridge Univ. Press, Cambridge, 1990.

[186] LEMAITRE, J., AND DESMORAT, R. *Engineering Damage Mechanics. Ductile, Creep, Fatigue and Brittle Failures*. Springer, Berlin, 2005.

[187] LIU, I.-S. Method of lagrange multipliers for exploitation of the entropy principle. *Arch. Rat. Mech. Anal. 46* (1972), 131–148.

[188] LIU, I.-S. An extended fields theory of viscoelastic materials. *Int. J. Engng. Sci. 26* (1988), 331–342.

[189] LIU, I.-S. Extended thermodynamics of viscoelastic materials. *Continuum Mech. Thermodyn. 1* (1989), 143–164.

[190] LIU, I.-S. On entropy flux-heat flux relation in thermodynamics with lagrange multipliers. *CMT 8* (1996), 247–256.

[191] LIU, I.-S. *Continuum Mechanics.* Springer, Berlin, 2002.

[192] LIU, I.-S. On well-posedness of classical boundary conditions in extended thermodynamics. In *Trends in Applications of Mathematics to Mechanics, STAMM 2004* (Aachen, 2005), K. H. Y. Wang, Ed., Shaker, pp. 225–233.

[193] LIU, I.-S. A method of differential iteration for boundary value problems in extended thermodynamics. *Nonlinear Analysis: Real World Application 8*, 4 (2007), 1113–1131.

[194] LIU, I.-S., AND MÜLLER, I. Extended thermodynamics of classical and degenerate gases. *Arch. Rat. Mech. Anal. 83*, 4 (1983), 285–332.

[195] LIU, I.-S., MÜLLER, I., AND RUGGERI, T. Relativistic thermodynamics of gases. *Ann. Phys. 169*, 1 (1986), 191–219.

[196] LIU, I.-S., AND RINCON, M. A. A boundary value problem in extended thermodynamics. one-dimensional steady flows with heat conduction. *Continuum Mech. Thermodyn. 16* (2004), 109–124.

[197] LIU, I.-S., RINCON, M. A., AND MÜLLER, I. Iterative approximation of stationary heat conduction in extended thermodynamics. *Continuum Mech. Thermodyn. 14* (2002), 483–493.

[198] LOPATNIKOV, S., AND CHENG, A.-D. Variational formulation of fluid infiltrated porous material in thermal and mechanical equilibrium. *Mech. Mater. 34* (2002), 685–704.

[199] MALVERN, L. *Introduction to the Mechanics of a Continuous Medium.* Prentice-Hall, New Jersey, 1969.

[200] MARSDEN, J. E., AND HUGHES, T. J. R. *Mathematical Foundations of Elasticity.* Dover, N.Y., 1983.

[201] MAUGIN, G. *Thermomechanics of Plasticity and Fracture.* Cambridge University Press, U.K., 1992.

[202] MAUGIN, G. *Thermomechanics of Nonlinear Irreversible Behavior.* World Scientific Publ, Singapore, 1999.

[203] MAXWELL, J. C. *A Treatise on Electricity and Magnetism, Vols. I and II.* Clarendon Press Series. University of Oxford, London, 1873.

[204] MAYER, J. R. Bemerkungen über die Kräfte der unbelebten Natur. *Annalen der Chemie und Pharmacie 42* (1842), 233–240.

[205] MCCONNELL, A. J. *Application of Tensor Analysis.* Dover, N.Y.,

1957.

[206] MENDOZA, E., Ed. *Reflections on the Motive Power of Fire by Sadi Carnot and other Papers on the Second Law of Thermodynamics by E. Clapeyron and R. Clausius.* Dover, N.Y., 1960.

[207] MIKELIĆ, A. *Recent Developments in Multiscale Problems Coming from Fluid Mechanics.* Springer, Heidelberg, 2002, ch. 5, pp. 225–267. M. Kirkilionis, S. Krömker, R. Rannacher, F. Tomi (eds.).

[208] MIKELIĆ, A., AND PRIMICERIO, M. Homogenization of a problem modelling remediation of porous media. *Far East J. Appl. Math. 15* (2004), 365–380.

[209] MISNER, C., THORNE, K., AND WHEELER, J., Eds. *Gravitation.* W. H. Freeman and Co., San Francisco, 1973.

[210] MOECKEL, G. Thermodynamics of an interface. *Arch. Rat. Mech. Anal. 57*, 3 (1975), 255–280.

[211] MOLINARI, A. Bandes de cisaillement dans un monocristal viscoplastique en traction simple. *C. R. Acad. Sci. Paris 306* (1988), 841–845.

[212] MOLINARI, A. Shear band analysis. *Solid State Phenomena 3 and 4* (1988), 447–468.

[213] MOLINARI, A., CANOVA, G., AND AHZI, S. A self consistent approach of the large deformation polycrystal viscoplasticity. *Acta. Metall. 35*, 12 (1987), 2983–2994.

[214] MOONEY, M. The thermodynamics of a strained elastomer. i. general analysis. *J. Appl. Phys. 19* (1948), 434–444.

[215] MORLAND, J. L. W. A simple constitutive theory for a fluid-saturated porous solid. *J. Geophys. Res. 77* (1972), 890–900.

[216] MURNAGHAN, F. D. *Finite Deformation of an Elastic Solid.* J. Wiley, N.Y., London, 1951.

[217] MÜLLER, I. A thermodynamic theory of mixtures of fluids. *Arch. Rat. Mech. Anal. 28* (1968), 1–39.

[218] MÜLLER, I. The coldness, a universal function in thermoelastic bodies. *Arch. Rat. Mech. Anal. 41* (1971), 319–332.

[219] MÜLLER, I. Die Kältefunktion, eine universelle Funktion in der Thermodynamik viskoser wärmeleitender Flüssigkeit. *Arch. Rat. Mech. Anal. 40* (1971), 1–36.

[220] MÜLLER, I. A new approach to thermodynamics of simple mixtures. *Z. Naturforschung 28*, 2A (1973), 1801–1813.

[221] MÜLLER, I. *Thermodynamik. Die Grundlagen der Materialtheorie.* Bertelsmann Universitätsverlag, Düsseldorf, 1973.

[222] MÜLLER, I. *Thermodynamics.* Pitman, N.Y., 1985.

[223] MÜLLER, I. *A History of Thermodynamics.* Springer, Berlin, 2007.

[224] MÜLLER, I., REITEBUCH, D., AND WEISS, W. Extended thermodynamics - consistent in order of magnitude. *Cont. Mech. Thermodyn.*

15, 2 (2003).

[225] Müller, I., and Ruggeri, T. *Rational Extended Thermodynamics (Second Edition).* Springer, N.Y., 1998.

[226] Müller, I., and Strehlow, P. *Rubber and Rubber Balloons. Paradigms of Thermodynamics.* Springer, Berlin, 2004.

[227] Müller, I., and Wilmanski, K. Extended thermodynamics of a non-newtonian fluid. *Rheologica Acta 22* (1986), 335–349.

[228] Needleman, A. Dynamic shear band development in palne strain. *J. Appl. Mech. 56* (1989), 1–9.

[229] Neimitz, A. *Mechanika pekania.* PWN, Warsaw, 1998.

[230] Nemat-Nasser, S. *Micromechanically Based Finite Plasticity.* Elsevier, 1985, pp. 85–96. Sawczuk, A. and G. Bianchi, G. (eds.).

[231] Noll, W., Toupin, R. A., and Wang, C.-C., Eds. *Continuum Theory of Inhomogeneities in Simple Bodies.* Springer, N.Y., 1968.

[232] Nowacki, W. *Thermoelasticity.* PWN and Pergamon Press, Oxford, 1986.

[233] Nunziato, J. W., and Walsh, E. K. On the influence of void compaction and material non-uniformity on the propagation of one-dimensional acceleration waves in granular materials. *Arch. Rat. Mech. Anal. 64*, 4 (1977), 299–316.

[234] Ogden, R. W. *Non-Linear Elastic Deformations.* Dover, Mineola, N.Y., 1984.

[235] Ogden, R. W. Nonlinear elasticity with application to material modelling. Tech. Rep. Lecture Notes 6, Centre of Excellence for Advanced Materials and Structures, Warsaw, 2003.

[236] Oldroyd, J. Finite strains in an anisotropic elastic continuum. *Proc. Roy. Soc. Lond., A 202* (1950), 345–358.

[237] Oldroyd, J. On the formulation of rheological equations of state. *Proc. Roy. Soc. Lond., A 200* (1950), 523–541.

[238] Oldroyd, J. Non-newtonian effects in steady motion of some idealized elastico-viscous liquids. *Proc. Roy. Soc. Lond., A 245* (1958), 278.

[239] Onsager, L. Reciprocal relations in irreversible processes. i. *Phys. Rev. 37* (1931), 405–426.

[240] Onsager, L. Reciprocal relations in irreversible processes. ii. *Phys. Rev. 38* (1931), 2265–2279.

[241] Oppenheim, I., and Williams, W. Geometric effects in an elastic tensegrity structure. *J. of Elasticity 59*, 1 (2000), 51–65.

[242] Panagiotopoulos, P. D. *Inequality Problems in Mechanics and Applications. Convex and Nonconvex Energy Functions.* Birkhäuser, Boston, 1985.

[243] Passman, S. L., Nunziato, J. W., and Walsh, E. K. A theory

of multiphase mixtures. In *Truessdellt's Rational Thermodynamics*, C. Truesdell, Ed., vol. App. 5C. Springer, 1984, pp. 286–325.

[244] PERZYNA, P. The constitutive equations for rate sensitive plastic materials. *Quart. Appl. Math. 20* (1963), 321–332.

[245] PERZYNA, P. Fundamental problems in viscoplasticity. *Adv. Appl. Mech. 9* (1966), 343–377.

[246] PERZYNA, P. Thermodynamic theory of viscoplasticity. *Adv. Appl. Mech. 11* (1971), 313–354.

[247] PERZYNA, P. *Thermodynamics of inelastic materials (in Polish).* PWN, Warszawa, 1978.

[248] PERZYNA, P. Thermodynamics of dissipative materials. In *Recent Developments in Thermomechanics of Solids*, G. Lebon and P. Perzyna, Eds., CISM Courses and Lectures No. 262. Springer, Wien, New York, 1980.

[249] PERZYNA, P. The thermodynamical theory of elasto-viscoplasticity. *Engineering Transactions 53* (2005), 235–316.

[250] PIOLA, G. *Sull' applicazione de' proncipj della meccanica analitica del Lagrange ai principali problemi*, vol. VII+. Regia Stamparia, Milano, 1825.

[251] PIPKIN, A. *Lectures on Viscoelasticity Theory.* Springer, N.Y., 1972.

[252] PLANCK, M. *Treatise on Thermodynamics (Third Edition).* Dover Publ., N.Y., 1945.

[253] PONTE CASTANEDA, P., TELEGA, J. J., AND GAMBIN, B. E., Eds. *Nonlinear Homogenization and its Applications to Composites, Polycrystals and Smart Materials, Proceedings of the NATO Advanced Research Workshop, held in Warsaw, Poland, 23-26 June 2003* (N.Y., 2004), vol. 170 of *NATO Science Series II: Mathematics, Physics and Chemistry*, Springer.

[254] PONTER, A. R. S., BATAILLE, J., AND KESTIN, J. Thermodynamic model for the time dependent plastic deformation of solids. *J. Mec. 18*, 3 (1979), 511–539.

[255] PRIGOGINE, I. *Introduction to Thermodynamics of Irreversible Processes.* Interscience, New York, 1967.

[256] RADKEVICH, E. V., AND WILMANSKI, K. On dispersion in the mathematical model of poroelastic materials with the balance equation for porosity. *J. Math. Sci. 114*, 4 (2003), 1431–1449.

[257] RAJAGOPAL, K. R. On the stability of third-grade fluids. *Arch. Mech. 3* (1980), 867–875.

[258] RAJAGOPAL, K. R., AND SACCOMANDI, G. On internal constraints in continuum mechanics. *Differential Equations and Non-Linear Mechanics* (2006). Article ID 18572.

[259] RAJAGOPAL, K. R., AND TAO, L. *Mechanics of Mixtures.* World

Scientific, Singapore, 1995.

[260] RANKINE, W. J. M. Laws of the elasticity of solid bodies. *Cambr. Dubl. Math. J. 6* (1851), 67–101. papers (pages 41-80, 178-181, 185-186).

[261] RESIBOIS, P., AND DE LEENER, M. *Classical Kinetic Theory of Fluids.* John Wiley, N.Y., 1977.

[262] RICCI, G., AND LEVI-CIVITA, T. Méthodes de calcul différentiel absolut et leurs applications. *Math. Ann. 54* (1901), 125–201.

[263] RICE, J. R. Continuum mechanics and thermodynamics of plasticity in relation of microscale deformation mechanisms. In *Constitutive Equations of Plasticity,* A. Argon, Ed. MIT Press, 1975, pp. 23–79.

[264] RIVLIN, R. Large elastic deformations of isotropic materials, II: Some uniqueness theorems for pure homogenous deformation. *Philos. Trans. Roy. Soc.,A. A240* (1948), 491–508.

[265] RIVLIN, R. Large elastic deformations of isotropic materials, V: The problem of flexture. *Philos. Trans. Roy. Soc.,A. 195* (1949), 463–473.

[266] RIVLIN, R. Large elastic deformations of isotropic materials, VI: Further results in the theory of torsion, shear and flexture. *Philos. Trans. Roy. Soc.,A. A242* (1949), 173–195.

[267] RIVLIN, R., AND ERICKSEN, J. Stress-deformation relations for isotropic materials. *J. Rat. Mech. Anal. 4* (1955), 323–425.

[268] RIVLIN, R., AND SAUNDERS, D. Large elastic deformations of isotropic materials, VII. Experiments on the deformation of rubber. *Philos. Trans. Roy. Soc.,A. 243* (1951), 251–258.

[269] RIVLIN, R., AND WILMANSKI, K. The passage from memory functionals to rivlin-ericksen constitutive equations. *ZAMP 38* (1987), 624–629.

[270] RIVLIN, R. S. Large elastic deformations of isotropic materials, I: Some uniqueness theorems for pure homogenous deformation. *Philos. Trans. Roy. Soc.,A. A240* (1948), 459–490.

[271] RIVLIN, R. S. Large elastic deformations of isotropic materials, III: Some uniqueness theorems for pure homogenous deformation. *Philos. Trans. Roy. Soc.,A. A240* (1948), 509–525.

[272] RIVLIN, R. S. Large elastic deformations of isotropic materials, IV: Further developments of the general theory. *Philos. Trans. Roy. Soc.,A. A241* (1948), 379–397.

[273] ROMANO, A., LANCELLOTTA, R., AND MARASCO, A. *Continuum Mechanics using Mathematica.* Birkhäuser, Boston, 2006.

[274] RUBIN, M. B. A time integration procedure for large plastic deformation in elastic-viscoplastic metals. *J. Math. Phys. (ZAMP) 40* (1989), 846–871.

[275] RUBIN, M. B. On the treatment of elastic deformation in finite

elastic-viscoplastic theory. *Int. J. Plasticity 12* (1996), 951–965.

[276] RUBIN, M. B. *Cosserat Theories: Shells, Rods and Points.* Kluwer, Dordrecht, 2000.

[277] RUBIN, M. B. Physical reasons for abandoning plastic deformation measures in finite plasticity and viscoplasticity theory. *Arch. Mech. 53* (2001), 519–539.

[278] RUBIN, M. B. Introduction to continuum mechanics. Tech. Rep., Technion, Haifa, 2005.

[279] RUGGERI, T. Galilean invariance and entropy principle for systems of balance laws. *Continuum Mech. Thermodyn. 1* (1989), 3–20.

[280] RUGGERI, T., AND SIMIC, S. *Mixture of Gases with Multi-Temperature: Maxwellian Iteration.* World Scientific, New Jersey, 2007, pp. 186–194. Ruggeri, T. and Sammartino, M. (eds.).

[281] SANCHEZ-PALENCIA, E. *Nonhomogeneous Media and Vibration Theory.* Springer, Berlin, 1980.

[282] SANDRI, G. The foundations of nonequilibrium statistical mechanics. i and ii. *Ann. Phys. (N.Y.) 24* (1963), 332–418.

[283] SANDRI, G. New kinetic equation. *Phys. Rev. Lett. 11*, 4 (1963), 178–179.

[284] SCHMID, E. Über die schubverfestigung von einkristallen bei plastischer deformation. *Z. Phys. 40* (1924), 54–74.

[285] SCHMID, E., AND BOAS, W. *Kristallplastizität mit besonderer Berücksichtigung der Metalle.* Springer, Berlin, 1935.

[286] SCHROETER, J. Kinetic theory of charged particles. In *Symposium on Kinetic Theory and Exrended Thermodynamics* (Bologna, 1987), I. Mueller and T. Ruggeri, Eds., Pitagora Editrice, pp. 295–311.

[287] SECCIA, L., AND STRUMIA, A. Wave propagation in relativistic extended thermodynamics. *Cont. Mech. Thermodyn. 2* (1990), 151–161.

[288] SEDOV, L. *A Course in Continuum Mechanics Vol.II: Physical Foundations and Formulations of Problems.* Springer, Berlin, 1987.

[289] SEDOV, L. *A Course in Continuum Mechanics Vol.III: Fluids, Gases and ... Plastic Solids and the Formation of Cracks.* Springer, Berlin, 1987.

[290] SEDOV, L. I. *A Course in Continuum Mechanics Vol.I: Basic Equations and Analytical Techniques.* Springer, Berlin, 1987.

[291] SERRIN, J. Conceptual analysis of the classical second law of thermodynamics. *Arch. Rat. Mech. Anal. 70* (1979), 355–371.

[292] SERRIN, J., Ed. *New Perspectives in Thermodynamics.* Springer, Berlin, 1986.

[293] SHAVIT, A., AND GUTFINGER, C. *Thermodynamics. From Concepts to Applications.* Prentice Hall, London, 1955.

[294] SIENIUTYCZ, S., AND SALAMON, P. E., Eds. *Extended Thermody-*

namic Systems. Taylor and Francis, New York, 1992.

[295] SIGNIORINI, A. Sulle deformazioni termoelastiche finite. In *Proc. 3rd Int. Congr. Appl. Mech.* (Stockholm, 1930), vol. 2, pp. 80–89.

[296] SIGNIORINI, A. Deformazioni elastiche finite: elasticita di 2 grado. In *Atti 2 Congr. Mat. Ital.* (1940), pp. 56–71.

[297] SILHAVY, M. *The Mechanics and Thermodynamics of Continuous Media.* Springer, Berlin, 1997.

[298] SLEBODZINSKI, W. *Exterior forms and their applications.* Polish Scientific Publishers, Warsaw, 1970.

[299] SMITH, G. F. On the minimality of integrity bases for symmetric 3×3 matrices. *Quart. Appl. Math. 5* (1960), 382–389.

[300] SMITH, G. F., AND RIVLIN, R. S. The anisotropic tensors. *Quart. Appl. Math. 15* (1957), 107–112.

[301] SOBCZYK, K. *Stochastic Wave Propagation.* Elsevier, Amsterdam, 1985.

[302] SOMMERFELD, A. *Thermodynamik und Statistik (2. Auflage).* Verlag Harri Deutsch, Frankfurt, 1977.

[303] SOO, S. L. *Fluid Dynamics of Multiphase Systems.* Blaisdell, Waltham, 1967. Particularly classical modeling of suspensions.

[304] SPANOS, T. J. T. *The Thermophysics of Porous Media.* Monographs and Surveys in pure and Appleid Mathematics, No. 126. Chapman and Hall/CRC, 2002.

[305] SPENCER, A. The static theory of finite elasticity. *J. Inst. of Mathematics and its Applications 6* (1970), 164–200.

[306] SPENCER, A. Theory of invariants. In *Continuum Physics*, A. Eringen, Ed., vol. 1. Academic Press, 1975, ch. 3.

[307] SPENCER, A., AND RIVLIN, R. Further results on the theory of matrix polynomials. *Arch. Rat. Mech. Anal. 4* (1960), 214–230.

[308] SPENCER, A. J. M. The invariants of six symmetric 3×3 matrices. *Arch. Rat. Mech. Anal. 7* (1961), 64–77.

[309] SPENCER, A. J. M., AND RIVLIN, R. S. Finite integrity bases for five or fewer symmetric 3×3 matrices. *Arch. Rat. Mech. Anal. 2* (1958/1959), 435–446.

[310] SPENCER, A. J. M., AND RIVLIN, R. S. Isotropic integrity bases for vectors and second-order tensors, i. *Arch. Rat. Mech. Anal. 9* (1962), 45–63.

[311] ST. VENANT, A.-J.-C. B. Sur les pressions qui se développent à l'intérieur des corps solides lorsque les dél'élasticité, ne peuvent cependant pas être considérés comme trés petits. *Bull. Soc. Philomath. 5* (1844), 26–28.

[312] STERNBERG, S. *Lectures on Differential Geometry, 2nd edition.* Chelsea Pub. Co., 1982.

[313] STOLL, R. D. *Sediment Acoustics*, vol. 26 of *Lecture Notes in Earth Sciences*. Springer, New York, 1989.

[314] STRUCHTRUP, H. *Macroscopic Transport Equations for Rarefied Gas Flows. Approximation Methods in Kinetic Theory*. Springer, Heidelberg, 2005.

[315] STRUCHTRUP, H., AND WEISS, W. Maximum of the local entropy production becomes minimal in stationary processes. *Phys. Rev. Lett. 80* (1998), 5048–5051.

[316] STRUMINSKII, V. V. Hilbert's method of solving the boltzmann kinetic equation. *Sov. Phys. - Dokl. 9*, 9 (1965), 733–735.

[317] STRUMINSKII, V. V. One method of solving the boltzmann kinetic equation. *Sov. Phys. - Dokl. 9*, 9 (1965), 736–739.

[318] STRUMINSKII, V. V. The structure of solutions of sequences of equations in the kinetic theory of gases. *Sov. Phys. - Dokl. 10*, 11 (1966), 999–1002.

[319] STRUMINSKII, V. V. Theory of systems of identical particles. *Sov. Phys. - Dokl. 25*, 6 (1979), 439–441.

[320] SYNGE, J., AND SCHILD, A. *Tensor Calculus*. University of Toronto Press, Toronto, 1959.

[321] SZERMEGOR, T. *Theory of Elasticity Microinhomogeneous Media, (in Russian)*. Nauka, Moscow, 1977.

[322] THOMAS, T. Y. Extended compatibility conditions for the study of surfaces of discontinuity in continuum mechanics. *J. Math. Mech. 6* (1957), 311–322, 907–908.

[323] THOMAS, T. Y. *Plastic Flow and Fracture in Solids*. Academic Press, N.Y., London, 1961.

[324] TOLMAN, R. *Relativity, Thermodynamics and Cosmology*. Oxfors University Press, 1934.

[325] TOLSTOY, I., Ed. *Acoustics, elasticity and thermodynamics of porous media: Twenty-one papers by M. A. Biot*. Acoustical Society of America, 1992.

[326] TRUESDELL, C. A. The mechanical foundations of elasticity and fluid dynamics. *J. Rat. Mech. Anal. 1* (1952), 125–300.

[327] TRUESDELL, C. A. Sulle basi della termomeccanica. *Accademia Nazionale dei Lincei, Rendiconti della Classe die Scienze Fisiche, Mathematiche e Naturali (8) 22*, 8 (1957), 33–38, 158–166.

[328] TRUESDELL, C. A. The rational mechanics of flexible or elastic bodies. *L. Euleri Opera Omnia (2) 11*, 2 (1960), 1638–1788.

[329] TRUESDELL, C. A. *The Elements of Continuum Mechanics*. Springer, N.Y., 1966.

[330] TRUESDELL, C. A. *A First Course in Rational Continuum Mechanics*. The Johns Hopkins University, Baltimore, Maryland, 1972.

[331] TRUESDELL, C. A. *A First Course in Rational Continuum Mechanics, part I, Fundamental Concepts.* Academic Press, New York, 1977.

[332] TRUESDELL, C. A. *Rational Thermodynamics (Second Edition).* Springer, 1984.

[333] TRUESDELL, C. A., AND MUNCASTER, R. G. *Fundamentals of Maxwell's Kinetic Theory of a Simple Monatomic Gas.* Academic Press, 1980.

[334] TRUESDELL, C. A., AND NOLL, W. *The Non-Linear Field Theories of Mechanics*, vol. 3. Springer, Berlin, 1965, ch. 3. Flügge, S. (ed.).

[335] TRUESDELL, C. A., AND TOUPIN, R. A. *The Classical Field Theories*, vol. 3. Springer, Berlin, 1960, ch. 1. Flügge, S. (ed.).

[336] TRUESDELL, C. A.AND BHARATHA, S. *Classical Thermodynamics as a Theory of Heat Engines.* Springer, N.Y., 1977.

[337] VIKTOROV, I. A. *Rayleigh and Lamb waves. Physical theory and applications.* Plenum Press, New York, 1967.

[338] VILLAGGIO, P. The tickness of roman arches. In *Contemporary Research in the Mechanics and Mathematics of Materials* (Barcelona, 1996), e. R. C. Batra, M. F. Beatty, Ed., CIMNE, pp. 472–479.

[339] VILLAGGIO, P. *Mathematical Models for Elastic Structures.* Cambridge Univ. Press, 1997.

[340] VISINTIN, A. *Differential Models of Hysteresis.* Springer, New York, 1994.

[341] VOLTERRA, V. Sull'equilibrio dei corpi elastici più volte connessi. *Rend. Accad. Lincei 14* (1905), 193–202.

[342] VON TERZAGHI, K. *Erdbaumechanik auf bodenphysikalischer Grundlage.* Franz Deuticke, Leipzig, 1925.

[343] VON TERZAGHI, K. *Theoretical Soil Mechanics.* John Wiley and Sons, New York, 1943.

[344] VON TERZAGHI, K., AND FROEHLICH, O. *Theorie der Setzung von Tonschichten.* Franz Deuticke, Leipzig, Wien, 1936.

[345] WANG, H. *Theory of Linear Poroelasticity with Applications to Geomechanics and Hydrogeology.* Princeton University Press, Princeton, 2000.

[346] WANG, H. F., AND TRUESDELL, C. *Introduction to Rational Elasticity.* Noordhoff, 1973.

[347] WEINGARTEN, G. Sulle superficie di discontinuitelasticità dei corpi solidi. *Rend. Accad. Lincei 10* (1901), 5, 57–60.

[348] WHITE, J. E. *Underground sound. Application of seismic waves, Methods in Geochemistry and Geophysics*, vol. 18. Elsevier, Amsterdam, New York, 1983.

[349] WHITHAM, G. B. *Linear and Nonlinear Waves.* John Wiley & Sons, New York, 1974.

[350] WILHELM, T., AND WILMANSKI, K. On the onset of flow instabilities in granular media due to porosity inhomogeneities. *Int. Jour. Multiphase Flow 28* (2002), 1929–1944.

[351] WILLIAMS, W. O. Snelson tensegrity dtructures. Tech. Rep. 18, WIAS, 2000.

[352] WILMANSKI, K. Dynamics of bodies with microstructure. *Arch. Mech. Stos. (Arch. Mech.) 45*, 4 (1968), 705–744.

[353] WILMANSKI, K. On thermodynamics and functions of states of non-isolated systems. *Arch. Rat. Mech. Anal. 45*, 4 (1972), 251–281.

[354] WILMANSKI, K. *Podstawy Termodynamiki Fenomenologicznej (in Polish)*. PWN, Warszawa, 1974.

[355] WILMANSKI, K. Thermodynamic foundations of thermoelasticity. In *Recent Developments in Thermomechanics of Solids*, P. P. G. Lebon, Ed., no. 262 in CISM Courses and Lectures. Springer, Wien-New York, 1980.

[356] WILMANSKI, K. Thermodynamics of non-newtonian fluids. Tech. rep., Wissenschaftskolleg zu Berlin, Institute for Advanced Study, Jahrbuch 1984/85, 1986.

[357] WILMANSKI, K. Thermodynamics of a heat conducting maxwellian fluid. *Arch. Mech. 40* (1988), 2–3, 217–232.

[358] WILMANSKI, K. Macroscopic theory of evolution of deformation textures. *Int. J. Plasticity 8* (1992), 959–975.

[359] WILMANSKI, K. Phenomenological thermodynamics. In *Foundations of Mechanics*, H. Zorski, Ed. PWN & Elsevier, Amsterdam, 1992, pp. 485–589.

[360] WILMANSKI, K. Symmetric model of stress-strain hysteresis loops in shape memory alloys. *Int. J. Engng. Sci. 31*, 8 (1993), 1121–1138.

[361] WILMANSKI, K. Lagrangean model of two-phase porous material. *J. Non-Equilibrium Thermodyn. 20* (1995), 50–77.

[362] WILMANSKI, K. Porous media at finite strains. the new model with the balance equation for porosity. *Arch. Mech. 48*, 4 (1996), 591–628.

[363] WILMANSKI, K. A thermodynamic model of compressible porous materials with the balance equation of porosity. *Transp. Porous Media 32* (1998), 21–47.

[364] WILMANSKI, K. *Thermomechanics of Continua*. Springer, Berlin, N.Y., 1998.

[365] WILMANSKI, K. Waves in porous and granular materials. In *Kinetic and continuum theories of granular and porous media*, K. Hutter and K. Wilmanski, Eds., no. 400 in CISM Courses and Lectures. Springer, Wien New York, 1999, pp. 131–186.

[366] WILMANSKI, K. Toward extended thermodynamics of porous and granular materials. In *Trends in Applications of Mathematics to Me-*

chanics (2000), A. N. G. Iooss, O. Guès, Ed., Chapman & Hall/CRC, pp. 147–160.

[367] WILMANSKI, K. On a homogeneous adsorption in porous materials. *ZAMM 81* (2001), 119–124.

[368] WILMANSKI, K. Thermodynamics of multicomponent continua. In *Earthquake Thermodynamics and Phase Transitions in the Earth's Interior*, R. Teisseyre and E. Majewski, Eds. Academic Press, San Diego, 2001, pp. 567–655.

[369] WILMANSKI, K. Thermodynamical admissibility of Biot's model of poroelastic saturated materials. *Arch. Mech. 54*, 5-6 (2002), 709–736.

[370] WILMANSKI, K. On thermodynamics of nonlinear poroelastic materials. *J. Elasticity 71* (2003), 247–261.

[371] WILMANSKI, K. Some questions on material objectivity arising in models of porous materials. In *Rational continua, classical and new*, P. Podio-Guidugli and M. Brocato, Eds. Springer, Italia Srl, Milano, 2003, pp. 183–196.

[372] WILMANSKI, K. On microstructural tests for poroelastic materials and corresponding Gassmann-type relations. *Géotechnique 54* (2004), 593–603.

[373] WILMANSKI, K. Elastic modelling of surface waves in single and multicomponent systems. In *Surface Waves in Geomechanics: Direct and Inverse Modelling for Soils and Rocks*, C. C. Lai and K. Wilmanski, Eds., CISM, No. 481. Springer, N.Y., 2005, pp. 203–276.

[374] WILMANSKI, K. Tortuosity and objective relative accelerations in the theory of porous materials. *Proc. R. Soc. A 461* (2005), 1533–1561.

[375] WILMANSKI, K. A few remarks on Biot's model and linear acoustics of poroelastic saturated materials. *Soil Dynamics & Earthquake Engineering 26*, 6-7 (2006), 509–536.

[376] WILMANSKI, K. A few remarks on micro/macro transitions and Gassmann relations for poroelastic materials. *Mechanics of Solids and Structures 6* (2007), 191–204.

[377] WILMANSKI, K., AND RADKEVICH, E. O dispersii w modeli dla poristo-uprugich sred s urawnieniem balansa poristosti (inrussian). *Trudy Seminara im. I. G. Petrowskogo 22* (2002), 74–104.

[378] WILMANSKI, K., AND RADKEVICH, E. O globalnoj ustoicziwosti w zadacze dla poristo-uprugich sred s urawnieniem bal-ansa poristosti (in russian). *Trudy Moskowskogo Matematiczeskogo Obszczestwa 63* (2002), 7–44.

[379] WOLFFERSDORFF, P.-A. A hypoplastic relation for granular materials with a predefined limit state surface. *Mechanics of Cohesive-Frictional Materials 1* (1996), 251–271.

[380] WOOD, D. M. *Soil Behaviour and Critical State Soil Mechanics.*

Cambridge Univ. Press, Cambridge, 1990.

[381] WOZNIAK, C. *Podstawy dynamiki cial odksztalcalnych (in Polish).* PWN, Warsaw, 1968.

[382] WOZNIAK, C. *Wiezy w mechanice cial odksztalcalnych (in Polish).* Ossolineum, Wroclaw, 1988.

[383] WU, T. Y. *Kinetic Equations of Gases and Plasmas.* Addison-Wesley Publ. Co., 1966.

[384] WU, W., AND KOLYMBAS, D. Numerical testing of the stability criterion for hypoplastic constitutive equations. *Mechanics of Materials 9* (1990), 245–253.

[385] YVON, J. *Théorie Statistique des Fluides et l'Equation d'Etat.* Hermann, Paris, 1935.

[386] ZORAWSKI, M. *La Théorie MathDunod, Paris, 1967.*

Index

Series on Advances in Mathematics for Applied Sciences

Aims and Scope

This Series reports on new developments in mathematical research relating to methods, qualitative and numerical analysis, mathematical modeling in the applied and the technological sciences. Contributions related to constitutive theories, fluid dynamics, kinetic and transport theories, solid mechanics, system theory and mathematical methods for the applications are welcomed.

This Series includes books, lecture notes, proceedings, collections of research papers. Monograph collections on specialized topics of current interest are particularly encouraged. Both the proceedings and monograph collections will generally be edited by a Guest editor.

High quality, novelty of the content and potential for the applications to modern problems in applied science will be the guidelines for the selection of the content of this series.

Instructions for Authors

Submission of proposals should be addressed to the editors-in-charge or to any member of the editorial board. In the latter, the authors should also notify the proposal to one of the editors-in-charge. Acceptance of books and lecture notes will generally be based on the description of the general content and scope of the book or lecture notes as well as on sample of the parts judged to be more significantly by the authors.

Acceptance of proceedings will be based on relevance of the topics and of the lecturers contributing to the volume.

Acceptance of monograph collections will be based on relevance of the subject and of the authors contributing to the volume.

Authors are urged, in order to avoid re-typing, not to begin the final preparation of the text until they received the publisher's guidelines. They will receive from World Scientific the instructions for preparing camera-ready manuscript.

SERIES ON ADVANCES IN MATHEMATICS FOR APPLIED SCIENCES

Published:*

Vol. 63 Lecture Notes on the Discretization of the Boltzmann Equation
eds. N. Bellomo and R. Gatignol

Vol. 64 Generalized Kinetic Models in Applied Sciences — Lecture Notes on Mathematical Problems
by L. Arlotti et al.

Vol. 65 Mathematical Methods for the Natural and Engineering Sciences
by R. E. Mickens

Vol. 66 Advanced Mathematical and Computational Tools in Metrology VI
eds. P. Ciarlini et al.

Vol. 67 Computational Methods for PDE in Mechanics
by B. D'Acunto

Vol. 68 Differential Equations, Bifurcations, and Chaos in Economics
by W. B. Zhang

Vol. 69 Applied and Industrial Mathematics in Italy
eds. M. Primicerio, R. Spigler and V. Valente

Vol. 70 Multigroup Equations for the Description of the Particle Transport in Semiconductors
by M. Galler

Vol. 71 Dissipative Phase Transitions
eds. P. Colli, N. Kenmochi and J. Sprekels

Vol. 72 Advanced Mathematical and Computational Tools in Metrology VII
eds. P. Ciarlini et al.

Vol. 73 Introduction to Computational Neurobiology and Clustering
by B. Tirozzi, D. Bianchi and E. Ferraro

Vol. 74 Wavelet and Wave Analysis as Applied to Materials with Micro or Nanostructure
by C. Cattani and J. Rushchitsky

Vol. 75 Applied and Industrial Mathematics in Italy II
eds. V. Cutello et al.

Vol. 76 Geometric Control and Nonsmooth Analysis
eds. F. Ancona et al.

Vol. 77 Continuum Thermodynamics
by K. Wilmanski

*To view the complete list of the published volumes in the series, please visit:
http://www.worldscibooks.com/series/samas_series.shtml